PROBLEMS OF ESTIMATION

Confidence Interval for

Mean (large sample, σ known or estimated by s)

$$\bar{x} - z_{\alpha/2} \cdot \frac{\sigma}{\sqrt{n}} < \mu < \bar{x} + z_{\alpha/2} \cdot \frac{\sigma}{\sqrt{n}}$$

Mean (small sample)

$$\bar{x} - t_{\alpha/2} \cdot \frac{s}{\sqrt{n}} < \mu < \bar{x} + t_{\alpha/2} \cdot \frac{s}{\sqrt{n}}$$

Proportion (large sample)

$$\hat{p} - z_{\alpha/2} \cdot \sqrt{\frac{\hat{p}(1-\hat{p})}{n}} < p < \hat{p} + z_{\alpha/2} \cdot \sqrt{\frac{\hat{p}(1-\hat{p})}{n}}$$

where $\hat{p} = \dfrac{x}{n}$

Sample Size

Estimation of mean

$$n = \left[\frac{z_{\alpha/2} \cdot \sigma}{E}\right]^2$$

Estimation of proportion

$$n = p(1-p)\left[\frac{z_{\alpha/2}}{E}\right]^2$$

TESTS OF HYPOTHESES

Statistics for Tests Concerning

Difference between means (σ_1 and σ_2 known or estimated by s_1 and s_2)

$$z = \frac{\bar{x}_1 - \bar{x}_2 - \delta}{\sqrt{\dfrac{\sigma_1^2}{n_1} + \dfrac{\sigma_2^2}{n_2}}}$$

Difference between means (small samples)

$$t = \frac{\bar{x}_1 - \bar{x}_2 - \delta}{\sqrt{\dfrac{(n_1-1)s_1^2 + (n_2-1)s_2^2}{n_1+n_2-2} \cdot \left(\dfrac{1}{n_1} + \dfrac{1}{n_2}\right)}}$$

Continued inside back cover

PROBABILITY AND STATISTICS FOR ENGINEERS

Fourth Edition

Irwin Miller

Vice-President
Jones Reilly & Associates, Inc.

John E. Freund

Arizona State University

Richard A. Johnson

University of Wisconsin-Madison

PRENTICE HALL / Englewood Cliffs, New Jersey 07632

Library of Congress Cataloging in Publication Data

Miller, Irwin,
Probability and statistics for engineers/Irwin Miller, John E. Freund, Richard A. Johnson.—4th ed.
p. cm.
Bibliography: p.
Includes index.
ISBN 0-13-712761-8
1. Engineering—Statistical methods. 2. Probabilities.
I. Freund, John E. II. Johnson, Richard A., . III. Title.
TA340.M5 1990
519.2′02462—dc20 89-8595
CIP

A Division of Simon & Schuster
Englewood Cliffs, New Jersey 07632

Manufacturing buyer: Paula Massenaro
Cover design: Suzanne Behnke

Printed in the United States of America

10 9 8 7 6 5 4 3 2 1

ISBN 0-13-712761-8

Prentice-Hall International (UK) Limited, *London*
Prentice-Hall of Australia Pty. Limited, *Sydney*
Prentice-Hall Canada Inc., *Toronto*
Prentice-Hall Hispanoamericana, S.A., *Mexico*
Prentice-Hall of India Private Limited, *New Delhi*
Prentice-Hall of Japan, Inc., *Tokyo*
Simon & Schuster Asia Pte. Ltd., *Singapore*
Editora Prentice-Hall do Brasil, Ltda., *Rio de Janeiro*

CONTENTS

PREFACE TO THE FOURTH EDITION

In the fourth edition we tried to preserve all the strengths of the earlier edition while at the same time bringing the material in line with the current application of statistics in scientific fields. Primarily, the additions have been of three forms.

First, because of the international surge of interest in quality improvement programs, the special chapter devoted to the key underlying statistical ideas has been expanded. Throughout the book, beginning with the first chapter, we have inserted examples that help emphasize the procedures for improving processes. Nevertheless, there must be a basic understanding of variation before any quality improvement can be undertaken. In a classroom setting, we feel that this is best achieved by introducing the most useful probability models and then teaching the elements of statistical inference. With this sound basis, the techniques for quality improvement are readily understood and easily implemented on specific processes.

Second, we have acknowledged the fact that scientists now have good computer support by including generic computer output in a number of examples. Several exercises, based on the statistical software package *MINITAB*, have also been added. Consequently, we are able to extend the treatment of plots for checking probability models, residual plots for checking models with predictor variables, and transformations. These are essential tools for the practitioner.

Third, we have included a number of examples that we have encountered while consulting with engineers and scientists from industry and research laboratories. Others were culled from our experience of teaching and interacting with engineering students over the past 22 years.

To give students an earlier preview of statistics, we have moved the discussion of descriptive statistics to Chapter 2. Also, throughout this edition, we have placed more emphasis on confidence estimation. Finally, to aid the student, we have also included a section of summary exercises at the end of each chapter.

We wish to thank Minitab (*State College, Penn.*) for permission to include commands and output from their *MINITAB* software package and the SAS Institute (*Cary, N.C.*) for permission to include output from their *SAS* package. Thanks also to Jim Evans, Greg Reinsel, and J. C. Lu for help in preparing some of the computer output. A. Mouhab and K. T. Wu provided valuable help in checking the manuscript.

Thank you to the reviewers: D. S. Gill, *North Dakota State University*; Iris B. Ibrahim, *Clemson University*; Robert L. Schaefer, *Miami University, Ohio*; U. Narayan Bhat, *Southern Methodist University*; Peter Wollan, *Michigan Technological University*; Bernard Gleimer, *New York Institute of Technology*, Bhushan L. Wadhwa, *Cleveland State University*; Joe H. Mize, *Ohio State University.*

All revisions in this edition were the responsibility of R. A. Johnson.

RICHARD A. JOHNSON

PREFACE TO THE THIRD EDITION

This book has been written for an introductory course in probability and statistics for students of engineering and the physical sciences. In the Third Edition, the authors have revised the exercise material, and many new exercises have been added. Exercises are numbered consecutively within each chapter. Each chapter is keynoted by an introductory statement and has a check list of key terms (with page references) at the end. Important formulas, theorems, and rules have been set out from the text; they are enclosed in boxes with a description of each in the margin. Notation and technical terms have been updated to conform with current usage. New material has been added on combinatorial methods, stem-and-leaf plots, and the distinction between "probability" and "confidence." The treatment of statistical inference has been extensively revised, including a more systematic treatment of tests of hypotheses and a discussion of tests of composite null hypotheses.

Throughout, the text has been revised with the aim of improving clarity and bringing the applications up to date. It has been tested extensively in courses for university students as well as by in-plant training of practicing engineers. The authors have found that the material in the book can be covered in a two-semester or three-quarter course consisting of three lectures a week. However, through a choice of topics the book also lends itself as a text for shorter courses with emphasis upon either theory or applications.

Chapters 2, 4, 5, and 6 provide a brief, though rigorous, introduction to the theory of statistics and, together with some of the material in Chapter 15, they are suitable for an introductory semester (or quarter) course on the mathematics of probability and statistics. Chapters 3, 7, 8, 9, and 10 contain conventional material on the elementary methods of statistical inference, including a treatment of nonparametric methods. Chapters 11, 12, and 13 comprise an introduction to some of the standard, though more advanced, topics of experimental statistics, while

Chapters 14 and 15 deal with special applications which have become increasingly important in recent years.

The mathematical background expected of the reader is a year course in calculus; actually, calculus is required mainly for Chapters 2 and 6, dealing with basic distribution theory in the continuous case; for Chapter 15, dealing with applications to reliability theory; and for the least-squares methods of Chapters 11 and 12. The treatment of probability in Chapter 3 is modern in the sense that it is based on the elementary theory of sets.

The authors would like to express their appreciation and indebtedness to the D. Van Nostrand Company for permission to reproduce the material in Table 2; to the Macmillan Publishing Co., Inc., for permission to reprint Table 4 from *Statistical Methods for Research Workers* by R. A. Fisher; to Professor E. S. Pearson and the *Biometrika* trustees for permission to reproduce the material in Tables 5, 6, and 9; to A. H. Bowker and G. J. Lieberman and Prentice-Hall, Inc., for permission to reproduce Table 8; to Donald B. Owen and Addison-Wesley, Inc., for permission to reproduce part of the table of random numbers from their *Handbook of Statistical Tables*; to Frank J. Massey, Jr., and *Journal of the American Statistical Association* for permission to reproduce the material in Table 10; to D. B. Duncan, H. L. Harter, and *Biometrics* for permission to reproduce Table 12; to the American Society for Testing and Materials for permission to reproduce Table 13; and to the McGraw-Hill Book Company for permission to reproduce the material in Table 14.

The authors would also like to express their appreciation to the editorial staff of Prentice Hall, Inc., for their courteous cooperation in the production of this book, to Judi Fabian, who helped in the typing of the manuscript, to Rita Ewer for her help with the proofreading, and above all to their wives for not complaining too much about the demands made on their husbands throughout the writing of this book.

IRWIN MILLER

JOHN E. FREUND

INTRODUCTION

Everything dealing with the collection, processing, analysis, and interpretation of numerical data belongs to the domain of statistics. In engineering, this includes such diversified tasks as calculating the average length of the downtimes of a computer, collecting and presenting data on the numbers of persons attending seminars on solar energy, evaluating the effectiveness of commercial products, predicting the reliability of a rocket, or studying the vibrations of airplane wings.

In Sections 1.1, 1.2, and 1.3 we discuss the recent growth of statistics and, in particular, its applications to problems of engineering. Statistics plays a major role in the improvement of quality of any product or service. An engineer using the techniques described in this book can become much more effective in all phases of work relating to research, development, or production.

1.1 MODERN STATISTICS

The origin of statistics can be traced to two areas of interest that, on the surface, have little in common: games of chance and what is now called political science. Mid-eighteenth century studies in probability, motivated largely by interest in games of chance, led to the mathematical treatment of errors of measurement and the theory that now forms the foundation of statistics. In the same century, interest in the numerical description of political units (cities, provinces, counties, etc.) led to what is now called **descriptive statistics**. At first, descriptive statistics consisted merely of the presentation of data in tables and charts; nowadays, it includes also the summarization of data by means of numerical descriptions.

In recent decades, the growth of statistics has made itself felt in almost every major phase of human activity, and the most important feature of its growth has been the shift in emphasis from descriptive statistics to **statistical inference**. Statistical inference concerns generalizations based on sample data; it applies to such problems as estimating an engine's average emission of pollutants from trial runs, testing a manufacturer's claim on the basis of measurements performed on samples of his product, and predicting the fidelity of an audio system on the basis of sample data pertaining to the performance of its components.

When one makes a statistical inference, namely, an inference that goes beyond the information contained in a set of data, one must always proceed with caution. One must decide carefully how far one can go in generalizing from a given set of data, whether such generalizations are at all reasonable or justifiable, whether it might be wise to wait until there are more data, and so forth. Indeed, some of the most important problems of statistical inference concern the appraisal of the risks and the consequences to which one might be exposed by making generalizations from sample data. This includes an appraisal of the probabilities of making wrong decisions, the chances of making incorrect predictions, and the possibility of obtaining estimates that do not lie within permissible limits.

In recent years, attempts have been made to treat all these problems within the framework of a unified theory called **decision theory**. Although this theory has many conceptual as well as theoretical advantages, its application still contains a strong element of subjectivity. It is at least partly a subjective decision whether to base an experiment (say, the determination of a specific heat) on 5 measurements, on 12 measurements, or on 25 or more. Also, subjective factors invariably enter the design of equipment, the hiring of personnel, and even one's deciding how to formulate a hypothesis and the alternative against which it is to be tested. Above all, subjective judgments are virtually unavoidable when one is asked to put "cash values" on the various risks to which one is exposed. After all, if scientists are asked to judge the safety of a piece of equipment, how can they possibly put a cash value on the possibility that they might make an error, when such an error may lead to the loss of human lives.

Whether or not statistical inference is viewed within the broader framework of decision theory, it depends heavily on the theory of probability. This is a mathematical theory, but the question of objectivity versus subjectivity arises in its application and in its interpretation. As we shall see in Chapter 3, statements such as "there is a fifty-fifty chance that a given fabric will ignite when exposed to a cigarette ash" or "the probability is 0.05 that a defective part will not be caught in the final inspection" can be looked upon as objective or subjective evaluations of the uncertainties involved.

We shall approach the subject of statistics as a science, developing each statistical idea insofar as possible from its probabilistic foundation, and applying each idea to problems of physical or engineering science as soon as it has been developed. The great majority of the methods we shall use in stating and solving these problems belong to the **classical approach**, because they do not formally take into account the various subjective factors mentioned above; in selected applications we present also the **Bayesian approach** (see Sections 7.3 and 9.2), which accounts formally for at least some of these subjective factors. In any case, we shall endeavor continually to make the reader aware that the subjective factors do exist, and to indicate whenever possible what role they might play in making the final decision. Subjectivity plays an important role in the choice among statistical methods or formulas to be used in a given situation, in deciding on the size of a sample, in specifying the probabilities with which we are willing to risk errors, and so forth. This "bread-and-butter" approach to statistics presents the subject in the form in which it has so successfully contributed to engineering science, as well as to the natural social sciences, in the last 30 years.

1.2 STATISTICS AND ENGINEERING

There are few areas where the impact of the recent growth of statistics has been felt more strongly than in engineering and industrial management. Indeed, it would be difficult to overestimate the contributions statistics has made to problems of production, to the effective use of materials and labor, to basic research, and to the development of new products. As in the other sciences, statistics has become a vital tool to engineers. It enables them to understand phenomena subject to variation and to effectively predict or control them.

In this text, our attention will be directed largely toward engineering applications, but we shall not hestitate to refer also to other areas to impress upon the reader the great generality of most statistical techniques. Thus, the reader will find that the statistical method which is used to estimate the coefficient of thermal expansion of a metal serves also to estimate the average time it takes a secretary to perform a given task, the average thickness of a pelican egg, or the average IQ of an immigrant arriving in the United States. Similarly, the statistical method that is used to compare the strength of two alloys serves also to compare the effectiveness of two teaching methods, the merits of two insect sprays, or the performance of men and women in a current-events test.

1.3 THE ROLE OF THE SCIENTIST AND ENGINEER IN QUALITY IMPROVEMENT

Since the 1960s, the United States has found itself in an increasingly competitive world market. At present, we are in the midst of an international revolution in quality improvement. The teaching and ideas of W. Edwards Deming were instrumental in the rejuvenation of Japan's industry. He now stresses that American industry, in order to survive, must mobilize with a continuing commitment to quality improvement. From design to production, processes need to be continually improved. The engineer and scientist, with their technical knowledge and armed with basic statistical skills in data collection and graphical display, can be main participants in attaining this goal.

The **quality improvement** movement is based on the philosophy of "make it right the first time." Furthermore, one should not be content with any process or product but should continue to look for ways of improving it. We emphasize the key statistical components of any modern quality improvement program. In Chapter 14, we outline the basic issues of quality improvement and present some of the specialized statistical techniques for studying production processes. The experimental designs discussed in Chapter 13 are also basic to the process of quality improvement.

Closely related to quality improvement techniques are the statistical techniques that have been developed to meet the **reliability** needs of the highly complex products of space-age technology. Chapter 15 provides an introduction to this area.

1.4 CHECK LIST OF KEY TERMS (with page references)

Bayesian approach to statistics 3
Classical approach to statistics 2
Decision theory 2
Descriptive statistics 1
Quality improvement 4
Reliability 4
Statistical inference 2

TREATMENT OF DATA

Statistical data, obtained from surveys, experiments, or any series of measurements, are often so numerous that they are virtually useless unless they are condensed, or reduced, into a more suitable form. We begin with the use of simple graphics. Next, Sections 2.2 and 2.3 deal with problems relating to the grouping of data and the presentation of such groupings in graphical form; in Section 2.4 we discuss a relatively new way of presenting grouped data.

Sometimes it may be satisfactory to present data just as they are and let them speak for themselves; on other occasions it may be necessary only to group the data and present the result in tabular or graphical form. However, most of the time data have to be summarized further, and in Sections 2.5 through 2.7 we introduce some of the most widely used kinds of statistical descriptions.

2.1 PARETO DIAGRAMS AND DOT DIAGRAMS

Data need to be collected to provide the vital information necessary to solve engineering problems. Once gathered, these data must be described and analyzed to produce summary information. Graphical presentations can often be the most effective way to communicate this information. To illustrate the power of graphical techniques, we first describe a **Pareto diagram**. This display, which orders each type of failure or defect according to its frequency, can help engineers identify important defects and their causes.

When a company identifies a process as a candidate for improvement, the first step is to collect data on the frequency of each type of failure. For example, for a computer-controlled lathe whose performance was below par, workers recorded the following causes and their frequencies:

power fluctuations	6
controller not stable	22
operator error	13
worn tool not replaced	2
other	5

These data are presented as a **Pareto diagram** in Figure 2.1. This diagram graphically depicts Pareto's empirical law that any assortment of events consists of a few major and many minor elements. Typically, two or three elements will account for more than half of the total frequency. In the context of quality improvement, we want to select the few vital major opportunities for improvement from the many trivial minor opportunities for improvement. This graph visually emphasizes the importance of reducing the frequency of controller misbehavior. An initial goal may be to cut it in half.

As a second step toward improvement of the process, data were collected on the deviations of cutting speed from the target value set by the controller. The seven observed values of (cutting speed) − (target),

$$3,\ 6,\ -2,\ 4,\ 7,\ 4,\ 3$$

are plotted as a **dot diagram** in Figure 2.2. The dot diagram visually summarizes the information that the lathe is, generally, running fast. In Chapters 13 and 14 we develop efficient experimental designs and methods for identifying primary causal factors that contribute to the variability in a response such as cutting speed.

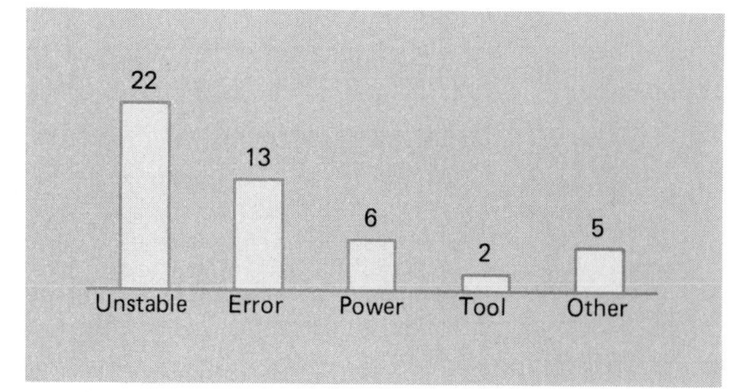

FIGURE 2.1
A Pareto diagram of failures.

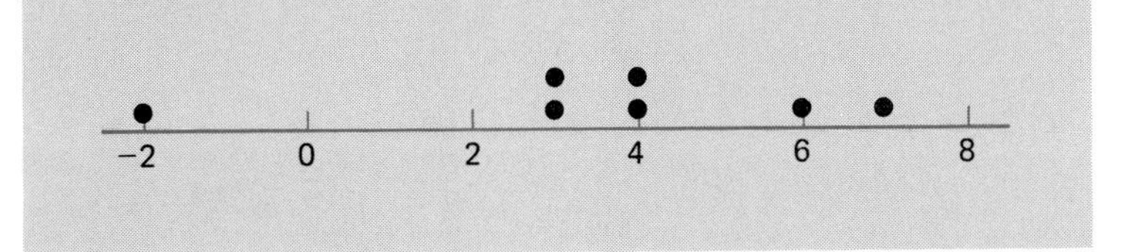

FIGURE 2.2
Dot diagram of cutting speed deviations.

When the number of observations is small, it is difficult to identify any pattern of variation. Still, it is a good idea to plot the data and look for unusual features.

EXAMPLE

In 1987, for the first time, physicists observed supernova neutrinos from outside of our solar system. At a site in Kamiokande, the following times (second) between neutrinos were recorded:

0.107, 0.196, 0.021, 0.283, 0.179, 0.854, 0.58, 0.19, 7.3, 1.18, 2.0

Draw a dot diagram.

Solution

We plot to the nearest 0.1 second to avoid crowding. (See Figure 2.3.) Note the extremely long gap between 2.0 and 7.3 seconds. Statisticians call such an unusual observation an **outlier**. Usually, outliers merit further attention. Was there a recording error, were neutrino's missed in that long time interval, or were there two separate explosions in the supernova?

■

EXAMPLE

The vessels that contain the reactions at some nuclear power plants consist of two components that are welded together. Copper in the welds could cause them to become brittle after years of service. Samples of welding material from one heat that were used in one plant had the copper contents 0.27, 0.35, 0.37. Samples from the next heat had values 0.23, 0.15, 0.25, 0.24, 0.30, 0.33, 0.26. Draw a dot diagram that highlights possible differences in the two production runs (heats) of welding material.

Solution

We plot the first group as solid circles and the second as open circles. (See Figure 2.4.) It seems unlikely that the two production runs are alike because the top two values are from the first run. (In Chapter 10 we confirm this fact.)

■

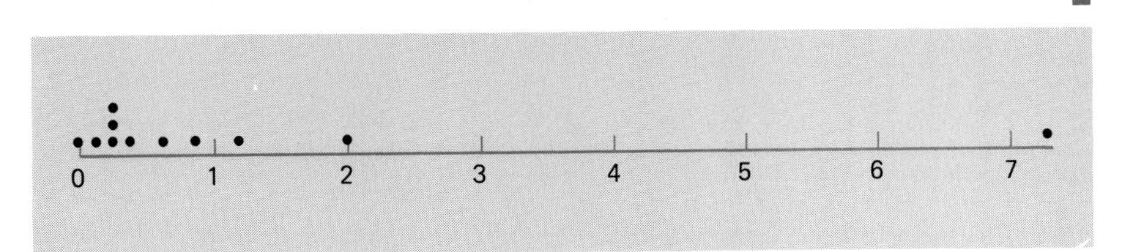

FIGURE 2.3
Dot diagram of times between neutrinos.

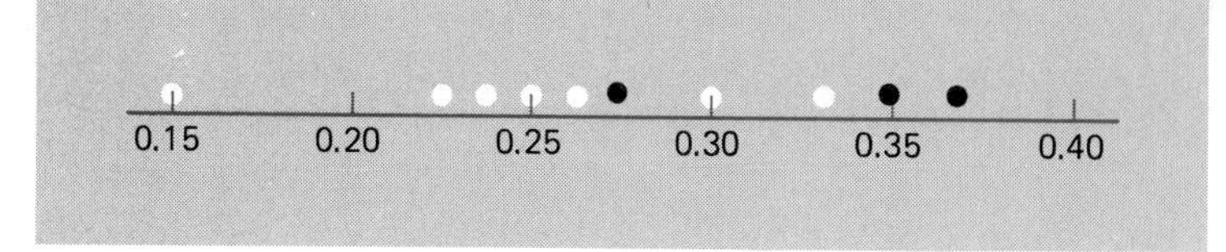

FIGURE 2.4
Dot diagram of copper content.

When a set of data consists of a large number of observations, we take the approach in the next section. The observations are first summarized in the form of a table.

2.2
FREQUENCY DISTRIBUTIONS

A **frequency distribution** is a table that divides a set of data into a suitable number of classes (categories), showing also the number of items belonging to each class. Such a table sacrifices some of the information contained in the data; instead of knowing the exact value of each item, we only know that it belongs to a certain class. On the other hand, the kind of grouping it represents often brings out important features of the data, and the gain in "legibility" usually more than compensates for the loss of information. In what follows, we shall consider mainly **numerical distributions**, that is, frequency distributions where the data are grouped according to size; if the data are grouped according to some quality, or attribute, we refer to such a distribution as a **categorical distribution**.

The first step in constructing a frequency distribution consists of deciding how many classes to use and choosing the **limits** for each class, that is, from where to where each class is to go. Generally speaking, the number of classes we use depends on the number of observations, but it is seldom profitable to use fewer than 5 or more than 15. It also depends on the range of the data, namely, the difference between the largest observation and the smallest. Then, we tally the observations and thus determine the **class frequencies**, namely, the number of observations in each class.

To illustrate the construction of a frequency distribution, let us consider the following 80 determinations of the daily emission (in tons) of sulfur oxides from an industrial plant:

15.8	26.4	17.3	11.2	23.9	24.8	18.7	13.9	9.0	13.2
22.7	9.8	6.2	14.7	17.5	26.1	12.8	28.6	17.6	23.7
26.8	22.7	18.0	20.5	11.0	20.9	15.5	19.4	16.7	10.7
19.1	15.2	22.9	26.6	20.4	21.4	19.2	21.6	16.9	19.0
18.5	23.0	24.6	20.1	16.2	18.0	7.7	13.5	23.5	14.5
14.4	29.6	19.4	17.0	20.8	24.3	22.5	24.6	18.4	18.1
8.3	21.9	12.3	22.3	13.3	11.8	19.3	20.0	25.7	31.8
25.9	10.5	15.9	27.5	18.1	17.9	9.4	24.1	20.1	28.5

Since the largest observation is 31.8, the smallest is 6.2, and the range is 25.6, we might choose the six classes having the limits 5.0–9.9, 10.0–14.9, ..., 30.0–34.9, we might choose the seven classes 5.0–8.9, 9.0–12.9, ..., 29.0–32.9, or we might choose the nine classes 5.0–7.9, 8.0–10.9, ..., 29.0–31.9. Note that in each case **the classes do not overlap, they accommodate all the data, and they are all of the same size.**

Deciding on the second of these classifications, we now tally the 80 observations and obtain the results shown in the following table:

Class limits	*Tally*	*Frequency*
5.0– 8.9	///	3
9.0–12.9	~~////~~ ~~////~~	10
13.0–16.9	~~////~~ ~~////~~ ////	14
17.0–20.9	~~////~~ ~~////~~ ~~////~~ ~~////~~ ~~////~~	25
21.0–24.9	~~////~~ ~~////~~ ~~////~~ //	17
25.0–28.9	~~////~~ ////	9
29.0–32.9	//	2
	Total	80

Note that the class limits are given to as many decimal places as the original data. Had the original data been given to two decimal places, we would have used the class limits 5.00–8.99, 9.00–12.99, ..., 29.00–32.99, and if they had been rounded to the nearest ton, we would have used the class limits 5–8, 9–12, ..., 29–32.

In the preceding example, the data may be thought of as values of a continuous random variable, but if we use classes such as 5.0–9.0, 9.0–13.0, ..., 29.0–33.0, there exists the possibility of ambiguities: 9.0 could go into the first class or into the second, 13.0 could go into the second class or into the third, and so on. To avoid this difficulty, we can let the first class go from 4.95 to 8.95, the second from 8.95 to 12.95, ..., and the last from 28.95 to 32.95. We refer to these new limits as the **class boundaries**, and it should be noted that there will be no ambiguities even though the boundaries overlap. These class boundaries are, so to speak, "impossible" values, since the data were given to only one decimal. In practice, we use the class boundaries rather than the original class limits chiefly when we want to stress the fact that we are dealing with continuous kinds of measurements.

As we pointed out earlier, once data have been grouped, each observation has lost its identity in the sense that its exact value is no longer known. This may lead to difficulties when we want to give further descriptions of the data, but we can avoid them by representing each observation in a class by its midpoint, called the **class mark**. In general, the class marks of a frequency distribution are obtained by averaging successive class limits or successive class boundaries. If the classes of a distribution are all of equal length, as in our example, we refer to the common interval between any successive class marks as the **class interval** of the distribution. Note that the class interval may also be obtained from the difference between any successive class boundaries, but not from the difference between successive class limits.

EXAMPLE With reference to the distribution of the sulfur oxides data, find (a) the class marks, and (b) the class interval.

Solution (a) The class marks are $\frac{5.0 + 8.9}{2} = 6.95$, $\frac{9.0 + 12.9}{2} = 10.95$, 14.95, 18.95, 22.95, 26.95, and 30.95. (b) The class interval is $10.95 - 6.95 = 4$.

■

There are several alternative forms of distributions into which data are sometimes grouped. Foremost among these are the "less than," "or less," "more than," and "or more" **cumulative distributions**. A cumulative "less than" distribution shows the total number of observations that are less than given values. These values must be class boundaries or appropriate class limits, but they may not be class marks.

EXAMPLE Convert the distribution of the sulfur oxides emission data into a distribution showing how many of the observations are less than 4.95, less than 8.95, less than 12.95, ..., and less than 32.95.

Solution Since none of the values is less than 4.95, 3 are less than 8.95, $3 + 10 = 13$ are less than 12.95, $3 + 10 + 14 = 27$ are less than 16.95, ..., and all 80 are less than 32.95, we have:

Tons of sulfur oxides	*Cumulative frequency*
less than 4.95	0
less than 8.95	3
less than 12.95	13
less than 16.95	27
less than 20.95	52
less than 24.95	69
less than 28.95	78
less than 32.95	80

Note that instead of "less than 4.95" we could also have written "less than 5.0" or "4.9 or less."

■

Cumulative "more than" and "or more" distributions are constructed similarly by adding the frequencies, one by one, starting at the other end of the frequency distribution. In practice, "less than" cumulative distributions are used most widely, and it is not uncommon to refer to "less than" cumulative distributions simply as cumulative distributions.

If it is desired to compare frequency distributions, it may be necessary (or at least advantageous) to convert them into **percentage distributions**. We simply divide each class frequency by the total frequency (the total number of observations in the distribution) and multiply by 100; in this way we indicate what percentage of the data falls into each class of the distribution. The same can also be done with cumulative distributions, thus converting them to **cumulative percentage distributions**.

EXAMPLE The following are the numbers of workers absent from a factory on 50 working days:

13	5	13	37	10	16	2	11	6	12
8	21	12	11	7	7	9	16	49	18
3	11	19	6	15	10	14	10	7	24
11	3	6	10	4	6	32	9	12	7
29	12	9	19	8	20	15	5	17	10

Use the six classes 0–4, 5–9, 10–14, 15–19, 20–24, and "25 or more" to construct (a) a frequency distribution, and (b) a percentage distribution.

Solution (a) Tallying the data into the six classes, we get

Number of absences	*Frequency*
0–4	4
5–9	15
10–14	16
15–19	8
20–24	3
25 or more	4

and (b) dividing each class frequency by 50 and multiplying by 100, that is, multiplying each class frequency by 2, we get

Number of absences	*Percentage*
0–4	8
5–9	30
10–14	32
15–19	16
20–24	6
25 or more	8

Note that in these tables we replaced the five classes 25–29, 30–34, 35–39, 40–44, and 45–49, which would ordinarily have been needed, by a single class having no upper limit. This is called an **open class**, and it avoids having to use a large number of classes that are empty or have very small frequencies. Note also that when distributions are shown in their final form, the tally is usually omitted.

■

2.3

GRAPHS OF FREQUENCY DISTRIBUTIONS

Properties of frequency distributions relating to their shape are best exhibited by means of graphs, and in this section we shall introduce some of the most widely used forms of graphical presentations of frequency distributions, percentage distributions, and cumulative distributions.

The most common form of graphical presentation of a frequency distribution is the **histogram**. The histogram of a frequency distribution is constructed of adjacent rectangles, the heights of the rectangles represent the class frequencies and the bases of the rectangles extend between successive class boundaries. A histogram of the sulfur oxides emission data is shown in Figure 2.5.

In connection with histograms, it is sometimes preferable to look on the areas of the rectangles, rather than their heights, as representing the class frequencies.

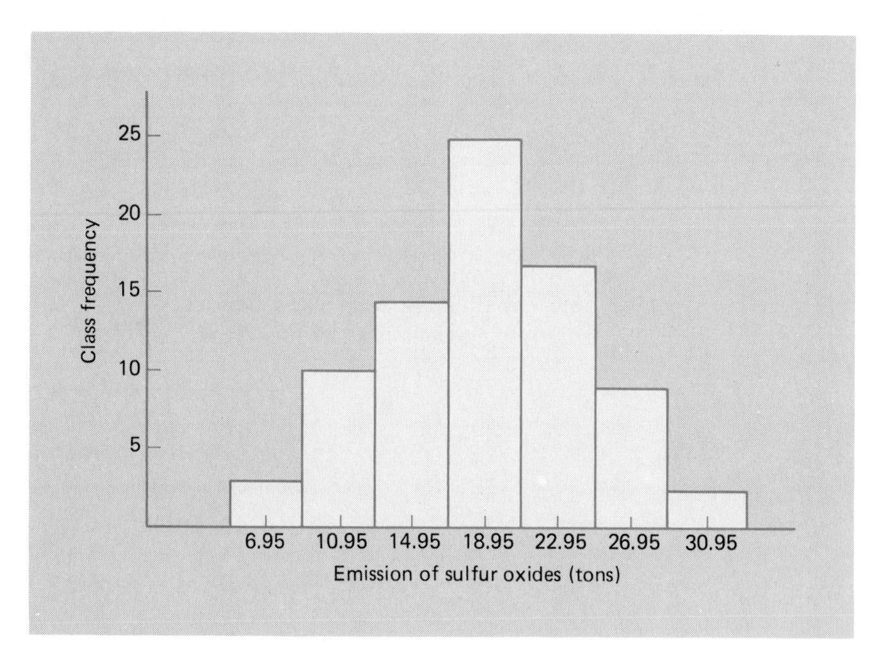

FIGURE 2.5
Histogram.

This applies in particular to situations where we wish to approximate histograms with smooth curves or where there are classes of unequal length (see Exercise 2.18 on page 20).

Similar to histograms are **bar charts**, such as the one shown in Figure 2.1; the heights of the rectangles, or bars, again represent the class frequencies, but there is no pretense of having a continuous horizontal scale. Another alternative way of presenting frequency distributions in graphical form is the **frequency polygon**. Here the class frequencies are plotted at the class marks, that is, we plot the points (x_i, f_i) where x_i is the class mark of the ith class and f_i is the corresponding frequency, and the successive points are connected by means of straight lines after having added classes with zero frequencies at both ends of the distribution.

Inspection of the graph of a frequency distribution often brings out features that are not immediately apparent from the data themselves. Aside from the fact that such a graph presents a good overall picture of the data, it can also emphasize irregularities and unusual features. For instance, outlying observations which somehow do not fit the overall picture, that is, the overall pattern of the data, may be due to errors of measurement, equipment failure, and similar causes. Also, the fact that a histogram or a frequency polygon exhibits two or more **modes** (maxima) can provide pertinent information. The appearance of two modes may imply, for example, a shift in the process that is being measured, or it may imply that the data come from several sources. With some experience one learns to spot such irregularities or anomalies, and an experienced engineer would find it just as surprising if the histogram of a distribution of integrated-circuit failure times were symmetrical as if a distribution of American men's hat sizes were bimodal.

Sometimes it can be enough to draw a histogram in order to solve an engineering problem.

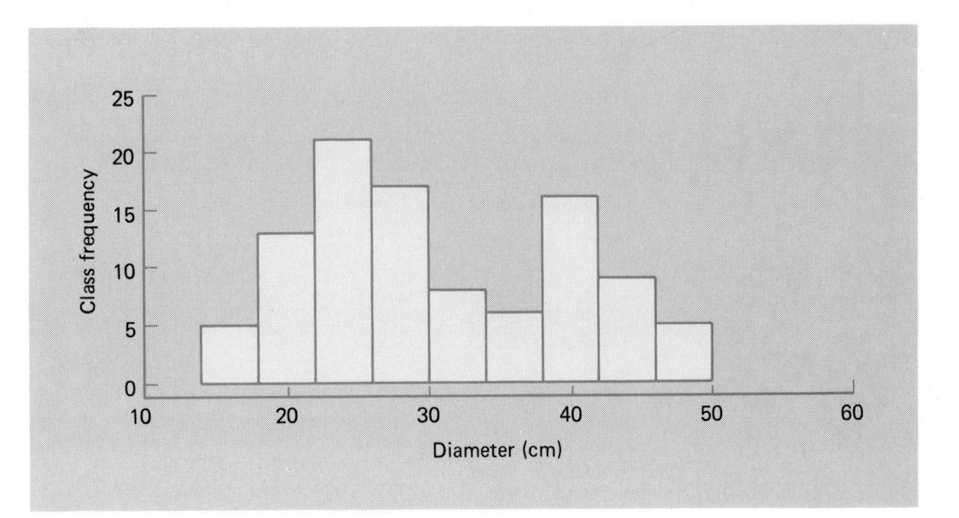

FIGURE 2.6
Histogram of pellet diameter.

EXAMPLE A metallurgical engineer was experiencing trouble with a grinding operation. After some thought he collected a sample of pellets used for grinding and measured their diameters. His histogram is displayed in Figure 2.6. It exhibits two distinct peaks, one for a group of pellets centered near 25 and the other centered near 40.

By getting his supplier to do a better sort, so all pellets would be essentially from the first group, the engineer completely solved his problem.

■

As illustrated by the next example, not all histograms are symmetric.

EXAMPLE A computer scientist, trying to optimize system performance, collected data on the time, in microseconds, between requests for a particular process service.

2,808	4,201	3,848	9,112	2,082	5,913	1,620	6,719	21,657
3,072	2,949	11,768	4,731	14,211	1,583	9,853	78,811	6,655
1,803	7,012	1,892	4,227	6,583	15,147	4,740	8,528	10,563
43,003	16,723	2,613	26,463	34,867	4,191	4,030	2,472	28,840
24,487	14,001	15,241	1,643	5,732	5,419	28,608	2,487	995
3,116	29,508	11,440	28,336	3,440				

Draw a histogram.

Solution The histogram of this interrequest time data, shown in Figure 2.7, has a long right-hand tail.

■

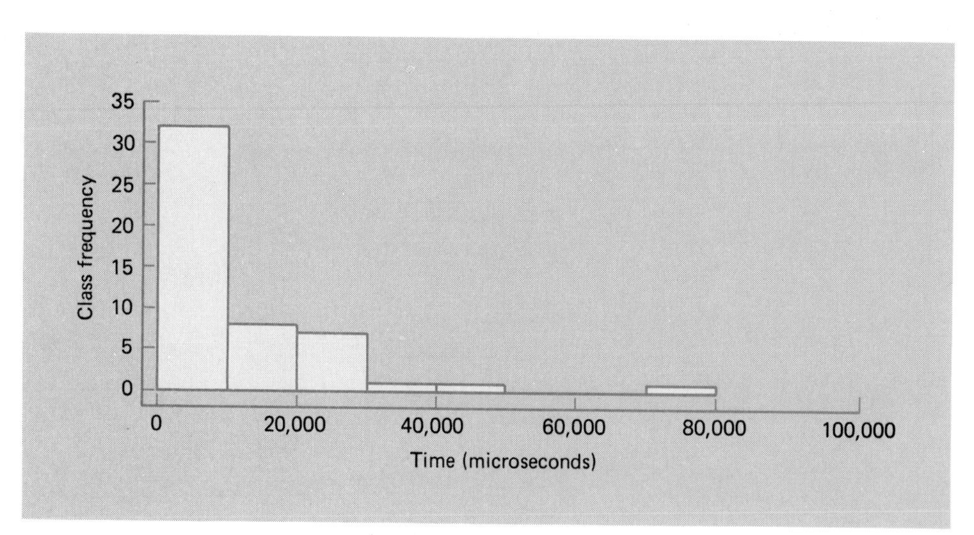

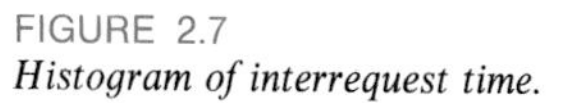
FIGURE 2.7
Histogram of interrequest time.

EXAMPLE Compressive strength was measured on 58 specimens of a new aluminum alloy undergoing development as a material for airplanes.

66.4	67.7	68.0	68.0	68.3	68.4	68.6	68.8	68.9	69.0	69.1
69.2	69.3	69.3	69.5	69.5	69.6	69.7	69.8	69.8	69.9	70.0
70.0	70.1	70.2	70.3	70.3	70.4	70.5	70.6	70.6	70.8	70.9
71.0	71.1	71.2	71.3	71.3	71.5	71.6	71.6	71.7	71.8	71.8
71.9	72.1	72.2	72.3	72.4	72.6	72.7	72.9	73.1	73.3	73.5
74.2	74.5	75.3								

Draw a histogram that is scaled to have a total area of 1 unit. For reasons to become apparent in Chapter 6, we call the vertical scale density.

Solution We make the height of each rectangle equal to *relative frequency/width*, so that its area equals the relative frequency. The resulting histogram, constructed by computer, has a nearly symmetric shape. (See Figure 2.8.) We have also graphed a continuous curve that approximates the overall shape. In Chapter 6, we will be introduced to this bell-shaped family of curves.

This example suggests that histograms, for observations that come from a continuous scale, can be approximated by smooth curves.

■

Cumulative distributions are usually presented graphically in the form of **ogives**, which are similar to frequency polygons, except that we plot the cumulative frequencies at the class boundaries instead of the ordinary frequencies at the class

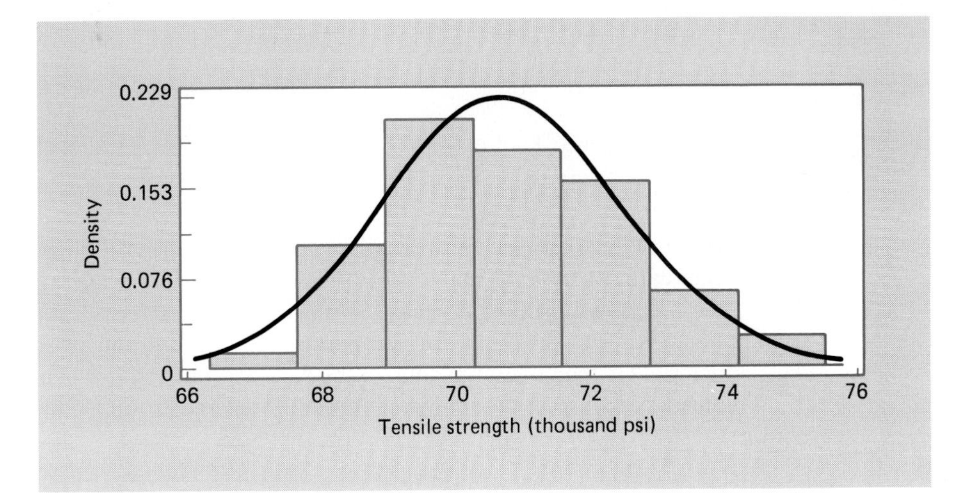

FIGURE 2.8
Histogram of aluminum alloy compressive strength.

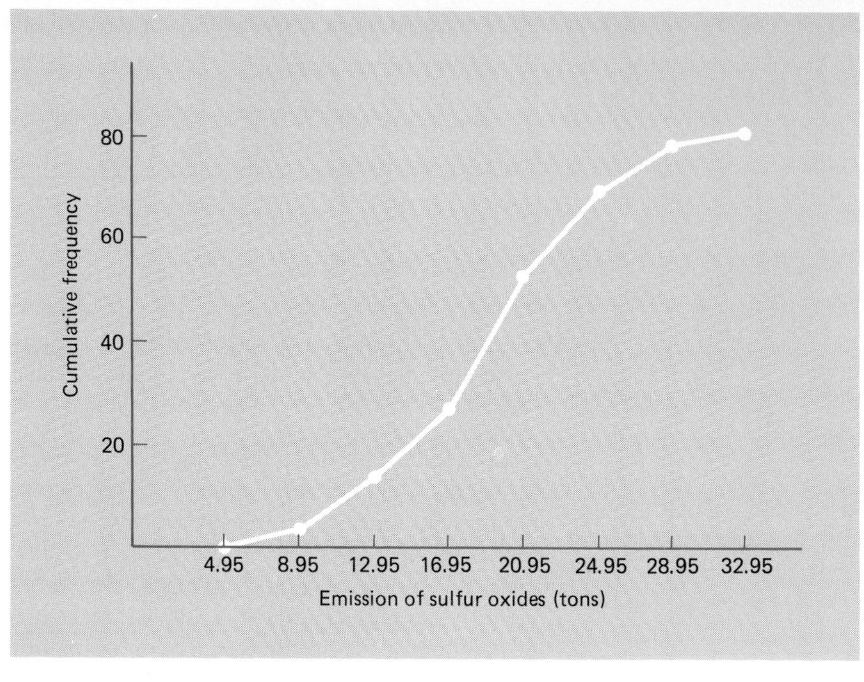

FIGURE 2.9
Ogive.

marks. The resulting points are again connected by means of straight lines, as shown in Figure 2.9, which represents the cumulative "less than" distribution of the sulfur oxides emission data on page 8.

2.4 STEM-AND-LEAF DISPLAYS

In the two preceding sections we directed our attention to the grouping of relatively large sets of data, with the objective of putting such data into a manageable form. As we saw, this entailed some loss of information. In recent years, similar techniques have been proposed for the preliminary exploration of small sets of data, which yield a good overall picture of the data without any loss of information.

To illustrate, consider the following humidity readings rounded to the nearest percent:

29	44	12	53	21	34	39	25	48	23
17	24	27	32	34	15	42	21	28	37

Proceeding as in Section 2.2, we might group these data into the following distribution:

Humidity readings	*Tally*	*Frequency*
10–19	///	3
20–29	~~////~~ ///	8
30–39	~~////~~	5
40–49	///	3
50–59	/	1

where the tally pictures the overall pattern of the data like a histogram (or bar chart) lying on its side.

If we wanted to avoid the loss of information inherent in the preceding table, we could replace the tally marks with the last digits of the corresponding readings, getting

```
10–19 | 2 7 5
20–29 | 9 1 5 3 4 7 1 8
30–39 | 4 9 2 4 7
40–49 | 4 8 2
50–59 | 3
```

This can also be written as

```
1* | 2 7 5
2* | 9 1 5 3 4 7 1 8
3* | 4 9 2 4 7
4* | 4 8 2
5* | 3
```

where * is a placeholder for 0, 1, 2, 3, 4, 5, 6, 7, 8, or 9, or simply as

```
1 | 2 7 5
2 | 9 1 5 3 4 7 1 8
3 | 4 9 2 4 7
4 | 4 8 2
5 | 3
```

In either of these final forms, the table is called a **stem-and-leaf display** (or simply a **stem-leaf display**)—each line is a **stem** and each digit on a stem to the right of the vertical line is a **leaf**. To the left of the vertical line are the **stem labels**, which, in our example, are 1*, 2*, . . . , 5*, or 1, 2, . . . , 5.

Essentially, a stem-and-leaf display presents the same picture as the corresponding tally, yet it retains all the original information. For instance, if a stem-and-leaf display has the stem

1.2∗ | 3 5 2 0 8

the corresponding data are 1.23, 1.25, 1.22, 1.20, and 1.28, and if a stem-and-leaf display has the stem

0.3∗∗ | 17 03 55 89

with two-digit leaves, the corresponding data are 0.317, 0.303, 0.355, and 0.389.

There are various ways in which stem-and-leaf displays can be modified to meet particular needs (see Exercises 2.23 and 2.24), but we shall not go into this here in any detail as it has been our objective to present only one of the relatively new techniques, which come under the general heading of **exploratory data analysis**.

EXERCISES

2.1 Accidents at a potato chip plant are categorized according to the area injured.

fingers	17
eyes	5
arm	2
leg	1

Draw a Pareto chart.

2.2 Damage at a paper mill (thousands dollars) due to breakage can be divided according to the product:

toilet paper	132
hand towels	85
napkins	43
12 other products	50

(a) Draw a Pareto chart.

What percent of the loss occurs in making

(b) toilet paper;

(c) toilet paper or hand towels?

2.3 The following are 15 measurements of the boiling point of a silicon compound (in degrees Celsius): 166, 141, 136, 153, 170, 162, 155, 146, 183, 157, 148, 132, 160, 175, and 150. Construct a dot diagram.

2.4 Measurements of the "fraction solar" of solar-assisted heat pumps in a city in the Southwest are grouped into the following classes: 0.50–0.59, 0.60–0.69, 0.65–0.79, and 0.85–0.94. If the measurements are all rounded to two decimals, none is less than 0.50 and none exceeds 0.94, explain where difficulties might arise.

2.5 The weights of certain mineral specimens, given to the nearest tenth of an ounce, are grouped into a table having the classes 10.5–11.4, 11.5–12.4, 12.5–13.4, and 13.5–14.4 ounces. Find

(a) the class marks; (b) the class boundaries;
(c) the class interval.

2.6 With reference to the preceding exercise, is it possible to determine from the grouped data how many of the mineral specimens weigh

(a) less than 11.5 ounces; (b) more than 11.5 ounces;
(c) at least 12.4 ounces; (d) at most 12.4 ounces;
(e) from 11.5 to 13.5 ounces?

2.7 The following are measurements of the breaking strength (in ounces) of a sample of 60 linen threads:

32.5	15.2	35.4	21.3	28.4	26.9	34.6	29.3	24.5	31.0
21.2	28.3	27.1	25.0	32.7	29.5	30.2	23.9	23.0	26.4
27.3	33.7	29.4	21.9	29.3	17.3	29.0	36.8	29.2	23.5
20.6	29.5	21.8	37.5	33.5	29.6	26.8	28.7	34.8	18.6
25.4	34.1	27.5	29.6	22.2	22.7	31.3	33.2	37.0	28.3
36.9	24.6	28.9	24.8	28.1	25.4	34.5	23.6	38.4	24.0

Group these measurements into a distribution having the classes 15.0–19.9, 20.0–24.9, . . . , 35.0–39.9 and construct a histogram.

2.8 Convert the distribution obtained in the preceding exercise into a cumulative "less than" distribution and graph its ogive.

2.9 The class marks of a distribution of temperature readings (given to the nearest degree Celsius) are 16, 25, 34, 43, 52, and 61. Find

(a) the class boundaries; (b) the class limits.

2.10 The following are the ignition times of certain upholstery materials exposed to a flame, given to the nearest hundredth of a second:

2.58	2.51	4.04	6.43	1.58	4.32	2.20	4.19
4.79	6.20	1.52	1.38	3.87	4.54	5.12	5.15
5.50	5.92	4.56	2.46	6.90	1.47	2.11	2.32
6.75	5.84	8.80	7.40	4.72	3.62	2.46	8.75
2.65	7.86	4.71	6.25	9.45	12.80	1.42	1.92
7.60	8.79	5.92	9.65	5.09	4.11	6.37	5.40
11.25	3.90	5.33	8.64	7.41	7.95	10.60	3.81
3.78	3.75	3.10	6.43	1.70	6.40	3.24	1.79
4.90	3.49	6.77	5.62	9.70	5.11	4.50	2.50
5.21	1.76	9.20	1.20	6.85	2.80	7.35	11.75

Group these figures into a table with a suitable number of equal classes and construct a histogram.

2.11 Convert the distribution obtained in Exercise 2.10 into a cumulative "less than" percentage distribution and plot its ogive.

2.12 In a 2-week study of the productivity of workers, the following data were obtained on the total number of acceptable pieces which 100 workers produced:

65	36	49	84	79	56	28	43	67	36
43	78	37	40	68	72	55	62	22	82
88	50	60	56	57	46	39	57	73	65
59	48	76	74	70	51	40	75	56	45
35	62	52	63	32	80	64	53	74	34
76	60	48	55	51	54	45	44	35	51
21	35	61	45	33	61	77	60	85	68
45	53	34	67	42	69	52	68	52	47
62	65	55	61	73	50	53	59	41	54
41	74	82	58	26	35	47	50	38	70

Group these figures into a distribution having the classes 20–29, 30–39, 40–49, . . . , and 80–89, and plot a histogram.

2.13 Convert the distribution obtained in Exercise 2.12 into a cumulative "less than" percentage distribution and plot its ogive.

2.14 The following are the number of automobile accidents that occurred at 60 major intersections in a certain city during the Fourth of July weekend:

0	2	5	0	1	4	1	0	2	1
5	0	1	3	0	0	2	1	3	1
1	4	0	2	4	1	2	4	0	4
3	5	0	1	3	6	4	2	0	2
0	2	3	0	4	2	5	1	1	2
2	1	6	5	0	3	3	0	0	4

Group these data into a frequency distribution showing how often each of the values occurs and draw a bar chart.

2.15 Convert the distribution obtained in Exercise 2.14 into a cumulative "or more" distribution and draw its ogive.

2.16 Categorical distributions are often presented graphically by means of **pie charts**, in which a circle is divided into sectors proportional in size to the frequencies (or percentages) with which the data are distributed among the categories. Draw a pie chart to represent the following data, obtained in a study in which 40 drivers were asked to judge the maneuverability of a certain make car: Very good, good, good, fair, excellent, good, good, good, very good, poor, good, good, good, good, very good, good, fair, good, good, very poor, very good, fair, good, good, excellent, very good, good, good, good, fair, fair, very good, good, very good, excellent, very good, fair, good, good, and very good.

2.17 The pictogram of Figure 2.10 is intended to illustrate the fact that per capita income in the United States doubled from $6,000 in 1977 to $12,000 in 1986. Does this pictogram convey a "fair" impression of the actual change? If not, state how it might be modified.

2.18 Convert the distribution of the sulfur oxides emission data on page 8 into a distribution having the classes 5.0–8.9, 9.0–20.9, 21.0–28.9, and 29.0–32.9. Draw two histograms of this distribution, one in which the class frequencies are given by the heights of the rectangles and one in which the class frequencies are given by the areas of the rectangles. Explain why the first of these histograms gives a very misleading picture.

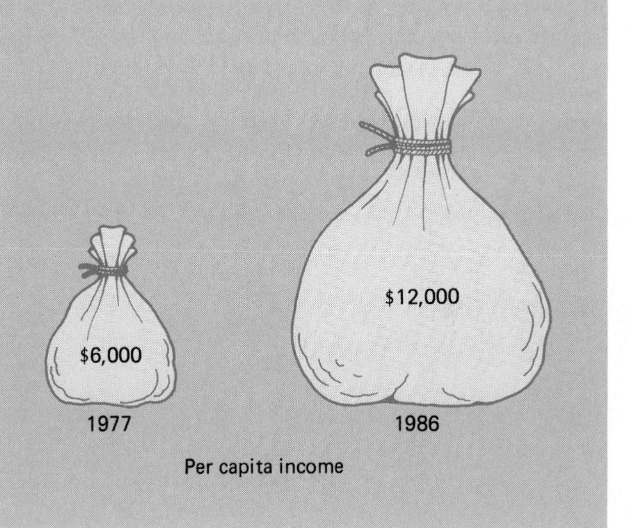

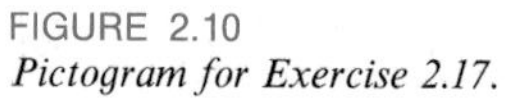
FIGURE 2.10
Pictogram for Exercise 2.17.

2.19 Given a set of observations $x_1, x_2, \ldots,$ and x_n, we define their **empirical cumulative distribution** as the function whose values $F(x)$ equal the proportion of the observations less than or equal to x. Graph the empirical cumulative distribution for the 15 measurements of Exercise 2.3.

2.20 The following are figures on a well's daily production of oil in barrels: 214, 203, 226, 198, 243, 225, 207, 203, 208, 200, 217, 202, 208, 212, 205, and 220. Construct a stem-and-leaf display with the stem labels 19∗, 20∗, ..., and 24∗.

2.21 The following are determinations of a river's annual maximum flow in cubic meters per second: 405, 355, 419, 267, 370, 391, 612, 383, 434, 462, 288, 317, 540, 295, and 508. Construct a stem-and-leaf display with two-digit leaves.

2.22 List the data that correspond to the following stems of stem-and-leaf displays:

(a) 1∗ | 3 2 5 7 1 4 8 (b) 23 | 4 0 0 1 6

(c) 2∗∗ | 35 18 57 03 (d) 3.2 | 1 7 4 4 3

2.23 If we want to construct a stem-and-leaf display with more stems than there would be otherwise, we might use ∗ as a placeholder for 0, 1, 2, 3, and 4, and • as a placeholder for 5, 6, 7, 8, and 9. For the humidity readings on page 16, we would thus get the **double-stem display**

Stem	Leaves
1∗	2
1•	7 5
2∗	1 3 4 1
2•	9 5 7 8
3∗	4 2 4
3•	9 7
4∗	4 2
4•	8
5∗	3

where we doubled the number of stems by cutting the interval covered by each stem in half. Construct a double-stem display with one-digit leaves for the data of Exercise 2.12.

2.24 If we want to construct a stem-and-leaf display equivalent to a distribution with a class interval of 2, we can use * as a placeholder for 0 and 1, *t* for 2 and 3, *f* for 4 and 5, *s* for 6 and 7, and • for 8 and 9. The resulting stem-and-leaf display is called a **five-stem display**.

(a) The following are the IQ's of 20 applicants to an undergraduate engineering program: 109, 111, 106, 106, 125, 108, 115, 109, 107, 109, 108, 110, 112, 104, 110, 112, 128, 106, 111, and 108. Construct a five-stem display with one-digit leaves.

(b) The following is part of a five-stem display:

Stem	Leaves
53*f*	5 4 4 4 5 4
53*s*	6 7 6 6
53•	9 8
54*	1

List the corresponding measurements.

2.5 DESCRIPTIVE MEASURES

Given a set of n measurements or observations, $x_1, x_2, \ldots, x_n$, there are many ways in which we can describe their center (middle, or central location). Most popular among these are the **arithmetic mean** and the **median**, although other kinds of "averages" are sometimes used for special purposes. The arithmetic mean—or, more succinctly, the **mean**—is defined by the formula

Mean

$$\bar{x} = \frac{\sum_{i=1}^{n} x_i}{n}$$

To emphasize that it is based on a set of observations, we often refer to $\bar{x}$ as the **sample mean**.

Sometimes it is preferable to use the **median** as a descriptive measure of the center, or location, of a set of data. This is true, particularly, if it is desired to minimize the calculations or if it is desired to eliminate the effect of extreme (very large or very small) values. The median of n observations $x_1, x_2, \ldots, x_n$ can be defined loosely as the "middlemost" value once the data are arranged according to size. More precisely, if the observations are arranged according to size and n is an odd number, the median is the value of the observation numbered $\frac{n+1}{2}$; if n is an even number, the median is defined as the mean (average) of the observations numbered $\frac{n}{2}$ and $\frac{n+2}{2}$.

EXAMPLE Find the mean and the median of 15, 14, 2, 27, and 13.

Solution The mean is

$$\bar{x} = \frac{15 + 14 + 2 + 27 + 13}{5} = 14.2$$

and the median is the third largest value, namely, 14.

■

EXAMPLE Find the mean and the median of 11, 9, 17, 19, 4, and 15.

Solution The mean is

$$\bar{x} = \frac{11 + 9 + 17 + 19 + 4 + 15}{6} = 12.5$$

and the median, the mean of the third and fourth largest values, is 13.

■

Although the mean and the median each provides a single number to represent an entire set of data, the mean is usually preferred in problems of estimation and other problems of statistical inference. An intuitive reason for preferring the mean is that the median does not utilize all the information contained in the observations. A related reason is that the median is generally subject to greater chance fluctuations, that is, it is apt to vary more from sample to sample. This important concept of "sampling variability" will be explored in detail in Chapter 6.

The following is an example where the median actually gives a more useful description of a set of data than the mean.

EXAMPLE A small company employs four young engineers, who each earn $24,000, and the owner (also an engineer), who gets $114,000. Comment on the claim that on the average the company pays $42,000 to its engineers and, hence, is a good place to work.

Solution The mean of the five salaries is $42,000, but it hardly describes the situation. The median, on the other hand, is $24,000 and it is at least representative of what a young engineer earns with this firm. Moneywise, the company is not such a good place for young engineers.

■

This example illustrates that there is always an inherent danger in summarizing a set of data by means of a single number.

One of the most important characteristics of almost any set of data is that the values are not all alike; indeed, the extent to which they are unalike, or vary among themselves, is of basic importance in statistics. Measures such as the mean and median describe one important aspect of a set of data—their "middle" or their "average"—but they tell us nothing about this other basic characteristic. We observe that the dispersion of a set of data is small if the values are closely bunched about their mean, and that it is large if the values are scattered widely about their mean. It would seem reasonable, therefore, to measure the variation of a set of data in terms of the amounts by which the values deviate from their mean. If a set of numbers $x_1, x_2, \ldots, x_n$ has mean $\bar{x}$, the differences $x_1 - \bar{x}, x_2 - \bar{x}, \ldots, x_n - \bar{x}$ are called the **deviations from the mean**. It suggests itself that we might use their average as a measure of variation in the data set. Unfortunately, this will not do. As the reader will be asked to show in Exercise 2.42 on page 33, the sum of the deviations is always zero. That is $\sum_{i=1}^{n} (x_i - \bar{x}) = 0$, so the mean of the deviations is always zero.

An alternative approach is to work with the squares of the deviations from the reader will be asked to show in Exercise 2.42 on page 33, the sum of the deviations $x_1, x_2, \ldots, x_n$ measures essentially the average of their squared deviations from their mean $\bar{x}$, and it is defined by the formula

Variance

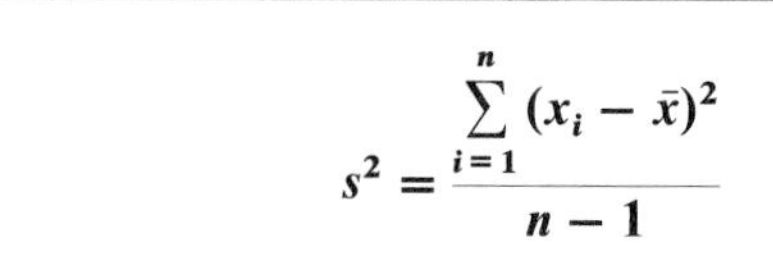

$$s^2 = \frac{\sum_{i=1}^{n} (x_i - \bar{x})^2}{n - 1}$$

Our reason for dividing by $n - 1$ instead of n is that there are only $n - 1$ independent deviations $x_i - \bar{x}$. Because their sum is always zero, the value of any particular one is always equal to the negative of the sum of the other $n - 1$ deviations.

Consistent with the terminology of Chapters 4 and 5, we define the **standard deviation** of n observations $x_1, x_2, \ldots, x_n$ as the square root of their variance, namely,

Standard deviation

$$s = \sqrt{\frac{\sum_{i=1}^{n} (x_i - \bar{x})^2}{n - 1}}$$

The standard deviation is by far the most generally useful measure of variation. Its advantage over the variance is that it is given in the same units as the observations.

EXAMPLE The delay times (handling, setting, and positioning the tools) for cutting six parts on an engine lathe are 0.6, 1.2, 0.9, 1.0, 0.6, and 0.8 minutes. Calculate s.

Solution First we calculate the mean:

$$\bar{x} = \frac{0.6 + 1.2 + 0.9 + 1.0 + 0.6 + 0.8}{6} = 0.85$$

Then we set up the work required to find $\sum (x_i - \bar{x})^2$ in the following table:

x_i	$x_i - \bar{x}$	$(x_i - \bar{x})^2$
0.6	−0.25	0.0625
1.2	0.35	0.1225
0.9	0.05	0.0025
1.0	0.15	0.0225
0.6	−0.25	0.0625
0.8	−0.05	0.0025
		0.2750

Then we divide 0.2750 by 6 − 1 = 5, take the square root, and get

$$s = \sqrt{\frac{0.2750}{5}} = 0.23$$

■

The standard deviation and the variance are measures of **absolute variation**, that is, they measure the actual amount of variation present in a set of data, and they depend on the scale of measurement. To compare the variation in several sets of data, it is generally desirable to use a measure of **relative variation**; for instance, the **coefficient of variation**, which gives the standard deviation as a percentage of the mean.

Coefficient of variation

$$V = \frac{s}{\bar{x}} \cdot 100$$

EXAMPLE Measurements made with one micrometer of the diameter of a ball bearing have a mean of 3.92 mm and a standard deviation of 0.015 mm, whereas measurements made with another micrometer of the unstretched length of a spring have a mean of 1.54 inches and a standard deviation of 0.008 inch. Which of these two measuring instruments is relatively more precise?

Solution For the first micrometer the coefficient of variation is

$$V = \frac{0.015}{3.92} \cdot 100 = 0.38\,\%$$

and for the second micrometer the coefficient of variation is

$$V = \frac{0.008}{1.54} \cdot 100 = 0.52\,\%$$

Thus, the measurements made with the first micrometer are relatively more precise.

■

In this section, we have limited the discussion to the mean, the median, the variance, and the standard deviation, but there are many other ways of describing sets of data.

2.6 QUARTILES AND OTHER PERCENTILES

In addition to the median, which divides a set of data into halves, we can consider other division points. When an ordered data set is divided into quarters, the resulting division points are called the sample **quartiles**. The **first quartile**, Q_1, is a value that has one-fourth, or 25%, of the observations below its value. The first quartile is also the sample 25th **percentile** $P_{0.25}$. More generally, we define the sample 100 pth percentile as follows.

> **The sample 100 pth percentile is a value such that at least 100 p % of the observations are at or below this value and at least 100 $(1 - p)$ % are at or above this value.**

As in the case of the median, which is the 50th percentile, this may not uniquely define a percentile. For simplicity, if more than one observation satisfies the definition, we will take their mean. (Most computer packages linearly interpolate between the two adjacent values.) For moderate or large sample sizes the convention used to locate the point in the interval is inconsequential.

The quartiles are the 25th, 50th, and 75th percentiles.

first quartile	Q_1 = 25th percentile
second quartile	Q_2 = 50th percentile
third quartile	Q_3 = 75th percentile

EXAMPLE Obtain the quartiles and the 95th percentile for the sulfur emission data on page 8.

Solution The ordered data are:

6.2	7.7	8.3	9.0	9.4	9.8	10.5	10.7	11.0	11.2	11.8
12.3	12.8	13.2	13.3	13.5	13.9	14.4	14.5	14.7	15.2	15.5
15.8	15.9	16.2	16.7	16.9	17.0	17.3	17.5	17.6	17.9	18.0
18.0	18.1	18.1	18.4	18.5	18.7	19.0	19.1	19.2	19.3	19.4
19.4	20.0	20.1	20.1	20.4	20.5	20.8	20.9	21.4	21.6	21.9
22.3	22.5	22.7	22.7	22.9	23.0	23.5	23.7	23.9	24.1	24.3
24.6	24.6	24.8	25.7	25.9	26.1	26.4	26.6	26.8	27.5	28.5
28.6	29.6	31.8								

The first quartile must have at least $\frac{1}{4} \times 80 = 20$ observations at or below its value and at least $\frac{3}{4} \times 80 = 60$ at or above. Both the 20th and 21st smallest values satisfy the criterion, so we take their mean.

$$Q_1 = \frac{14.7 + 15.2}{2} = 14.95$$

The second quartile, or median, is the mean of the 40th and 41st ordered observations

$$Q_2 = \frac{19.0 + 19.1}{2} = 19.05$$

while the third quartile is the mean of the 59th and 60th:

$$Q_3 = \frac{22.9 + 23.0}{2} = 22.95$$

To obtain the 95th percentile $P_{0.95}$, we determine that $0.95 \times 80 = 76$ of the observations must lie at or below and $0.05 \times 80 = 4$ must be at or above. Again, a mean is appropriate.

$$P_{0.95} = \frac{27.5 + 28.5}{2} = 28.0$$

The 95th percentile provides a useful description regarding days of high emission. On only 5% of the days are more than 28.0 tons of sulfur put into the air.

In the context of monitoring high values, we also record that the maximum emission was 31.8.

■

The **minimum** and **maximum** observations also convey information concerning the amount of variability present in a set of data. We encounter the **range** = *maximum* − *minimum* in Section 8.1. Care must be taken when interpreting the range since a single large or small observation can greatly inflate its value.

The amount of variation in the middle half of the data is described by the

$$\textbf{interquartile range} = \textit{third quartile} - \textit{first quartile} = Q_3 - Q_1$$

EXAMPLE Obtain the range and interquartile range for the sulfur emission data on page 8.

Solution From the previous example,

$$\text{range} = \text{maximum} - \text{minimum} = 31.8 - 6.2 = 25.6 \text{ tons}$$

$$\text{interquartile range} = Q_3 - Q_1 = 22.95 - 14.95 = 8.00 \text{ tons}$$

■

Boxplots

The summary information contained in the quartiles is highlighted in a graphic display called a **boxplot**. The center half of the data, extending from the first to the third quartile, is represented by a rectangle. The median is identified by a bar within this box. A line extends from the third quartile to the maximum and another line extends from the first quartile to the minimum. (For large data sets the lines may only extend to the 95th and 5th percentiles.)

Figure 2.11 gives the boxplot for the sulfur emission data on page 8. The symmetry seen in the histogram is also evident in this boxplot.

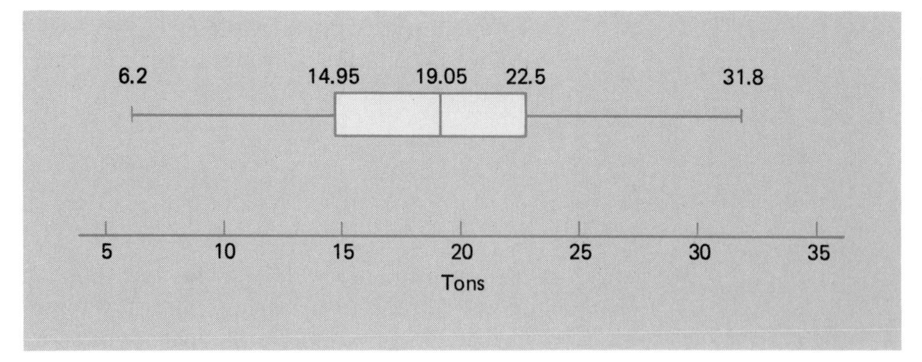

FIGURE 2.11
Boxplot of the sulfur emission data.

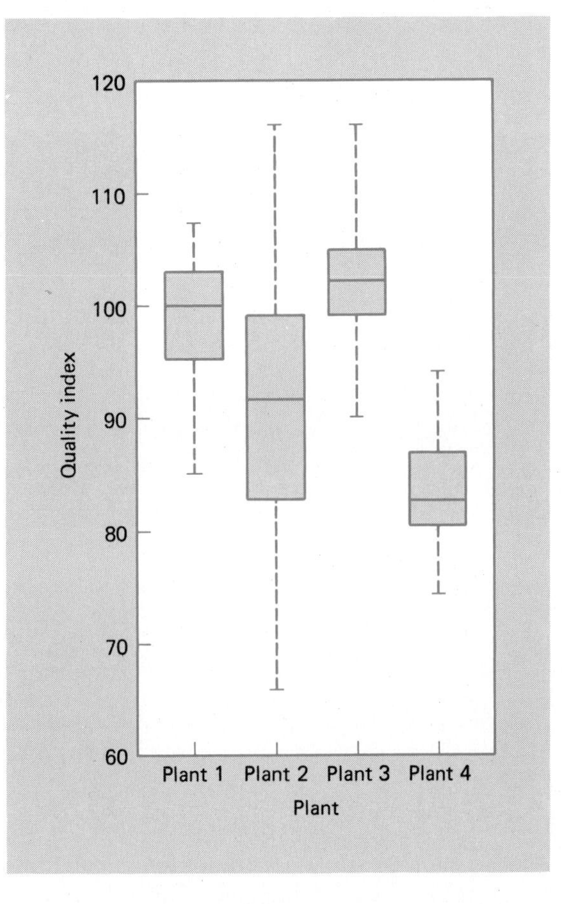

FIGURE 2.12
Boxplot of the quality index.

Boxplots are particularly effective for graphically portraying comparisons among sets of observations. They are easy to understand and have a high visual impact.

Sometimes, with rather complicated components like hard disk drives or RAM chips for computers, quality is quantified as an index with target value 100. Typically, a quality index will be based upon the deviations of several physical characteristics from their engineering specifications. Figure 2.12 shows the quality index at four manufacturing plants. It is clear from this graphic that plant 2 needs to reduce its variability and that plants 2 and 4 need to improve their quality level.

We conclude this section with a warning. Sometimes it is the trend over time that is the most important feature of the data. This feature would be lost entirely if the set of data were summarized in a dot diagram, stem-and-leaf display, or boxplot. Figure 2.13 illustrates this point by a time plot of the ozone in October, in Dobson units, over a region of the South Pole. The apparent downward trend, if real, is of major scientific interest and, maybe, vital to life.

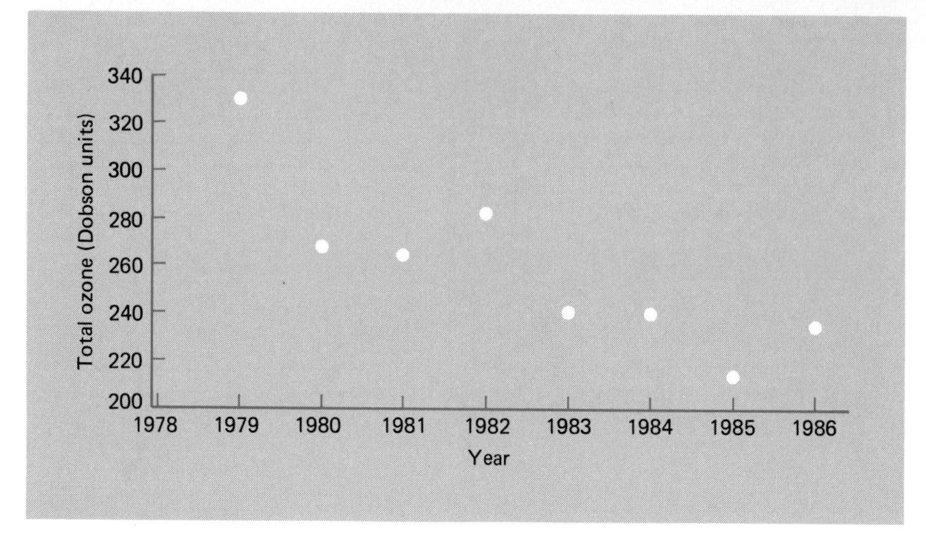

FIGURE 2.13
The monthly average total atmospheric ozone, for October, over the South Polar latitudes.

2.7
THE CALCULATION OF $\bar{x}$ AND s

In this section we discuss methods for calculating $\bar{x}$ and s for **raw** (ungrouped) as well as grouped data. These methods are particularly well suited for small hand-held calculators, and they are both rapid and accurate.

The calculation of $\bar{x}$ for ungrouped data does not pose any problems; we have only to add the values of the observations and divide by n. On the other hand, the calculation of s^2 is usually cumbersome if we directly use the formula defining s^2 on page 24. Instead, we shall use the algebraically equivalent form

Variance (computing formula)

$$s^2 = \frac{n \cdot \sum_{i=1}^{n} x_i^2 - \left(\sum_{i=1}^{n} x_i\right)^2}{n(n-1)}$$

which requires less labor to evaluate with a calculator. (In Exercise 2.43 on page 33 the reader will be asked to show that this formula is, in fact, equivalent to the one on page 24.)

EXAMPLE Find the mean and the standard deviation of the following miles per gallon obtained in 20 test runs performed on urban roads with an intermediate-size car:

19.7	21.5	22.5	22.2	22.6
21.9	20.5	19.3	19.9	21.7
22.8	23.2	21.4	20.8	19.4
22.0	23.0	21.1	20.9	21.3

Solution Using a calculator, we find that the sum of these figures is 427.7 and that the sum of their squares is 9,173.19. Consequently,

$$\bar{x} = \frac{427.7}{20} = 21.38$$

and

$$s^2 = \frac{20(9{,}173.19) - (427.7)^2}{20 \cdot 19} = 1.412$$

and it follows that $s = 1.19$. In computing the necessary sums we usually retain all decimal places, but as in this example, at the end we usually round to one more decimal than we had in the original data.

■

See Exercise 2.50 for a computer calculation.

Not too many years ago, one of the main reasons for grouping data was to simplify the calculation of descriptions such as the mean and the standard deviation. With easy access to statistical calculators and computers, this is no longer the case, but we shall nevertheless discuss here the calculation of $\bar{x}$ and s from grouped data, since some data (for instance, from government publications) may be available only in grouped form.

To calculate $\bar{x}$ and s from grouped data, we shall have to make some assumption about the distribution of the values within each class. If we represent all values within a class by the corresponding class mark, the sum of the x's and the sum of their squares can now be written

$$\sum_{i=1}^{k} x_i f_i \quad \text{and} \quad \sum_{i=1}^{k} x_i^2 f_i$$

where x_i is the class mark of the ith class, f_i is the corresponding class frequency, and k is the number of classes in the distribution. Substituting these sums into the formula for $\bar{x}$ and the computing formula for s^2, we get

Mean and variance (grouped data)

$$\bar{x} = \frac{\sum_{i=1}^{k} x_i f_i}{n}$$

$$s^2 = \frac{n \cdot \sum_{i=1}^{k} x_i^2 f_i - \left(\sum_{i=1}^{k} x_i f_i\right)^2}{n(n-1)}$$

EXAMPLE Use the distribution obtained on page 8 to calculate the mean and the variance of the sulfur oxides emission data.

Solution Recording the class marks and the class frequencies in the first two columns, and the products $x_i f_i$ and $x_i^2 f_i$ in the third and fourth columns, we obtain

x_i	f_i	$x_i f_i$	$x_i^2 f_i$
6.95	3	20.85	144.9075
10.95	10	109.50	1,199.0250
14.95	14	209.30	3,129.0350
18.95	25	473.75	8,977.5625
22.95	17	390.15	8,953.9425
26.95	9	242.55	6,536.7225
30.95	2	61.90	1,915.8050
	80	1,508.00	30,857.0000

Then, substitution into the formulas yields

$$\bar{x} = \frac{1{,}508}{80} = 18.85$$

and

$$s^2 = \frac{80(30{,}857) - (1{,}508)^2}{80 \cdot 79} = 30.77$$

■

EXERCISES

2.25 The following are the numbers of twists that were required to break 12 forged alloy bars: 33, 24, 39, 48, 26, 35, 38, 54, 23, 34, 29, and 37. Find
(a) the mean; (b) the median.

2.26 With reference to the preceding exercise, find s using
(a) the formula that defines s; (b) the computing formula for s.

2.27 If the mean annual salary paid to the chief executives of three engineering firms is \$125,000, can one of them receive \$400,000?

2.28 By mistake, an instructor erased the grade that one of ten students received. If the nine other students got 43, 66, 74, 90, 40, 52, 70, 78, and 92 and the mean of all ten grades is 67, what grade did the instructor erase?

2.29 The following are the numbers of minutes that a person had to wait for the bus to work on 15 working days: 10, 1, 13, 9, 5, 9, 2, 10, 3, 8, 6, 17, 2, 10, and 15. Find
(a) the mean; (b) the median.
(c) Draw a boxplot.

2.30 With reference to the preceding exercise, find s^2 using
(a) the formula that defines s^2; (b) the computing formula for s^2.

2.31 Records show that in Phoenix, Arizona, the normal daily maximum temperature for each month is 65, 69, 74, 84, 93, 102, 105, 102, 98, 88, 74, and 66 degrees Fahrenheit. Verify that the mean of these figures is 85 and comment on the claim that, in Phoenix, the average daily maximum temperature is a very comfortable 85 degrees.

2.32 With reference to Exercise 2.20 on page 21, find the mean and the median of the data on the well's daily production of oil.

2.33 With reference to Exercise 2.21 on page 21, find the standard deviation of the data on the river's annual maximum flow.

2.34 For the four observations 9, 7, 15, 5;
(a) calculate the deviations $(x_i - \bar{x})$ and check that they add to 0;
(b) calculate the variance and the standard deviation.

2.35 With reference to Exercise 2.12 on page 20, draw a boxplot.

2.36 With reference to Exercise 2.3 on page 18, calculate $\bar{x}$ and s.

2.37 Find the mean and the standard deviation of the 20 humidity readings on page 16 by using
(a) the raw (ungrouped) data;
(b) the distribution obtained on the next page.

2.38 Use the distribution obtained in Exercise 2.7 on page 19 to find the mean and the variance of the breaking strengths.

2.39 Use the distribution obtained in Exercise 2.10 on page 19 to find the mean and the standard deviation of the ignition times. Also determine the coefficient of variation.

2.40 Use the distribution obtained in Exercise 2.12 on page 20 to find the coefficient of variation of the productivity data.

2.41 In three recent years, the price of copper was 69.6, 66.8, and 66.3 cents per pound, and the price of bituminous coal was 19.43, 19.82, and 22.40 dollars per short ton. Which of the two sets of prices is relatively more variable?

2.42 Show that $\sum_{i=1}^{n} (x_i - \bar{x}) = 0$ for any set of observations $x_1, x_2, \ldots,$ and x_n.

2.43 Show that the computing formula for s^2 on page 30 is equivalent to the one used to define s^2 on page 24.

2.44 If data are coded so that $x_i = c \cdot u_i + a$, show that $\bar{x} = c \cdot \bar{u} + a$ and $s_x = c \cdot s_u$.

2.45 To find the **median** of a distribution obtained for n observations, we first determine the class into which the median must fall. Then, if there are j values in this class and k values below it, the median is located $\frac{(n/2) - k}{j}$ of the way into this class, and to obtain the median we multiply this fraction by the class interval and add the result to the lower class boundary of the class into which the median must fall. This method is based on the assumption that the observations in each class are "spread uniformly" throughout the class interval, and this is why we count $\frac{n}{2}$ of the observations instead of $\frac{n+1}{2}$ as on page 22. To illustrate, let us refer to the distribution of the sulfur oxides emission data on page 8. Since $n = 80$, it can be seen that the median must fall into the class 17.0–20.9, and since $j = 25$ and $k = 27$, it follows that the median is $16.95 + \frac{40 - 27}{25} \cdot 4 = 19.03$.

(a) Find the median of the distribution of the absenteeism data given on page 11.

(b) Use the distribution obtained in Exercise 2.7 on page 19 to find the median of the grouped breaking strengths.

(c) Use the distribution obtained in Exercise 2.10 on page 19 to find the median of the grouped ignition times.

2.46 For each of the following distributions, decide whether it is possible to find the mean and/or the median. Explain your answers.

(a)

Grade	*Frequency*
40–49	5
50–59	18
60–69	27
70–79	15
80–89	6

(b)

IQ	*Frequency*
less than 90	3
90– 99	14
100–109	22
110–119	19
more than 119	7

(c)

Weight	*Frequency*
100 or less	41
101–110	13
111–120	8
121–130	3
131–140	1

2.47 To find the quartiles Q_1 and Q_3 for grouped data, we proceed as in Exercise 2.45, but count $\frac{n}{4}$ and $\frac{3n}{4}$ of the observations instead of $\frac{n}{2}$.

(a) With reference to the distribution of the sulfur oxides emission data on page 8, find Q_1, Q_3, and the interquartile range.

(b) Find Q_1 and Q_3 for the distribution of the absenteeism data given on page 11.

2.48 If k sets of data consist, respectively, of $n_1, n_2, \ldots, n_k$ observations and have the means $\bar{x}_1, \bar{x}_2, \ldots, \bar{x}_k$, then the overall mean of all the data is given by the formula

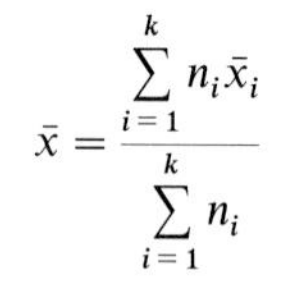

$$\bar{x} = \frac{\sum_{i=1}^{k} n_i \bar{x}_i}{\sum_{i=1}^{k} n_i}$$

(a) The average annual salaries paid to top-level management in three companies are \$84,000, \$92,000, and \$89,000. If the respective numbers of top-level executives in these companies are 4, 15, and 11, find the average salary paid to these 30 executives.

(b) In a nuclear engineering class there are 22 juniors, 18 seniors, and 10 graduate students. If the juniors averaged 71 in the midterm examination, the seniors averaged 78, and the graduate students averaged 89, what is the mean for the entire class?

2.49 The formula of the preceding exercise is a special case of the following formula for the **weighted mean**

$$\bar{x}_w = \frac{\sum_{i=1}^{k} w_i x_i}{\sum_{i=1}^{k} w_i}$$

where w_i is a weight indicating the relative importance of the ith observation.

(a) If an instructor counts the final examination in a course four times as much as each 1-hour examination, what is the weighted average grade of a student who received grades of 69, 75, 56, and 72 in four 1-hour examinations and a final examination grade of 78?

(b) From 1975 to 1980 the cost of food increased by 53% in a certain city, the cost of housing increased by 40%, and the cost of transportation increased by 34%. If the average salaried worker spent 28% of his or her income on food, 35% on housing, and 14% on transportation, what is the combined percentage increase in the cost of these items?

2.50 Modern computer software packages have come a long way towards removing the tedium of calculating statistics. *MINITAB* is one common and easy to use package. We illustrate the use of the computer using *MINITAB* commands. Other easy-to-use packages have a quite similar command structure.

The command **SET C1** (set data in column 1) places the data in column 1 of worksheet within the computer.

```
SET C1
3 7 2 9 4
END
```

The command **DESCRIBE C1** produces several statistics, including $\bar{x}$ and s along with the sample size n. (This package uses a convention slightly different from ours when calculating quartiles.)

We use this command with the five observations above in C1.

```
DESCRIBE C1
```

Part of the output is

```
          N     MEAN    MEDIAN     STDEV
C1        5     5.00      4.00      2.92

        MIN      MAX        Q1        Q3
C1     2.00     9.00      2.50      8.00
```

Use *MINITAB*, or some other statistical package, to find $\bar{x}$ and s for

(a) the decay times on page 157; (b) the interrequest times on page 14.

2.51 (Further *MINITAB* commands and calculations) With the observations on the strength (in pounds per square inch) of 2×4 pieces of lumber already set in C1, we can order and then print the data using the two commands

```
SORT C1 SET C2
PRINT C2

1325  1419  1490  1633  1645  1655  1710  1712  1725  1727  1745
1828  1840  1856  1859  1867  1889  1899  1943  1954  1976  2046
2061  2104  2168  2199  2276  2326  2403  2983
```

Descriptive statistics may be obtained from individual commands

```
     MEAN C1
MEAN     =          1908.8
     STAND C1
ST.DEV. =            327.12
     MEDIAN C1
MEDIAN =            1863.0
```

MINITAB also creates histograms.

```
          HISTOGRAM C1

Histogram of C1       N = 30

Midpoint     Count
    1400         3    ***
    1600         3    ***
    1800        12    ************
    2000         5    *****
    2200         4    ****
    2400         2    **
    2600         0
    2800         0
    3000         1    *
```

And, it creates stem-and-leaf displays.

```
          STEM-AND-LEAF C1

Stem-and-leaf of C1          N = 30
Leaf Unit = 100

     1    1 3
     3    1 44
    11    1 66677777
  (10)    1 8888888999
     9    2 00111
     4    2 23
     2    2 4
     1    2
     1    2 9
```

Here the first column gives the cumulative number of points, counting in from each end, as an aid to locating percentiles. The second column is the stem to which the leaves are attached. From the ordered data

(a) obtain the quartiles;
(b) construct a histogram with five cells and locate the mean, median, Q_1, and Q_3 on the horizontal axes;
(c) repeat parts (a) and (b) with the aluminum alloy data on page 15.

2.8 REVIEW EXERCISES

2.52 From 2000 computer chips inspected by the manufacturer, the following numbers of defects were recorded.

holes not open	182
holes too large	55
poor connections	31
incorrect size chip	5
other	7

Draw a Pareto chart.

2.53 Draw
(a) a frequency table of the aluminum alloy strength data on page 15 using the classes 66.0–67.4, 67.5–68.9, 69.0–70.4, 70.5–71.9, 72.0–73.4, 73.5–74.9, 75.0–76.4;
(b) a histogram using the frequency table in part (a).

2.54 Draw
(a) a frequency table of the interrequest time data on page 14 using the intervals 0–2,499, 2,500–4,999, 5,000–9,999, 10,000–19,999, 20,000–39,999, 40,000–59,999, 60,000–79,999;
(b) a histogram using the frequency table in part (a) (note that the intervals are unequal, so make the height of the rectangle equal relative frequency/width).

2.55 Direct evidence of Newton's universal law of gravitation was provided from a renowned experiment by Henry Cavendish (1731–1810). In the experiment, masses of objects were determined by weighing and the measured force of attraction was used to calculate the density of the earth. The values of the earth's density, in time order by row, are

5.36	5.29	5.58	5.65	5.57	5.53	5.62	5.29
5.44	5.34	5.79	5.10	5.27	5.39	5.42	5.47
5.63	5.34	5.46	5.30	5.75	5.68	5.85	

(Source: *Philosophical Transactions* **17** (1798): 469.)
(a) Find the mean and standard deviation.
(b) Find the median, Q_1, and Q_3.
(c) Plot the observations versus time order. Is there any obvious trend?

2.56 J. J. Thomson (1856–1940) discovered the electron by isolating negatively charged particles for which he could measure the mass-charge ratio. This ratio appeared to be constant over a wide range of experimental conditions and, consequently, could be a characteristic of a new particle. His observations, from two different cathode-ray tubes that used air as the gas, are: (Source: *Philosophical Magazine* **44**; 5 (1897): 293.)

Tube 1	0.57	0.34	0.43	0.32	0.48	0.40	0.40
Tube 2	0.53	0.47	0.47	0.51	0.63	0.61	0.48

(a) Draw a dot diagram with solid dots for Tube 1 observations and circles for Tube 2 observations.
(b) Calculate the mean and standard deviation for the Tube 1 observations.
(c) Calculate the mean and standard deviation for the Tube 2 observations.

2.57 With reference to Exercise 2.56,
(a) calculate the median, maximum, minimum, and range for the Tube 1 observations;
(b) calculate the median, maximum, minimum, and range for the Tube 2 observations.

2.58 A. A. Michelson (1852–1931) made many series of measurements of the speed of light. Using a revolving mirror technique, he obtained

12, 30, 30, 27, 30, 39, 18, 27, 48, 24, 18

for the differences (velocity of light in air) $-(299{,}700)$ km/s. (Source: *The Astrophysical Journal* **65** (1927): 11)
(a) Draw a dot diagram.
(b) Find the median and the mean. Locate both on the dot diagram.
(c) Find the variance and standard deviation.

2.59 With reference to Exercise 2.58
(a) find the quartiles;
(b) find the minimum, maximum, range, and interquartile range;
(c) draw a boxplot.

2.60 A civil engineer monitors water quality by measuring the amount of suspended solids in a sample of river water. Over 11 weekdays, she observed

14, 12, 21, 28, 30, 63, 29, 63, 55, 19, 20

suspended solids (parts per million).
(a) Draw a dot diagram.
(b) Find the median and the mean. Locate both on the dot diagram.
(c) Find the variance and standard deviation.

2.61 With reference to Exercise 2.60
(a) find the quartiles;
(b) find the minimum, maximum, range, and the interquartile range;
(c) construct a boxplot.

2.62 With reference to the aluminum alloy strength data in the example on page 15,
(a) find the quartiles;
(b) find the minimum, maximum, range, and interquartile range;
(c) find the 10th percentile and the 20th percentile.

```
                                              UNIVARIATE
VARIABLE= STRENGTH

                MOMENTS                                QUANTILES(DEF=4)

N              30  SUM WGTS       30    100% MAX     2983       99% 2983
MEAN      1908.77  SUM         57263     75% Q3   2071.75       95% 2664
STD DEV   327.115  VARIANCE   107004     50% MED     1863       90% 2321
SKEWNESS  1.11841  KURTOSIS  2.88335     25% Q1    1711.5       10% 1504.3
USS     112404829  CSS       3103123      0% MIN     1325        5% 1376.7
CV        17.1375  STD MEAN  59.7228                             1% 1325

                                        RANGE           1658
                                        Q3-Q1         360.25
```

FIGURE 2.14
Selected SAS output to describe the lumber strength data from Exercise 2.51.

2.63 With reference to Exercise 2.62, draw a boxplot.

2.64 With reference to the aluminum alloy strength data in the example on page 15, make a stem-and-leaf display.

2.65 Measurements of the ignition temperature of a gas vary from 1,161 to 1,319 degrees Fahrenheit. Construct a table with eight equal classes into which these data might be grouped. Give

(a) the class limits; (b) the class marks;
(c) the class boundaries; (d) the class interval.

2.66 The following are 12 temperature readings at various locations in a large kiln (in degrees Fahrenheit): 475, 500, 460, 425, 460, 410, 470, 475, 460, 510, 450, and 415. Find

(a) the mean; (b) the median;
(c) the standard deviation.

2.67 In five tests, one student averaged 63.2 with a standard deviation of 3.3, whereas another student averaged 78.8 with a standard deviation of 5.3. Which student is relatively more consistent?

2.68 With reference to the lumber strength data in Exercise 2.51, the statistical software package *SAS* produced the output in Figure 2.14. Using this output

(a) identify the mean and standard deviation and compare these answers with the values given in Exercise 2.51;
(b) draw a boxplot.

2.9
CHECK LIST OF KEY TERMS (with page references)

Absolute variation 25
Arithmetic mean 22
Bar chart 13
Boxplot 28
Categorical distribution 8
Class boundary 9
Class frequency 8
Class interval 9
Class limit 8
Class mark 9
Coefficient of variation 25
Cumulative distribution 10

PROBABILITY

In the study of probability there are basically three kinds of questions: (1) What do we mean when we say that the probability of an event is, say, 0.50, 0.02, or 0.81? (2) How are the numbers we call probabilities determined, or measured in actual practice? (3) What are the mathematical rules which probabilities must obey?

After some mathematical preliminaries in Sections 3.1 and 3.2, we study the first two kinds of questions in Section 3.3 and the third kind of question in Sections 3.4 through 3.7. The related concept of a mathematical expectation is introduced in Section 3.8.

3.1 SAMPLE SPACES AND EVENTS

In statistics, a set of all possible outcomes of an experiment is called a **sample space**, owing to the fact that it usually consists of all the things that can happen when one takes a sample. Sample spaces are usually denoted by the letter S. To avoid misunderstandings about the words "experiment" and "outcome" as we have used them here, it should be understood that statisticians use these terms in a very wide sense. An **experiment** may consist of the simple process of noting whether a switch is turned on or off; it may consist of determining the time it takes a car to accelerate to 30 miles per hour; or it may consist of the complicated process of finding the mass of an electron. Thus, the **outcome** of an experiment may be a simple choice between two alternatives; it may be the result of a direct measurement or count; or it may be an answer obtained after extensive measurements and calculations.

When we study the outcomes of an experiment, we usually identify the various possibilities with numbers, points, or some other kinds of symbols. For instance, if four contractors bid on a highway construction job and we let a, b, c, and d denote that it is awarded to Mr. Adam, Mrs. Brown, Mr. Clark, or Ms. Dean, then the sample space for this experiment is the set $S = \{a, b, c, d\}$.

Also, if a government agency must decide where to locate two new computer research facilities and that (for a certain purpose) it is of interest to indicate how many of them will be located in Texas and how many in California, we can write the sample space as

$$S = \{(0, 0), (1, 0), (0, 1), (2, 0), (1, 1), (0, 2)\}$$

where the first coordinate is the number of research facilities that will be located in Texas and the second coordinate is the number that will be located in California. Geometrically, this sample space may be pictured as in Figure 3.1, from which it is apparent, for example, that in two of the six possibilities Texas and California will get equally many of the new research facilities. The use of points rather than letters or numbers has the advantage that it makes it easier to visualize the various possibilities, and perhaps discover some special features which several of the outcomes may have in common.

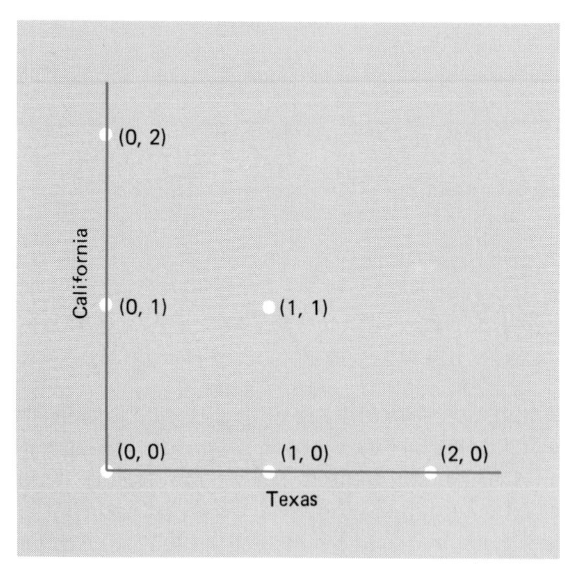

FIGURE 3.1
Sample space for the number of new computer research facilities to be located in Texas and in California.

Generally, sample spaces are classified according to the number of elements (points) that they contain. In the two preceding examples, the sample spaces had four and six elements, and they are both referred to as **finite**. Other examples of finite sample spaces are the one for the various ways in which a president and a vice-president can be selected from among the 25 members of a union local and the one for the various ways in which a student can answer the 12 questions on a true-false test. As we see on page 48, the first of these sample spaces has 600 elements and the other has 4,096.

The following are examples of sample spaces that are not finite. If persons checking the nitrogen oxide emission of cars are interested in the number of cars they have to inspect before they observe the first one that does not meet government regulations, it could be the first, the second, ..., the fiftieth, ..., and for all we know they may have to check thousands of cars before they find one that does not meet government regulations. Not knowing how far they may have to go, it is appropriate in an example like this to take as the sample space the whole set of natural numbers, of which there is a countable infinity. To go one step further, if they were interested in the nitrogen oxide emission of a given car in grams per mile, the sample space would have to consist of all the points on a continuous scale (a certain interval on the line of real numbers), of which there is a continuum.

In general, a sample space is said to be **discrete** if it has finitely many or a countable infinity of elements. If the elements (points) of a sample space constitute a continuum, for example, all the points on a line, all the points on a line segment, or all the points in a plane, the sample space is said to be **continuous**. In the remainder of this chapter we shall consider only discrete and mainly finite sample spaces.

In statistics, any subset of a sample space is called an **event**. By subset we mean any part of a set, including the whole set and, trivially, a set called the **empty set** and denoted by $\emptyset$, which has no elements at all. For instance, with reference to Figure 3.1,

$$C = \{(1, 0), (0, 1)\}$$

is the event that, between them, Texas and California will get one of the two research facilities,

$$D = \{(0, 0), (0, 1), (0, 2)\}$$

is the event that Texas will not get either of the two research facilities, and

$$E = \{(0, 0), (1, 1)\}$$

is the event that Texas and California will get equally many of the facilities. Note that events C and E have no elements in common—they are **mutually exclusive events**.

In many probability problems we are interested in events which can be expressed in terms of two or more events by forming **unions**, **intersections**, and

complements. Although the reader must surely be familiar with these terms, let us review briefly that if A and B are any two sets in a sample space S, their union $A \cup B$ is the subset of S that contains all elements that are either in A, in B, or in both; their intersection $A \cap B$ is the subset of S that contains all elements that are in both A and B; and the complement A' of A is the subset of S that contains all the elements of S that are not in A.

EXAMPLE With reference to the sample space of Figure 3.1 and the events C, D, and E just defined, list the outcomes comprising each of the following events and also express the events in words:

(a) $C \cup E$;
(b) $C \cap D$;
(c) D'.

Solution (a) Since $C \cup E$ contains all the elements that are in C, in E, or in both,

$$C \cup E = \{(1, 0), (0, 1), (0, 0), (1, 1)\}$$

is the event that neither Texas nor California will get both of the new research facilities. (b) Since $C \cap D$ contains all the elements that are in both C and D,

$$C \cap D = \{(0, 1)\}$$

is the event that Texas will not get either of the two new facilities and California will get one. (c) Since D' contains all the elements of the sample space that are not in D,

$$D' = \{(1, 0), (1, 1), (2, 0)\}$$

is the event that Texas will get at least one of the new computer research facilities.

■

Sample spaces and events, particularly relationships among events, are often depicted by means of **Venn diagrams** like those of Figures 3.2 and 3.4. In each case the sample space is represented by a rectangle, whereas events are represented by regions within the rectangle, usually by circles or parts of circles. The shaded regions of the four Venn diagrams of Figure 3.2 represent event A, the complement of event A, the union of events A and B, and the intersection of events A and B.

EXAMPLE If A is the event that a certain student is taking a course in calculus and B is the event that the student is taking a course in applied mechanics, what events are represented by the shaded regions of the four Venn diagrams of Figure 3.2?

Solution The shaded region of the first diagram represents the event that the student is taking a course in calculus; that of the second diagram represents the event that the student is not taking a course in calculus; that of the third diagram represents the

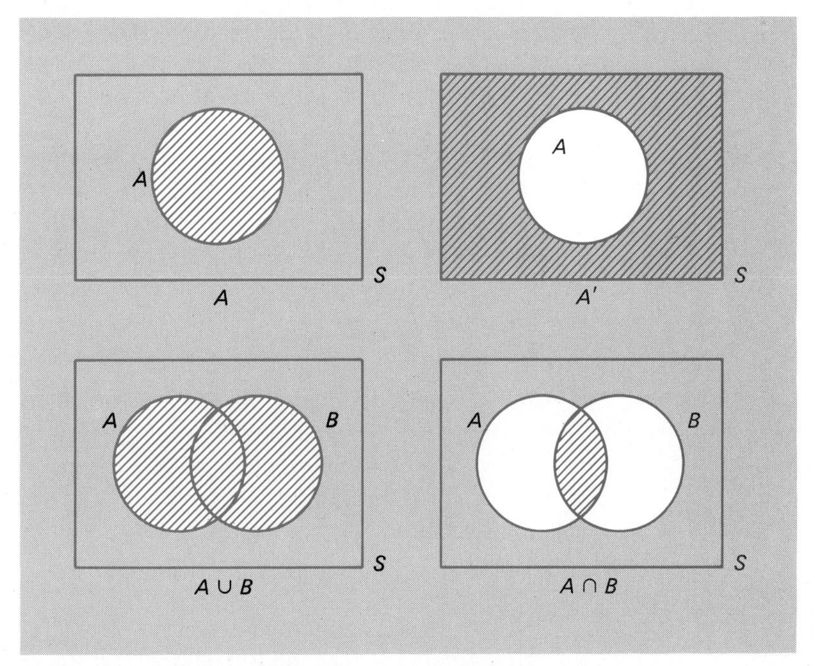

FIGURE 3.2
Venn diagrams.

event that the student is taking a course in calculus and/or a course in applied mechanics; and that of the fourth diagram represents the event that the student is taking a course in calculus as well as a course in applied mechanics.

■

Venn diagrams are often used to verify relationships among sets, thus making it unnecessary to give formal proofs based on the algebra of sets. To illustrate, let us show that $(A \cup B)' = A' \cap B'$, which expresses the fact that the complement of the union of two sets equals the intersection of their complements. To begin, note that the shaded region of the first Venn diagram of Figure 3.3 represents the set $(A \cup B)'$

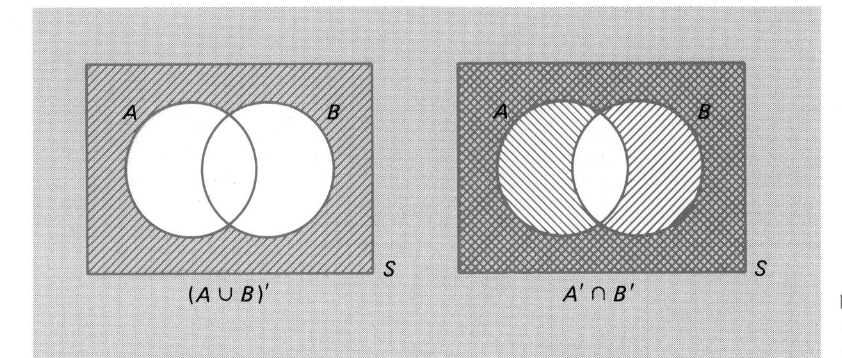

FIGURE 3.3
Use of Venn diagrams to show that $(A \cup B)' = A' \cap B'$.

(compare this diagram with the third diagram of Figure 3.2). The crosshatched region of the second Venn diagram of Figure 3.3 was obtained by shading the region representing A' with lines going in one direction and that representing B' with lines going in another direction. Thus, the crosshatched region represents the intersection of A' and B', and it can be seen that it is identical with the shaded region of the first Venn diagram of Figure 3.3.

When we deal with three events, we draw the circles as in Figure 3.4. In this diagram, the circles divide the sample space into eight regions, numbered 1 through 8, and it is easy to determine whether the corresponding events are parts of A or A', B or B', and C or C'.

EXAMPLE A manufacturer of small motors is concerned with three major types of defects. If A is the event that the shaft size is too large, B is the event that the windings are improper, and C is the event that the electrical connections are unsatisfactory, express in words what events are represented by the following regions of the Venn diagram of Figure 3.4.

(a) region 2
(b) regions 1 and 3 together
(c) regions 3, 5, 6, and 8 together

Solution (a) Since this region is contained in A and B but not in C, it represents the event that the shaft is too large and the windings improper, but the electrical connections are satisfactory. (b) Since this region is common to B and C, it represents the event that the windings are improper and the electrical connections are unsatisfactory. (c) Since this is the entire region outside A, it represents the event that the shaft size is not too large.

■

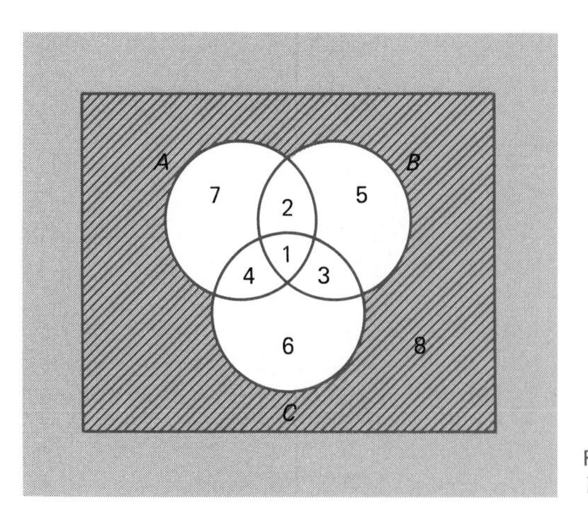

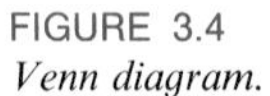
FIGURE 3.4
Venn diagram.

3.2 COUNTING

At times it can be quite difficult, or at least tedious, to determine the number of elements in a finite sample space by direct enumeration. To illustrate, suppose that a consumer testing service rates lawn mowers as being easy, average, or difficult to operate; as being expensive or inexpensive; and as being costly, average, or cheap to repair. In how many different ways can a lawn mower be rated by this testing service?

Clearly, there are many possibilities: a lawn mower may be rated easy to operate, inexpensive, but costly to repair; it may be rated difficult to operate, expensive, and cheap to repair; it may be rated as neither easy nor difficult to operate, inexpensive, with average cost of repairs; and so on. Continuing this way, we may be able to list all 18 possibilities, but the chances are that we will omit at least one or two.

To handle this kind of problem systematically, it helps to draw a **tree diagram** like that of Figure 3.5, where the three alternatives for ease of operation are

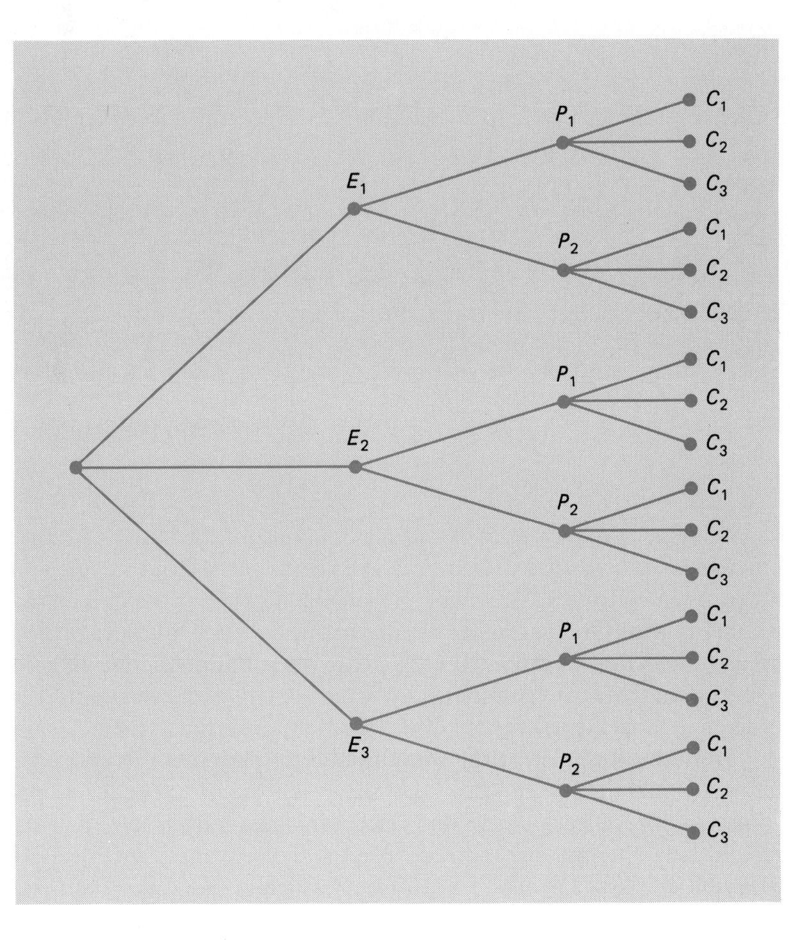

FIGURE 3.5
Tree diagram for rating of lawn mowers.

denoted by E_1, E_2, and E_3; the price is either P_1 or P_2; and the three alternatives for the cost of repairs are denoted by C_1, C_2, and C_3. Following a given path from left to right along the branches of the tree, we obtain a particular rating, namely, a particular element of the sample space, and it can be seen that altogether there are 18 possibilities.

This result could also have been obtained by observing that there are three E-branches, that each E-branch forks into two P-branches, and that each P-branch forks into three C-branches. Thus, there are $3 \cdot 2 \cdot 3 = 18$ combinations of branches, or paths. This result is a special case of the following theorem:

Multiplication of choices

> ***Theorem 3.1*** ***If sets $A_1, A_2, \ldots, A_k$ contain, respectively, $n_1, n_2, \ldots, n_k$ elements, there are $n_1 \cdot n_2 \cdots n_k$ ways of choosing first an element of A_1, then an element of $A_2, \ldots,$ and finally an element of A_k.***

In our example we had $n_1 = 3$, $n_2 = 2$, and $n_3 = 3$, and, hence, $3 \cdot 2 \cdot 3 = 18$ possibilities.

EXAMPLE In how many different ways can a union local with a membership of 25 choose a vice president and a president?

Solution Since the vice president can be chosen in 25 ways and, subsequently, the president in 24 ways, there are altogether $25 \cdot 24 = 600$ ways in which the whole choice can be made.

■

EXAMPLE If a test consists of 12 true-false questions, in how many different ways can a student mark the test paper with one answer to each question?

Solution Since each question can be answered in two ways, there are altogether

$$2 \cdot 2 \cdot 2 \cdot 2 \cdot 2 \cdot 2 \cdot 2 \cdot 2 \cdot 2 \cdot 2 \cdot 2 \cdot 2 = 2^{12} = 4{,}096 \text{ possibilities}$$

■

As in the first of these two examples, the rule for the multiplication of choices is often used when several choices are made from one set and we are concerned with the order in which they are made. In general, if r objects are chosen from a set of n distinct objects, any particular arrangement, or order, of these objects is called a **permutation**. For instance, 4 1 2 3 is a permutation of the first four positive integers, and Maine, Vermont, and Connecticut is a permutation, a particular ordered arrangement, of three of the six New England states.

To find a formula for the total number of permutations of r objects selected from a set of n distinct objects, we observe that the first selection is made from the whole set of n objects, the second selection is made from the $n - 1$ objects which remain after the first selection has been made, ..., and the rth selection is made from the $n - (r - 1) = n - r + 1$ objects which remain after the first $r - 1$ selections have been made. Therefore, by the rule for the multiplication of choices, the total number of permutations of r objects selected from a set of n distinct objects is

$$_nP_r = n(n-1)(n-2)\cdots(n-r+1)$$

for $r = 1, 2, \ldots, n$.

Since products of consecutive integers arise in many problems relating to permutations or other kinds of special selections, it will be convenient to introduce here the **factorial notation**, where $1! = 1$, $2! = 2\cdot 1 = 2$, $3! = 3\cdot 2\cdot 1 = 6$, $4! = 4\cdot 3\cdot 2\cdot 1 = 24, \ldots$, and in general $n! = n(n-1)(n-2)\cdots 3\cdot 2\cdot 1$. Also, to make various formulas more generally applicable, we let $0! = 1$ by definition.

To express the formula for $_nP_r$ in terms of factorials, we multiply and divide by $(n-r)!$ getting

$$_nP_r = \frac{n(n-1)(n-2)\cdots(n-r+1)(n-r)!}{(n-r)!} = \frac{n!}{(n-r)!}$$

To summarize,

Number of permutations of n objects taken r at a time

> ***Theorem 3.2** The number of permutations of r objects selected from a set of n distinct objects is*
>
> $$_nP_r = n(n-1)(n-2)\cdots(n-r+1)$$
>
> *or, in factorial notation,*
>
> $$_nP_r = \frac{n!}{(n-r)!}$$

Note that the second formula also holds for $r = 0$.

EXAMPLE In how many different ways can one make a first, second, third, and fourth choice among 12 firms leasing construction equipment?

Solution For $n = 12$ and $r = 4$, the first formula yields

$$_{12}P_4 = 12\cdot 11\cdot 10\cdot 9 = 11{,}880$$

and the second formula yields

$$_{12}P_4 = \frac{12!}{(12-4)!} = \frac{12!}{8!} = \frac{12\cdot 11\cdot 10\cdot 9\cdot 8!}{8!} = 11{,}880$$

■

EXAMPLE An electronic controlling mechanism requires five identical memory chips. In how many ways can this mechanism be assembled using 5 given chips?

Solution For $n = 5$ and $r = 5$, the first formula yields

$$_5P_5 = 5 \cdot 4 \cdot 3 \cdot 2 \cdot 1 = 120$$

and the second formula yields

$$_5P_5 = \frac{5!}{(5-5)!} = \frac{5!}{0!} = 5! = 120$$

■

The first formula for $_nP_r$ is generally easier to use unless we can refer to a table of factorials or use a calculator which directly yields factorials and/or ratios of factorials.

There are many problems in which we must find the number of ways in which r objects can be selected from a set of n objects, but we do not care about the order in which the selection is made. For instance, we may want to know in how many ways 3 of 20 laboratory assistants can be chosen to assist with an experiment. In general, there are $r!$ permutations of any r objects we select from a set of n distinct objects. So, the $_nP_r$ permutations of r objects selected from a set of n objects contain each set of r objects $r!$ times. Therefore, to find the number of ways in which r objects can be selected from a set of n distinct objects, also called the number of **combinations** of n objects taken r at a time and denoted by $_nC_r$ or $\binom{n}{r}$, we divide $_nP_r$ by $r!$ and get

Number of combinations of n objects taken r at a time

***Theorem 3.3** The number of ways in which r objects can be selected from a set of n distinct objects is*

$$\binom{n}{r} = \frac{n(n-1)(n-2)\cdots(n-r+1)}{r!}$$

or, in factorial notation,

$$\binom{n}{r} = \frac{n!}{r!\,(n-r)!}$$

EXAMPLE In how many different ways can 3 of 20 laboratory assistants be chosen to assist with an experiment?

Solution For $n = 20$ and $r = 3$, the first formula for $\binom{n}{r}$ yields

$$\binom{20}{3} = \frac{20 \cdot 19 \cdot 18}{3!} = 1{,}140$$

■

EXAMPLE In how many different ways can the director of a research laboratory choose two chemists from among seven applicants and three physicists from among nine applicants?

Solution The two chemists can be chosen in $\binom{7}{2} = 21$ ways, the three physicists can be chosen in $\binom{9}{3} = 84$ ways, so that the whole selection can be made in $21 \cdot 84 = 1{,}764$ ways.

■

EXERCISES

3.1 A technician has to check the suitability of three solid crystal lasers and two carbon dioxide lasers for a given task.

(a) Using two coordinates so that (2, 1), for example, represents the event that the technician will find two of the solid crystal lasers and one of the carbon dioxide lasers suitable for the task, draw a diagram similar to that of Figure 3.1 showing the 12 points of the sample space.

(b) If R is the event that equally many solid crystal lasers and carbon dioxide lasers are suitable for the task, T is the event that none of the carbon dioxide lasers is suitable for the task, and U is the event that fewer solid crystal lasers than carbon dioxide lasers are suitable for the task, express each of these events symbolically by listing its elements.

3.2 With reference to Exercise 3.1, which of the three pairs of events, R and T, R and U, and T and U, are mutually exclusive?

3.3 With reference to Exercise 3.1, list the outcomes comprising each of the following events, and also express the events in words.

(a) $R \cup U$ (b) $R \cap T$ (c) T'

3.4 With reference to the sample space of Figure 3.1, express each of the following events in words.

(a) $F = \{(1, 0), (1, 1)\}$ (b) $G = \{(0, 2), (1, 1), (2, 0)\}$
(c) $F \cap G$

3.5 To construct sample spaces for experiments in which we deal with nonnumerical data, we often code the various alternatives by assigning them numbers. For instance, if a mechanic is asked whether work on a certain model car is very easy, easy, average, difficult, or very difficult, we might assign these alternatives the codes 1, 2, 3, 4, and 5. If $A = \{3, 4\}$, $B = \{2, 3\}$, and $C = \{4, 5\}$, express each of the following symbolically by listing its elements and also in words.

(a) $A \cup B$ (b) $A \cap B$
(c) $A \cup B'$ (d) C'

3.6 With reference to Exercise 3.5, which of the three pairs of events, A and B, A and C, and B and C, are mutually exclusive?

3.7 Two professors and three graduate assistants are responsible for the supervision of a physics lab, and at least one professor and one graduate assistant has to be present at all times.

(a) Using two coordinates so that (1, 3), for example, represents the event that one professor and three graduate assistants are present, draw a diagram similar to that of Figure 3.1 showing the points of the corresponding sample space.

(b) Describe in words the events which are represented by $B = \{(1, 3), (2, 3)\}$, $C = \{(1, 1), (2, 2)\}$, and $D = \{(1, 2), (2, 1)\}$.

(c) With reference to part (b), express $C \cup D$ symbolically by listing its elements, and also express this event in words.

(d) With reference to part (b), are B and D mutually exclusive?

3.8 For each of the following experiments decide whether it would be appropriate to use a sample space which is finite, countably infinite, or continuous.

(a) The amount of cosmic radiation to which passengers are exposed during a transcontinental jet flight is measured by means of a suitable counter.

(b) Five of the members of a professional society with 12,600 members are chosen to serve on a nominating committee.

(c) An experiment is conducted to measure the heat of vaporization of water.

(d) A study is made to determine in how many of 450 airplane accidents the main cause is pilot error.

(e) Measurements are made to determine the uranium content of a certain ore.

(f) In a torture test, a watch is dropped from a tall building until it stops running.

3.9 In Figure 3.6, C is the event that an ore contains copper and U is the event that it contains uranium. Explain in words what events are represented by regions 1, 2, 3, and 4.

3.10 With reference to Exercise 3.9, what events are represented by

(a) regions 1 and 3 together; (b) regions 3 and 4 together;

(c) regions 1, 2, and 3 together?

3.11 With reference to Figure 3.4, what events are represented by

(a) region 5; (b) regions 4 and 6 together;

(c) regions 7 and 8 together; (d) regions 1, 2, 3, and 5 together?

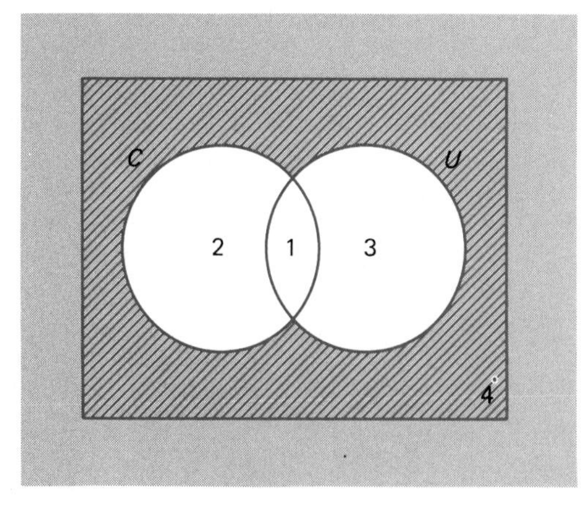

FIGURE 3.6
Venn diagram for Exercises 3.9 and 3.10.

3.12 With reference to Figure 3.4, what regions or combinations of regions represent the events that a motor will have

(a) none of the major defects;
(b) shaft that is large and windings improper;
(c) shaft that is large and/or windings improper but the electrical connections are satisfactory;
(d) the shaft is large and the windings improper and/or the electrical connections are unsatisfactory.

3.13 Use Venn diagrams to verify that

(a) $(A \cap B)' = A' \cup B'$;
(b) $A \cup (A \cap B) = A$;
(c) $(A \cap B) \cup (A \cap B') = A$;
(d) $A \cup B = (A \cap B) \cup (A \cap B') \cup (A' \cap B)$;
(e) $A \cup (B \cap C) = (A \cup B) \cap (A \cup C)$.

3.14 A building inspector has to check the wiring in a new apartment building either on Monday, Tuesday, Wednesday, or Thursday, and at 8 A.M., 1 A.M., or 2 P.M. Draw a tree diagram which shows the various ways in which the inspector can schedule the inspection of the wiring of the new apartment building.

3.15 If the five finalists in an international volleyball tournament are Spain, U.S.A., Uruguay, Portugal and Japan, draw a tree diagram which shows the various possible first and second place finishers.

3.16 A student can study either 0, 1, or 2 hours for a test in computer programming on any given night. Construct a tree diagram to show that there are 10 different ways in which the student can study altogether 6 hours for the test on four consecutive nights.

3.17 In an optics kit there are six concave lenses, four convex lenses, and three prisms. In how many ways can one choose one of the concave lenses, one of the convex lenses, and one of the prisms?

3.18 A questionnaire sent through the mail as part of a market study consists of eight questions, each of which can be answered in three different ways. In how many different ways can a person answer the eight questions on this questionnaire?

3.19 In a small geology class, each of the four students must write a report on one of eight field trips. In how many different ways can they each choose one of the field trips if

(a) no two students may choose the same field trip;
(b) there is no restriction on their choice?

3.20 If there are nine cars in a race, in how many different ways can they place first, second, and third?

3.21 In how many ordered ways can a television director schedule six different commercials during the six time slots allocated to commercials during the telecast of the first period of a hockey game?

3.22 If among n objects k are alike and the others are all distinct, the number of permutations of these n objects taken all together is $\frac{n!}{k!}$.

(a) How many permutations are there of the letters of the word *class*?
(b) In how many ways can the television director of Exercise 3.21 fill the six time slots allocated to commercials, if there are four different commercials, of which a given one is to be shown three times while each of the others is to be shown once?

3.23 Determine the number of ways in which a manufacturer can choose 2 of 15 locations for a new warehouse.

3.24 If the order does not matter, in how many different ways can 4 of 18 tax returns be chosen for a special audit?

3.25 A carton of 12 transistor batteries contains one that is defective. In how many ways can an inspector choose three of the batteries and
(a) get the one that is defective; (b) not get the one that is defective?

3.26 With reference to Exercise 3.25, suppose that two of the batteries are defective. In how many ways can the inspector choose three of the batteries and get
(a) none of the defective batteries; (b) one of the defective batteries;
(c) both of the defective batteries?

3.27 A major appliance store carries eight kinds of refrigerators, six kinds of washer-driers, and five kinds of microwave ovens. In how many different ways can two of each kind be chosen for a special sale?

3.3 PROBABILITY

So far we have studied only what is possible in a given situation; now we shall go one step further and judge also what is probable and what is improbable. Historically, the oldest way of measuring uncertainties is the **classical probability concept**, which was developed originally in connection with games of chance. It applies when all possible outcomes are equally likely, in which case we say that

The classical probability concept

> ***If there are n equally likely possibilities, of which one must occur and s are regarded as favorable, or as a "success," then the probability of a "success" is given by*** $\frac{s}{n}$.

In the application of this rule, the terms favorable and success are used rather loosely—favorable may mean that a television set does not work and success may mean that someone catches the flu.

EXAMPLE What is the probability of drawing an ace from a well-shuffled deck of 52 playing cards?

Solution There are $s = 4$ aces among the $n = 52$ cards, so we get

$$\frac{s}{n} = \frac{4}{52} = \frac{1}{13}$$

■

Although equally likely possibilities are found mostly in games of chance, the classical probability concept applies also to a great variety of situations where gambling devices are used to make random selections; say, when offices are assigned to research assistants by lot, when laboratory animals are chosen for an experiment so that each one has the same chance of being selected, or when washing machine parts are chosen for inspection so that each part produced has the same chance of being selected.

EXAMPLE If 3 of 20 tires are defective and 4 of them are randomly chosen for inspection (that is, each tire has the same chance of being selected), what is the probability that only one of the defective tires will be included?

Solution There are $\binom{20}{4} = 4{,}845$ equally likely ways of choosing 4 of the 20 tires, so $n = 4{,}845$. The number of favorable outcomes is the number of ways in which one of the defective tires and three of the nondefective tires can be selected, or $s = \binom{3}{1}\binom{17}{3} = 3 \cdot 680 = 2{,}040$. It follows that the probability is

$$\frac{s}{n} = \frac{2{,}040}{4{,}845} = \frac{8}{19}$$

or approximately 0.42.

■

A major shortcoming of the classical probability concept is its limited applicability, for there are many situations in which the various possibilities cannot all be regarded as equally likely. This would be the case, for example, if we are concerned with the question of whether it will rain the next day, whether a missile launching will be a success, whether a newly designed engine will function for at least 1,000 hours, or whether a certain candidate will win an election.

Among the various probability concepts, most widely held is the **frequency interpretation**, according to which

The frequency interpretation of probability

> ***The probability of an event (happening or outcome) is the proportion of times the event would occur in a long run of repeated experiments.***

If we say that the probability is 0.78 that a jet from New York to Boston will arrive on time, we mean that such flights arrive on time 78% of the time. Also, if the Weather Service predicts that there is a 40% chance for rain (that the probability is 0.40), this means that under the same weather conditions it will rain 40% of the time. More generally, we say that an event has a probability of, say, 0.90, in the same sense in which we might say that in cold weather our car will start 90% of the time. We cannot guarantee what will happen on any particular occasion—the car may start and then it may not—but if we kept records over a long period of time, we should find that the proportion of "successes" is very close to 0.90.

In accordance with the frequency interpretation of probability, we estimate the probability of an event by observing what fraction of the time similar events have occurred in the past.

EXAMPLE If records show that 294 of 300 ceramic insulators tested were able to withstand a certain thermal shock, what is the probability that any one such insulator will be able to withstand the thermal shock?

Solution Among the insulators tested, $\frac{294}{300} = 0.98$ were able to withstand the thermal shock, and we use this figure as an estimate of the probability.

■

An alternative point of view which is currently gaining in favor is to interpret probabilities as **personal** or **subjective evaluations**. Such probabilities express the strength of one's belief with regard to the uncertainties that are involved, and they apply especially when there is little or no direct evidence, so that there is no choice but to consider collateral (indirect) evidence, educated guesses, and perhaps intuition and other subjective factors. Subjective probabilities are best determined by referring to risk taking, or betting situations, as will be explained in Exercise 3.53.

3.4 THE AXIOMS OF PROBABILITY

In this section we define probabilities mathematically as the values of **additive set functions**. Since the reader is probably most familiar with functions for which the elements of the domain and the range are all numbers, let us first give a very simple example where the elements of the domain are sets while the elements of the range are nonnegative integers, namely, a **set function** that assigns to each subset A of a finite sample space S the number of elements in A, written $N(A)$. Suppose that 500 machine parts are inspected before they are shipped, that I denotes that a machine part is improperly assembled, D denotes that it contains one or more defective components, and the distribution of the 500 machine parts among the various categories is as shown in the Venn diagram of Figure 3.7. The numbers in Figure 3.7 are $N(I \cap D') = 20$, $N(I \cap D) = 10$, $N(I' \cap D) = 5$, and $N(I' \cap D') = 465$. Using these values and the fact that the set function is **additive** (meaning that the number which it assigns to the union of two subsets which have no elements in common is

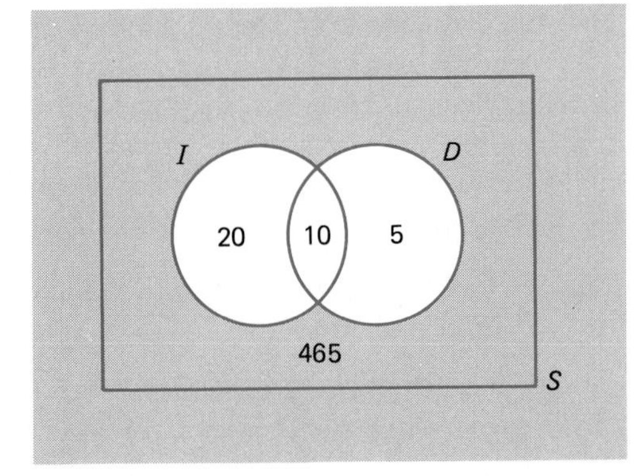

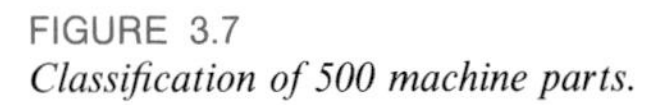
FIGURE 3.7
Classification of 500 machine parts.

the sum of the numbers assigned to the individual subsets), we can determine the value of $N(X)$ for any other subset X of S. For instance,

$$N(I') = N(I' \cap D) + N(I' \cap D') = 5 + 465 = 470$$

$$\begin{aligned} N(I \cup D) &= N(I \cap D') + N(I \cap D) + N(I' \cap D) \\ &= 20 + 10 + 5 = 35 \end{aligned}$$

$$\begin{aligned} N(I' \cup D) &= N(I \cap D) + N(I' \cap D) + N(I' \cap D') \\ &= 10 + 5 + 465 = 480 \end{aligned}$$

and

$$N(D) = N(I \cap D) + (I' \cap D) = 10 + 5 = 15$$

Using the concept of an additive set function, let us now explain what we mean by the probability of an event. Given a finite sample space S and an event A in S, we define $P(A)$, the probability of A, to be a value of an additive set function that satisfies the following three conditions:

The axioms of probability

Axiom 1 **$0 \le P(A) \le 1$ *for each event A in S.***
Axiom 2 **$P(S) = 1$.**
Axiom 3 ***If A and B are any mutually exclusive events in S, then*** **$P(A \cup B) = P(A) + P(B)$.**

The first axiom states that probabilities are real numbers on the interval from 0 to 1. The second axiom states that the sample space as a whole is assigned a probability of 1 and this expresses the idea that the probability of a certain event, an event which must happen, is equal to 1. The third axiom states that probability functions must be additive.

Axioms for a mathematical theory require no proof, but if such a theory is to be applied to the physical world, we must show somehow that the axioms are "realistic." Thus, let us show that the three postulates are consistent with the classical probability concept and the frequency interpretation. The situation is more complicated when it comes to subjective probabilities, as is explained in Exercise 3.54.

So far as the first axiom is concerned, fractions of the form $\frac{s}{n}$, where $0 \le s \le n$ and n is a positive integer, cannot be negative or exceed 1, and the same is true also for the proportion of the time that an event will occur. To show that the second axiom is consistent with the classical probability concept and the frequency interpretation, we have only to observe that for the whole sample space $P(S) = \frac{n}{n} = 1$ and that an event which is certain to occur must happen 100% of the time. So far as the third axiom is concerned, if $P(A) = \frac{s_1}{n}$, $P(B) = \frac{s_2}{n}$, and A and B are mutually exclusive, then $P(A \cup B) = \frac{s_1 + s_2}{n} = P(A) + P(B)$; also, if one event occurs 36% of the time, another event occurs 41% of the time, and the two events are mutually exclusive, then one or the other will occur $36 + 41 = 77\%$ of the time.

Before we go any further, it is important to stress the point that the axioms of probability do not tell us how to assign probabilities to the various outcomes of an experiment; they merely restrict the ways in which it can be done. In actual practice, probabilities are assigned on the basis of past experience, on the basis of a careful analysis of conditions underlying the experiment, on the basis of subjective evaluations, or on the basis of assumptions—say, the common assumption that all the outcomes are equiprobable.

EXAMPLE If an experiment has the three possible and mutually exclusive outcomes A, B, and C, check in each case whether the assignment of probabilities is permissible:

(a) $P(A) = \frac{1}{3}$, $P(B) = \frac{1}{3}$, and $P(C) = \frac{1}{3}$
(b) $P(A) = 0.64$, $P(B) = 0.38$, and $P(C) = -0.02$
(c) $P(A) = 0.35$, $P(B) = 0.52$, and $P(C) = 0.26$
(d) $P(A) = 0.57$, $P(B) = 0.24$, and $P(C) = 0.19$

Solution (a) The assignment of probabilities is permissible because the values are all on the interval from 0 to 1 and their sum is $\frac{1}{3} + \frac{1}{3} + \frac{1}{3} = 1$. (b) The assignment is not permissible because $P(C)$ is negative. (c) The assignment is not permissible because $0.35 + 0.52 + 0.26 = 1.13$, which exceeds 1. (d) The assignment is permissible because the values are all on the interval from 0 to 1 and their sum is $0.57 + 0.24 + 0.19 = 1$.

■

3.5
SOME ELEMENTARY THEOREMS

With the use of mathematical induction, the third axiom of probability can be extended to include any number of mutually exclusive events; in other words, the following can be shown.

Generalization of third axiom of probability

> ***Theorem 3.4*** *If $A_1, A_2, \ldots, A_n$ are mutually exclusive events in a sample space S, then*
>
> $$P(A_1 \cup A_2 \cup \cdots \cup A_n) = P(A_1) + P(A_2) + \cdots + P(A_n)$$

In the next chapter we shall see how the third axiom of probability must be modified so that the axioms apply also to sample spaces which are not finite.

EXAMPLE The probability that a consumer testing service will rate a new antipollution device for cars very poor, poor, fair, good, very good, or excellent are 0.07, 0.12, 0.17, 0.32, 0.21, and 0.11. What are the probabilities that it will rate the device

(a) very poor, poor, fair, or good;
(b) good, very good, or excellent?

Solution Since the possibilities are all mutually exclusive, direct substitution into the formula of Theorem 3.4 yields

$$0.07 + 0.12 + 0.17 + 0.32 = 0.68$$

for part (a) and

$$0.32 + 0.21 + 0.11 = 0.64$$

for part (b).

■

As it can be shown that a sample space of n points (outcomes) has 2^n subsets, it would seem that the problem of specifying a probability function (namely, a probability for each subset or event) can easily become very tedious. Indeed, for $n = 20$ there are already more than 1 million possible events. Fortunately, this task can be simplified considerably by the use of the following theorem:

Rule of calculating probability of an event

> ***Theorem 3.5*** *If A is an event in the finite sample space S, then P(A) equals the sum of the probabilities of the individual outcomes comprising A.*

To prove this theorem, let $E_1, E_2, \ldots, E_n$ be the n outcomes comprising event A, so that we can write $A = E_1 \cup E_2 \cup \cdots \cup E_n$. Since the E's are individual outcomes they are mutually exclusive, and by Theorem 3.4 we have

$$P(A) = P(E_1 \cup E_2 \cup \cdots \cup E_n)$$
$$= P(E_1) + P(E_2) + \cdots + P(E_n)$$

which completes the proof.

EXAMPLE With reference to the lawn-mower-rating example on page 47, suppose that the probabilities of the 18 outcomes are as shown in Figure 3.8 (which, except for the probabilities, is identical with Figure 3.5). Find $P(E_1)$, $P(P_1)$, $P(C_1)$, $P(E_1 \cap P_1)$, and $P(E_1 \cap C_1)$.

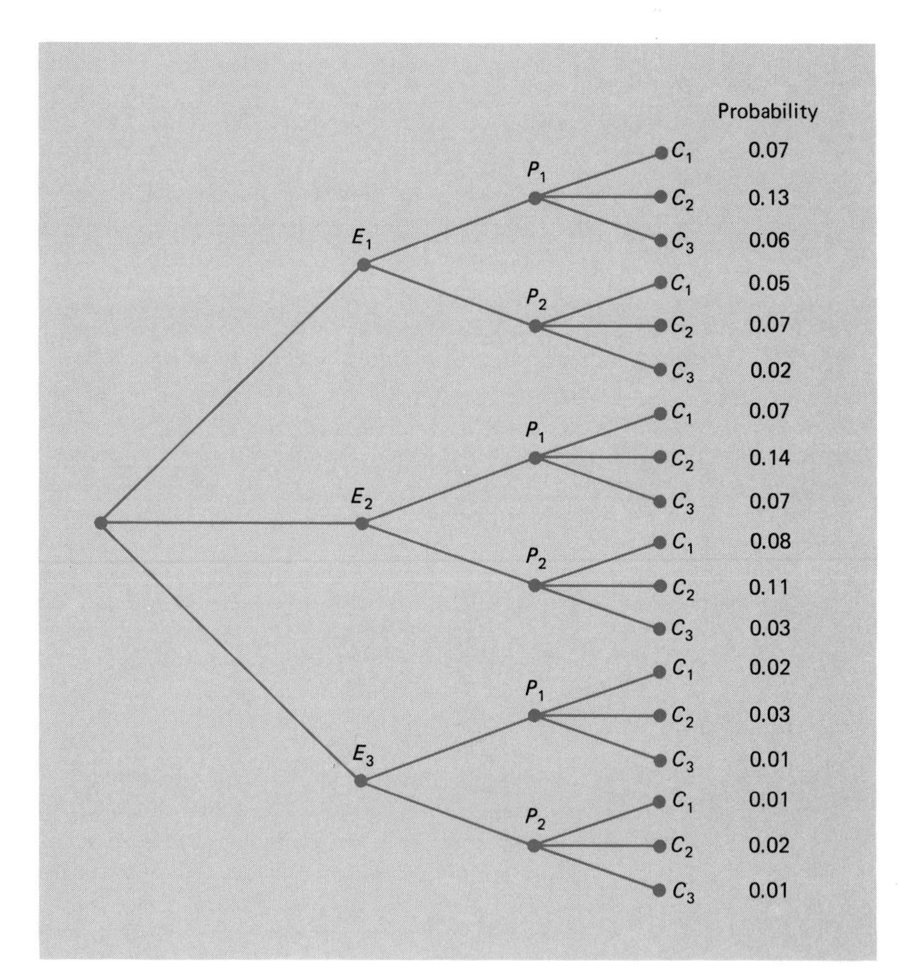

FIGURE 3.8
Ratings of lawn mowers and their probabilities.

Solution Adding the probabilities of the outcomes comprising the respective events, we get

$$P(E_1) = 0.07 + 0.13 + 0.06 + 0.05 + 0.07 + 0.02 = 0.40$$

$$P(P_1) = 0.07 + 0.13 + 0.06 + 0.07 + 0.14 + 0.07 + 0.02 + 0.03 + 0.01 = 0.60$$

$$P(C_1) = 0.07 + 0.05 + 0.07 + 0.08 + 0.02 + 0.01 = 0.30$$

$$P(E_1 \cap P_1) = 0.07 + 0.13 + 0.06 = 0.26$$

and

$$P(E_1 \cap C_1) = 0.07 + 0.05 = 0.12$$

■

In Theorem 3.4 we saw that the third axiom of probability can be extended to include more than two mutually exclusive events. Another useful and important extension of this axiom allows us to find the probability of the union of any two events in S regardless of whether they are mutually exclusive. To motivate the theorem which follows, let us consider the Venn diagram of Figure 3.9, which concerns the job applications of recent engineering school graduates. The letters I and G stand for getting a job in industry or getting a job with the government, and it follows from the Venn diagram that

$$P(I) = 0.18 + 0.12 = 0.30$$

$$P(G) = 0.12 + 0.24 = 0.36$$

and

$$P(I \cup G) = 0.18 + 0.12 + 0.24 = 0.54$$

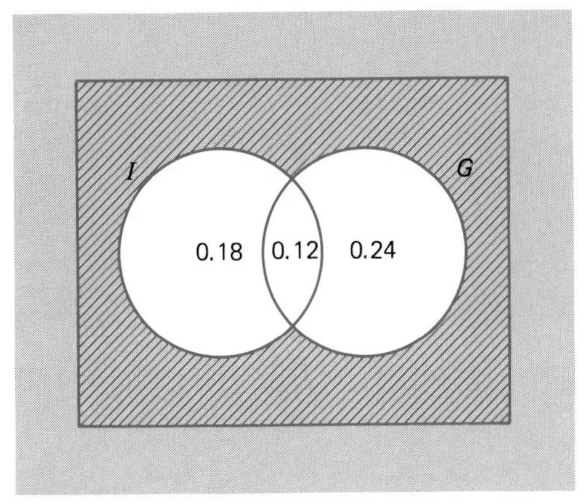

FIGURE 3.9
Venn diagram for example.

We were able to add the various probabilities because they represent mutually exclusive events.

Had we erroneously used the third axiom of probability to calculate $P(I \cup G)$, we would have obtained $P(I) + P(G) = 0.30 + 0.36$, which exceeds the correct value by 0.12. This error results from adding in $P(I \cap G)$ twice, once in $P(I) = 0.30$ and once in $P(G) = 0.36$, and we could correct for it by subtracting 0.12 from 0.66. Thus, we would get

$$\begin{aligned} P(I \cup G) &= P(I) + P(G) - P(I \cap G) \\ &= 0.30 + 0.36 - 0.12 \\ &= 0.54 \end{aligned}$$

and this agrees, as it should, with the result obtained before.

In line with this motivation, let us now state and prove the following theorem:

General addition rule

***Theorem 3.6** If A and B are any events in S, then*

$$P(A \cup B) = P(A) + P(B) - P(A \cap B)$$

To prove this theorem, we make use of the identities of parts (c) and (d) of Exercise 3.13, getting

$$\begin{aligned} P(A \cup B) &= P(A \cap B) + P(A \cap B') + P(A' \cap B) \\ &= [P(A \cap B) + P(A \cap B')] + [P(A \cap B) + P(A' \cap B)] - P(A \cap B) \\ &= P(A) + P(B) - P(A \cap B) \end{aligned}$$

Note that when A and B are mutually exclusive so that $P(A \cap B) = 0$, Theorem 3.6 reduces to the third axiom of probability. For this reason, we sometimes refer to the third axiom of probability as the **special addition rule**.

EXAMPLE With reference to the lawn-mower-rating example, find the probability that a lawn mower will be rated easy to operate and/or having a high average cost of repairs, namely, $P(E_1 \cup C_1)$.

Solution Making use of the results obtained on page 61, $P(E_1) = 0.40$, $P(C_1) = 0.30$, and $P(E_1 \cap C_1) = 0.12$, we substitute into the formula of Theorem 3.6 and get

$$\begin{aligned} P(E_1 \cup C_1) &= 0.40 + 0.30 - 0.12 \\ &= 0.58 \end{aligned}$$

■

EXAMPLE If the probabilities are 0.87, 0.36, and 0.29 that a family, randomly chosen as part of a sample survey in a large metropolitan area, owns a color television set, a black-and-white set, or both, what is the probability that a family in this area will own one or the other or both kinds of sets?

Solution Substituting these given values into the formula of Theorem 3.6, we get

$$0.87 + 0.36 - 0.29 = 0.94$$

■

Note that the general addition rule, Theorem 3.6, can be generalized further so that it applies to more than two events (see Exercise 3.49).

Using the axioms of probability, we can derive many other theorems which play important roles in applications. For instance, let us show the following:

Probability of complement

Theorem 3.7 ***If A is any event in S, then $P(A') = 1 - P(A)$.***

To prove this theorem, we make use of the fact that A and A' are mutually exclusive by definition, and that $A \cup A' = S$ (namely, that among them A and A' contain all the elements in S). Hence, we can write

$$\begin{aligned} P(A) + P(A') &= P(A \cup A') \\ &= P(S) \\ &= 1 \end{aligned}$$

so that $P(A') = 1 - P(A)$. As a special case we find that $P(\varnothing) = 1 - P(S) = 0$, since the empty set $\varnothing$ is the complement of S.

EXAMPLE Referring again to the lawn-mower-rating example and the results on page 61, find

(a) the probability that a lawn mower will not be rated easy to operate;

(b) the probability that a lawn mower will be rated as either not being easy to operate or not having a high average repair cost.

Solution (a) $P(E_1') = 1 - P(E_1) = 1 - 0.40 = 0.60$. (b) Since $E_1' \cup C_1' = (E_1 \cap C_1)'$ according to the identity of part (a) of Exercise 3.13, we get $P(E_1' \cup C_1') = P[(E_1 \cap C_1)'] = 1 - P(E_1 \cap C_1) = 1 - 0.12 = 0.88$.

■

EXERCISES

3.28 If a card is drawn from a well-shuffled deck of 52 playing cards, what is the probability of drawing

(a) a red king; (b) a 3, 4, 5, or 6;

(c) a black card; (d) a red ace or a black queen?

3.29 When we roll a pair of balanced dice, what are the probabilities of getting

(a) 7; (b) 11;
(c) 7 or 11; (d) 3;
(e) 2 or 12; (f) 2, 3, or 12?

3.30 A lottery sells tickets numbered from 00001 through 50000. What is the probability of drawing a number that is divisible by 200?

3.31 A car rental agency has 18 compact cars and 12 intermediate-size cars. If four of the cars are randomly selected for a safety check, what is the probability of getting two of each kind?

3.32 Among 842 armed robberies in a certain city, 143 were never solved. Estimate the probability that an armed robbery in this city will be solved.

3.33 In a group of 160 graduate engineering students, 92 are enrolled in an advanced course in statistics, 63 are enrolled in a course in operations research, and 40 are enrolled in both. How many of these students are not enrolled in either course?

3.34 Among 150 persons interviewed as part of an urban mass transportation study, some live more than 3 miles from the center of the city (A), some now regularly drive their own car to work (B), and some would gladly switch to public mass transportation if it were available (C). Use the information given in Figure 3.10 to find

(a) $N(A)$; (b) $N(B)$;
(c) $N(C)$; (d) $N(A \cap B)$;
(e) $N(A \cap C)$; (f) $N(A \cap B \cap C)$;
(g) $N(A \cup B)$; (h) $N(B \cup C)$;
(i) $N(A' \cup B' \cup C)$; (j) $N[B \cap (A \cup C)]$.

3.35 An experiment has the four possible mutually exclusive outcomes A, B, C, and D. Check whether the following assignments of probability are permissible:

(a) $P(A) = 0.38$, $P(B) = 0.16$, $P(C) = 0.11$, $P(D) = 0.35$;
(b) $P(A) = 0.31$, $P(B) = 0.27$, $P(C) = 0.28$, $P(D) = 0.16$;
(c) $P(A) = 0.32$, $P(B) = 0.27$, $P(C) = -0.06$, $P(D) = 0.47$;
(d) $P(A) = \frac{1}{2}$, $P(B) = \frac{1}{4}$, $P(C) = \frac{1}{8}$, $P(D) = \frac{1}{16}$;
(e) $P(A) = \frac{5}{18}$, $P(B) = \frac{1}{6}$, $P(C) = \frac{1}{3}$, $P(D) = \frac{2}{9}$.

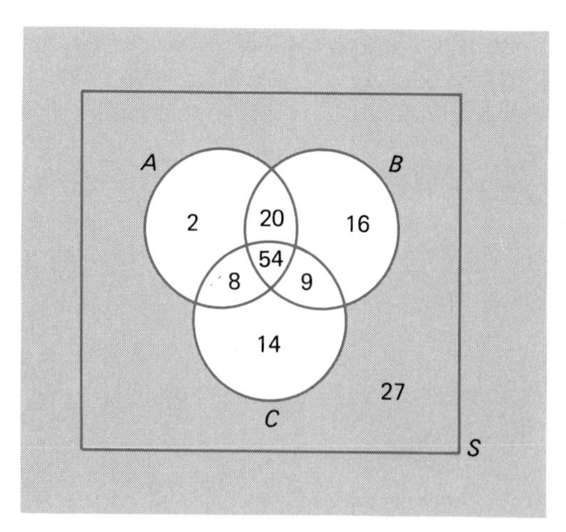

FIGURE 3.10
Diagram for Exercise 3.34.

3.36 With reference to Exercise 3.1, suppose that the points (0, 0), (0, 1), (0, 2), (1, 0), (1, 1), (1, 2), (2, 0), (2, 1), (2, 2), (3, 0), (3, 1), and (3, 2) have the probabilities 0.001, 0.002, 0.006, 0.001, 0.017, 0.078, 0.004, 0.069, 0.311, 0.004, 0.092, and 0.415.

(a) Verify that this assignment of probabilities is permissible.
(b) Find the probabilities of events R, T, and U given in part (b) of that exercise.
(c) Calculate the probabilities that zero, one, or two carbon dioxide lasers will be suitable for the given task.

3.37 With reference to Exercise 3.7, suppose that each point (i, j) of the sample space is assigned the probability $\frac{15/28}{i+j}$.

(a) Verify that this assignment of probabilities is permissible.
(b) Find the probabilities of events B, C, and D described in part (b) of that exercise.
(c) Find the probabilities that one, two, or three of the graduate students will be supervising the physics lab.

3.38 Explain why there must be a mistake in each of the following statements:

(a) The probability that a mineral sample will contain silver is 0.38 and the probability that it will not contain silver is 0.52.
(b) The probability that a drilling operation will be a success is 0.34 and the probability that it will not be a success is -0.66.
(c) An air-conditioning repairperson claims that the probability is 0.82 that the compressor is all right, 0.64 that the fan motor is all right, and 0.41 that they are both all right.

3.39 Refer to parts (c) and (d) of Exercise 3.13 to show that

(a) $P(A \cap B) \leq P(A)$; (b) $P(A \cup B) \geq P(A)$.

3.40 Explain why there must be a mistake in each of the following statements:

(a) The probability that a student will get an A in a geology course is 0.32, and the probability that he or she will get either an A or a B is 0.27.
(b) A company is working on the construction of two shopping centers; the probability that the larger one will be completed on time is 0.35 and the probability that both will be completed on time is 0.42.

3.41 If A and B are mutually exclusive events, $P(A) = 0.29$, and $P(B) = 0.43$, find

(a) $P(A')$;
(b) $P(A \cup B)$;
(c) $P(A \cap B')$;
(d) $P(A' \cap B')$.

3.42 With reference to Exercise 3.34, suppose that the questionnaire filled in by one of the 150 persons is to be double-checked. If it is chosen in such a way that each questionnaire has a probability of $\frac{1}{150}$ of being selected, find the probabilities that the person

(a) lives more than 3 miles from the center of the city;
(b) regularly drives his or her car to work;
(c) does not live more than 3 miles from the center of the city and would not want to switch to public mass transportation if it were available;
(d) regularly drives his or her car to work but would gladly switch to public mass transportation if it were available.

3.43 A police department needs new tires for its patrol cars and the probabilities are 0.17, 0.22, 0.03, 0.29, 0.21, and 0.08 that it will buy Uniroyal tires, Goodyear tires, Michelin tires, General tires, Goodrich tires, or Armstrong tires. Find the probabilities that it will buy

(a) Goodyear or Goodrich tires;
(b) Uniroyal, General, or Goodrich tires;
(c) Michelin or Armstrong tires;
(d) Goodyear, General, or Armstrong tires.

3.44 The probabilities that a TV station will receive 0, 1, 2, 3, ..., 8, or at least 9 complaints after showing a controversial program are, respectively, 0.01, 0.03, 0.07, 0.15, 0.19, 0.18, 0.14, 0.12, 0.09, and 0.02. What are the probabilities that after showing such a program the station will receive

(a) at most 4 complaints;
(b) at least 6 complaints;
(c) from 5 to 8 complaints?

3.45 If each point of the sample space of Figure 3.11 represents an outcome having the probability $\frac{1}{32}$, find

(a) $P(A)$; (b) $P(B)$;
(c) $P(A \cap B)$; (d) $P(A \cup B)$;
(e) $P(A' \cap B)$; (f) $P(A' \cap B')$.

3.46 The probability that a person stopping at a gas station will ask to have the tires checked is 0.12, the probability that he or she will ask to have the oil checked is 0.29, and the probability that he or she will ask to have them both checked is 0.07.

(a) What is the probability that a person stopping at this gas station will have either the tires or the oil checked?
(b) What is the probability that a person stopping at this gas station will have neither the tires nor the oil checked?

3.47 The probability that a new airport will get an award for its design is 0.16, the probability that it will get an award for the efficient use of materials is 0.24, and the probability that it will get both awards is 0.11.

(a) What is the probability that it will get at least one of the two awards?
(b) What is the probability that it will get only one of two awards?

3.48 Given $P(A) = 0.35$, $P(B) = 0.73$, and $P(A \cap B) = 0.14$, find

(a) $P(A \cup B)$; (b) $P(A' \cap B)$;
(c) $P(A \cap B')$; (d) $P(A' \cup B')$.

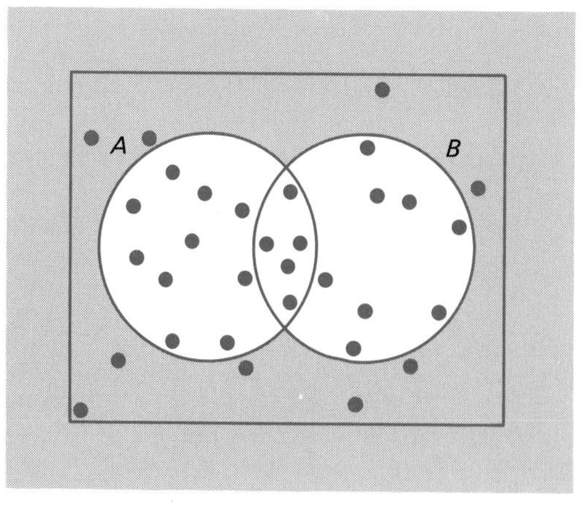

FIGURE 3.11
Diagram for Exercise 3.45.

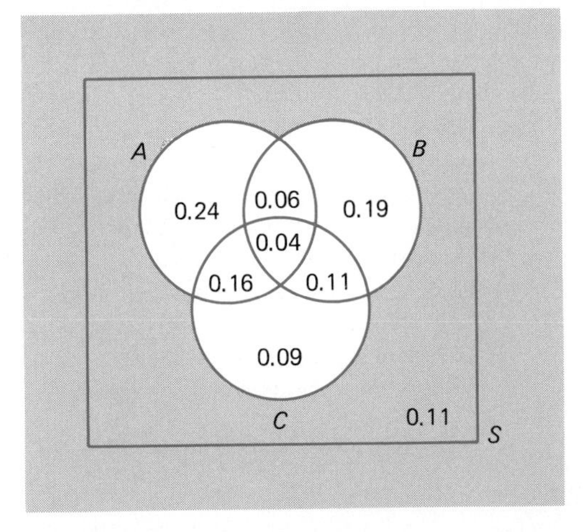

FIGURE 3.12
Diagram for Exercise 3.49.

3.49 It can be shown that for any three events A, B, and C, the probability that at least one of them will occur is given by

$$P(A \cup B \cup C) = P(A) + P(B) + P(C) - P(A \cap B) - P(A \cap C) - P(B \cap C) + P(A \cap B \cap C)$$

Verify that this formula holds for the probabilities of Figure 3.12.

3.50 Suppose that in the maintenance of a large medical-records file for insurance purposes the probability of an error in processing is 0.0010, the probability of an error in filing is 0.0009, the probability of an error in retrieving is 0.0012, the probability of an error in processing as well as filing is 0.0002, the probability of an error in processing as well as retrieving is 0.0003, the probability of an error in filing as well as retrieving is 0.0003, and the probability of an error in processing, filing, as well as retrieving is 0.0001. What is the probability of making at least one of these errors?

3.51 If the probability of event A is p, then the **odds** that it will occur are given by the ratio of p to $1 - p$. Odds are usually given as a ratio of two positive integers having no common factor, and if an event is more likely not to occur than to occur, it is customary to give the odds that it will not occur rather than the odds that it will occur. What are the odds for or against the occurrence of an event if its probability is

(a) $\frac{4}{7}$; (b) 0.05; (c) 0.80?

3.52 Use the definition of Exercise 3.51 to show that if the odds for the occurrence of event A are a to b, where a and b are positive integers, then $p = \frac{a}{a+b}$.

3.53 The formula of Exercise 3.52 is often used to determine subjective probabilities. For instance, if an applicant for a job "feels" that the odds are 7 to 4 of getting the job, the subjective probability the applicant assigns to getting the job is $p = \frac{7}{7+4} = \frac{7}{11}$.

(a) If a business person feels that the odds are 3 to 2 that a new venture will succeed (say, by betting \$300 against \$200 that it will succeed), what subjective probability is he or she assigning to its success?

(b) If a student is willing to bet \$30 against \$10, but not \$40 against \$10 that he or she will get a passing grade in a certain course, what does this tell us about the subjective probability the student assigns to getting a passing grade in the course?

3.54 Subjective probabilities may or may not satisfy the third axiom of probability. When they do, we say that they are **consistent**; when they do not, they ought not to be taken too seriously.

(a) The supplier of delicate optical equipment feels that the odds are 7 to 5 against a shipment arriving late and 11 to 1 against it not arriving at all. Furthermore, he feels that there is a fifty-fifty chance (the odds are 1 to 1) that such a shipment will either arrive late or not at all. Are the corresponding probabilities consistent?

(b) There are two Ferraris in a race, and an expert feels that the odds against their winning are, respectively, 2 to 1 and 3 to 1. Furthermore, she claims that there is a less-than-even chance that either of the two Ferraris will win. Discuss the consistency of these claims.

3.6 CONDITIONAL PROBABILITY

As we have defined probability, it is meaningful to ask for the probability of an event only if we refer to a given sample space S. To ask for the probability that an engineer earns at least \$40,000 a year is meaningless unless we specify whether we are referring to all engineers in the western hemisphere, all engineers in the United States, all those in a particular industry, all those affiliated with a university, and so forth. Thus, when we use the symbol $P(A)$ for the probability of A, we really mean the probability of A given some sample space S. Since the choice of S is by no means always evident, and since there are problems in which we are interested in the probabilities of A with respect to more sample spaces than one, the notation $P(A|S)$ is used to make it clear that we are referring to a particular sample space S. We read $P(A|S)$ as the conditional probability of A relative to S, and every probability is thus a conditional probability. Of course, we use the simplified notation $P(A)$ whenever the choice of S is clearly understood.

To illustrate some of the ideas connected with conditional probabilities, let us consider again the 500 machine parts of which some are improperly assembled and some contain one or more defective components as shown in Figure 3.7. Assuming equal probabilities in the selection of one of the machine parts for inspection, it can be seen that the probability of getting one with one or more defective components is $P(D) = \frac{10 + 5}{500} = \frac{3}{100}$. To check whether the probability is the same if the choice is restricted to the machine parts that are improperly assembled, we have only to look at the reduced sample space of Figure 3.13 and assume that each of the 30 improperly assembled parts has the same chance of being selected. We thus get

$$P(D|I) = \frac{N(D \cap I)}{N(I)} = \frac{10}{30} = \frac{1}{3}$$

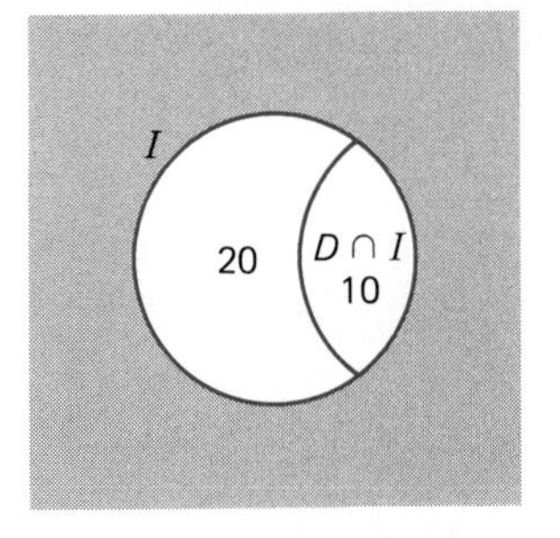

FIGURE 3.13
Reduced sample space.

and it can be seen that the probability of getting a machine part with one or more defective components has increased from $\frac{3}{100}$ to $\frac{1}{3}$. Note that if we divide the numerator and denominator of the preceding formula for $P(D|I)$ by $N(S)$, we get

$$P(D|I) = \frac{\frac{N(D \cap I)}{N(S)}}{\frac{N(I)}{N(S)}} = \frac{P(D \cap I)}{P(I)}$$

where $P(D|I)$ is given by the ratio of $P(D \cap I)$ to $P(I)$.

Looking at this example in another way, note that with respect to the whole sample space S we have

$$P(D \cap I) = \frac{10}{500} = \frac{1}{50} \quad \text{and} \quad P(D' \cap I) = \frac{20}{500} = \frac{2}{50}$$

assuming, as before, that each of the 500 machine parts has the same chance of being selected. Thus, the probabilities that the machine part selected will or will not contain one or more defective components, given that it is improperly assembled, should be in the ratio 1 to 2. Since the probabilities of D and D' in the reduced sample space must add up to 1, it follows that

$$P(D|I) = \frac{1}{3} \quad \text{and} \quad P(D'|I) = \frac{2}{3}$$

and this agrees with the result obtained before. This explains why we had to divide by $P(I)$ when we wrote

$$P(D|I) = \frac{P(D \cap I)}{P(I)}$$

above—division by $P(I)$, or multiplication by $\frac{1}{P(I)}$, takes care of the proportionality factor which makes the sum of the probabilities over the reduced sample space equal to 1.

Following these observations, let us now make the following general definition:

Conditional probability

> ***If A and B are any events in S and P(B) ≠ 0, the conditional probability of A given B is***
>
> $$P(A|B) = \frac{P(A \cap B)}{P(B)}$$

EXAMPLE If the probability that a communication system will have high fidelity is 0.81 and the probability that it will have high fidelity and high selectivity is 0.18, what is the probability that a system with high fidelity will also have high selectivity?

Solution If A is the event that a communication system has high selectivity and B is the event that it has high fidelity, we have $P(B) = 0.81$ and $P(A \cap B) = 0.18$, and substitution into the formula yields

$$P(A|B) = \frac{0.18}{0.81} = \frac{2}{9}$$

■

EXAMPLE If the probability that a research project will be well planned is 0.80 and the probability that it will be well planned and well executed is 0.72, what is the probability that a research project that is well planned will also be well executed?

Solution Substitution into the formula for a conditional probability yields $\frac{0.72}{0.80} = 0.90$.

■

EXAMPLE Referring to the lawn-mower-rating example, for which the probabilities of the individual outcomes are given in Figure 3.8, use the results obtained on page 61 to find $P(E_1|C_1)$.

Solution Since we had $P(E_1 \cap C_1) = 0.12$ and $P(C_1) = 0.30$, substitution into the formula for a conditional probability yields

$$P(E_1|C_1) = \frac{0.12}{0.30} = 0.40$$

■

It is of interest to note that the value obtained here for $P(E_1|C_1)$ equals that obtained on page 61 for $P(E_1)$. This means that the probability of a lawn mower being rated easy to repair is the same regardless of whether or not it is given that it has a high average cost of repairs, and we say that E_1 is **independent** of C_1. As the reader will be asked to verify this in Exercise 3.59 on page 77, it also follows from the results on page 61 that E_1 is not independent of P_1, namely, that a lawn mower's ease of operation is related to its retail price.

In general, if A and B are any two events in a sample space S, we say that A is independent of B if and only if $P(A|B) = P(A)$, but as it can be shown that B is independent of A whenever A is independent of B, it is customary to say simply that ***A* and *B* are independent**.

The following theorem is an immediate consequence of the definition of conditional probability:

General multiplication rule

> ***Theorem 3.8*** ***If A and B are any events in S, then***
>
> $$P(A \cap B) = P(A) \cdot P(B|A) \qquad \text{if } P(A) \neq 0$$
> $$= P(B) \cdot P(A|B) \qquad \text{if } P(B) \neq 0$$

The second of these rules is obtained directly from the definition of conditional probability by multiplying both sides by $P(B)$; the first is obtained from the second by interchanging the letters A and B.

EXAMPLE The supervisor of a group of 20 construction workers wants to get the opinion of 2 of them (to be selected at random) about certain new safety regulations. If 12 of them favor the new regulations and the other 8 are against it, what is the probability that both of the workers chosen by the supervisor will be against the new safety regulations?

Solution Assuming equal probabilities for each selection (which is what we mean by the selections being random), the probability that the first worker selected will be against the new safety regulations is $\frac{8}{20}$, and the probability that the second worker selected will be against the new safety regulations given that the first one is against them is $\frac{7}{19}$. Thus, the desired probability is $\frac{8}{20} \cdot \frac{7}{19} = \frac{14}{95}$.

■

In the special case where A and B are independent, Theorem 3.8 leads to the following result:

Special multiplication rule

> ***Theorem 3.9*** ***If A and B are independent events, then***
>
> $$P(A \cap B) = P(A) \cdot P(B)$$

Thus, the probability that two independent events will both occur is simply the product of their probabilities. This rule is sometimes used as the definition of independence; in any case, it may be used to determine whether two given events are independent.

EXAMPLE What is the probability of getting two heads in two flips of a balanced coin?

Solution Since the probability of heads is $\frac{1}{2}$ for each flip and the two flips are independent, the probability is $\frac{1}{2} \cdot \frac{1}{2} = \frac{1}{4}$.

■

EXAMPLE Two cards are drawn at random from an ordinary deck of 52 playing cards. What is the probability of getting two aces if

(a) the first card is replaced before the second card is drawn;
(b) the first card is not replaced before the second card is drawn?

Solution (a) Since there are four aces among the 52 cards, we get $\frac{4}{52} \cdot \frac{4}{52} = \frac{1}{169}$. (b) Since there are only three aces among the 51 cards that remain after one ace has been removed from the deck, we get $\frac{4}{52} \cdot \frac{3}{51} = \frac{1}{221}$.

Note that $\frac{1}{221} \neq \frac{4}{52} \cdot \frac{4}{52}$, so independence is violated when the sampling is without replacement.

■

EXAMPLE If $P(C) = 0.65$, $P(D) = 0.40$, and $P(C \cap D) = 0.24$, are the events C and D independent?

Solution Since $P(C) \cdot P(D) = (0.65)(0.40) = 0.26$ and not 0.24, the two events are not independent.

In the preceding examples we have used the assigned probabilities to check if two events are independent. The concept of independence can be—and frequently is—employed when probabilities are assigned to events that concern unrelated parts of an experiment.

■

EXAMPLE Let A be the event that raw material is available and B be the event that the machining time is less than 1 hour. If $P(A) = 0.8$ and $P(B) = 0.7$, assign probability to the event $A \cap B$.

Solution Since the events A and B concern unrelated steps in the manufacturing process, we invoke independence and make the assignment

$$P(A \cap B) = P(A)P(B) = (0.8)(0.7) = 0.56$$

■

The special multiplication rule can easily be extended so that it applies to more than two independent events—again, we multiply together all the individual probabilities.

EXAMPLE What is the probability of not rolling any 6's in four rolls of a balanced die?

Solution The probability is $\frac{5}{6} \cdot \frac{5}{6} \cdot \frac{5}{6} \cdot \frac{5}{6} = \frac{625}{1,296}$.

■

For three or more dependent events the multiplication rule becomes more complicated, as is illustrated in Exercise 3.70.

3.7 BAYES' THEOREM

The general multiplication rules are useful in solving many problems in which the ultimate outcome of an experiment depends on the outcomes of various intermediate stages. Suppose, for instance, that an assembly plant receives its voltage regulators from three different suppliers, 60% from supplier B_1, 30% from supplier B_2, and 10% from supplier B_3. In other words, the probabilities that any one voltage regulator received by the plant comes from these three suppliers are 0.60, 0.30, and 0.10. If 95% of the voltage regulators from B_1, 80% of those from B_2, and 65% of those from B_3 perform according to specifications, what we would like to know is the probability that any one voltage regulator received by the plant will perform according to specifications.

If A denotes the event that a voltage regulator received by the plant performs according to specifications, and B_1, B_2, and B_3 are the events that it comes from the respective suppliers, we can write

$$\begin{aligned} A &= A \cap [B_1 \cup B_2 \cup B_3] \\ &= (A \cap B_1) \cup (A \cap B_2) \cup (A \cap B_3) \end{aligned}$$

and

$$P(A) = P(A \cap B_1) + P(A \cap B_2) + P(A \cap B_3)$$

since $A \cap B_1$, $A \cap B_2$, and $A \cap B_3$ are mutually exclusive. Then, if we apply the second of the general multiplication rules to $P(A \cap B_1)$, $P(A \cap B_2)$, and $P(A \cap B_3)$, we get

$$P(A) = P(B_1) \cdot P(A|B_1) + P(B_2) \cdot P(A|B_2) + P(B_3) \cdot P(A|B_3)$$

and substitution of the given numerical values yields

$$\begin{aligned} P(A) &= (0.60)(0.95) + (0.30)(0.80) + (0.10)(0.65) \\ &= 0.875 \end{aligned}$$

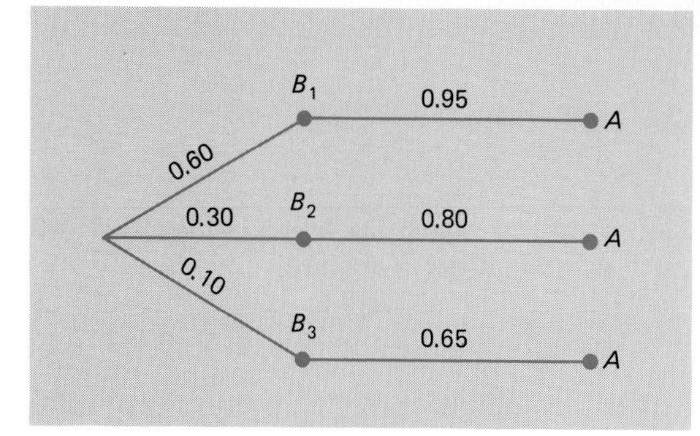

FIGURE 3.14
Tree diagram for example dealing with three suppliers of voltage regulators.

for the probability that any one voltage regulator received by the given plant will perform according to specifications.

To visualize this result, we have only to construct a tree diagram like that of Figure 3.14, where the probability of the final outcome is given by the sum of the products of the probabilities corresponding to each branch of the tree.

In the preceding example there were only three alternatives at the intermediate stage, but if there are n mutually exclusive alternatives $B_1, B_2, \ldots, B_n$ at the intermediate stage, a similar argument will lead to the following result, sometimes called the **rule of elimination** or the **rule of total probability**:

Rule of elimination

> ***Theorem 3.10** If $B_1, B_2, \ldots, B_n$ are mutually exclusive events of which one must occur, then*
>
> $$P(A) = \sum_{i=1}^{n} P(B_i) \cdot P(A|B_i)$$

To visualize this result, we have only to construct a tree diagram like that of Figure 3.15, where the probability of the final outcome is again given by the sum of the products of the probabilities corresponding to each branch of the tree.

To consider a problem that is closely related to the one we have just discussed, suppose we want to know the probability that a particular voltage regulator, which is known to perform according to specifications, came from supplier B_3. Symbolically, we want to know the value of $P(B_3|A)$, and to find a formula for this probability we first write

$$P(B_3|A) = \frac{P(A \cap B_3)}{P(A)}$$

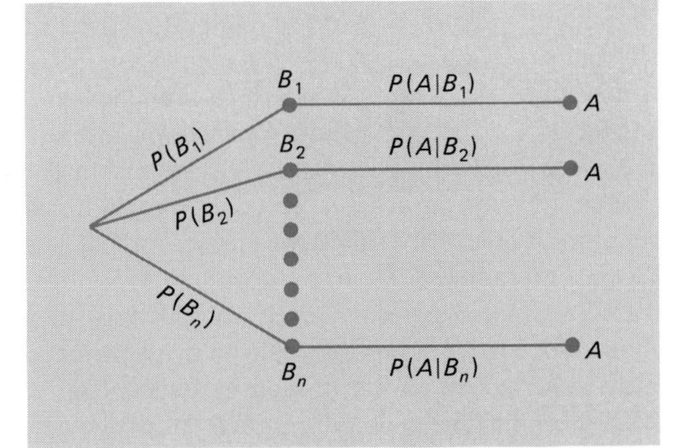

FIGURE 3.15
Tree diagram for rule of elimination.

Then, substituting $P(B_3) \cdot P(A|B_3)$ for $P(A \cap B_3)$ and $\sum_{i=1}^{3} P(B_i) \cdot P(A|B_i)$ for $P(A)$ in accordance with Theorems 3.8 and 3.10, we get the formula

$$P(B_3|A) = \frac{P(B_3) \cdot P(A|B_3)}{\sum_{i=1}^{3} P(B_i) \cdot P(A|B_i)}$$

which expresses $P(B_3|A)$ in terms of given probabilities. Substituting the values from page 73 (or from Figure 3.14), we finally obtain

$$P(B_3|A) = \frac{(0.10)(0.65)}{(0.60)(0.95) + (0.30)(0.80) + (0.10)(0.65)}$$
$$= 0.074$$

Note that the probability that a voltage regulator is supplied by B_3 decreases from 0.10 to 0.074 once it is known that it performs according to specifications.

The method used to solve the preceding example can easily be generalized to yield the following formula:

Bayes' theorem

***Theorem 3.11** If $B_1, B_2, \ldots, B_n$ are mutually exclusive events of which one must occur, then*

$$P(B_r|A) = \frac{P(B_r) \cdot P(A|B_r)}{\sum_{i=1}^{n} P(B_i) \cdot P(A|B_i)}$$

for $r = 1, 2, \ldots, n$.

Note that the expression in the numerator is the probability of reaching A via the ith branch of the tree and that the expression in the denominator is the sum of the probabilities of reaching A via the n branches of the tree.

Bayes' theorem provides a formula for finding the probability that the "effect" A was "caused" by the event B_r. For instance, in our example we found the probability that an acceptable voltage regulator was made by supplier B_3. The probabilities $P(B_i)$ are called the "prior," or "a priori," probabilities of the "causes" B_i, and in practice it is often difficult to assign them numerical values. For many years Bayes' theorem was looked upon with suspicion because it was used with the often-erroneous assumption that the prior probabilities are all equal. A good deal of the controversy once surrounding Bayes' theorem has been cleared up with the realization that the probabilities $P(B_3)$ must be determined separately in each case from the nature of the problem, preferably on the basis of past experience. We shall return to this problem as it relates to Bayesian inference in Chapters 7 and 9.

EXAMPLE The four attendants of a gasoline service station are supposed to wash the windshield of each customer's car. Janet, who services 20% of all cars, fails to wash the windshield one time in 20; Tom, who services 60% of all cars, fails to wash the windshield one time in 10; Georgia, who services 15% of all cars, fails to wash the windshield one time in 10; and Peter, who services 5% of all cars, fails to wash the windshield one time in 20. If a customer complains later that her windshield was not washed, what is the probability that her car was serviced by Janet?

Solution Substituting the various probabilities into the formula of Theorem 3.11, we get

$$P(B_1|A) = \frac{(0.20)(0.05)}{(0.20)(0.05) + (0.60)(0.10) + (0.15)(0.10) + (0.05)(0.05)}$$
$$= 0.114$$

and it is of interest to note that although Janet fails to wash the windshield of only 1 car in 20, namely, 5% of the cars, more than 11% of the windshields that are not washed at this service station are her responsibility.

■

EXERCISES

3.55 With reference to Figure 3.7, find $P(I|D)$ and $P(I|D')$, assuming that originally each of the 500 machine parts has the same chance of being chosen for inspection.

3.56 With reference to the example on page 70, the probability is 0.75 that the research project will be well executed. What is the probability that the project was well planned given that it is well executed?

3.57 With reference to Exercise 3.34 and Figure 3.10, assume that each of 150 persons has the same chance of being selected and find the probabilities that he or she

(a) lives more than 3 miles from the center of the city given that he or she would gladly switch to public mass transportation;

(b) regularly drives his or her car to work given that he or she lives more than 3 miles from the center of the city;

(c) would not want to switch to public mass transportation given that he or she does not regularly drive his or her car to work.

3.58 With reference to Figure 3.12, find

(a) $P(A|B)$;
(b) $P(B|C')$;
(c) $P(A \cap B|C)$;
(d) $P(B \cup C|A')$;
(e) $P(A|B \cup C)$;
(f) $P(A|B \cap C)$;
(g) $P(A \cap B \cap C|B \cap C)$;
(h) $P(A \cap B \cap C|B \cup C)$.

3.59 With reference to the lawn-mower-rating example and the probabilities given in Figure 3.8, find

(a) $P(E_1|P_1)$ and compare its value with that of $P(E_1)$;
(b) $P(C_2|P_2)$ and compare its value with that of $P(C_2)$;
(c) $P(E_1|P_1 \cap C_1)$ and compare its value with that of $P(E_1)$.

3.60 With reference to Exercise 3.47, find the probabilities that the airport will get the design award given that

(a) it got the award for the efficient use of materials;
(b) it did not get the award for the efficient use of materials.

3.61 Prove that $P(A|B) = P(A)$ implies that $P(B|A) = P(B)$ provided that $P(A) \neq 0$ and $P(B) \neq 0$.

3.62 If the probabilities are 0.58, 0.25, and 0.19 that a person in a certain income bracket will invest in money market funds, common stocks, or both, find the probabilities that a person in that income bracket

(a) who invests in money market funds will also invest in common stocks;
(b) who invests in common stocks will also invest in money market funds.

3.63 Given $P(A) = 0.50$, $P(B) = 0.30$, and $P(A \cap B) = 0.15$, verify that

(a) $P(A|B) = P(A)$;
(b) $P(A|B') = P(A)$;
(c) $P(B|A) = P(B)$;
(d) $P(B|A') = P(B)$.

3.64 Among the 24 invoices prepared by a billing department, 4 contain errors while the others do not. If we randomly check 2 of these invoices, what are the probabilities that

(a) both will contain errors;
(b) neither will contain an error?

3.65 Among 60 automobile repair parts loaded on a truck in San Francisco, 45 are destined for Seattle and 15 for Vancouver. If two of the parts are unloaded in Portland by mistake and the "selection" is random, what are the probabilities that

(a) both parts should have gone to Seattle;
(b) both parts should have gone to Vancouver;
(c) one should have gone to Seattle and one to Vancouver?

3.66 For two rolls of a balanced die, find the probabilities of getting

(a) two 4's;
(b) first a 4 and then a number less than 4.

3.67 If $P(A) = 0.80$, $P(C) = 0.35$, and $P(A \cap C) = 0.28$, are events A and C independent?

3.68 If the odds are 5 to 3 that event M will not occur, 2 to 1 that event N will occur, and 4 to 1 that they will not both occur, are the two events M and N independent?

3.69 Find the probabilities of getting

(a) eight heads in a row with a balanced coin;
(b) three 3's and then a 4 or a 5 in four rolls of a balanced die;
(c) five multiple-choice questions answered correctly, if for each question the probability of answering it correctly is $\frac{1}{3}$.

3.70 For three or more events which are not independent, the probability that they will all occur is obtained by multiplying the probability that one of the events will occur times the probability that a second of the events will occur given that the first event has occurred times the probability that a third of the events will occur given that the first two events have occurred, and so on. For instance, for three events we can write

$$P(A \cap B \cap C) = P(A) \cdot P(B|A) \cdot P(C|A \cap B)$$

and we find that the probability of drawing without replacement three aces in a row from an ordinary deck of 52 playing cards is

$$\frac{4}{52} \cdot \frac{3}{51} \cdot \frac{2}{50} = \frac{1}{5{,}525}$$

(a) If six bullets, of which three are blanks, are randomly inserted into a gun, what is the probability that the first three bullets fired will all be blanks?

(b) In a certain city during the month of May, the probability that a rainy day will be followed by another rainy day is 0.80 and the probability that a sunny day will be followed by a rainy day is 0.60. Assuming that each day is classified as being either rainy or sunny and that the weather on any given day depends only on the weather the day before, find the probability that in the given city a rainy day in May is followed by two more rainy days, then a sunny day, and finally another rainy day.

(c) A department store which bills its charge-account customers once a month has found that if a customer pays promptly one month, the probability is 0.90 that he will also pay promptly the next month; however, if a customer does not pay promptly one month, the probability that he will pay promptly the next month is only 0.50. What is the probability that a customer who has paid promptly one month will not pay promptly the next three months?

(d) If 5 of a company's 12 delivery trucks do not meet emission standards and 4 of the 12 trucks are randomly picked for inspection, what is the probability that none of them meets emission standards?

3.71 At an electronics plant, it is known from past experience that the probability is 0.86 that a new worker who has attended the company's training program will meet the production quota and that the corresponding probability is 0.35 for a new worker who has not attended the company's training program. If 80% of all new workers attend the training program, what is the probability that a new worker will meet the production quota?

3.72 With reference to the preceding exercise, find the probability that a new worker who meets the production quota will have attended the company's training program.

3.73 A consulting firm rents cars from three agencies, 20% from agency D, 20% from agency E, and 60% from agency F. If 10% of the cars from D, 12% of the cars from E, and 4% of the cars from F have bad tires, what is the probability that the firm will get a car with bad tires?

3.74 With reference to Exercise 3.73, what is the probability that a car with bad tires rented by the firm came from agency F?

3.75 Use the information on the tree diagram of Figure 3.16 to determine the value of
(a) $P(A)$; (b) $P(B|A)$; (c) $P(B|A')$.

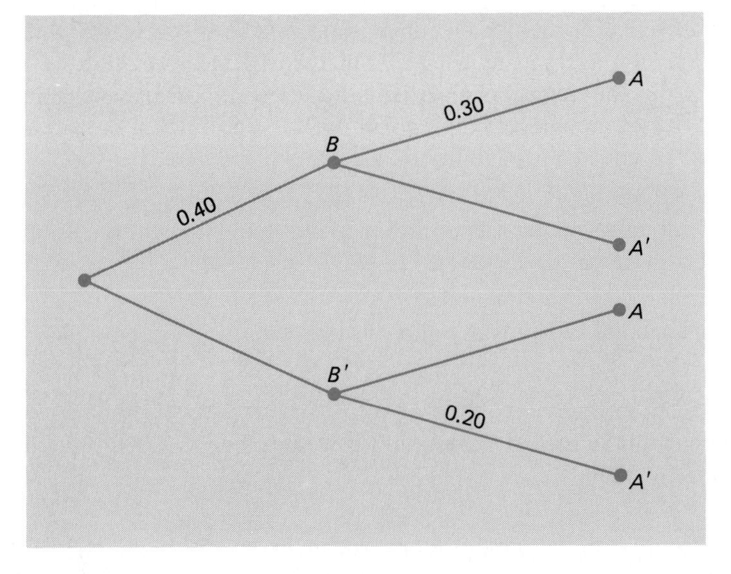

FIGURE 3.16
Diagram for Exercise 3.75.

3.76 With reference to the example on page 73, find the probabilities that a voltage regulator which performs according to specifications came
(a) from supplier B_1; (b) from supplier B_2.

3.77 With reference to the example on page 76, find the probabilities that the car of the complaining customer was serviced by
(a) Tom; (b) Georgia; (c) Peter.

3.78 Two firms V and W consider bidding on a road-building job, which may or may not be awarded depending on the amounts of the bids. Firm V submits a bid and the probability is $\frac{3}{4}$ that it will get the job provided firm W does not bid. The odds are 3 to 1 that W will bid, and if it does, the probability that V will get the job is only $\frac{1}{3}$.
(a) What is the probability that V will get the job?
(b) If V gets the job, what is the probability that W did not bid?

3.8 MATHEMATICAL EXPECTATION AND DECISION MAKING

If an insurance agent tells us that in the United States a 40-year-old woman can expect to live 38 more years, this does not mean that anyone really expects a 40-year-old woman to live until her seventy-eighth birthday and then die the next day. Similarly, if we read that a radial tire can be expected to last 40,000 miles, that in the coal mining industry we can expect 41.5 nonfatal injuries per million work-hours, or that the U.S. Patent Office can expect to issue 27,000 certificates of trademarks each year, it must be understood that the word "expect" is not used in its colloquial sense. So far as the first statement is concerned, some 40-year-old women will live another 15 years, some will live another 30 years, some will live another 50 years, ..., and the life expectance of "38 more years" will have to be interpreted as an average, namely, as a **mathematical expectation**.

Originally, the concept of a mathematical expectation arose in connection with games of chance, and in its simplest form it is the product of the amount a player stands to win and the probability that he or she will win.

EXAMPLE What is our mathematical expectation if we stand to win \$8 if and only if a balanced coin comes up heads?

Solution The probability of heads is $\frac{1}{2}$ and our mathematical expectation is $8 \cdot \frac{1}{2} = \$4$.

■

EXAMPLE What is our mathematical expectation if we buy one of 1,000 raffle tickets for a grand prize of \$500.

Solution The probability of winning the grand prize is $\frac{1}{1{,}000}$ and the mathematical expectation is $500 \cdot \frac{1}{1{,}000} = \$0.50$.

■

In both of these examples there was a single prize, but in each case there were two possible payoffs—\$8 or \$0 in the first example and \$500 or \$0 in the other. Indeed, in the second example we can argue that 999 of the tickets will pay \$0 and one of the tickets will pay \$500; altogether, the 1,000 tickets will thus pay \$500, or on the average \$0.50 per ticket, and this is the mathematical expectation.

To generalize the concept of a mathematical expectation, let us consider the following example:

EXAMPLE A distributor makes a profit of \$20 on an item if it is shipped from the factory in perfect condition and arrives on time, but it is reduced by \$2 if it does not arrive on time, and by \$12 regardless of whether it arrives on time if it is not shipped from the factory in perfect condition. If 70% of such items are shipped in perfect condition and arrive on time, 10% are shipped in perfect condition but do not arrive on time, and 20% are not shipped in perfect condition, what is the distributor's expected profit per item?

Solution We can argue that the distributor will make the \$20 profit 70% of the time, the \$18 profit 10% of the time, and the \$8 profit 20% of the time, so that the average profit per item (the expected profit) is

$$20(0.70) + 18(0.10) + 8(0.20) = \$17.40$$

This result is the sum of the products obtained by multiplying each amount by the corresponding proportion or probability.

■

Generalizing from this example, let us now make the following definition:

Mathematical expectation

> ***If the probabilities of obtaining the amounts $a_1, a_2, \ldots,$ or a_k are $p_1, p_2, \ldots,$ and p_k, then the mathematical expectation is***
>
> $$E = a_1p_1 + a_2p_2 + \cdots + a_kp_k$$

As far as the a's are concerned, it is important to keep in mind that they are positive when they represent profits, winnings, or gains and that they are negative when they represent losses, deficits, or penalties.

EXAMPLE An engineering firm is faced with the task of preparing a proposal for a research contract. The cost of preparing the proposal is \$5,000 and the probabilities for potential gross profits of \$50,000, \$30,000, \$10,000, or \$0 are 0.20, 0.50, 0.20, and 0.10, provided that the proposal is accepted. If the probability is 0.30 that the firm's proposal will be accepted, what is its expected net profit?

Solution The probability that the firm will make a net profit of \$45,000 (\$50,000 minus the cost of the proposal) is (0.30)(0.20) = 0.06. Similarly, the probabilities that the firm will make net profits of \$25,000 or \$5,000 are (0.30)(0.50) = 0.15 and (0.30)(0.20) = 0.06, whereas the probability of a \$5,000 loss is (0.30)(0.10) + 0.70 = 0.73, allowing for the possibility that the proposal will not be accepted. Thus, the expected net profit is

$$45{,}000(0.06) + 25{,}000(0.15) + 5{,}000(0.06) - 5{,}000(0.73) = \$3{,}100$$

Whether or not it is wise to risk \$5,000 to make an expected profit of \$3,100 is not a question of statistics. If a company has little capital and can make only few such proposals, the 0.73 chance of losing \$5,000 may well make the venture unattractive. On the other hand, if a company is well capitalized and prepares many such proposals, the risk may be worth taking, for it promises an expected return of 62% on the original investment.

■

The following is a typical inventory problem in which expected values are used to determine optimum conditions:

EXAMPLE It is known from past experience that the daily demand for a perishable product is as shown in the following table:

Number of orders	3	4	5	6	7	8	9
Probability	0.05	0.12	0.20	0.24	0.17	0.14	0.08

If each item costs \$35 (including the cost of carrying it in stock), it sells for \$50 provided that it is in stock, and it represents a total loss if it remains in stock at the end of a day, how many items should be stocked each day so as to maximize the expected profit?

Solution If three items are stocked, the profit is 150 − 105 = \$45, since there is a probability of 1 that there will be a demand for three or more. If four items are stocked, there is a probability of 0.05 that three items will be sold, a probability of 0.95 that there will be a demand for four or more, and the expected profit is

$$150(0.05) + 200(0.95) - 140 = \$57.50$$

Similarly, if five items are stocked, there is a probability of 0.05 that three items will be sold, a probability of 0.12 that four items will be sold, a probability of 0.83 that there will be a demand for five or more, and the expected profit is

$$150(0.05) + 200(0.12) + 250(0.83) - 175 = \$64.00$$

Continuing in this way (see Exercise 3.85), it can be shown that if six items are stocked the expected profit is \$60.50, if seven items are stocked the expected profit is \$45.00, and that the expected profit is a maximum when the number of items stocked is five.

■

In recent years, mathematical expectations have been playing an increasingly important role in scientific decision making, as it is generally considered rational to select whichever alternative has the most promising mathematical expectation: the one which maximizes the expected profit, minimizes the expected cost, maximizes tax advantages, minimizes the expected loss, and so forth.

In applying the methods of decision making illustrated in this section, there are essentially two practical limitations: first, we must be able to assign "cash values" to the various outcomes, and second, we must have adequate information concerning all relevant probabilities. The problem of assigning cash values to the various outcomes can be much more difficult than it was in our examples. For instance, if a government agency were faced with the decision of whether to require crash barriers for interstate highway construction, it would be extremely difficult to assign a cash value to the savings in life and reductions in injuries that might result.

The specification of probabilities concerning the various outcomes can be equally difficult. In our examples, the probabilities were presumably based on past experience, and, hence, are really only estimates (or approximations). This may be questioned with respect to the example dealing with the preparation of the research proposal, where it would be extremely difficult even to estimate the probability that a given proposal will be accepted. This would have to depend not only on the quality of the proposal itself, but on such unknown factors as the number and quality of competing proposals as well as the policy of the sponsoring agency. The estimate of 0.30 in our example was purely subjective, being based on some executive's "feeling" about the situation measured against the background of his personal experience. While such estimates of probabilities are not without value, they do leave open the possibility that different individuals (acting in the same situation and having the same information) would arrive at different "optimum" decisions.

EXERCISES

3.79 If a service club sells 4,000 raffle tickets for a cash prize of $800, what is the mathematical expectation of a person who buys one of the tickets?

3.80 A charitable organization raises funds by selling 2,000 raffle tickets for a first prize worth $500 and a second prize worth $100. What is the mathematical expectation of a person who buys one of the tickets?

3.81 What is our mathematical expectation if we win $10 if a balanced coin comes up heads and lose $10 if it comes up tails?

3.82 A game between two players is fair if each player has the same mathematical expectation. If someone gives us $5 each time we roll a 1 or a 2 with a balanced die, how much must we pay that person each time we roll a 3, 4, 5, or 6 to make the game fair?

3.83 When the American League and National League champions are evenly matched, the probabilities that a World Series will end in 4, 5, 6, or 7 games are, respectively, $\frac{1}{8}$, $\frac{1}{4}$, $\frac{5}{16}$, and $\frac{5}{16}$. What is the expected length of a World Series when the two teams are evenly matched?

3.84 The two finalists in a golf tournament play 18 holes, with the winner getting $50,000 and the runner-up getting $30,000. What are the two players' mathematical expectations if

(a) they are evenly matched;
(b) one of the two players should be favored by odds of 3 to 1?

3.85 With reference to the example on page 81, show that for six and seven items stocked, the expected profits are $60.50 and $45.00.

3.86 To handle a liability suit, a lawyer has to decide whether to charge a straight fee of $3,000 or a contingent fee of $12,000, which she will get only if her client wins. What does the lawyer think about her client's chances if she feels that

(a) taking the straight fee will give her a higher mathematical expectation;
(b) taking the contingent fee will give her a higher mathematical expectation?

3.87 A union wage negotiator feels that the probabilities are 0.40, 0.30, 0.20, and 0.10 that the union members will get a $1.50 an hour raise, a $1.00 an hour raise, a $.50 an hour raise, or no raise at all. What is their expected raise?

3.88 An importer is offered a shipment of machine tools for $140,000, and the probabilities that he will be able to sell them for $180,000, $170,000, or $150,000 are 0.32, 0.55, and 0.13. What is the importer's expected gross profit?

3.89 The manufacturer of a new battery additive has to decide whether to sell her product for $0.80 a can or for $1.20 with a "double-your-money-back-if-not-satisfied" guaranty. How does she feel about the chances that a person will ask for double his or her money back if

(a) she decides to sell the product for $0.80;
(b) she decides to sell the product for $1.20 with the guarantee;
(c) she cannot make up her mind?

3.90 A merchant can buy an item for \$2.10 and sell it for \$4.50. The probabilities for a demand of 0, 1, 2, 3, 4, or "5 or more" items are, respectively, 0.05, 0.15, 0.30, 0.25, 0.15, and 0.10. Calculate the expected profit resulting from stocking 0, 1, 2, 3, 4, or 5 items and determine how many the merchant should stock so as to maximize expected profit.

3.91 Mr. Brown and Mr. Jones are betting on repeated flips of a balanced coin. At the beginning Mr. Brown has m dollars, Mr. Jones has n dollars; at each flip the loser pays the winner one dollar, and they continue playing until one of them has lost all the money with which he began. To find p, the probability that Mr. Brown will win Mr. Jones' n dollars before he loses his m dollars, set up an equation in m, n, and p which makes use of the fact that in a fair game each player's mathematical expectation is the same. What is the probability that Mr. Brown will win if he begins with \$12 and Mr. Jones begins with \$4?

3.92 The management of a mining company must decide whether to continue an operation at a certain location. If they continue and are successful, this will be worth \$1,000,000; if they continue and are not successful, this will entail a loss of \$600,000; if they do not continue but would have been successful, this will entail a loss of \$400,000 (for competitive reasons); and if they do not continue and would not have been successful anyhow, this will be worth \$100,000 to the company (because funds allocated to the operation remain unspent). What decision would maximize the company's expected profit if the probabilities for and against success are 0.40 and 0.60?

3.93 With reference to Exercise 3.92, what decision would maximize the company's expected profit if the probabilities for and against success are 0.20 and 0.80?

3.94 With reference to Exercise 3.92, suppose that it is possible to continue the operation for two months at a cost of \$300,000, after which it will be possible to know for certain whether the operation will be a success. At that point, the only relevant costs are the \$1,000,000 and \$100,000 gains. Would it be worthwhile to spend these additional funds?

3.95 A power company is faced with the choice of whether to construct a light-water reactor (LWR) or a fossil-fuel power plant (FF). The LWR plant will cost \$300 per kilowatt to construct, and the FF plant will cost \$150 per kilowatt. Due to uncertainties concerning fuel availability and the impact of future air- and water-quality regulations, the useful operating life of each plant is unknown, but the following probabilities have been estimated:

Useful life (years)		10	20	30	40
Probabilities	*LWR plant*	0.05	0.25	0.50	0.20
	FF plant	0.10	0.50	0.30	0.10

(a) Which plant can be expected to cost less per year of useful life?
(b) What relative fuel cost per kilowatt would favor construction of the LWR plant?
(c) A further uncertainty concerning safety regulations makes it a fifty-fifty bet that \$50 per kilowatt will be added to the actual construction cost of the LWR plant. What effect does this have on these results?

3.9
REVIEW EXERCISES

3.96 A salesperson of industrial chemicals has four customers in Sacramento, whom he may or may not be able to visit on a 2-day trip to this city. He will not visit any of these customers more than once.

(a) Using two coordinates so that (2, 1), for example, represents the event that he will visit two of his customers on the first day and one on the second day, draw a diagram similar to that of Figure 3.1 showing the points of the corresponding sample space.

(b) List the points of the sample space that constitute the events X, Y, and Z that he will visit all four of his customers, that he will visit more of his customers on the first day than on the second day, and that he will visit at least three of his customers on the second day.

(c) Which of the three pairs of events, X and Y, X and Z, and Y and Z, are mutually exclusive?

3.97 With reference to the preceding exercise, express each of the following events symbolically by listing its elements, and also express it in words:

(a) X'; (b) $X \cup Y$;
(c) $X \cap Z$; (d) $X' \cap Y$.

3.98 Use Venn diagrams to verify that

(a) $(A' \cup B')' = A \cap B$; (b) $(A \cap B')' = A' \cup B$.

3.99 If the tone quality of four audio systems is to be rated superior, average, or inferior and we are interested only in how many of the systems get each of these ratings, draw a tree diagram which shows the 15 different possibilities.

3.100 Alarm units are to be connected at four fixed positions along a pipeline. In how many ways can the four available alarm units be connected to the line?

3.101 In how many ways can two out of seven chemical engineers be assigned to a new project?

3.102 An experiment has three possible mutually exclusive outcomes, A, B, and C. Check whether the following assignments of probability are possible.

(a) $P(A) = 0.3$, $P(B) = 0.4$, and $P(C) = 0.34$
(b) $P(A) = 0.5$, $P(B) = 0.1$, and $P(C) = 0.4$
(c) $P(A) = 0.4$, $P(B) = 0.64$, and $P(C) = -0.04$

3.103 Given $P(A) = 0.3$, $P(B) = 0.5$, and $P(A \cap B) = 0.24$. Find

(a) $P(A \cup B)$;
(b) $P(A' \cap B)$;
(c) $P(A \cap B')$;
(d) $P(A' \cup B')$.
(e) Are A and B independent?

3.104 In a sample of 446 cars stopped at a roadblock, only 67 of the drivers had their seatbelts fastened. Estimate the probability that a driver stopped on that road will have his or her seatbelt fastened.

3.105 The personnel manager of a manufacturing plant claims that among the 400 employees 312 got a raise in 1989, 248 got increased pension benefits, 173 got both, and 43 got neither. Explain why this claim should be questioned.

3.106 If the probabilities that a satellite launching rocket will explode during lift-off or have its guidance system fail in flight are 0.0002 and 0.0005, find the probabilities that such a rocket will

(a) not explode during lift-off;
(b) explode during lift-off or have its guidance system fail in flight;
(c) neither explode during lift-off nor have its guidance system fail in flight.

3.107 Given $P(A) = 0.4$, $P(B) = 0.5$, and $P(A \cap B) = 0.2$, verify that

(a) $P(A|B) = P(A)$; (b) $P(A|B') = P(A)$;
(c) $P(B|A) = P(B)$; (d) $P(B|A') = P(B)$.

3.108 If events A and B are independent and $P(A) = 0.25$ and $P(B) = 0.40$, find

(a) $P(A \cap B)$; (b) $P(A|B)$;
(c) $P(A \cup B)$; (d) $P(A' \cap B')$.

3.109 An explosion in an LNG storage tank in the process of being repaired could have occurred as the result of static electricity, malfunctioning electrical equipment, an open flame in contact with the liner, or purposeful action (industrial sabotage). Interviews with engineers who were analyzing the risks involved led to estimates that such an explosion would occur with probability 0.25 as a result of static electricity, 0.20 as a result of malfunctioning electric equipment, 0.40 as a result of an open flame, and 0.75 as a result of purposeful action. These interviews also yielded subjective estimates of the prior probabilities of these four causes of 0.30, 0.40, 0.15, and 0.15, respectively. What was the most likely cause of the explosion?

3.110 It costs \$60 to test a certain component of a machine. If a defective component is installed, it costs \$1,200 to repair the resulting damage to the machine. Is it more profitable to install the component without testing if it is known that

(a) 3% of all components produced are defective;
(b) 5% of all components produced are defective;
(c) 8% of all components produced are defective?

3.111 A contractor has to choose between two jobs. The first job promises a profit of \$240,000 with a probability of 0.75 or a loss of \$60,000 (due to strikes and other delays) with a probability of 0.25; the second job promises a profit of \$360,000 with a probability of 0.50 or a loss of \$90,000 with a probability of 0.50.

(a) Which job should the contractor choose if she wants to maximize her expected profit?
(b) Which job would the contractor probably choose if her business is in bad shape and she will go broke unless she can make a profit of at least \$300,000 on her next job?

3.10
CHECK LIST OF KEY TERMS (with page references)

Additive set function 56
Axioms of probability 57
Bayes' theorem 75
Classical probability concept 54
Combination 50
Complement 44
Conditional probability 70
Consistency of probabilities 68
Continuous sample space 43
Discrete sample space 43

Empty set 43
Event 43
Experiment 42
Factorial notation 49
Finite sample space 43
Frequency interpretation 55
General addition rule 62
General multiplication rule 71
Independent events 71
Intersection 43
Mathematical expectation 80
Multiplication of choices 48
Mutually exclusive events 43
Odds 67
Outcome 42
Permutation 48
Rule of elimination 74
Rule of total probability 74
Sample space 41
Set function 56
Special addition rule 62
Special multiplication rule 71
Subjective probability 56
Tree diagram 47
Union 43
Venn diagram 44

PROBABILITY DISTRIBUTIONS

In most statistical problems we are concerned with one number or a few numbers that are associated with the outcomes of experiments. In the inspection of a manufactured product we may be interested only in the number of defectives; in the analysis of a road test we may be interested only in the average speed and the average fuel consumption; and in the study of the performance of a rotary switch we may be interested only in its lubrication, the current, and the humidity. All these numbers are associated with situations involving an element of chance—in other words, they are values of random variables.

In the study of random variables we are usually interested in their probability distributions, namely, in the probabilities with which they take on the various values in their range. The introduction to random variables and probability distributions in Section 4.1 is followed by a discussion of various special probability distributions in Sections 4.2, 4.3, 4.7, 4.8, and 4.9, and descriptions of the most important features of probability distributions in Sections 4.4 and 4.5.

4.1
RANDOM VARIABLES

To be more explicit about the concept of a random variable, let us refer again to the lawn-mower-rating example on page 47, and the corresponding probabilities shown in Figure 3.8. Now let us suppose that we refer to E_1 (easy to operate), P_2 (inexpensive), and C_3 (low average cost of repairs) as preferred ratings, and that we are interested only in the number of preferred ratings a lawn mower will get. To find the probabilities that a lawn mower will get 0, 1, 2, or 3 preferred ratings, let us refer to Figure 4.1, which is like Figure 3.8 except that we indicate for each outcome

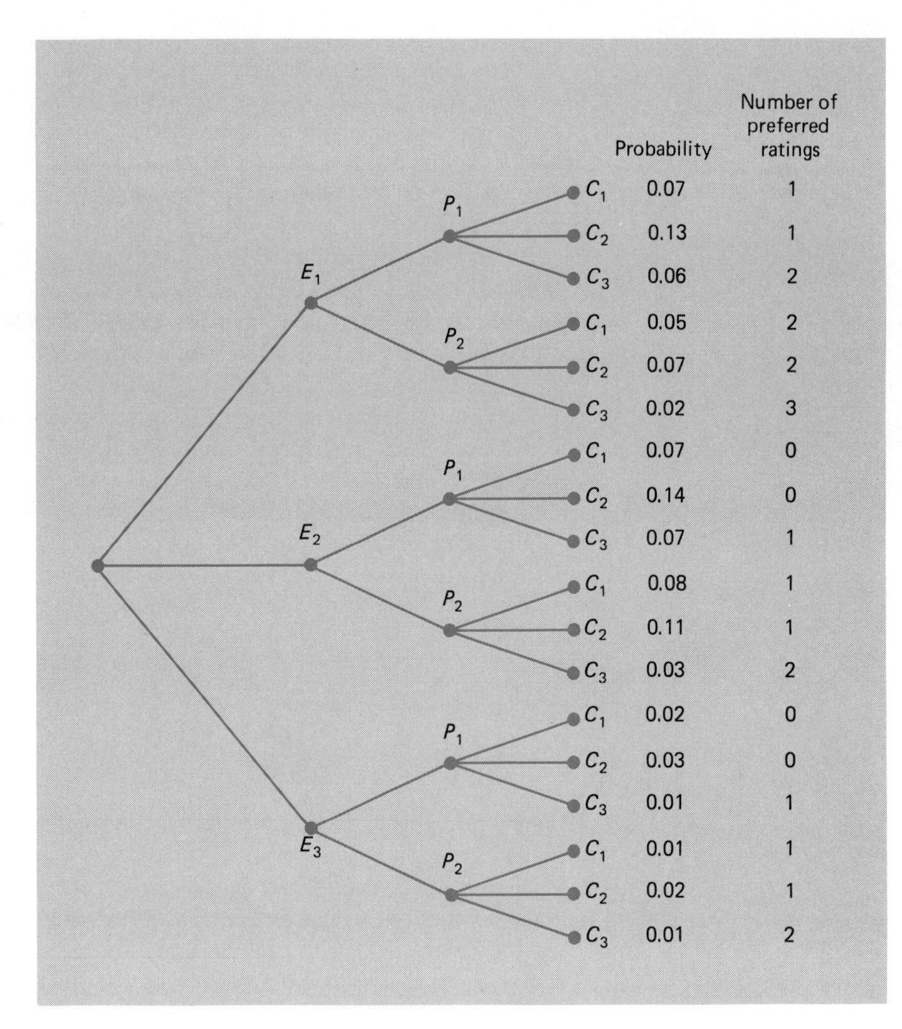

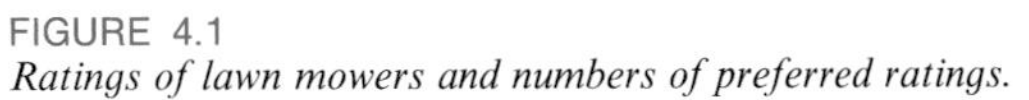
FIGURE 4.1
Ratings of lawn mowers and numbers of preferred ratings.

the number of preferred ratings. Adding the respective probabilities, we find that for zero preferred ratings the probability is $0.07 + 0.14 + 0.02 + 0.03 = 0.26$, for one preferred rating the probability is $0.07 + 0.13 + 0.07 + 0.08 + 0.11 + 0.01 + 0.01 + 0.02 = 0.50$, for two preferred ratings the probability is $0.06 + 0.05 + 0.07 + 0.03 + 0.01 = 0.22$, and for three preferred ratings the probability is 0.02. These results may be summarized as in the following table, where x denotes the number of preferred ratings:

x	0	1	2	3
Probability	0.26	0.50	0.22	0.02

The numbers 0, 1, 2, and 3 in this table are values of a **random variable**—the number of preferred ratings. Corresponding to each outcome in the sample space there is one and only one value x of this random variable, so that the random variable may be thought of as a function defined over the elements of the sample space. This is how we define random variables in general; they are functions defined over the elements of a sample space.

To find the probability that a random variable will take on any one value within its range, we proceed as in the above example; indeed, the table which we obtained displays another function, called the **probability distribution** of the random variable. To denote the values of a probability distribution we shall use such symbols as $f(x)$, $g(x)$, $\varphi(y)$, $h(z)$, and so on.

Random variables are usually classified according to the number of values which they can assume. In this chapter we shall limit our discussion to **discrete random variables**, which can take on only a finite number, or a countable infinity of values; **continuous random variables** are taken up in Chapter 5.

Whenever possible, we try to express probability distributions by means of equations; otherwise we must give a table that actually exhibits the correspondence between the values of the random variable and the associated probabilities. For instance,

$$f(x) = \frac{1}{6} \qquad \text{for } x = 1, 2, 3, 4, 5, 6$$

gives the probability distribution for the number of points we roll with a balanced die.

Of course, not every function defined for the values of a random variable can serve as a probability distribution. Since the values of probability distributions are probabilities and one value of a random variable must always occur, it follows that if $f(x)$ is a value of a probability distribution, then

$$f(x) \geq 0 \qquad \text{for all } x$$

and

$$\sum_{\text{all } x} f(x) = 1$$

EXAMPLE Check whether the following can serve as probability distributions:

(a) $f(x) = \dfrac{x-2}{2}$ for $x = 1, 2, 3, 4$

(b) $h(x) = \dfrac{x^2}{25}$ for $x = 0, 1, 2, 3, 4$

Solution (a) This function cannot serve as a probability distribution because $f(1)$ is negative. (b) This function cannot serve as a probability distribution because the sum of the five probabilities is $\frac{6}{5}$ and not 1.

■

It is often helpful to visualize probability distributions by means of graphs like those of Figure 4.2. The one on the left is called a **probability histogram**; the heights of the rectangles are proportional to the corresponding probabilities and the bases touch so that there are no gaps between the rectangles representing the successive values of the random variable. The one on the right is called a **bar chart**; the heights of the rectangles are also proportional to the corresponding probabilities, but they are narrow and their width is of no significance.

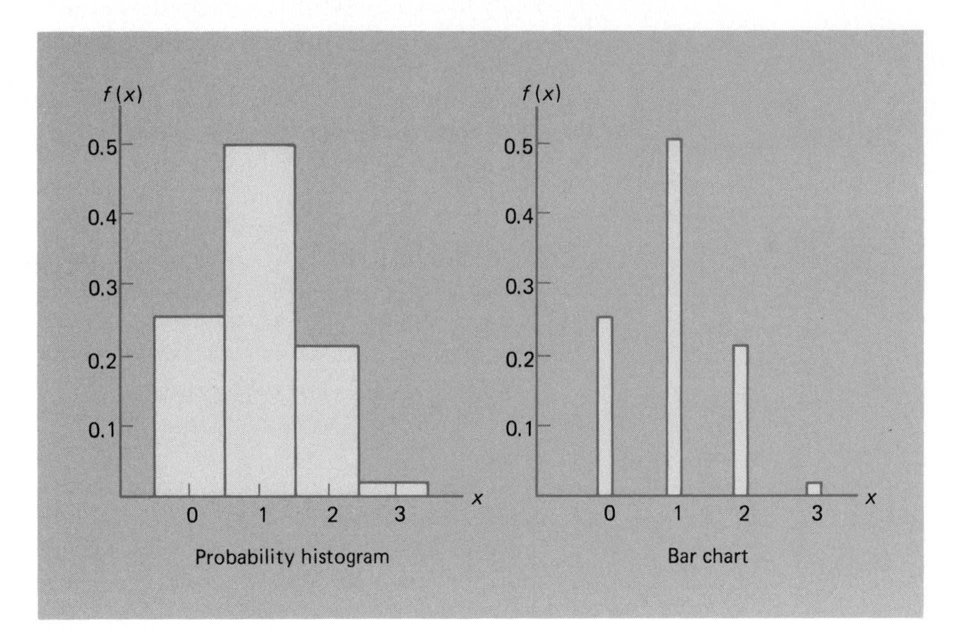

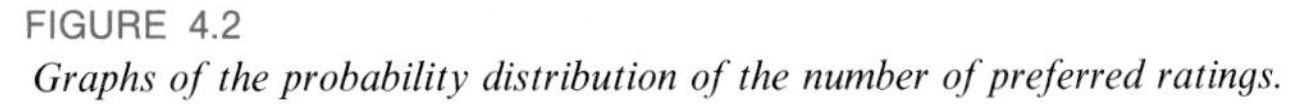

FIGURE 4.2
Graphs of the probability distribution of the number of preferred ratings.

As we see later in this chapter, there are many problems in which we are interested not only in the probability $f(x)$ that the value of a random variable is x, but also in the probability $F(x)$ that the value of a random variable is less than or equal to x. We refer to the function that assigns a value $F(x)$ to each x within the range of a random variable as the **distribution function** of the random variable. Referring again to the lawn-mower-rating example and basing our calculations on the table on page 90, we get

x	0	1	2	3
$F(x)$	0.26	0.76	0.98	1.00

for the distribution function of the number of preferred ratings.

4.2 THE BINOMIAL DISTRIBUTION

Many statistical problems deal with situations referred to as **repeated trials**. For example, we may want to know the probability that one of five rivets will rupture in a tensile test, the probability that 9 of 10 VCR's will run at least 1,000 hours, the probability that 45 of 300 drivers stopped at a road block will be wearing seatbelts, or the probability that 66 of 200 television viewers (interviewed by a rating service) will recall what products were advertised on a given program. To borrow from the language of games of chance, we might say that in each of these examples we are interested in the probability of getting x **successes** in n **trials**, or in other words, x successes and $n - x$ failures in n attempts.

In the problems we study in this section, we shall always make the following assumptions:

1. **There are only two possible outcomes for each trial** (arbitrarily called "success" and "failure," without inferring that a success is necessarily desirable).
2. **The probability of a success is the same for each trial.**
3. **There are *n* trials, where *n* is a constant.**
4. **The *n* trials are independent.**

Trials satisfying these assumptions are referred to as **Bernoulli trials**; if the assumptions cannot be met, the theory we shall develop here does not apply.

To solve problems that do meet the conditions listed in the preceding paragraph, we use a formula obtained in the following way: If p and $1 - p$ are the probabilities of success and failure on any one trial, then the probability of getting x successes and $n - x$ failures *in some specific order* is $p^x(1 - p)^{n-x}$; clearly, in this product of p's and $(1 - p)$'s there is one factor p for each success, one factor $1 - p$

for each failure, and the x factors p and $n - x$ factors $1 - p$ are all multiplied together by virtue of the generalized multiplication rule for more than two independent events. Since this probability applies to any point of the sample space that represents x successes and $n - x$ failures (in any specific order), we have only to count how many points of this kind there are, and then multiply $p^x(1 - p)^{n-x}$ by this number. Clearly, the number of ways in which we can select the x trials on which there is to be a success is $\binom{n}{x}$, the number of combinations of x objects selected from a set of n objects, and we thus arrive at the following result:

Binomial distribution

$$b(x; n, p) = \binom{n}{x} p^x (1 - p)^{n-x} \qquad \text{for } x = 0, 1, 2, \ldots, n$$

This probability distribution is called the **binomial distribution** because for $x = 0, 1, 2, \ldots$, and n, the values of the probabilities are the successive terms of the binomial expansion of $[p + (1 - p)]^n$; for the same reason, the combinatorial quantities $\binom{n}{x}$ are referred to as **binomial coefficients**. Actually, the preceding equation defines a family of probability distributions with each member characterized by given values of the **parameters** n and p.

EXAMPLE It has been claimed that in 60% of all solar heat installations the utility bill is reduced by at least one-third. Accordingly, what are the probabilities that the utility bill will be reduced by at least one-third in

(a) four of five installations;

(b) at least four of five installations?

Solution (a) Substituting $x = 4$, $n = 5$, and $p = 0.60$ into the formula for the binomial distribution, we get

$$b(4; 5, 0.60) = \binom{5}{4}(0.60)^4(1 - 0.60)^{5-4}$$
$$= 0.259$$

(b) Substituting $x = 5$, $n = 5$, and $p = 0.60$ into the formula for the binomial distribution, we get

$$b(5; 5, 0.60) = \binom{5}{5}(0.60)^5(1 - 0.60)^{5-5}$$
$$= 0.078$$

and the answer is $b(4; 5, 0.60) + b(5; 5, 0.60) = 0.259 + 0.078 = 0.337$.

■

If n is large, the calculation of binomial probabilities can become quite tedious. Many statistical software packages have binomial distribution commands (see Exercises 4.28 and 4.29). Otherwise it is convenient to refer to special tables. Values of the binomial distribution and the corresponding cumulative distribution function have been tabulated for $n = 2$ to $n = 49$ by the National Bureau of Standards and for $n = 50$ to $n = 100$ by H. G. Romig (see the bibliography at the end of the book). Table 1 at the end of this book gives the values of

$$B(x; n, p) = \sum_{k=0}^{x} b(k; n, p) \qquad \text{for } x = 0, 1, 2, \ldots, n$$

for $n = 2$ to $n = 20$ and $p = 0.05, 0.10, 0.15, \ldots, 0.90, 0.95$. We tabulated the **cumulative probabilities** rather than the values of $b(x; n, p)$, because the values of $B(x; n, p)$ are the ones needed more often in statistical applications. Note, however, that the values of $b(x; n, p)$ can be obtained by subtracting adjacent entries in Table 1. Because the two cumulative probabilities $B(x; n, p)$ and $B(x - 1; n, p)$ differ by the single term $b(x; n, p)$

$$b(x; n, p) = B(x; n, p) - B(x - 1; n, p)$$

The examples that follow illustrate the direct use of Table 1 and the use of this relation.

EXAMPLE If the probability is 0.05 that a certain wide-flange column will fail under a given axial load, what are the probabilities that among 16 such columns

(a) at most two will fail;
(b) at least four will fail?

Solution (a) Table 1 shows that $B(2; 16, 0.05) = 0.9571$. (b) Since

$$\sum_{x=4}^{16} b(x; 16, 0.05) = 1 - B(3; 16, 0.05)$$

Table 1 yields $1 - 0.9930 = 0.0070$.

■

EXAMPLE If the probability is 0.20 that any one person will dislike the taste of a new toothpaste, what is the probability that 5 of 18 randomly selected persons will dislike it?

Solution Using the relationship to cumulative probabilities and then looking up these probabilities in Table 1, we get

$$\begin{aligned} b(5; 18, 0.20) &= B(5; 18, 0.20) - B(4; 18, 0.20) \\ &= 0.8671 - 0.7164 \\ &= 0.1507 \end{aligned}$$

■

The following example illustrates the use of the binomial distribution in a problem of decision making.

EXAMPLE A washing machine manufacturer claims that only 10% of his machines require repairs within the warranty period of 12 months. If 5 of 20 of his machines required repairs within the first year, does this tend to support or refute the claim?

Solution Let us first find the probability that five or more of 20 of the washing machines will require repairs within a year when the probability that any one will require repairs within a year is 0.10. Using Table 1, we get

$$\begin{aligned}\sum_{x=5}^{20} b(x; 20, 0.10) &= 1 - B(4; 20, 0.10)\\ &= 1 - 0.9568\\ &= 0.0432\end{aligned}$$

Since this value is very small, it would seem reasonable to reject the washing machine manufacturer's claim.

■

Important information about the shape of binomial distributions is shown in Figures 4.3 and 4.4. First, if $p = 0.50$ the equation for the binomial distribution is

$$b(x; n, 0.50) = \binom{n}{x}(0.5)^n$$

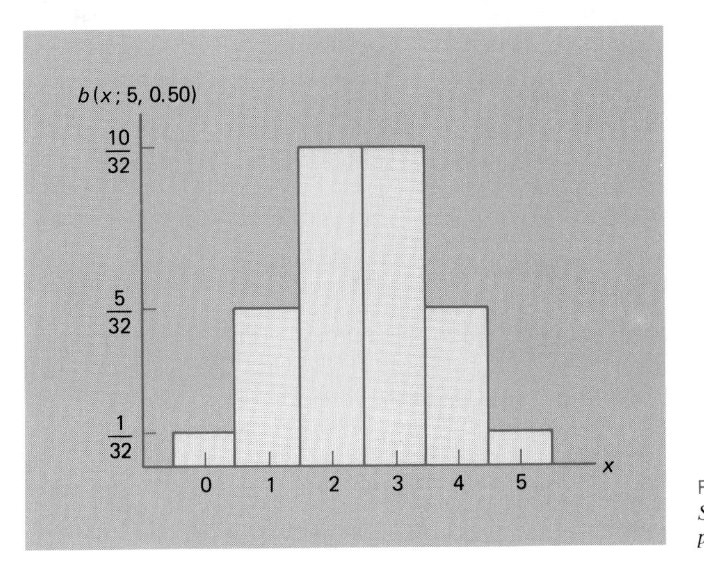

FIGURE 4.3
Symmetrical binomial distribution with $n = 5$ and $p = 0.50$.

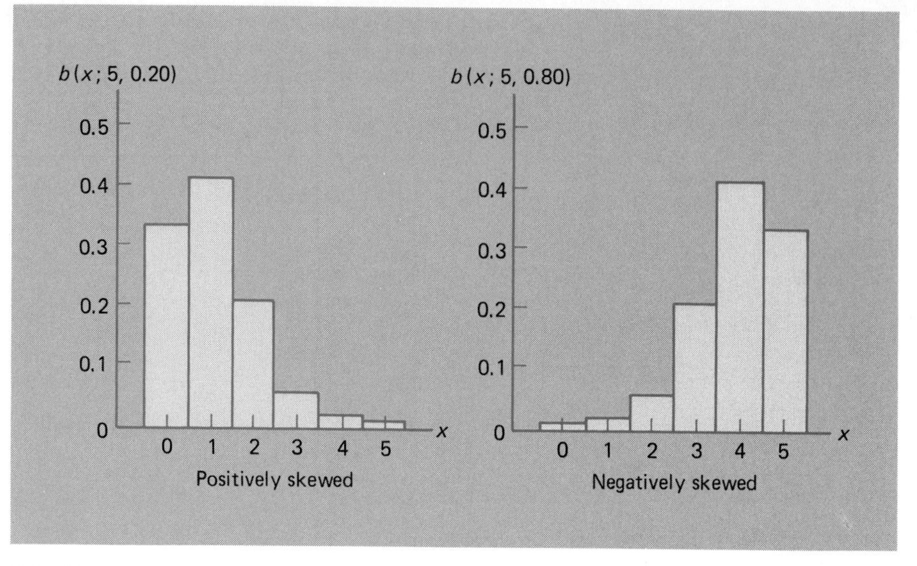

FIGURE 4.4
Skewed binomial distributions with $n = 5$ and $p = 0.20$ and $p = 0.80$.

and since $\binom{n}{n-x} = \binom{n}{x}$, it follows that $b(x; n, 0.50) = b(n - x; n, 0.50)$. This means that the probability histograms of such binomial distributions are **symmetrical**, as is illustrated in Figure 4.3. Note, however, that if p is less than 0.50, it is more likely that x will be small rather than large compared to n and that the opposite is true if p is greater than 0.50. This is illustrated by Figure 4.4, showing the histograms of binomial distributions with $n = 5$ and $p = 0.20$ and $p = 0.80$. A probability distribution that has a probability histogram like either of those of Figure 4.4 is said to be **skewed** (or long-tailed); it is said to be **positively skewed** if the tail is on the right, and it is said to be **negatively skewed** if the tail is on the left.

4.3 THE HYPERGEOMETRIC DISTRIBUTION

Suppose that we are interested in the number of defectives in a sample of n units drawn from a lot containing N units, of which a are defective. If the sample is drawn in such a way that at each successive drawing whatever units are left in the lot have the same chance of being selected, the probability that the first drawing will yield a defective unit is $\frac{a}{N}$, but for the second drawing it is $\frac{a-1}{N-1}$ or $\frac{a}{N-1}$, depending on whether or not the first unit drawn was defective. Thus, the trials are not independent, the fourth assumption underlying the binomial distribution is not met, and the binomial distribution does not apply. Note that the binomial distribution would apply if we **sample with replacement**, namely, if each unit selected for the sample is replaced before the next one is drawn.

To solve the problem of **sampling without replacement** (that is, as we originally formulated the problem), let us proceed as follows: The x successes (defectives) can be chosen in $\binom{a}{x}$ ways, the $n - x$ failures (nondefectives) can be chosen in $\binom{N-a}{n-x}$ ways, and, hence, x successes and $n - x$ failures can be chosen in $\binom{a}{x}\binom{N-a}{n-x}$ ways. Also, n objects can be chosen from a set of N objects in $\binom{N}{n}$, and if we consider all these possibilities as equally likely, it follows that for sampling without replacement the probability of getting "x successes in n trials" is

Hypergeometric distribution

$$h(x; n, a, N) = \frac{\binom{a}{x}\binom{N-a}{n-x}}{\binom{N}{n}} \quad \text{for } x = 0, 1, \ldots, n$$

where x cannot exceed a and $n - x$ cannot exceed $N - a$. This equation defines the **hypergeometric distribution**, whose parameters are the sample size n, the lot size (or population size) N, and the number of "successes" in the lot a.

EXAMPLE A shipment of 20 tape recorders contains 5 that are defective. If 10 of them are randomly chosen for inspection, what is the probability that 2 of the 10 will be defective?

Solution Substituting $x = 2$, $n = 10$, $a = 5$, and $N = 20$ into the formula for the hypergeometric distribution, we get

$$h(2; 10, 5, 20) = \frac{\binom{5}{2}\binom{15}{8}}{\binom{20}{10}} = \frac{10 \cdot 6{,}435}{184{,}756} = 0.348$$

■

In the preceding example, n was large compared to N, and if we had made the mistake of using the binomial distribution with $n = 10$ and $p = \frac{5}{20} = 0.25$ to calculate the probability of two defectives, the result would have been 0.282, which is much too small. However, when n is small compared to N, the composition of the lot is not seriously affected by drawing the sample, and the binomial distribution with the parameters n and $p = \frac{a}{N}$ will yield a good approximation.

EXAMPLE Repeat the preceding example for a lot of 100 tape recorders, of which 25 are defective, by using

(a) the formula for the hypergeometric distribution;
(b) the formula for the binomial distribution as an approximation.

Solution (a) Substituting $x = 2$, $n = 10$, $a = 25$, and $N = 100$ into the formula for the hypergeometric distribution, we get

$$h(2; 10, 25, 100) = \frac{\binom{25}{2}\binom{75}{8}}{\binom{100}{10}} = 0.292$$

(b) Substituting $x = 2$, $n = 10$, and $p = \frac{25}{100} = 0.25$ into the formula for the binomial distribution, we get

$$\begin{aligned} b(2; 10, 0.25) &= \binom{10}{2}(0.25)^2(1 - 0.25)^{10-2} \\ &= 0.282 \end{aligned}$$

■

Observe that the difference between the two values is only 0.010. In general, it can be shown that $h(x; n, a, N)$ approaches $b(x; n, p)$ with $p = \frac{a}{N}$ when $N \to \infty$, and a good rule of thumb is to use the binomial distribution as an approximation to the hypergeometric distribution if $n \leq \frac{N}{10}$.

Although we have introduced the hypergeometric distribution in connection with a problem of sampling inspection, it has many other applications. For instance, it can be used to find the probability that three of 12 housewives prefer Brand A detergent to Brand B, if they are selected from among 200 housewives among whom 40 actually prefer Brand A to Brand B. Also, it can be used in connection with a problem of selecting industrial diamonds, some of which have superior qualities and some do not, in connection with a problem of sampling income tax returns, where a among N returns filed contain questionable deductions, and so on.

EXERCISES

4.1 Suppose that a probability of $\frac{1}{12}$ is assigned to each point of the sample space of part (a) of Exercise 3.1 on page 51. Find the probability distribution of the total number of lasers that are suitable for the given task.

4.2 An experiment consists of four tosses of a coin. Denoting the outcomes $HHTH$, $THTT, \ldots$, and assuming that all 16 outcomes are equally likely, find the probability distribution for the total number of heads.

4.3 Determine whether the following can be probability distributions of a random variable which can take on only the values 1, 2, 3, and 4.

(a) $f(1) = 0.26, f(2) = 0.26, f(3) = 0.26$, and $f(4) = 0.26$
(b) $f(1) = 0.15, f(2) = 0.28, f(3) = 0.29$, and $f(4) = 0.28$
(c) $f(1) = 0.33, f(2) = 0.37, f(3) = -0.03$, and $f(4) = 0.33$

4.4 Check whether the following can define probability distributions, and explain your answers.

(a) $f(x) = \dfrac{x}{15}$ for $x = 0, 1, 2, 3, 4, 5$

(b) $f(x) = \dfrac{5 - x^2}{6}$ for $x = 0, 1, 2, 3$

(c) $f(x) = \dfrac{1}{4}$ for $x = 3, 4, 5, 6$

(d) $f(x) = \dfrac{x + 1}{25}$ for $x = 1, 2, 3, 4, 5$.

4.5 Given that $f(x) = \dfrac{k}{2^x}$ is a probability distribution for a random variable that can take on the values $x = 0, 1, 2, 3$, and 4, find k.

4.6 With reference to Exercise 4.5, find an expression for the values $F(x)$ of the distribution function of the random variable.

4.7 Prove that $b(x; n, p) = b(n - x; n, 1 - p)$.

4.8 Prove that $B(x; n, p) = 1 - B(n - x - 1; n, 1 - p)$.

4.9 Prove that

$$\frac{b(x + 1; n, p)}{b(x; n, p)} = \frac{p(n - x)}{(1 - p)(x + 1)}$$

for $x = 0, 1, 2, \ldots, n - 1$, and use this recursion formula to calculate the values of the binomial distribution with $n = 6$ and $p = 0.30$. Also, verify the results by means of Table 1.

4.10 Use Table 1 to find

(a) $B(8; 16, 0.40)$;
(b) $b(8; 16, 0.40)$;
(c) $B(9; 12, 0.60)$;
(d) $b(9; 12, 0.60)$;
(e) $\sum_{k=6}^{20} b(k; 20, 0.15)$;
(f) $\sum_{k=6}^{9} b(k; 9, 0.70)$.

4.11 Use Table 1 to find

(a) $B(7; 19, 0.45)$;
(b) $b(7; 19, 0.45)$;
(c) $B(8; 10, 0.95)$;
(d) $b(8; 10, 0.95)$;
(e) $\sum_{k=4}^{10} b(k; 10, 0.35)$;
(f) $\sum_{k=2}^{4} b(k; 9, 0.30)$.

4.12 Rework the decision problem on page 95, supposing that only 3 of the 20 washing machines required repairs within the first year.

4.13 In a given city, medical expenses are given as the reason for 75% of all personal bankruptcies. Use the formula for the binomial distribution to find the probability that medical expenses will be given as the reason for two of the next four personal bankruptcies filed in that city.

4.14 If the probability is 0.40 that steam will condense in a thin-walled aluminum tube at 10 atm pressure, use the formula for the binomial distribution to find the probability that under the stated conditions steam will condense in 4 of 12 such tubes.

4.15 Suppose that a civil service examination is designed so that 70% of all persons with an IQ of 90 can pass. Use Table 1 to find the probabilities that among 15 persons with an IQ of 90 who take the test

(a) at least 12 will pass;
(b) at most six will pass;
(c) 10 will pass.

4.16 The probability that the noise level of a wide-band amplifier will exceed 2 dB is 0.05. Use Table 1 to find the probabilities that among 12 such amplifiers the noise level of

(a) one will exceed 2 dB;
(b) at most two will exceed 2 dB;
(c) two or more will exceed 2 dB.

4.17 An agricultural cooperative claims that 90% of the watermelons shipped out are ripe and ready to eat. Find the probabilities that among 18 watermelons shipped out

(a) all 18 are ripe and ready to eat;
(b) at least 16 are ripe and ready to eat;
(c) at most 14 are ripe and ready to eat.

4.18 A quality-control engineer wants to check whether (in accordance with specifications) 95% of the electronic components shipped by his company are in good working condition. To this end, he randomly selects 15 from each large lot ready to be shipped and passes the lot if the selected components are all in good working condition; otherwise, each of the components in the lot is checked. Find the probabilities that the quality control engineer will commit the error of

(a) holding a lot for further inspection even though 95% of the components are in good working condition;
(b) letting a lot pass through without further inspection even though only 90% of the components are in good working condition;
(c) letting a lot pass through without further inspection even though only 80% of the components are in good working condition.

4.19 A food processor claims that at most 10% of her jars of instant coffee contain less coffee than claimed on the label. To test this claim, 16 jars of her instant coffee are randomly selected and the contents are weighed; her claim is accepted if fewer than 3 of the jars contain less coffee than claimed on the label. Find the probabilities that the food processor's claim will be accepted when the actual percentage of her jars containing less coffee than claimed on the label is

(a) 5%; (b) 10%;
(c) 15%; (d) 20%.

4.20 A study shows that a computer firm answers 70% of all inquiries within 6 days. Find the probabilities that the firm will answer 0, 1, 2, ..., or 10 of 10 inquiries within 6 days, and draw a probability histogram of this probability distribution.

4.21 What is the probability that an IRS auditor will catch only 2 income tax returns with illegitimate deductions, if she randomly selected 6 returns from among 18 returns, of which 8 contain illegitimate deductions?

4.22 Among the 12 solar collectors on display at a trade show, 9 are flat-plate collectors and the others are concentrating collectors. If a person visiting the show randomly selects four of the solar collectors to check out, what is the probability that 3 of them will be flat-plate collectors?

4.23 A quality-control engineer inspects a random sample of 3 batteries from each lot of 24 car batteries that is ready to be shipped. If such a lot contains six batteries with slight defects, what are the probabilities that the inspector's sample will contain

(a) none of the batteries with defects;
(b) only one of the batteries with defects;
(c) at least two of the batteries with defects?

4.24 If 6 of 18 new buildings in a city violate the building code, what is the probability that a building inspector, who randomly selects 4 of the new buildings for inspection, will catch

(a) none of the new buildings that violate the building code;
(b) one of the new buildings that violate the building code;
(c) two of the new buildings that violate the building code;
(d) at least three of the new buildings that violate the building code?

4.25 Among the 16 cities that a professional society is considering for its next three annual conventions, 7 are in the western part of the United States. To avoid arguments, the selection is left to chance. If none of the cities can be chosen more than once, what are the probabilities that

(a) none of the conventions will be held in the western part of the United States;
(b) all of the conventions will be held in the western part of the United States?

4.26 A shipment of 120 burglar alarms contains 5 that are defective. If three of these alarms are randomly selected and shipped to a customer, find the probability that the customer will get one bad unit by using

(a) the formula for the hypergeometric distribution;
(b) the formula for the binomial distribution as an approximation.

4.27 Among the 300 employees of a company, 240 are union members, while the others are not. If eight of the employees are chosen by lot to serve on the committee which administers the pension fund, find the probability that five of them will be union members while the others are not, by using

(a) the formula for the hypergeometric distribution;
(b) the formula for the binomial distribution as an approximation.

4.28 Binomial probabilities can be calculated using the *MINITAB* command

```
PDF;
  BINOMIAL with n=7, p=.33.
```

which produces the output

```
BINOMIAL WITH N =  7 P = 0.330000
    K           P( X = K)
    0           0.0606
    1           0.2090
    2           0.3088
    3           0.2535
    4           0.1248
    5           0.0369
    6           0.0061
    7           0.0004
```

Find the binomial probabilities with $n = 27$ and $p = 0.47$.

4.29 Cumulative binomial probabilities can be calculated using the *MINITAB* command

```
CDF ;
  BINOMIAL with n=7 ,  p=.33.
```

The output is

```
BINOMIAL WITH N =  7 P = Ø.33ØØØØ
  K P( X LESS        OR =K)
  Ø                  Ø.Ø6Ø6
  1                  Ø.2696
  2                  Ø.5783
  3                  Ø.8318
  4                  Ø.9566
  5                  Ø.9935
  6                  Ø.9996
  7                  1.ØØØØ
```

Find the cumulative binomial probabilities with $n = 27$ and $p = 0.47$.

4.4

THE MEAN AND THE VARIANCE OF A PROBABILITY DISTRIBUTION

Besides the binomial and hypergeometric distributions, there are many other probability distributions that have important engineering applications. However, before we go any further, let us discuss some general characteristics of probability distributions.

One such characteristic, that of the symmetry or skewness of a probability distribution, was illustrated by means of Figures 4.3 and 4.4; two others will be presented here with reference to Figure 4.5, which shows the probability histograms of two binomial distributions. One of these binomial distributions has the parameters $n = 4$ and $p = \frac{1}{2}$, and the other has the parameters $n = 16$ and $p = \frac{1}{2}$. Essentially, these two probability distributions differ in two respects. The first probability distribution is centered about $x = 2$, whereas the other (whose histogram is shaded) is centered about $x = 8$, and we say that the two distributions differ in their **location**. Another distinction is that the histogram of the second distribution is more spread out, and we say that the two distributions differ in **variation**. To make such comparisons more specific, we shall introduce in this section two of the most important statistical measures, describing, respectively, the location and the variation of a probability distribution—the **mean** and the **variance**.

The mean of a probability distribution is simply the mathematical expectation of a corresponding random variable. If a random variable takes on the values $x_1, x_2, \ldots,$ or x_k, with the probabilities $f(x_1), f(x_2), \ldots,$ and $f(x_k)$, its mathematical expectation is

$$x_1 \cdot f(x_1) + x_2 \cdot f(x_2) + \cdots + x_k \cdot f(x_k)$$

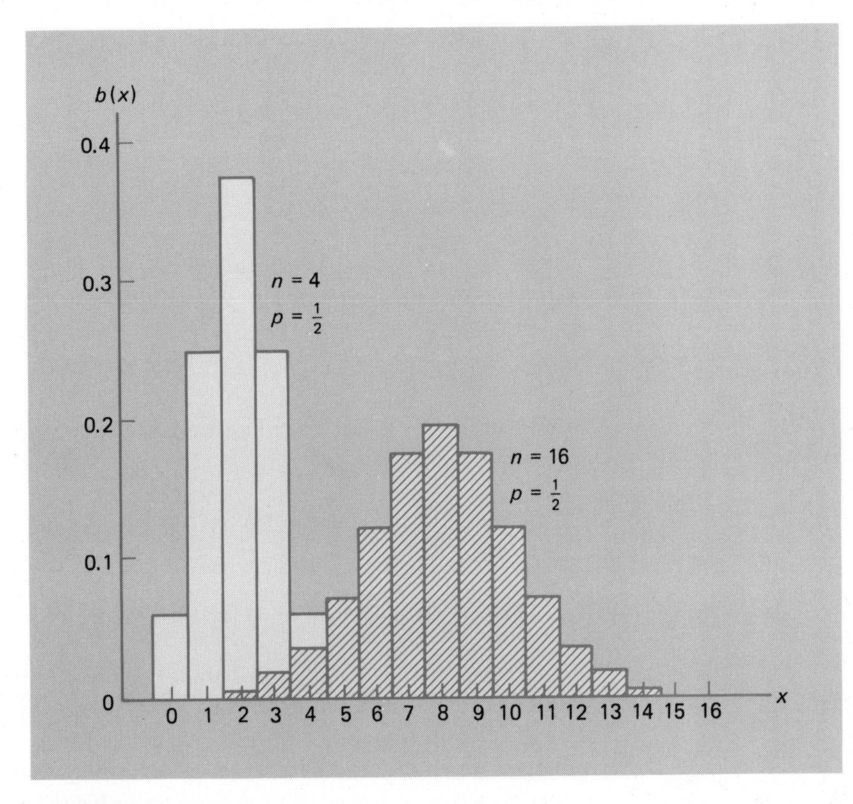

FIGURE 4.5
Probability histograms of two binomial distributions.

and, using the $\sum$ notation, we can thus write

Mean of discrete probability distribution

$$\mu = \sum_{\text{all } x} x \cdot f(x)$$

where the **mean** is denoted by the Greek letter μ (*mu*). The mean of a probability distribution measures its center in the sense of an average, or better, in the sense of a center of gravity. Note that the above formula for μ is, in fact, that for the **first moment about the origin** of a discrete system of masses $f(x)$ arranged on a straight line at distances x from the origin. We do not have to divide here by $\sum_{\text{all } x} f(x)$, as we do in the usual formula for the x-coordinate of the center of gravity, since this sum equals 1 by definition.

EXAMPLE Find the mean of the probability distribution of the number of heads obtained in three flips of a balanced coin.

Solution The probabilities for 0, 1, 2, or 3 heads are $\frac{1}{8}$, $\frac{3}{8}$, $\frac{3}{8}$, and $\frac{1}{8}$, as can easily be verified by counting equally likely possibilities or by using the formula for the binomial distribution with $n = 3$ and $p = \frac{1}{2}$. Thus,

$$\mu = 0\cdot\frac{1}{8} + 1\cdot\frac{3}{8} + 2\cdot\frac{3}{8} + 3\cdot\frac{1}{8} = \frac{3}{2}$$

■

EXAMPLE With reference to the lawn-mower-rating example and the probabilities given on page 90, find the mean of the probability distribution of the number of preferred ratings.

Solution Substituting $x = 0$, 1, 2, and 3, and the corresponding probabilities, into the formula for μ, we get

$$\begin{aligned}\mu &= 0(0.26) + 1(0.50) + 2(0.22) + 3(0.02)\\ &= 1.00\end{aligned}$$

■

Returning to the second probability distribution of Figure 4.5, we could find its mean by calculating all the necessary probabilities (or by looking them up in Table 1) and substituting them into the formula for μ. However, if we reflect for a moment, we might argue that there is a fifty-fifty chance for a success on each trial, there are 16 trials, and it would seem reasonable to expect 8 heads and 8 tails (in the sense of a mathematical expectation). Similarly, we might argue that if a binomial distribution has the parameters $n = 200$ and $p = 0.20$, we can expect a success 20% of the time and, hence, on the average $200(0.20) = 40$ successes in 200 trials. These two values are, indeed, correct, and it can be shown in general that

Mean of binomial distribution

$$\boldsymbol{\mu = n \cdot p}$$

for the mean of a binomial distribution. To prove this formula, we substitute the expression that defines $b(x; n, p)$ into the formula for μ, and we get

$$\mu = \sum_{x=0}^{n} x\cdot\frac{n!}{x!\,(n-x)!}\,p^x(1-p)^{n-x}$$

Then, making use of the fact that $\frac{x}{x!} = \frac{1}{(x-1)!}$ and $n! = n(n-1)!$, and factoring out n and p, we obtain

$$\mu = np\sum_{x=1}^{n}\frac{(n-1)!}{(x-1)!\,(n-x)!}\,p^{x-1}(1-p)^{n-x}$$

where the summation starts with $x = 1$ since the original summand is zero for $x = 0$. If we now let $y = x - 1$ and $m = n - 1$, we obtain

$$\mu = np \sum_{y=0}^{m} \frac{m!}{y!(m-y)!} p^y (1-p)^{m-y}$$

and this last sum can easily be recognized as that of all the terms of the binomial distribution with the parameters m and p. Hence, this sum equals 1 and it follows that $\mu = np$.

EXAMPLE Find the mean of the probability distribution of the number of heads obtained in three flips of a balanced coin.

Solution For a binomial distribution with $n = 3$ and $p = \frac{1}{2}$, we get $\mu = 3 \cdot \frac{1}{2} = \frac{3}{2}$, and this agrees with the result obtained on page 104.

■

The formula $\mu = np$ applies, of course, only to binomial distributions, but for other special distributions, we can, similarly, express the mean in terms of their parameters. For instance, for the mean of the hypergeometric distribution with the parameters n, a, and N, we can write

Mean of hypergeometric distribution

$$\mu = n \cdot \frac{a}{N}$$

In Exercise 4.41 on page 113, the reader will be asked to derive this formula by a method similar to the one we used to derive the formula for the mean of a binomial distribution.

EXAMPLE With reference to the example on page 97, where 5 of 20 tape recorders were defective, find the mean of the probability distribution of the number of defectives in a sample of 10 randomly chosen for inspection.

Solution Substituting $n = 10$, $a = 5$, and $N = 20$ into the above formula for μ, we get

$$\mu = 10 \cdot \frac{5}{20} = 2.5$$

In other words, if we inspect 10 of the tape recorders, we can expect 2.5 defectives, where "expect" is to be interpreted in the sense of a mathematical expectation.

■

To study the second of the two properties of probability distributions mentioned on page 102, their variation, let us refer again to the two probability distributions of Figure 4.5. For the one where $n = 4$ there is a high probability of getting values close to the mean, but for the one where $n = 16$ there is a high probability of getting values scattered over considerable distances away from the mean. Using this property, it may seem reasonable to measure the variation of a probability distribution with the quantity

$$\sum_{\text{all } x} (x - \mu) \cdot f(x)$$

namely, the average amount by which the values of the random variable deviate from the mean. Unfortunately,

$$\begin{aligned}\sum_{\text{all } x} (x - \mu) \cdot f(x) &= \sum_{\text{all } x} x \cdot f(x) - \sum_{\text{all } x} \mu \cdot f(x) \\ &= \mu - \mu \cdot \sum_{\text{all } x} f(x) = \mu - \mu = 0\end{aligned}$$

so that this expression is always equal to zero. However, since we are really interested in the magnitude of the deviations $x - \mu$ and not in their signs, it suggests itself that we average the absolute values of these deviations from the mean. This would, indeed, provide a measure of variation, but on purely theoretical grounds we prefer to work instead with the squares of the deviations from the mean. These quantities are also nonnegative, and their average is indicative of the spread or dispersion of a probability distribution. We thus define the **variance** of a probability distribution with values $f(x)$ as

Variance of probability distribution

$$\boldsymbol{\sigma^2 = \sum_{\text{all } x} (x - \mu)^2 \cdot f(x)}$$

where σ is the lowercase Greek letter for *s*. This measure is not in the same units (or dimension) as the values of the random variable, but we can adjust for this by taking the square root, thus defining the **standard deviation** as

Standard deviation of probability distribution

$$\boldsymbol{\sigma = \sqrt{\sum_{\text{all } x} (x - \mu)^2 \cdot f(x)}}$$

EXAMPLE Compare the standard deviations of the two probability distributions of Figure 4.5.

Solution Since $\mu = 4 \cdot \frac{1}{2} = 2$ for the binomial distribution with $n = 4$ and $p = \frac{1}{2}$, we find that the variance of this probability distribution is

$$\begin{aligned}\sigma^2 &= (0 - 2)^2 \cdot \tfrac{1}{16} + (1 - 2)^2 \cdot \tfrac{4}{16} + (2 - 2)^2 \cdot \tfrac{6}{16} \\ &\qquad + (3 - 2)^2 \cdot \tfrac{4}{16} + (4 - 2)^2 \cdot \tfrac{1}{16} \\ &= 1\end{aligned}$$

and, hence, that its standard deviation is $\sigma = 1$. Similarly, it can be shown that for the other distribution $\sigma = 2$, and we find that the second (shaded) distribution with the greater spread also has the greater standard deviation.

■

Given any probability distribution, we can always calculate σ^2 by substituting the corresponding probabilities $f(x)$ into the formula which defines the variance. As in the case of the mean, however, this work can be simplified to a considerable extent when we deal with special kinds of distributions. For instance, it can be shown that the variance of the binomial distribution with the parameters n and p is given by the formula

Variance of binomial distribution

$$\sigma^2 = n \cdot p \cdot (1 - p)$$

and that the variance of the hypergeometric distribution with the parameters n, a, and N is given by the formula

Variance of hypergeometric distribution

$$\sigma^2 = \frac{n \cdot a \cdot (N - a) \cdot (N - n)}{N^2 \cdot (N - 1)}$$

EXAMPLE Verify the result stated in the preceding example, that $\sigma = 2$ for the binomial distribution with $n = 16$ and $p = \frac{1}{2}$.

Solution Substituting $n = 16$ and $p = \frac{1}{2}$ into the formula for the variance of a binomial distribution, we get

$$\sigma^2 = 16 \cdot \tfrac{1}{2} \cdot \tfrac{1}{2} = 4$$

and, hence, $\sigma = \sqrt{4} = 2$.

■

EXAMPLE With reference to the example on page 97, where 5 of 20 tape recorders were defective, find the standard deviation of the probability distribution of the number of defectives in a sample of 10 randomly chosen for inspection.

Solution Substituting $n = 10$, $a = 5$, and $N = 20$ into the formula for the variance of a hypergeometric distribution, we get

$$\sigma^2 = \frac{10 \cdot 5 \cdot (20 - 5) \cdot (20 - 10)}{20^2 \cdot 19} = \frac{75}{76}$$

and, hence, $\sigma = \sqrt{75/76} = 0.99$.

■

When we first defined the variance of a probability distribution, it may have occurred to the reader that the formula looked exactly like the one which we use in physics to define second moments, or moments of inertia. Indeed, it is customary in statistics to define the **kth moment about the origin** as

$$\mu_k' = \sum_{\text{all } x} x^k \cdot f(x)$$

and the **kth moment about the mean** as

$$\mu_k = \sum_{\text{all } x} (x - \mu)^k \cdot f(x)$$

Thus, the mean μ is the first moment about the origin, and the variance σ^2 is the second moment about the mean. Higher moments are often used in statistics to give further descriptions of probability distributions. For instance, the third moment about the mean (divided by σ^3 to make this measure independent of the scale of measurement) is used to describe the symmetry or skewness of a distribution; the fourth moment about the mean (divided by σ^4) is, similarly, used to describe its "peakedness," or **kurtosis**. To determine moments about the mean, it is usually easiest to express them in terms of moments about the origin and then to calculate the necessary moments about the origin. For the second moment about the mean we thus have the important formula

Computing formula for the variance

$$\boldsymbol{\sigma^2 = \mu_2' - \mu^2}$$

which the reader will be asked to prove in part (a) of Exercise 4.47 on page 113; part (b) of that exercise pertains to a similar formula for expressing μ_3 in terms of moments about the origin.

EXAMPLE Use the preceding computing formula to determine the variance of the probability distribution of the number of points rolled with a balanced die.

Solution Since $f(x) = \frac{1}{6}$ for $x = 1, 2, 3, 4, 5$, and 6, we get

$$\begin{aligned}\mu &= 1 \cdot \tfrac{1}{6} + 2 \cdot \tfrac{1}{6} + 3 \cdot \tfrac{1}{6} + 4 \cdot \tfrac{1}{6} + 5 \cdot \tfrac{1}{6} + 6 \cdot \tfrac{1}{6} \\ &= \tfrac{7}{2} \\ \mu_2' &= 1^2 \cdot \tfrac{1}{6} + 2^2 \cdot \tfrac{1}{6} + 3^2 \cdot \tfrac{1}{6} + 4^2 \cdot \tfrac{1}{6} + 5^2 \cdot \tfrac{1}{6} + 6^2 \cdot \tfrac{1}{6} \\ &= \tfrac{91}{6}\end{aligned}$$

and, hence, $\sigma^2 = \frac{91}{6} - (\frac{7}{2})^2 = \frac{35}{12}$.

■

4.5
CHEBYSHEV'S THEOREM

Earlier in this chapter we used examples to show how the standard deviation measures the variation of a probability distribution, that is, how it controls the concentration of probability in the neighborhood of the mean. If σ is small, there is a high probability of getting values close to the mean; if σ is large there is a correspondingly higher probability for getting values farther away from the mean. Formally, this idea is expressed by the following theorem.

Chebyshev's theorem

> ***Theorem 4.2*** ***If a probability distribution has mean*** $\boldsymbol{\mu}$ ***and standard deviation*** $\boldsymbol{\sigma}$***, the probability of getting a value which deviates from*** $\boldsymbol{\mu}$ ***by at least*** $\boldsymbol{k\sigma}$ ***is at most*** $\boldsymbol{\frac{1}{k^2}}$.

Symbolically,

$$P(|x - \mu| \geq k\sigma) \leq \frac{1}{k^2}$$

where $P(|x - \mu| \geq k\sigma)$ is the probability associated with the set of outcomes for which x, the value of a random variable having the given probability distribution, is such that $|x - \mu| \geq k\sigma$.

Thus, the probability that a random variable will take on a value which deviates (differs) from the mean by at least 2 standard deviations is at most $\frac{1}{4}$, the probability that it will take on a value which deviates from the mean by at least 5 standard deviations is at most $\frac{1}{25}$, and the probability that it will take on a value which deviates from the mean by 10 standard deviations or more is less than or equal to $\frac{1}{100}$.

To prove this theorem, consider any probability distribution with the values $f(x)$, the mean μ, and the variance σ^2. Dividing the sum defining the variance into three parts as indicated in Figure 4.6, we have

$$\begin{aligned}\sigma^2 &= \sum_{\text{all } x} (x - \mu)^2 f(x) \\ &= \sum_{R_1} (x - \mu)^2 f(x) + \sum_{R_2} (x - \mu)^2 f(x) + \sum_{R_3} (x - \mu)^2 f(x)\end{aligned}$$

where R_1 is the region for which $x \leq \mu - k\sigma$, R_2 is the region for which $\mu - k\sigma < x < \mu + k\sigma$, and R_3 is the region for which $x \geq \mu + k\sigma$. Since $(x - \mu)^2 f(x)$ cannot be negative, the above sum over R_2 is nonnegative, and

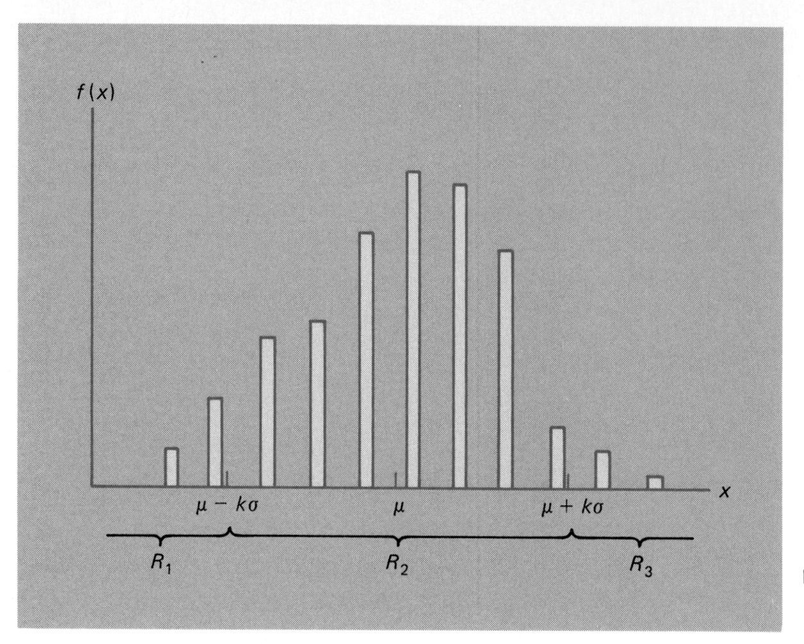

FIGURE 4.6
Diagram for proof of Chebyshev's theorem.

without it the sum of the summations over R_1 and R_3 is less than or equal to σ^2, that is,

$$\sigma^2 \geq \sum_{R_1} (x-\mu)^2 f(x) + \sum_{R_3} (x-\mu)^2 f(x)$$

But $x - \mu \leq -k\sigma$ in the region R_1 and $x - \mu \geq k\sigma$ in the region R_3, so that in either case $|x - \mu| \geq k\sigma$; hence, in both regions $(x-\mu)^2 \geq k^2\sigma^2$. If we now replace $(x-\mu)^2$ in each sum by $k^2\sigma^2$, a number less than or equal to $(x-\mu)^2$, we obtain the inequality

$$\sigma^2 \geq \sum_{R_1} k^2\sigma^2 f(x) + \sum_{R_3} k^2\sigma^2 f(x)$$

or

$$\frac{1}{k^2} \geq \sum_{R_1} f(x) + \sum_{R_3} f(x)$$

Since $\sum_{R_1} f(x) + \sum_{R_3} f(x)$ represents the probability that x is in the region $R_1 \cup R_3$, namely, that $|x - \mu| \geq k\sigma$, this completes the proof of Theorem 4.2.

To obtain an alternative form of Chebyshev's theorem, note that the event $|x - \mu| < k\sigma$ is the complement of the event $|x - \mu| \geq k\sigma$; hence, the probability of getting a value which deviates from μ by less than $k\sigma$ is at least $1 - \frac{1}{k^2}$.

EXAMPLE The number of customers who visit a car dealer's showroom on a Saturday morning is a random variable with $\mu = 18$ and $\sigma = 2.5$. With what probability can we assert that there will be between 8 and 28 customers?

Solution Since $k = \frac{28 - 18}{2.5} = \frac{18 - 8}{2.5} = 4$, the probability is at least $1 - \frac{1}{4^2} = \frac{15}{16}$.

■

Theoretically speaking, the most important feature of Chebyshev's theorem is that it applies to any probability distribution for which μ and σ exist. However, so far as applications are concerned, this generality is also its greatest weakness—it provides only an upper limit (often a very poor one) to the probability of getting a value that deviates from the mean by k standard deviations or more. For instance, we can assert in general that the probability of getting a value which differs from the mean by at least 2 standard deviations is at most 0.25, whereas the corresponding exact probability for the binomial distribution with $n = 16$ and $p = \frac{1}{2}$ is only 0.0768—"at most 0.25" is correct, but it does not tell us that the actual probability may be as small as 0.0768.

An important result is obtained if we apply Chebyshev's theorem to the binomial distribution. To illustrate this result, consider the following example.

EXAMPLE Show that for 40,000 flips of a balanced coin, the probability is at least 0.99 that the proportion of heads will fall between 0.475 and 0.525.

Solution Since $\mu = 40{,}000 \cdot \frac{1}{2} = 20{,}000$, $\sigma = \sqrt{40{,}000 \cdot \frac{1}{2} \cdot \frac{1}{2}} = 100$, and $1 - \frac{1}{k^2} = 0.99$ yields $k = 10$, the alternative form of Chebyshev's theorem tells us that the probability is at least 0.99 that we will get between $20{,}000 - 10(100) = 19{,}000$ and $20{,}000 + 10(100) = 21{,}000$ heads. Hence, the probability is at least 0.99 that the proportion of heads will fall between $\frac{19{,}000}{40{,}000} = 0.475$ and $\frac{21{,}000}{40{,}000} = 0.525$.

■

Correspondingly, the reader will be asked to show in Exercise 4.45 that for 1,000,000 flips of a balanced coin the probability is at least 0.99 that the proportion of heads will fall between 0.495 and 0.505, and these results suggest that when n is large, the chances are that the proportion of heads will be very close to $p = \frac{1}{2}$. When formulated for any binomial distribution with the parameters n and p, this result is referred to as the **law of large numbers**.

EXERCISES

4.30 Suppose that the probabilities are 0.4, 0.3, 0.2, and 0.1 that there will be 0, 1, 2, or 3 power failures in a certain city during the month of July. Use the formulas which define μ and σ^2 to find

(a) the mean of this probability distribution;
(b) the variance of this probability distribution.

4.31 Use the computing formula for σ^2 to rework part (b) of the preceding exercise.

4.32 The following table gives the probabilities that a certain computer will malfunction 0, 1, 2, 3, 4, 5, or 6 times on any one day:

Number of malfunctions: x	0	1	2	3	4	5	6
Probability: $f(x)$	0.17	0.29	0.27	0.16	0.07	0.03	0.01

Use the formulas which define μ and σ to find

(a) the mean of this probability distribution;

(b) the standard deviation of this probability distribution.

4.33 Use the computing formula for σ^2 to rework part (b) of the preceding exercise.

4.34 Find the mean and the variance of the uniform probability distribution given by

$$f(x) = \frac{1}{n} \qquad \text{for } x = 1, 2, 3, \ldots, n$$

[*Hint*: The sum of the first n positive integers is $\frac{1}{2}n(n+1)$, and the sum of their squares is $\frac{1}{6}n(n+1)(2n+1)$.]

4.35 Find the mean and the variance of the binomial distribution with $n = 4$ and $p = 0.70$ by using

(a) Table 1 and the formulas defining μ and σ^2;

(b) the special formulas for the mean and the variance of a binomial distribution.

4.36 As can easily be verified by means of the formula for the binomial distribution (or by listing all 32 possibilities), the probabilities of getting 0, 1, 2, 3, 4, or 5 heads in five flips of a balanced coin are $\frac{1}{32}$, $\frac{5}{32}$, $\frac{10}{32}$, $\frac{10}{32}$, $\frac{5}{32}$, and $\frac{1}{32}$. Find the mean of this probability distribution using

(a) the formula that defines μ;

(b) the special formula for the mean of a binomial distribution.

4.37 With reference to Exercise 4.36, find the variance of the probability distribution using

(a) the formula that defines σ^2;

(b) the computing formula for σ^2;

(c) the special formula for the variance of a binomial distribution.

4.38 If 95% of certain radial tires last at least 30,000 miles, find the mean and the standard deviation of the distribution of the number of these tires, among 20 selected at random, that last at least 30,000 miles, using

(a) Table 1, the formula which defines μ, and the computing formula for σ^2;

(b) the special formulas for the mean and the variance of a binomial distribution.

4.39 Find the mean and the standard deviation of the distribution of each of the following random variables (having binomial distributions):

(a) The number of heads obtained in 676 flips of a balanced coin.

(b) The number of 4's obtained in 720 rolls of a balanced die.

(c) The number of defectives in a sample of 600 parts made by a machine, when the probability is 0.04 that any one of the parts is defective.

(d) The number of students among 800 interviewed who do not like the food served at the university cafeteria, when the probability is 0.65 that any one of them does not like the food.

4.40 Find the mean and the standard deviation of the hypergeometric distribution with the parameters $n = 3$, $a = 4$, and $N = 8$

(a) by first calculating the necessary probabilities and then using the formulas which define μ and σ;

(b) by using the special formulas for the mean and the variance of a hypergeometric distribution.

4.41 Prove the formula for the mean of the hypergeometric distribution with the parameters n, a, and N, namely, $\mu = n \cdot \frac{a}{N}$. [*Hint*: Make use of the identity

$$\sum_{r=0}^{k} \binom{m}{r}\binom{s}{k-r} = \binom{m+s}{k}$$

which can be obtained by equating the coefficients of x^k in $(1 + x)^m(1 + x)^s$ and in $(1 + x)^{m+s}$.]

4.42 Construct a table showing the upper limits provided by Chebyshev's theorem for the probabilities of obtaining values differing from the mean by at least 1, 2, and 3 standard deviations and also the corresponding probabilities for the binomial distribution with $n = 16$ and $p = \frac{1}{2}$.

4.43 If a student answers the 144 questions of a true-false test by flipping a balanced coin—heads is "true" and tails is "false"—what does Chebyshev's theorem with $k = 4$ tell us about the number of correct answers he or she will get?

4.44 What does Chebyshev's theorem tell us about the probability of getting at most 30 or at least 105 sixes in 405 rolls of a balanced die?

4.45 Show that for 1,000,000 flips of a balanced coin the probability is at least 0.99 that the proportion of heads will fall between 0.495 and 0.505.

4.46 How many times do we have to flip a balanced coin to be able to assert with a probability of at most 0.01 that the difference between the proportion of tails and 0.50 will be at least 0.04?

4.47 Prove that

(a) $\sigma^2 = \mu_2' - \mu^2$;

(b) $\mu_3 = \mu_3' - 3\mu_2' \cdot \mu + 2\mu^3$.

4.6
THE POISSON APPROXIMATION TO THE BINOMIAL DISTRIBUTION

When n is large and p is small, binomial probabilities are often approximated by means of the formula

Poisson distribution

$$f(x; \lambda) = \frac{\lambda^x e^{-\lambda}}{x!} \quad \text{for } x = 0, 1, 2, \ldots$$

with λ (*lambda*) equal to the product np. Before we justify this approximation, let us point out that $x = 0, 1, 2, \ldots$ means that there is a countable infinity of possibilities, and this requires that we modify the third axiom of probability given on page 57. In its place we substitute the following axiom.

Modification of third axiom of probability

> ***Axiom 3'*** ***If $A_1, A_2, A_3, \ldots$ is a finite or infinite sequence of mutually exclusive events in S, then***
>
> $$P(A_1 \cup A_2 \cup A_3 \cup \cdots) = P(A_1) + P(A_2) + P(A_3) + \cdots$$

The other postulates remain unchanged. To verify that $P(S) = 1$ for this formula, we make use of Axiom 3′ and write

$$\sum_{x=0}^{\infty} f(x; \lambda) = \sum_{x=0}^{\infty} \frac{e^{-\lambda}\lambda^x}{x!} = e^{-\lambda} \sum_{x=0}^{\infty} \frac{\lambda^x}{x!}$$

Since the infinite series in the expression on the right is the Maclaurin's series for e^{λ}, it follows that

$$\sum_{x=0}^{\infty} f(x; \lambda) = e^{-\lambda} \cdot e^{\lambda} = 1$$

Let us now show that when $n \to \infty$ and $p \to 0$, while $np = \lambda$ remains constant, the limiting form of the binomial distribution is as given in the beginning of the preceding paragraph. First let us substitute $\frac{\lambda}{n}$ for p into the formula for the binomial distribution and simplify the resulting expression; thus, we get

$$\begin{aligned} b(x; n, p) &= \frac{n!}{x!\,(n-x)!}\left(\frac{\lambda}{n}\right)^x\left(1-\frac{\lambda}{n}\right)^{n-x} \\ &= \frac{n(n-1)(n-2)\cdots(n-x+1)}{x!\,n^x}(\lambda)^x\left(1-\frac{\lambda}{n}\right)^{n-x} \\ &= \frac{\left(1-\frac{1}{n}\right)\left(1-\frac{2}{n}\right)\cdots\left(1-\frac{x-1}{n}\right)}{x!}(\lambda)^x\left(1-\frac{\lambda}{n}\right)^{n-x} \end{aligned}$$

If we now let $n \to \infty$, we find that

$$\left(1-\frac{1}{n}\right)\left(1-\frac{2}{n}\right)\cdots\left(1-\frac{x-1}{n}\right) \to 1$$

that

$$\left(1-\frac{\lambda}{n}\right)^{n-x} = \left[\left(1-\frac{\lambda}{n}\right)^{n/\lambda}\right]^{\lambda}\left(1-\frac{\lambda}{n}\right)^{-x} \to e^{-\lambda}$$

and, hence, that the binomial distribution approaches

$$\frac{\lambda^x e^{-\lambda}}{x!} \qquad \text{for } x = 0, 1, 2, \ldots$$

This completes our proof; the distribution at which we arrived is called the **Poisson distribution**, as we already indicated on page 113.

Since the Poisson distribution has many important applications besides approximating binomial probabilities (see Section 4.7), it has been extensively tabulated. Table 2 at the end of the book gives the values of the probabilities $F(x; \lambda) = \sum_{k=0}^{x} f(k; \lambda)$ for values of λ in varying increments from 0.02 to 25, and its use is very similar to that of Table 1. Poisson probabilities are also calculated by many statistical software packages (see Exercises 4.70 and 4.71).

The following are some examples illustrating the Poisson approximation of binomial probabilities. An acceptable rule of thumb is to use this approximation if $n \geq 20$ and $p \leq 0.05$; if $n \geq 100$, the approximation is generally excellent so long as $np \leq 10$.

EXAMPLE

It is known that 5% of the books bound at a certain bindery have defective bindings. Find the probability that 2 of 100 books bound by this bindery will have defective bindings using

(a) the formula for the binomial distribution;

(b) the Poisson approximation to the binomial distribution.

Solution

(a) Substituting $x = 2$, $n = 100$, and $p = 0.05$ into the formula for the binomial distribution, we get

$$b(2; 100, 0.05) = \binom{100}{2}(0.05)^2(0.95)^{98} = 0.081$$

(b) Substituting $x = 2$ and $\lambda = 100(0.05) = 5$ into the formula for the Poisson distribution, we get

$$f(2; 5) = \frac{5^2 \cdot e^{-5}}{2!} = 0.084$$

It is of interest to note that the difference between the two values we obtained (the error we would make by using the Poisson approximation) is only 0.003. [Had we used Table 2 instead of looking up the value of e^{-5} in a mathematical table, we would have obtained $f(2; 5) = F(2; 5) - F(1; 5) = 0.125 - 0.040 = 0.085$.]

■

EXAMPLE A fire insurance company has 3,840 policyholders. If the probability is $\frac{1}{1,200}$ that any one of the policyholders will file at least one claim in any given year, find the probabilities that 0, 1, 2, 3, 4, ... of the policyholders will file at least one claim in a given year.

Solution For practical reasons we cannot use the formula for the binomial distribution, but use of the Poisson distribution with

$$\lambda = 3{,}840 \cdot \frac{1}{1{,}200} = 3.2$$

Table 2, and the identity $f(x; \lambda) = F(x; \lambda) - F(x - 1; \lambda)$ yields the results shown in the probability histogram of Figure 4.7.

■

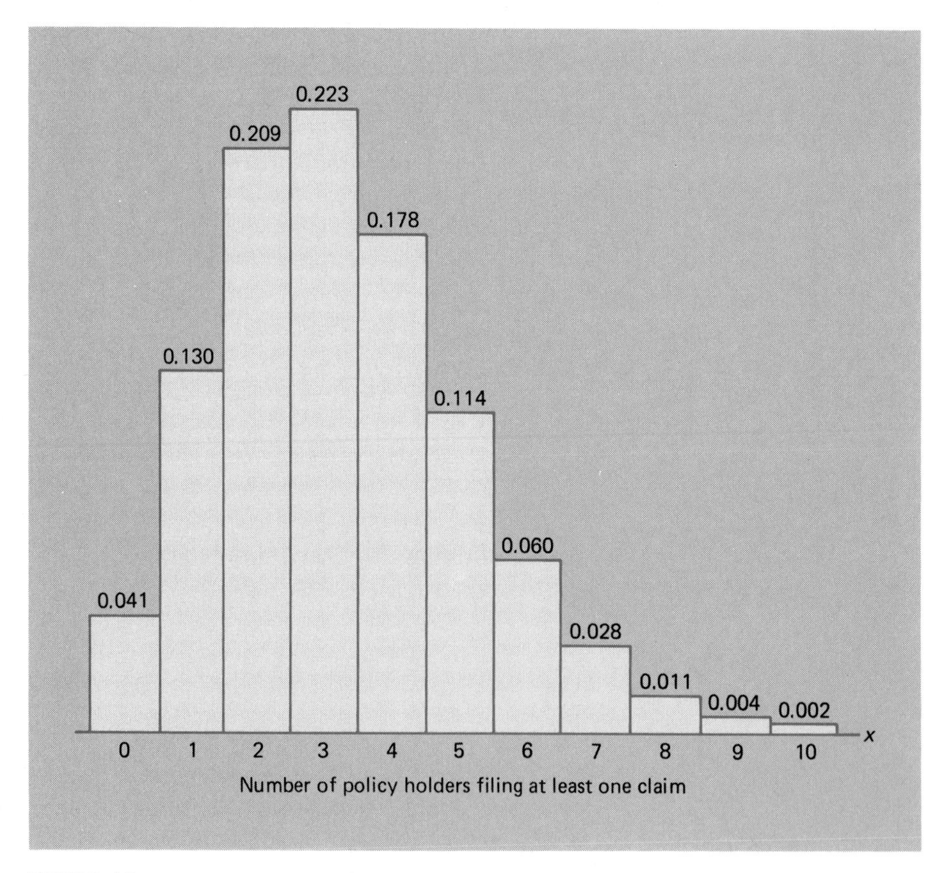

FIGURE 4.7
Probability histogram of Poisson distribution with $\lambda = 3.2$.

Using a method similar to that employed on page 104 to derive the formula for the mean of the binomial distribution, we can show that the mean and the variance of the Poisson distribution with the parameter λ are given by

Mean and variance of Poisson distribution

$$\mu = \lambda \quad \textit{and} \quad \sigma^2 = \lambda$$

The first of these results should really have been expected, for in our justification of the Poisson approximation to the binomial distribution we let $\lambda = np$. For the variance we can write $\sigma^2 = np(1 - p) = \lambda(1 - p)$, which approaches λ as $p \to 0$.

4.7 POISSON PROCESSES

In general, a **random process** is a physical process that is wholly or in part controlled by some sort of chance mechanism. It may be a sequence of repeated flips of a coin, measurements of the quality of manufactured products coming off an assembly line, the vibrations of airplane wings, the noise in a radio signal, or any one of numerous other phenomena. What characterizes such processes is their time dependence, namely, the fact that certain events do or do not take place (depending on chance) at regular intervals of time or throughout continuous intervals of time.

In this section we shall be concerned with processes taking place over continuous intervals of time, such as the occurrence of imperfections on a continuously produced bolt of cloth, the recording of radiation by means of a Geiger counter, the arrival of telephone calls at a switchboard, or the passing by cars of an electronic checking device. We will now show that the mathematical model which we can use to describe many situations like these is that of the Poisson distribution. To find the probability of x successes during a time interval of length T, we divide the interval into n equal parts of length Δt, so that $T = n \cdot \Delta t$, and we assume that:

1. The probability of a success during a very small interval of time Δt is given by $\alpha \cdot \Delta t$.
2. The probability of more than one success during such a small time interval Δt is negligible.
3. The probability of a success during such a time interval does not depend on what happened prior to that time.

This means that the assumptions underlying the binomial distribution are satisfied, and the probability of x successes in the time interval T is given by the binomial probability $b(x; n, p)$ with $n = \frac{T}{\Delta t}$ and $p = \alpha \cdot \Delta t$. Then, following the argument on page 114, we find that when $n \to \infty$ the probability of x successes during the time interval T is given by the corresponding Poisson probability with the parameter $\lambda = n \cdot p = \frac{T}{\Delta t} \cdot (\alpha \cdot \Delta t) = \alpha T$. Since λ is the mean of this Poisson distribution, note that α is the average (mean) number of successes per unit time.

EXAMPLE If a bank receives on the average $\lambda = 6$ bad checks per day, what are the probabilities that it will receive

(a) four bad checks on any given day;
(b) 10 bad checks on any two consecutive days?

Solution (a) Substituting $x = 4$ and $\lambda = 6$ into the formula for the Poisson distribution, we get

$$f(4; 6) = \frac{6^4 \cdot e^{-6}}{4!} = \frac{1{,}296(0.00248)}{24} = 0.134$$

(b) Here we want to find $f(10; 12)$, and we write

$$\begin{aligned} f(10; 12) &= F(10; 12) - F(9; 12) \\ &= 0.347 - 0.242 \\ &= 0.105 \end{aligned}$$

where the values of $F(10; 12)$ and $F(9; 12)$ were obtained from Table 2.

■

EXAMPLE In the inspection of tin plate by a continuous electrolytic process, 0.2 imperfection is spotted on the average per minute. Find the probabilities of spotting

(a) one imperfection in 3 minutes;
(b) at least two imperfections in 5 minutes;
(c) at most one imperfection in 15 minutes.

Solution Referring in each case to Table 2, we get

$$\begin{aligned} F(1; 0.6) - F(0; 0.6) &= 0.878 - 0.549 \\ &= 0.329 \end{aligned}$$

for part (a),

$$\begin{aligned} 1 - F(1; 1.0) &= 1 - 0.736 \\ &= 0.264 \end{aligned}$$

for part (b), and

$$F(1; 3.0) = 0.199$$

for part (c).

■

The Poisson distribution has many important applications in **queueing** problems, where we may be interested, for example, in the number of customers arriving for service at a cafeteria, the number of ships or trucks arriving to be unloaded at a receiving dock, the number of aircraft arriving at an airport, and so forth. Thus, if on the average 0.3 customer arrive per minute at a cafeteria, then the probability that exactly 3 customers will arrive during a 5-minute span is

$$F(3; 1.5) - F(2; 1.5) = 0.934 - 0.809 = 0.125$$

and if on the average 3 trucks arrive per hour to be unloaded at a warehouse, then the probability that at most 20 will arrive during an 8-hour day is

$$F(20; 24) = 0.243$$

Also, if on the average 14.5 planes arrive per day at a private airport, then the probability that the number of arrivals on any one day will be anywhere from 12 to 15, inclusive, is

$$F(15; 14.5) - F(11; 14.5) = 0.619 - 0.220 = 0.399$$

4.8 THE GEOMETRIC DISTRIBUTION

On page 43 we indicated that a countably infinite sample space would be needed if we are interested in the number of cars persons have to inspect until they find one whose nitrogen oxide emission does not meet government standards. To treat this kind of problem in general, suppose that in a sequence of trials we are interested in the number of the trial on which the first success occurs, and that all but the third of the assumptions underlying the binomial distribution are satisfied; in other words, n is not fixed.

Clearly, if the first success is to come on the xth trial, it has to be preceded by $x - 1$ failures, and if the probability of a success is p, the probability of $x - 1$ failures on $x - 1$ trials is $(1 - p)^{x-1}$. Then, if we multiply this expression by the probability p of a success on the xth trial, we find that the probability of getting the first success on the xth trial is given by

Geometric distribution

$$g(x; p) = p(1 - p)^{x-1} \qquad \text{for } x = 1, 2, 3, 4, \ldots$$

This probability distribution is called the **geometric distribution**, and, as the reader will be asked to verify in Exercise 4.67 on page 124, its mean is

Mean of geometric distribution

$$\mu = \frac{1}{p}$$

EXAMPLE If the probability is 0.20 that a burglar will get caught on any given job, what is the probability of being caught for the first time on the fourth job?

Solution Substituting $x = 4$ and $p = 0.20$ into the formula for the geometric distribution, we get

$$\begin{aligned} g(4; 0.20) &= (0.20)(1 - 0.20)^{4-1} \\ &= 0.102 \end{aligned}$$

■

EXAMPLE If the probability is 0.05 that a certain kind of measuring device will show excessive drift, what is the probability that the sixth of these measuring devices tested will be the first to show excessive drift?

Solution Substituting $x = 6$ and $p = 0.05$ into the formula for the geometric distribution, we get

$$\begin{aligned} g(6; 0.05) &= (0.05)(1 - 0.05)^{6-1} \\ &= 0.039 \end{aligned}$$

■

In **queueing theory**, the geometric distribution has important applications in connection with the number of units (customers, trucks, airplanes, etc.) that are being served or are waiting to be served at any given time. Suppose, for instance, that the arrival of cars at a gas station with one pump is a Poisson process with a **mean arrival rate** (average number of arrivals per unit time) of α cars per hour, and that the cars are served at a **mean service rate** of β cars per hour. Since the arrivals constitute a Poisson process, the probability of more than one arrival during a small time interval Δt is negligible, and we shall assume that the same is true also

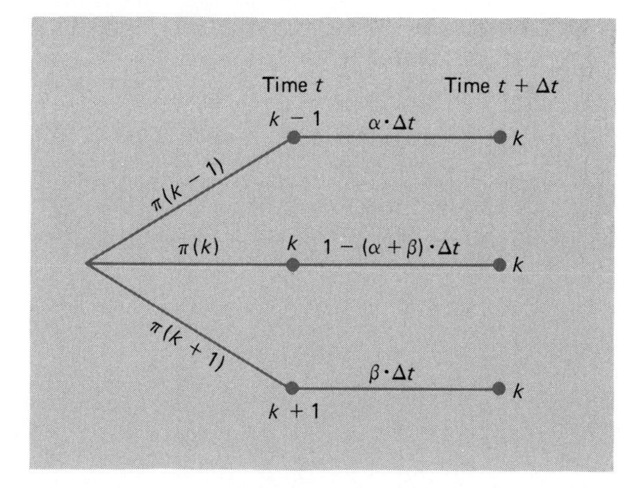

FIGURE 4.8
Tree diagram for queueing problem.

for the completions of service.† Furthermore, we shall assume that $\alpha < \beta$, namely, that the mean number of arrivals per unit time is less than the mean number of services that can be completed per unit time.

To arrive at an expression for $\pi(k)$, the probability that there are k cars in the gas station at any given time being served or waiting to be served, we shall first obtain a set of recursion relations expressing $\pi(k)$ in terms of $\pi(k-1)$ and $\pi(k-2)$. To illustrate how these relations are obtained, let us find the probability that there are no cars in the gas station (that is, $k = 0$) at time $t + \Delta t$. If $k = 0$ at time $t + \Delta t$, the only values that k could have had at time t are 0 and 1 by the assumption of Poisson arrivals. If $k = 0$ at time t, then no cars arrived in the interval from t to $t + \Delta t$ (with probability $1 - \alpha \cdot \Delta t$), but if $k = 1$ at time t, then a service was completed in this interval (with probability $\beta \cdot \Delta t$). Using the rule of elimination (see page 74), we thus conclude that

$$\pi(0) = \pi(0)(1 - \alpha \cdot \Delta t) + \pi(1)\beta \cdot \Delta t$$

or

$$\pi(1) = \frac{\alpha}{\beta} \cdot \pi(0)$$

In general, the situation is typified by the tree diagram of Figure 4.8, and the rule of elimination leads to the following recursion relation:

$$\pi(k) = \pi(k-1)\alpha \cdot \Delta t + \pi(k)[1 - (\alpha + \beta)\Delta t] + \pi(k+1)\beta \cdot \Delta t$$

† In fact, it will be assumed that so long as the service station is not empty, completions of service constitute a Poisson process.

for $k \geq 1$. In Exercise 4.66 on page 124, the reader will be asked to verify that a solution of this set of equations is given by the geometric distribution

$$\pi(k) = \left(1 - \frac{\alpha}{\beta}\right)\left(\frac{\alpha}{\beta}\right)^k \qquad \text{for } k = 0, 1, 2, \ldots$$

which may be obtained from the expression on page 120 by letting $p = 1 - \frac{\alpha}{\beta}$ and $x = k + 1$.

This distribution reveals many important properties of queues. Note first that since $\alpha < \beta$ by assumption, we have

$$\lim_{k \to \infty} \pi(k) = 0$$

so that the probability is 1 that the waiting line remains finite. (In fact, it can be shown that if $\alpha \geq \beta$ there is a nonzero probability that the waiting line will eventually grow beyond all bounds.) The mean number of cars in the station (being served and waiting to be served) is given by the mean of the above distribution, which is

$$\frac{1}{p} - 1 = \frac{1}{1 - \frac{\alpha}{\beta}} - 1 = \frac{\alpha}{\beta - \alpha}$$

in accordance with the formula for μ on page 120. (We subtracted the 1 because the values of the random variable stated with $k = 0$ instead of $x = 1$.)

Also, the distribution of the number of cars in the queue waiting to be served (thus, excluding the car being served) is given by $\pi(k + 1)$ for $k \geq 1$ and $\pi(0) + \pi(1)$ for $k = 0$. Thus, the mean number of cars in the queue is

$$\left(1 - \frac{\alpha}{\beta}\right)\sum_{k=1}^{\infty} k\left(\frac{\alpha}{\beta}\right)^{k+1} = \frac{\alpha^2}{\beta(\beta - \alpha)}$$

To find the mean time a customer spends waiting in the queue, it is necessary first to find the distribution of waiting times, and then to find its mean. We shall omit this work, stating only that the resulting mean waiting time is $\frac{\alpha}{\beta(\beta - \alpha)}$.

EXERCISES

4.48 Prove that for the Poisson distribution

$$\frac{f(x + 1; \lambda)}{f(x; \lambda)} = \frac{\lambda}{x + 1}$$

for $x = 0, 1, 2, \ldots$.

4.49 Use the recursion formula of Exercise 4.48 to calculate the values of the Poisson distribution with $\lambda = 3$ for $x = 0, 1, 2, \ldots$, and 9, and draw the probability histogram of this distribution. Verify your results by referring to Table 2.

4.50 Use Table 2 to find

(a) $F(4; 7)$; (b) $f(4; 7)$;

(c) $\sum_{k=6}^{19} f(4; 8)$.

4.51 Use Table 2 to find

(a) $F(9; 12)$; (b) $f(9; 12)$;

(c) $\sum_{k=3}^{12} f(k; 7.5)$.

4.52 Use the Poisson distribution to approximate the binomial probability $b(3; 100, 0.03)$.

4.53 In a given city, 6% of all drivers get at least one parking ticket per year. Use the Poisson approximation to the binomial distribution to determine the probabilities that among 80 drivers (randomly chosen in this city)

(a) four will get at least one parking ticket in any given year;

(b) at least three will get at least one parking ticket in any given year;

(c) anywhere from three to six, inclusive, will get at least one parking ticket in any given year.

4.54 If 0.8% of the fuses delivered to an arsenal are defective, use the Poisson approximation to determine the probability that four fuses will be defective in a random sample of 400.

4.55 The number of gamma rays emitted per second by a certain radioactive substance is a random variable having the Poisson distribution with $\lambda = 5.8$. If a recording instrument becomes inoperative when there are more than 12 rays per second, what is the probability that this instrument becomes inoperative during any given second?

4.56 Given that the switchboard of a consultant's office receives on the average 0.6 call per minute, find the probabilities that

(a) in a given minute there will be at least one call;

(b) in a 4-minute interval there will be at least three calls.

4.57 At a checkout counter customers arrive at an average rate of 1.5 per minute. Find the probabilities that

(a) at most four will arrive in any given minute;

(b) at least three will arrive during an interval of 2 minutes;

(c) at most 15 will arrive during an interval of 6 minutes.

4.58 A company rents out time on a computer for periods of t hours, for which it receives \$600 an hour. The number of times the computer breaks down during t hours is a random variable having the Poisson distribution with $\lambda = (0.8)t$, and if the computer breaks down x times during t hours it costs $50x^2$ dollars to fix it. How should the company select t in order to maximize its expected profit?

4.59 The determination of geometric probabilities is often simplified by making use of the identity

$$g(x; p) = \frac{1}{x} \cdot b(1; x, p)$$

and looking up $b(1; x, p)$ in a table of binomial probabilities. Verify this identity and use it (and Table 1) to evaluate

(a) $g(12; 0.10)$; (b) $g(10; 0.30)$.

4.60 An expert shot hits a target 95% of the time. What is the probability that the expert will miss the target for the first time on the fifteenth shot?

4.61 In a "torture test" a light switch is turned on and off until it fails. If the probability that the switch will fail any time it is turned on or off is 0.001, what is the probability that the switch will fail *after* it has been turned on or off 1,200 times. Assume that the conditions underlying the geometric distribution are met. [*Hint*: Use the formula for the value of an infinite geometric progression and logarithms.]

4.62 A pool company's records show that the probability is 0.20 that one of its new pools will require repairs within a year. What is the probability that the sixth pool it builds in a given year will be the first one to require repairs within a year?

4.63 In the example on page 119 we considered customers arriving at a cafeteria at an average rate of 0.3 per minute. If they are being served at an average rate of 0.5 per minute, find

(a) the average number of customers being served or waiting to be served at any given time;
(b) the average number of customers waiting to be served at any given time;
(c) the average time a customer spends waiting in line.

4.64 The arrival of trucks at a receiving dock is a Poisson process with a mean arrival rate of two per hour. The trucks can be unloaded at a mean rate of three per hour, and, so long as the receiving dock is not empty, completions of service constitute a Poisson process.

(a) What is the average number of trucks being unloaded and waiting to be unloaded?
(b) What is the mean number of trucks in the queue?
(c) What is the mean time a truck spends waiting in the queue?
(d) What is the probability that there are no trucks waiting to be unloaded?

4.65 With reference to Exercise 4.64, suppose that the cost of keeping a truck in the system is \$15 per hour. If it were possible to increase the mean loading rate to 3.5 trucks per hour at a cost of \$100 per day, would this be worthwhile?

4.66 Verify (by substitution) that the geometric distribution is a solution of the recursion relation given on page 121.

4.67 Differentiating with respect to p both sides of the equation

$$\sum_{x=1}^{\infty} p(1-p)^{x-1} = 1$$

show that the geometric distribution

$$f(x) = p(1-p)^{x-1} \qquad \text{for } x = 1, 2, 3, \ldots$$

has the mean $1/p$.

4.68 The following is an alternative way of introducing the Poisson distribution. Suppose that $f(x, t)$ is the probability of getting x successes during a time interval of length t when (1) the probability of a success during a very small time interval from t to $t + \Delta t$ is $\alpha \cdot \Delta t$, (2) the probability of more than one success occurring during such a time interval is negligible, and (3) the probability of a success during such a time interval does not depend on what happened prior to time t.

(a) Show that

$$f(x, t + \Delta t) = f(x, t)[1 - \alpha\Delta t] + f(x-1, t)\,\alpha\Delta t$$

(b) Using the result of part (a), show that

$$\frac{d[f(x, t)]}{dt} = \alpha[f(x-1, t) - f(x, t)]$$

(c) Verify by substitution that the differential equation obtained in (b) is satisfied by the formula for the Poisson distribution with $\lambda = \alpha t$.

4.69 Use the formulas defining μ and σ^2 to show that the mean and the variance of the Poisson distribution are both equal to λ.

4.70 Poisson probabilities can be calculated using the *MINITAB* command

```
PDF;
  POISSON mean = 1.64.
```

which produces the output

```
POISSON WITH MEAN =    1.640
   K          P( X = K)
   0            0.1940
   1            0.3181
   2            0.2609
   3            0.1426
   4            0.0585
   5            0.0192
   6            0.0052
   7            0.0012
   8            0.0003
   9            0.0000
```

Find the Poisson probabilities with $\lambda = 2.73$.

4.71 Cumulative Poisson probabilities can be calculated using the *MINITAB* command

```
CDF;
  POISSON mean= 1.64.
```

which produces the output

```
POISSON WITH MEAN =    1.640
   K  P( X LESS  OR = K)
   0             0.1940
   1             0.5121
   2             0.7730
   3             0.9156
   4             0.9740
   5             0.9932
   6             0.9985
   7             0.9997
   8             0.9999
   9             1.0000
```

Find the cumulative Poisson probabilities with $\lambda = 2.73$.

4.9

THE MULTINOMIAL DISTRIBUTION

An immediate generalization of the binomial distribution arises when each trial can have more than two possible outcomes. This happens, for example, when a manufactured product is classified as superior, average, or poor, when a student's performance is graded as an A, B, C, D, or F, or when an experiment is judged successful, unsuccessful, or inconclusive. To treat this kind of problem in general, let us consider the case where there are n independent trials, with each trial permitting k mutually exclusive outcomes whose respective probabilities are $p_1, p_2, \ldots, p_k$ $\left(\text{with } \sum_{i=1}^{k} p_i = 1\right)$. Referring to the outcomes as being of the first kind, the second kind, ..., and the kth kind, we shall be interested in the probability $f(x_1, x_2, \ldots, x_k)$ of getting x_1 outcomes of the first kind, x_2 outcomes of the second kind, ..., and x_k outcomes of the kth kind, with $\sum_{i=1}^{k} x_i = n$. Using arguments similar to those which we employed in deriving the equation for the binomial distribution in Section 4.2, it can be shown that the desired probability is given by

Multinomial distribution

$$f(x_1, x_2, \ldots, x_k) = \frac{n!}{x_1!\, x_2! \cdots x_k!} p_1^{x_1} \cdot p_2^{x_2} \cdots p_k^{x_k}$$

for $x_i = 0, 1, \ldots, n$ for each i, subject to the restriction $\sum_{i=1}^{k} x_i = n$. The **joint probability distribution** whose values are given by these probabilities is called the **multinomial distribution**; it owes its name to the fact that for the various values of the x_i the probabilities are given by the corresponding terms of the multinomial expansion of $(p_1 + p_2 + \cdots + p_k)^n$.

EXAMPLE The probabilities that the light bulb of a certain kind of slide projector will last fewer than 40 hours of continuous use, anywhere from 40 to 80 hours of continuous use, or more than 80 hours of continuous use, are 0.30, 0.50, and 0.20. Find the probability that among eight such bulbs two will last fewer than 40 hours, five will last anywhere from 40 to 80 hours, and one will last more than 80 hours.

Solution Substituting $n = 8$, $x_1 = 2$, $x_2 = 5$, $x_3 = 1$, $p_1 = 0.30$, $p_2 = 0.50$, and $p_3 = 0.20$ into the formula, we get

$$f(2, 5, 1) = \frac{8!}{2!\, 5!\, 1!}(0.30)^2(0.50)^5(0.20)^1$$
$$= 0.0945$$

■

EXERCISES

4.72 Suppose that the probabilities are, respectively, 0.40, 0.40, and 0.20 that in city driving a certain kind of imported car will average less than 22 miles per gallon, anywhere from 22 to 25 miles per gallon, or more than 25 miles per gallon. Find the probability that among 12 such cars tested, four will average less than 22 miles per gallon, six will average anywhere from 22 to 25 miles per gallon, and two will average more than 25 miles per gallon.

4.73 As can easily be shown, the probabilities of getting 0, 1, or 2 heads with a pair of balanced coins are $\frac{1}{4}$, $\frac{1}{2}$, and $\frac{1}{4}$. What is the probability of getting two tails twice, one head and one tail three times, and two heads once in six tosses of a pair of balanced coins?

4.74 Suppose that the probabilities are, respectively, 0.60, 0.20, 0.10, and 0.10 that an income tax form will be filled in correctly, that it will contain an error favoring the taxpayer, that it will contain an error favoring the government, or that it will contain both kinds of errors. Find the probability that among 10 of the income tax forms randomly selected for audit five will be correct, three will contain an error favoring the taxpayer, one will contain an error favoring the government, and one will contain both kinds of errors.

4.75 Using the same sort of reasoning as in the derivation of the formula for the hypergeometric distribution, we can derive a formula which is analogous to the multinomial distribution but applies to sampling without replacement. If a set of N objects contains a_1 objects of the first kind, a_2 objects of the second kind, ..., and a_k objects of the kth kind, so that $a_1 + a_2 + \cdots + a_k = N$, the number of ways in which we can select x_1 objects of the first kind, x_2 objects of the second kind, ..., and x_k objects of the kth kind is given by the product of the number of ways in which we can select x_1 of the a_1 objects of the first kind, x_2 of the a_2 objects of the second kind, ..., and x_k of the a_k objects of the kth kind. Thus, the probability of getting that many objects of each kind is simply this product divided by the total number of ways in which $x_1 + x_2 + \cdots + x_k = n$ objects can be selected from the whole set of N objects.

(a) Write a formula for the probability of thus obtaining x_1 objects of the first kind, x_2 objects of the second kind, ..., and x_k objects of the kth kind.

(b) If 20 defective glass bricks include 10 that have cracks but no discoloration, 7 that are discolored but have no cracks, and 3 that have cracks and discoloration, what is the probability that among 6 of the bricks chosen at random for further checks 3 will have cracks only, 2 will only be discolored, and 1 will have cracks as well as discoloration?

4.10 SIMULATION

In recent years, simulation techniques have been applied to many problems in the various sciences, and if the processes which are being simulated involve an element of chance, these techniques are referred to as **Monte Carlo methods**. Very often, the use of Monte Carlo simulation eliminates the cost of building and operating expensive equipment; it is used, for instance, in the study of collisions of photons with electrons, the scattering of neutrons, and similar complicated phenomena. Monte Carlo methods are also useful in situations where direct experimentation is impossible—say, in studies of the spread of cholera epidemics, which, of course, are not induced experimentally on human populations. In addition, Monte Carlo techniques are sometimes applied to the solution of mathematical problems which actually cannot be solved by direct means, or where a direct solution is too costly or requires too much time.

A classical example of the use of Monte Carlo methods in the solution of a problem of pure mathematics is the determination of π (the ratio of the circumference of a circle to its diameter) by probabilistic means. Early in the eighteenth century George de Buffon, a French naturalist, proved that if a very fine needle of length a is thrown at random on a board ruled with equidistant parallel lines, the probability that the needle will intersect one of the lines is $2a/\pi b$, where b is the distance between the parallel lines. What is remarkable about this fact is that it involves the constant $\pi = 3.1415926\ldots$, which in elementary geometry is approximated by the circumferences of regular polygons enclosed in a circle of radius $\frac{1}{2}$. Buffon's result implies that if such a needle is actually tossed a great many times, the proportion of the time it crosses one of the lines gives an estimate of $2a/\pi b$ and, hence, an estimate of π since a and b are known. Early experiments of this kind yielded an estimate of 3.1519 (based on 5,000 trials) and an estimate of 3.155 (based on 3,204 trials) in the middle of the nineteenth century.

Although Monte Carlo methods are sometimes based on actual gambling devices (for example, the needle tossing in the estimation of π), it is usually expedient to use so-called tables of **random digits** or **random numbers**. Tables of random numbers consist of many pages on which the digits 0, 1, 2, ..., and 9 are set down in a "random" fashion, much as they would appear if they were generated one at a time by a gambling device giving each digit an equal probability of being selected. Actually, we could construct such tables ourselves—say, by repeatedly drawing numbered slips out of a hat or by using a perfectly constructed spinner—but in practice such tables are usually generated by means of electronic computers (see Exercise 4.81).

Although tables of random numbers are constructed so that the digits can be looked upon as values of a random variable having the discrete uniform distribution $f(x) = \frac{1}{10}$ for $x = 0, 1, 2, \ldots,$ or 9, they can be used to simulate values of any discrete random variable, and even continuous random variables.

To illustrate the use of a table of random numbers, let us show how we can play "heads or tails" without actually flipping a coin. Letting 0, 2, 4, 6, and 8 represent heads and 1, 3, 5, 7, and 9 represent tails, we might arbitrarily choose the 4th column of the first page of Table 7, start at the top and go down the page. Thus, we get 6, 2, 7, 5, 5, 0, 1, 8, 6, 3, ..., and we interpret this as head, head, tail, tail, tail, head, tail, head, head, tail, ...

Repeated flips of any number of coins, say, three coins, can be simulated in the same way. If we arbitrarily choose the 9th, 10th, and 11th columns of the second page of Table 7, start at the top and go down the page, we get 480, 280, 085, 265, 303, 288, 295, 388, 127, 222, ..., and, counting the number of even digits in each case, we interpret this as 3, 3, 2, 2, 1, 3, 1, 2, 1, 3, ..., heads. If we did not want to count even digits, we could make use of the fact that the probabilities of getting 0, 1, 2, or 3 heads are $\frac{1}{8}$, $\frac{3}{8}$, $\frac{3}{8}$, and $\frac{1}{8}$, and use the following scheme:

Number of heads	*Probability*	*Random digits*
0	$\frac{1}{8}$	0
1	$\frac{3}{8}$	1, 2, 3
2	$\frac{3}{8}$	4, 5, 6
3	$\frac{1}{8}$	7

Ignoring the digits 8 and 9 wherever they may occur, we would thus interpret the random digits 4, 8, 1, 1, 1, 9, 3, 3, 3, 7, 6, 3, 6, 2, 6, 5, 8, 9, 3, 1, 0, and 6 (in the 5th row of the third page of Table 7) as getting 2, 1, 1, 1, 1, 1, 1, 3, 2, 1, 2, 1, 2, 2, 1, 1, 0, and 2 heads in 18 tosses of three coins.

Of the two methods used in the preceding example, the first has the disadvantage that we have to count how many even digits there are in each case, and the second has the disadvantage that we have to ignore 8 and 9. To avoid such waste of effort and time, we could have used the following scheme:

Number of heads	*Probability*	*Cumulative probability*	*Random numbers*
0	0.125	0.125	000–124
1	0.375	0.500	125–499
2	0.375	0.875	500–874
3	0.125	1.000	875–999

Here we used three-digit random numbers because the probabilities are given to three decimals, and we allocated 125 (or one-eighth) of the 1,000 random numbers from 000 to 999 to 0 heads, 375 (or three-eighths) to 1 head, 375 (or three-eighths) to 2 heads, and 125 (or one-eighth) to 3 heads. The column of "or less" **cumulative probabilities** was added to facilitate the assignment of the random numbers. Observe that in each case the last random digit is one less than the number formed by the three decimal digits of the corresponding cumulative probability.

With this scheme, if we arbitrarily use the 22nd, 23rd, and 24th columns of the first page of Table 7, starting with the 6th row and going down the page, we get 197, 565, 157, 520, 946, 951, 948, 568, 586, and 089, and we interpret this as 1, 2, 1, 2, 3, 3, 3, 2, 2, and 0 heads.

The method we have illustrated here with reference to a game of chance, can be used to simulate observations of any random variable with a given probability distribution.

EXAMPLE Suppose that the probabilities are 0.082, 0.205, 0.256, 0.214, 0.134, 0.067, 0.028, 0.010, 0.003, and 0.001 that 0, 1, 2, 3, ..., or 9 cars will arrive at a toll booth of a turnpike during any one-minute interval in the early afternoon.

(a) Distribute the three-digit random numbers from 000 to 999 among the 10 values of this random variable, so that they can be used to simulate the arrival of cars at the toll booth.

(b) Use the 5th, 6th, and 7th columns of the fourth page of Table 7, starting with the 11th row and going down the page, to simulate the arrival of cars at the toll booth during 20 one-minute intervals in the early afternoon.

Solution (a) Calculating the cumulative probabilities and following the suggestion given above, we arrive at the following scheme:

Number of cars	*Probability*	*Cumulative probability*	*Random numbers*
0	0.082	0.082	000–081
1	0.205	0.287	082–286
2	0.256	0.543	287–542
3	0.214	0.757	543–756
4	0.134	0.891	757–890
5	0.067	0.958	891–957
6	0.028	0.986	958–985
7	0.010	0.996	986–995
8	0.003	0.999	996–998
9	0.001	1.000	999

(b) Following the instructions, we get the random numbers 036, 417, 962, 458, 778, 541, 869, 379, 973, 553, 325, 674, 907, 710, 709, 499, 493, 384, 346, and 301, and this means that 0, 2, 6, 2, 4, 2, 4, 2, 6, 3, 2, 3, 5, 3, 3, 2, 2, 2, 2, and 2 cars arrived at the toll booth during the 20 one-minute intervals.

■

EXERCISES

4.76 Letting any five digits represent heads and the other five digits represent tails, use random numbers to simulate 100 flips of a balanced coin.

4.77 Using the digits 1, 2, 3, 4, 5, and 6 to represent the corresponding faces of a die (and omitting 0, 7, 8, and 9), simulate 120 rolls of a balanced die.

4.78 The probabilities that a computer software salesperson will make 0, 1, 2, 3, 4, or 5 sales on any one day are 0.14, 0.28, 0.27, 0.18, 0.09, and 0.04.

(a) Distribute the two-digit random numbers from 00 to 99 among the six values of this random variable, so that the corresponding random numbers can be used to simulate the salesman's sales.

(b) Use the result of part (a) to simulate the salesperson's sales on 25 days.

4.79 With reference to the example on page 116 and the probabilities shown in Figure 4.7, distribute the three-digit random numbers from 000 to 999 among the 11 values of the random variable, and simulate the fire insurance company's experience (that is, the number of policyholders who file at least one claim) over a period of 20 years.

4.80 Depending on the availability of parts, a company can manufacture 3, 4, 5, or 6 units of a certain item per week with corresponding probabilities of 0.10, 0.40, 0.30, and 0.20. The probabilities that there will be a weekly demand for 0, 1, 2, 3, ..., or 8 units are, respectively, 0.05, 0.10, 0.30, 0.30, 0.10, 0.05, 0.05, 0.04, and 0.01. If a unit is sold during the week that it is made, it will yield a profit of $100; this profit is reduced by $20 for each week that a unit has to be stored. Use random numbers to simulate the operations of this company for 50 consecutive weeks and estimate its expected weekly profit.

4.81 The statistical package *MINITAB* has random number generators. The command

```
RANDOM 5 observations into C1;
  INTEGER 1 to 1000.
```

produces five random integers between 1 and 1000 inclusive. These are placed in column 1 of the MINITAB worksheet. In one instance, this command produced the output

863 518 629 397 77

Use this command to select 10 observations between 50 and 100.

4.11 REVIEW EXERCISES

4.82 Determine whether the following can be probability distributions of a random variable that can take on only the values 0, 1, and 2.

(a) $f(0) = 0.34$, $f(1) = 0.34$, and $f(2) = 0.34$
(b) $f(0) = 0.2$, $f(1) = 0.6$, and $f(2) = 0.2$
(c) $f(0) = 0.7$, $f(1) = 0.4$, and $f(2) = -0.1$

4.83 Check whether the following can define probability distributions, and explain your answers.

(a) $f(x) = \frac{x}{10}$, for $x = 0, 1, 2, 3, 4$

(b) $f(x) = \frac{1}{3}$, for $x = -1, 0, 1$

(c) $f(x) = \frac{(x-1)^2}{4}$, for $x = 0, 1, 2, 3$

4.84 A basketball player makes 90% of her free throws. What is the probability she will miss for the first time on the 7th shot?

4.85 If the probability is 0.20 that a downtime of an automated production process will exceed 2 minutes, find the probability that three of eight downtimes of the process will exceed 2 minutes using

(a) the formula for the binomial distribution;
(b) Table 1.

4.86 If the probability that a fluorescent light has a useful life of at least 500 hours is 0.85, find the probabilities that among 20 such lights

(a) 18 will have a useful life of at least 500 hours;
(b) at least 15 will have a useful life of at least 500 hours;
(c) at least two will not have a useful life of at least 500 hours.

4.87 In 15 experiments studying the electrical behavior of single cells, 11 use microelectrodes made of metal and the other 4 use microelectrodes made from glass tubing. If two of the experiments are to be terminated for financial reasons, and they are selected at random, what are the probabilities that

(a) neither uses microelectrodes made from glass tubing;
(b) only one uses microelectrodes made from glass tubing;
(c) both use microelectrodes made from glass tubing?

4.88 As can be easily verified by means of the formula for the binomial distribution, the probabilities of getting 0, 1, 2, or 3 heads, in three flips of a coin whose probability of heads is 0.4, are 0.216, 0.432, 0.288 and 0.064. Find the mean of this probability distribution using

(a) the formula that defines μ;
(b) the special formula for the mean of a binomial distribution.

4.89 With reference to Exercise 4.88, find the variance of the probability distribution using

(a) the formula that defines σ^2;
(b) the special formula for the variance of a binomial distribution.

4.90 Find the mean and the standard deviation of the distribution of each of the following random variables (having binomial distributions).

(a) The number of heads in 555 flips of a balanced coin.
(b) The number of 6's in 300 rolls of a balanced die.
(c) The number of defectives in a sample of 700 parts made by a machine, when the probability is 0.03 that any one of the parts is defective.

4.91 Use the Poisson distribution to approximate the binomial probability $b(2; 100, 0.02)$.

4.92 With reference to Exercise 4.87, find the mean and the variance of the distribution of the number of microelectrodes made from glass tubing using

(a) the probabilities obtained in that exercise;
(b) the special formulas for the mean and the variance of a hypergeometric distribution.

4.93 The daily number of orders filled by the parts department of a repair shop is a random variable with $\mu = 142$ and $\sigma = 12$. According to Chebyshev's theorem, with what probability can we assert that on any one day it will fill between 82 and 202 orders?

4.94 Records show that the probability is 0.00004 that a car will have a flat tire while driving through a certain tunnel. Use the formula for the Poisson distribution to approximate the probability that at least 2 of 10,000 cars passing through the tunnel will have a flat tire.

4.95 The number of weekly breakdowns of a computer is a random variable having a Poisson distribution with $\lambda = 0.3$. What is the probability that the computer will operate without a breakdown for two consecutive weeks?

4.96 A manufacturer determines that a color TV set has probabilities 0.8, 0.15, 0.05, respectively, of being placed in the categories acceptable, minor defect, or major defect. If three TV's are inspected,

(a) find the probability that 2 are acceptable and 1 is a minor defect;
(b) find the marginal distribution of the number in minor defect;

(c) Compare your answer in (b) with the binomial probabilities $b(x; 3, 0.15)$. Comment.

4.97 Suppose that the probabilities are 0.2466, 0.3452, 0.2417, 0.1128, 0.0395, 0.0111, 0.0026, and 0.0005 that there will be 0, 1, 2, 3, 4, 5, 6, or 7 polluting spills in the Great Lakes on any one day.

(a) Distribute the four-digit random numbers from 0000 to 9999 to the eight values of this random variable, so that the corresponding random numbers can be used to simulate daily polluting spills in the Great Lakes.

(b) Use the result of part (a) to simulate the numbers of polluting spills in the Great Lakes on 30 days.

4.12
CHECK LIST OF KEY TERMS (with page references)

Bar chart 91
Bernoulli trials 92
Binomial coefficient 93
Binomial distribution 93
Chebyshev's theorem 109
Cumulative probability 94
Discrete random variable 90
Distribution function 92
Geometric distribution 120
Hypergeometric distribution 97
Joint probability distribution 126
kth moment about the mean 108
kth moment about the origin 108
Kurtosis 108
Law of large numbers 111
Mean 103
Mean arrival rate 120
Mean service rate 120
Monte Carlo methods 127
Multinomial distribution 126
Negatively skewed distribution 96
Parameter 93
Poisson distribution 113
Positively skewed distribution 96
Probability distribution 90
Probability histogram 91
Queueing 119
Random numbers 128
Random process 117
Random variable 90
Repeated trials 92
Sampling with replacement 97
Sampling without replacement 97
Skewed distribution 96
Standard deviation 106
Symmetrical distribution 96
Variance 106

5 PROBABILITY DENSITIES

Continuous sample spaces and continuous random variables arise when we deal with quantities that are measured on a continuous scale—for instance, when we measure the speed of a car, the amount of alcohol in a person's blood, the efficiency of a solar collector, or the amount of tar in a cigarette.

In this chapter we shall learn how to determine, and work with, probabilities relating to continuous sample spaces and continuous random variables. The introduction to probability densities in Section 5.1 is followed by a discussion of the normal distribution in Sections 5.2 and 5.3, and various other special probability densities in Sections 5.4 through 5.9. Problems involving more than one continuous random variable are discussed in Section 5.10. A method for checking whether a data set appears to be generated by a normal distribution is introduced in Section 5.11.

5.1

CONTINUOUS RANDOM VARIABLES

When we first introduced the concept of a random variable in Chapter 4, we presented it as a real-valued function defined over the sample space of an experiment, and we illustrated this idea with the random variable giving the number of preferred ratings which a lawn mower received, assigning the numbers 0, 1, 2, or 3 (whichever was appropriate) to the 18 possible outcomes of the experiment. In the continuous case, where random variables can assume values on a continuous scale, the procedure is very much the same. The outcomes of an experiment are represented by the points on a line segment or a line, and the value of a random variable is a number appropriately assigned to each point by means of some rule or equation. When the value of a random variable is given directly by a measurement or observation, we usually do not bother to differentiate between the value of the random variable, the measurement which we obtain, and the outcome of the experiment, the corresponding point on the real axis. Thus, if an experiment consists of determining what force is required to break a given tensile-test specimen, the result itself, say, 138.4 pounds, is the value of the random variable with which we are concerned. There is no real need in that case to add that the sample space of the experiment consists of all (or part of) the points on the positive real axis. In general, we shall write $P(a \leq x \leq b)$ for the probability associated with the points of the sample space for which the value of a random variable falls on the interval from a to b.

The problem of defining probabilities in connection with continuous sample spaces and continuous random variables involves some complications. To illustrate the nature of these complications, let us consider the following situation: Suppose we want to know the probability that if an accident occurs on a freeway whose length is 200 miles, it will happen at some given location or, perhaps, some particular stretch of the road. The outcomes of this experiment can be looked upon as a continuum of points, namely, those on the continuous interval from 0 to 200. Suppose the probability that the accident occurs on any interval of length L is $L/200$, with L measured in miles. Note that this arbitrary assignment of probability is consistent with Axioms 1 and 2 on page 57, since the probabilities are all nonnegative and less than or equal to 1, and $P(S) = \frac{200}{200} = 1$. Of course, we are considering so far only events represented by intervals which form part of the line segment from 0 to 200. Using Axiom 3′ on page 114, we can also obtain probabilities of events which are not intervals but which can be represented by the union of finitely many or countably many intervals. Thus, for two nonoverlapping intervals of length L_1 and L_2 we have a probability of

$$\frac{L_1 + L_2}{200}$$

and for an infinite sequence of nonoverlapping intervals of length $L_1, L_2, L_3, \ldots$, we have a probability of

$$\frac{L_1 + L_2 + L_3 + \cdots}{200}$$

Note that the probability that the accident occurs at any given point is equal to zero because we can look upon a point as an interval of zero length. However, the probability that the accident occurs in a very short interval is positive; for instance, for an interval of length 1 foot the probability is $9.5(10)^{-7}$.

Thus, in extending the concept of probability to the continuous case, we again use Axioms 1, 2, and 3′, but we shall have to restrict the meaning of the term *event*. So far as practical considerations are concerned, this restriction is of no consequence; we simply do not assign probabilities to some rather abstruse point sets, which cannot be expressed as the unions or intersections of finitely many or countably many intervals.

The way in which we assigned probabilities in the preceding example is, of course, very special; it is similar in nature to the way in which we assign equal probabilities to the six faces of a die, heads and tails, the 52 cards in a standard deck, and so forth. To treat the problem of associating probabilities with continuous random variables generally, suppose we are interested in the probability that a given random variable will take on a value on the interval from a to b, where a and b are constants with $a \leq b$. Suppose, furthermore, that we divide the interval from a to b into n equal subintervals of width Δx containing, respectively, the points $x_1, x_2, \ldots, x_n$, and that the probability that the random variable will take on a value on the subinterval containing x_i is given by $f(x_i) \cdot \Delta x$. Then, the probability that the random variable with which we are concerned will take on a value on the interval from a to b is given by

$$P(a \leq x \leq b) = \sum_{i=1}^{n} f(x_i) \cdot \Delta x$$

Now, if f is an integrable function defined for all values of the random variable with which we are concerned, we shall define the probability that the value of the random variable falls between a and b by letting $\Delta x \to 0$, namely, as

$$P(a \leq x \leq b) = \int_a^b f(x)\, dx$$

This definition of probability in the continuous case presupposes the existence of an appropriate function f which, integrated from any constant a to any constant b (with $a \leq b$), gives the probability that the corresponding random variable takes on a value on the interval from a to b. Note that a value $f(x)$ of f does not give the probability that the corresponding random variable takes on the value x; in the continuous case, probabilities are given by integrals and not by the values of f.

To obtain the probability that a random variable will actually take on a given value x, we might first determine the probability that it will take on a value on the interval from $x - \Delta x$ to $x + \Delta x$, and then let $\Delta x \to 0$. However, if we did this it would become apparent that the result is always zero. The fact that the probability is always zero that a continuous random variable will take on any given value x should not be disturbing. Indeed, our definition of probability for the continuous case provides a remarkably good model for dealing with measurements or observations. Owing to the limits of our ability to measure, experimental data never seem to come from a continuous sample space. Thus, while temperatures are fruitfully thought of as points on a continuous scale, any temperature measurement actually represents an interval on this scale. If we report a temperature measurement of 74.8 degrees centigrade, we really mean that the temperature lies in the interval from 74.75 to 74.85 degrees centigrade, and not that it is exactly 74.8000. . . .

It is important to add that when we say that there is a zero probability that a random variable will take on any given value x, this does not mean that it is impossible that the random variable will take on the value x. In the continuous case, a zero probability does not imply logical impossibility, but the whole matter is largely academic since, owing to the limitations of our ability to measure and observe, we are always interested in probabilities connected with intervals and not with isolated points.

As an immediate consequence of the fact that in the continuous case probabilities associated with individual points are always zero, we find that if we speak of the probability associated with the interval from a to b, it does not matter whether either endpoint is included. Symbolically,

$$P(a \leq x \leq b) = P(a \leq x < b) = P(a < x \leq b) = P(a < x < b)$$

Drawing an analogy with the concept of a density function in physics, we call the functions f, whose existence we stipulated in extending our definition of probability to the continuous case, **probability density functions**, or simply **probability densities**. Whereas density functions are integrated to obtain weights, probability density functions are integrated to obtain probabilities.

Since a probability density, integrated between any two constants a and b, gives the probability that a random variable assumes a value between these limits, f cannot be just any real-valued integrable function. However, if we impose the conditions that

$$f(x) \geq 0 \qquad \text{for all } x \text{ within the domain of } f$$

and

$$\int_{-\infty}^{\infty} f(x)\, dx = 1$$

it can be shown that the axioms of probability (with the modification about events discussed on page 136) are satisfied. Note the similarity between these conditions and those for probability distributions given on page 90.

As in the discrete case, we shall write as $F(x)$ the probability that a random variable with the probability density f takes on a value less than or equal to x, and we shall again refer to the corresponding function F as the **distribution function** of the random variable. Thus, if a random variable with values x has the probability density f, the values of its distribution function are given by

$$F(x) = \int_{-\infty}^{x} f(t)\, dt$$

Consequently, the probability that the random variable will take on a value on the interval from a to b is $F(b) - F(a)$, and according to the fundamental theorem of integral calculus it follows that

$$\frac{dF(x)}{dx} = f(x)$$

wherever this derivative exists.

EXAMPLE If a random variable has the probability density

$$f(x) = \begin{cases} 2e^{-2x} & \text{for } x > 0 \\ 0 & \text{for } x \leq 0 \end{cases}$$

find the probabilities that it will take on a value
(a) between 1 and 3;
(b) greater than 0.5.

Solution Evaluating the necessary integrals, we get

$$\int_{1}^{3} 2e^{-2x}\, dx = e^{-2} - e^{-6} = 0.133$$

for part (a) and

$$\int_{0.5}^{\infty} 2e^{-2x}\, dx = e^{-1} = 0.368$$

for part (b).

■

Note that in the preceding example we made the domain of f include all the real numbers even though the probability is zero that x will be negative. This is a practice we shall follow throughout this book. It is also apparent from the graph of this function in Figure 5.1 that it has a discontinuity at $x = 0$; indeed, a probability

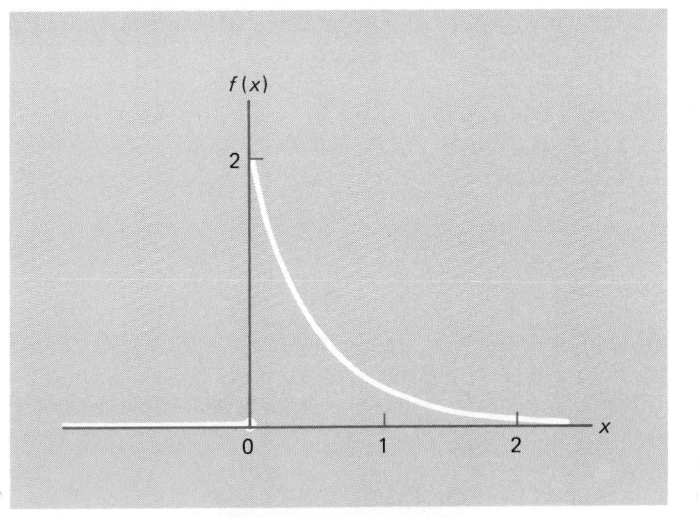

FIGURE 5.1
Graph of probability density.

density need not be everywhere continuous, as long as it is integrable between any two limits a and b (with $a \leq b$).

EXAMPLE With reference to the preceding example, find the distribution function and use it to determine the probability that the random variable will take on a value less than or equal to 1.

Solution Performing the necessary integrations, we get

$$F(x) = \begin{cases} 0 & \text{for } x \leq 0 \\ \displaystyle\int_0^x 2e^{-2t}\,dt = 1 - e^{-2x} & \text{for } x > 0 \end{cases}$$

and substitution of $x = 1$ yields

$$F(1) = 1 - e^{-2} = 0.865$$

■

Note that the distribution function of this example is nondecreasing and that $F(-\infty) = 0$ and $F(\infty) = 1$. Indeed, it follows by definition that these properties are shared by all distribution functions.

EXAMPLE Find k so that the following can serve as the probability density of a random variable:

$$f(x) = \begin{cases} 0 & \text{for } x \leq 0 \\ kxe^{-4x^2} & \text{for } x > 0 \end{cases}$$

Solution To satisfy the first of the two conditions on page 137, k must be nonnegative, and to satisfy the second condition we must have

$$\int_{-\infty}^{\infty} f(x)\,dx = \int_{0}^{\infty} kxe^{-4x^2}\,dx = \int_{0}^{\infty} \frac{k}{8} \cdot e^{-u}\,du = \frac{k}{8} = 1$$

so that $k = 8$.

■

The statistical measures that are used to describe probability densities are very similar to the ones which we used to describe probability distributions. Replacing summations with integrals, we define the **kth moment about the origin** as

$$\mu'_k = \int_{-\infty}^{\infty} x^k \cdot f(x)\,dx$$

analogous to the definition we gave on page 108. The first moment about the origin is again referred to as the **mean**, and it is denoted by μ; as before, it is the expected, or average, value of a random variable having the probability density f. We also define the **kth moment about the mean** as

$$\mu_k = \int_{-\infty}^{\infty} (x - \mu)^k \cdot f(x)\,dx$$

In particular, the second moment about the mean is again referred to as the **variance** and it is written as σ^2; as before, it measures the spread of a probability density in the sense that it gives the expected value of the squared deviation from the mean. Also, σ is again referred to as the **standard deviation**.

EXAMPLE With reference to the example on page 138, find the mean and the variance of the given probability density.

Solution Performing the necessary integrations, we get

$$\mu = \int_{-\infty}^{\infty} xf(x)\,dx = \int_{0}^{\infty} x \cdot 2e^{-2x}\,dx = \tfrac{1}{2}$$

and

$$\sigma^2 = \int_{-\infty}^{\infty} (x - \mu)^2 f(x)\,dx = \int_{0}^{\infty} (x - \tfrac{1}{2})^2 \cdot 2e^{-2x}\,dx = \tfrac{1}{4}$$

■

EXERCISES

5.1 Verify that the function of the example on page 138 is, in fact, a probability density.

5.2 If the probability density of a random variable is given by

$$f(x) = \begin{cases} kx^3 & 0 < x < 1 \\ 0 & \text{elsewhere} \end{cases}$$

Find the value k and the probability that the random variable takes on a value
(a) between $\frac{1}{4}$ and $\frac{3}{4}$; (b) greater than $\frac{2}{3}$.

5.3 With reference to the preceding exercise, find the corresponding distribution function and use it to determine the probabilities that a random variable having this distribution function will take on a value
(a) greater than 0.8; (b) between 0.2 and 0.4.

5.4 If the probability density of a random variable is given by

$$f(x) = \begin{cases} x & \text{for } 0 < x < 1 \\ 2 - x & \text{for } 1 \le x < 2 \\ 0 & \text{elsewhere} \end{cases}$$

find the probabilities that a random variable having this probability density will take on a value
(a) between 0.2 and 0.8; (b) between 0.6 and 1.2.

5.5 With reference to the preceding exercise, find the corresponding distribution function, and use it to determine the probabilities that a random variable having this distribution function will take on a value
(a) greater than 1.8; (b) between 0.4 and 1.6.

5.6 Given the probability density $f(x) = \dfrac{k}{1 + x^2}$ for $-\infty < x < \infty$, find k.

5.7 If the distribution function of a random variable is given by

$$F(x) = \begin{cases} 1 - \dfrac{4}{x^2} & \text{for } x > 2 \\ 0 & \text{for } x \le 2 \end{cases}$$

find the probabilities that this random variable will take on a value
(a) less than 3; (b) between 4 and 5.

5.8 Find the probability density that corresponds to the distribution function of Exercise 5.7. Are there any points at which it is undefined? Also sketch the graphs of the distribution function and the probability density.

5.9 Let the phase error in a tracking device have probability density

$$f(x) = \begin{cases} \cos x & 0 < x < \pi/2 \\ 0 & \text{elsewhere} \end{cases}$$

Find the probability that the phase error is
(a) between 0 and $\pi/4$; (b) greater than $\pi/3$.

5.10 The mileage (in thousands of miles) that car owners get with a certain kind of tire is a random variable having the probability density

$$f(x) = \begin{cases} \frac{1}{20}e^{-x/20} & \text{for } x > 0 \\ 0 & \text{for } x \leq 0 \end{cases}$$

Find the probabilities that one of these tires will last
(a) at most 10,000 miles;
(b) anywhere from 16,000 to 24,000 miles;
(c) at least 30,000 miles.

5.11 In a certain city, the daily consumption of electric power (in millions of kilowatt-hours) is a random variable having the probability density

$$f(x) = \begin{cases} \frac{1}{9}xe^{-x/3} & \text{for } x > 0 \\ 0 & \text{for } x \leq 0 \end{cases}$$

If the city's power plant has a daily capacity of 12 million kilowatt-hours, what is the probability that this power supply will be inadequate on any given day?

5.12 Prove that the identity $\sigma^2 = \mu_2' - \mu^2$ holds for any probability density for which these moments exist.

5.13 Find μ and σ^2 for the probability density of Exercise 5.2.

5.14 Find μ and σ^2 for the probability density of Exercise 5.4.

5.15 Find μ and σ for the probability density obtained in Exercise 5.8.

5.16 Find μ and σ for the distribution of the errors of measurement of Exercise 5.9.

5.17 Find μ for the distribution of the mileages of Exercise 5.10.

5.18 Show that μ_2' and, hence, σ^2 do not exist for the probability density of Exercise 5.6.

5.2 THE NORMAL DISTRIBUTION

Among the special probability densities we shall study in this chapter, the **normal probability density**, usually referred to simply as the **normal distribution**, is by far the most important.† It was studied first in the eighteenth century when scientists observed an astonishing degree of regularity in errors of measurement. They found that the patterns (distributions) they observed were closely approximated by a continuous distribution which they referred to as the "normal curve of errors" and

† The words *density* and *distribution* are often used interchangeably in the literature of applied statistics.

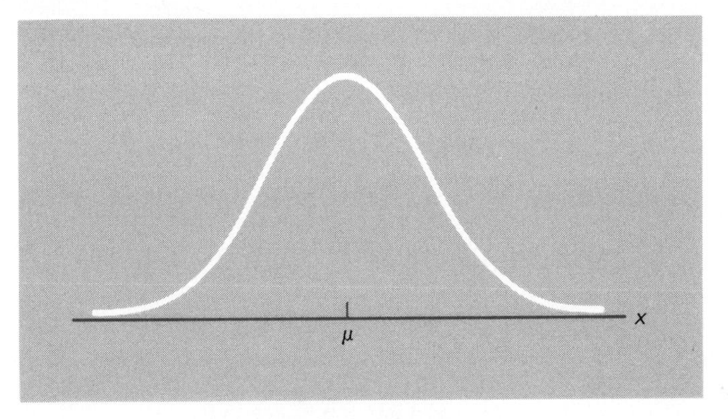

FIGURE 5.2
Graph of normal probability density.

attributed to the laws of chance. The equation of the normal probability density, whose graph (shaped like the cross section of a bell) is shown in Figure 5.2, is

Normal distribution

$$f(x; \mu, \sigma^2) = \frac{1}{\sqrt{2\pi\sigma^2}} e^{-(x-\mu)^2/2\sigma^2} \qquad -\infty < x < \infty$$

and in Exercises 5.42 and 5.43 on page 150, the reader will be asked to verify that its parameters μ and σ are, indeed its mean and its standard deviation.

Since the normal probability density cannot be integrated in closed form between every pair of limits a and b, probabilities relating to normal distributions are usually obtained from special tables, such as Table 3 at the end of this book. This table pertains to the **standard normal distribution**, namely, the normal distribution with $\mu = 0$ and $\sigma^2 = 1$, and its entries are the values of

$$F(z) = \frac{1}{\sqrt{2\pi}} \int_{-\infty}^{z} e^{-t^2/2}\, dt$$

for $z = 0.00, 0.01, 0.02, \ldots, 3.49$, and also $z = 4.00$, $z = 5.00$, and $z = 6.00$. To find the probability that a random variable having the standard normal distribution will take on a value between a and b, we use the equation $P(a < x < b) = F(b) - F(a)$, and if either a or b is negative, we also make use of the identity $F(-z) = 1 - F(z)$, which the reader will be asked to verify in Exercise 5.41 on page 150.

EXAMPLE Find the probabilities that a random variable having the standard normal distribution will take on a value

(a) between 0.87 and 1.28;
(b) between -0.34 and 0.62;
(c) greater than 0.85;
(d) greater than -0.65.

Solution Looking up the necessary values in Table 3, we get

$$F(1.28) - F(0.87) = 0.8997 - 0.8078$$
$$= 0.0919$$

for part (a),

$$F(0.62) - F(-0.34) = 0.7324 - (1 - 0.6331)$$
$$= 0.3655$$

for part (b),

$$1 - F(0.85) = 1 - 0.8023$$
$$= 0.1977$$

for part (c), and

$$1 - F(-0.65) = 1 - [1 - F(0.65)]$$
$$= F(0.65)$$
$$= 0.7422$$

for part (d).

■

There are also problems in which we are given probabilities relating to standard normal distributions and asked to find the corresponding values of z. The results of the example that follows will be used extensively in subsequent chapters.

EXAMPLE If z_α is such that the probability is α that it will be exceeded by a random variable having the standard normal distribution, find

(a) $z_{0.01}$;
(b) $z_{0.05}$.

Solution (a) Since $F(z_{0.01}) = 0.99$, we look for the entry in Table 3 which is closest to 0.99 and get 0.9901 corresponding to $z = 2.33$. Thus, $z_{0.01} = 2.33$. (b) Since $F(z_{0.05}) = 0.95$, we look for the entry in Table 3 which is closest to 0.95 and get 0.9495 and 0.9505 corresponding to $z = 1.64$ and $z = 1.65$. Thus, $z_{0.05} = 1.645$.

■

To use Table 3 in connection with a random variable which has the values x and a normal distribution with the mean μ and the variance σ^2, we refer to the corresponding **standardized random variable**, which has the values

$$z = \frac{x - \mu}{\sigma}$$

and the standard normal distribution. Thus, to find the probability that the original random variable will take on a value less than or equal to a, we look up $F\left(\frac{a-\mu}{\sigma}\right)$ in Table 3.

Also, if we want to find the probability that a random variable having the normal distribution with the mean μ and the variance σ^2 will take on a value between a and b, we have only to calculate the probability that a random variable having the standard normal distribution will take on a value between $\frac{a-\mu}{\sigma}$ and $\frac{b-\mu}{\sigma}$; symbolically,

$$P(a < x < b) = F\left(\frac{b-\mu}{\sigma}\right) - F\left(\frac{a-\mu}{\sigma}\right)$$

EXAMPLE If the amount of cosmic radiation to which a person is exposed while flying by jet across the United States is a random variable having the normal distribution with $\mu = 4.35$ mrem and $\sigma = 0.59$ mrem, find the probabilities that the amount of cosmic radiation to which a person will be exposed on such a flight is

(a) between 4.00 and 5.00 mrem;

(b) at least 5.50 mrem.

Solution Looking up the necessary values in Table 3, we get

$$\begin{aligned} F\left(\frac{5.00-4.35}{0.59}\right) - F\left(\frac{4.00-4.35}{0.59}\right) &= F(1.10) - F(-0.59) \\ &= 0.8643 - (1 - 0.7224) \\ &= 0.5867 \end{aligned}$$

for part (a) and

$$\begin{aligned} 1 - F\left(\frac{5.50-4.35}{0.59}\right) &= 1 - F(1.95) \\ &= 1 - 0.9744 \\ &= 0.0256 \end{aligned}$$

for part (b).

■

EXAMPLE The actual amount of instant coffee that a filling machine puts into "4-ounce" jars may be looked upon as a random variable having a normal distribution with $\sigma = 0.04$ ounce. If only 2% of the jars are to contain less than 4 ounces, what is the mean fill of these jars?

Solution To find μ such that $F\left(\frac{4-\mu}{0.04}\right) = 0.02$ and, hence, $F\left(-\frac{4-\mu}{0.04}\right) = 0.98$, we look for the entry in Table 3 closest to 0.98 and get 0.9798 corresponding to $z = 2.05$. Thus,

$$-\frac{4-\mu}{0.04} = 2.05$$

and, solving for μ, we find that $\mu = 4.082$ ounces. ■

According to Figure 2.8, the observations on the strength of an aluminum alloy appear to be normally distributed. The normal distribution is often used to model variation when the distribution is symmetric.

Although the normal distribution applies to continuous random variables, it is often used to approximate distributions of discrete random variables. Quite often, this yields satisfactory results, provided that we make the **continuity correction** illustrated in the following example.

EXAMPLE In a certain city, the number of power outages per month is a random variable having a distribution with $\mu = 11.6$ and $\sigma = 3.3$. If this distribution can be approximated closely with a normal distribution, what is the probability that there will be at least eight outages in any one month?

Solution The answer is given by the area of the white region of Figure 5.3; the area to the right of 7.5, not 8. The reason for this is that the number of outages is a discrete random variable, and if we want to approximate its distribution with a normal distribution, we must "spread" its values over a continuous scale. We do this by representing each integer k by the interval from $k - \frac{1}{2}$ to $k + \frac{1}{2}$. For instance, 3 is represented by the interval from 2.5 to 3.5, 10 is represented by the interval from 9.5

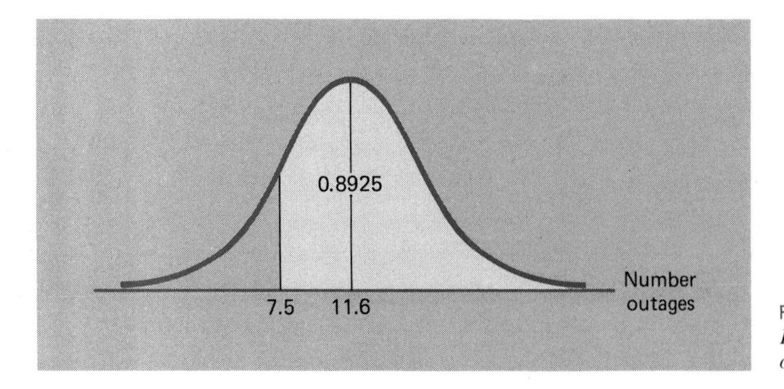

FIGURE 5.3
Diagram for example dealing with power outages.

to 10.5, and "at least 8" is represented by the interval to the right of 7.5. Thus, the desired probability is approximated by

$$1 - F\left(\frac{7.5 - 11.6}{3.3}\right) = 1 - F(-1.24)$$
$$= F(1.24)$$
$$= 0.8925$$

■

5.3
THE NORMAL APPROXIMATION TO THE BINOMIAL DISTRIBUTION

In some books the normal distribution is introduced as a probability density which can be used to approximate the binomial distribution when n is large and p, the probability of a success, is close to 0.50 and, hence, not small enough to use the Poisson approximation. Thus, let us state, without proof, the following theorem:

Normal approximation to binomial distribution

> ***Theorem 5.1** If x is a value of a random variable having the binomial distribution with the parameters n and p, and if*
>
> $$z = \frac{x - np}{\sqrt{np(1-p)}}$$
>
> *then the limiting form of the distribution function of this standardized random variable as $n \to \infty$ is given by*
>
> $$F(z) = \int_{-\infty}^{z} \frac{1}{\sqrt{2\pi}} e^{-t^2/2}\, dt \qquad -\infty < z < \infty$$

Note that although x takes on only the values $0, 1, 2, \ldots, n$, in the limit as $n \to \infty$ the distribution of the corresponding standardized random variable is continuous, and the corresponding probability density is the standard normal density. Note that in the following example, which is an application of Theorem 5.1, we use again the continuity correction given on page 146.

EXAMPLE If 20% of the diodes made in a certain plant are defective, what are the probabilities that in a lot of 100 randomly chosen for inspection

(a) at most 15 will be defective;
(b) exactly 15 will be defective?

Solution Since $\mu = 100(0.20) = 20$ and $\sigma = \sqrt{100(0.20)(0.80)} = 4$ for the binomial distribution with $n = 100$ and $p = 0.20$, we find that the normal approximation to the binomial distribution yields

$$\begin{aligned} F\left(\frac{15.5 - 20}{4}\right) &= F(-1.13) \\ &= 1 - F(1.13) \\ &= 1 - 0.8708 \\ &= 0.1292 \end{aligned}$$

for part (a) and

$$\begin{aligned} F\left(\frac{15.5 - 20}{4}\right) - F\left(\frac{14.5 - 20}{4}\right) &= F(-1.13) - F(-1.38) \\ &= F(1.38) - F(1.13) \\ &= 0.9162 - 0.8708 \\ &= 0.0454 \end{aligned}$$

for part (b).

■

Had we done the exact binomial calculation on a computer or used the Romig table (referred to in the bibliography) instead of using normal approximation in the preceding example, we would have obtained 0.1285 instead of 0.1292 for part (a) and 0.0481 instead of 0.0454 for part (b). Thus, it can be seen that both approximations are very close. A good rule of thumb is to use the normal approximation to the binomial distribution only when np and $n(1 - p)$ are both greater than 5.

EXERCISES

5.19 If a random variable has the standard normal distribution, find the probability that it will take on a value

(a) less than 1.50; (b) less than -1.20;
(c) greater than 2.16; (d) greater than -1.75.

5.20 If a random variable has the standard normal distribution, find the probability that it will take on a value

(a) between 0 and 2.7; (b) between 1.22 and 2.43;
(c) between -1.35 and -0.35; (d) between -1.70 and 1.35.

5.21 Find z if the probability that a random variable having the standard normal distribution will take on a value

(a) less than z is 0.9911; (b) greater than z is 0.1093;
(c) greater than z is 0.6443; (d) less than z is 0.0217;
(e) between $-z$ and z is 0.9298.

5.22 If a random variable has a normal distribution, what are the probabilities that it will take on a value within

(a) 1 standard deviation of the mean;
(b) 2 standard deviations of the mean;
(c) 3 standard deviations of the mean;
(d) 4 standard deviations of the mean?

5.23 Verify that

(a) $z_{0.005} = 2.575$;   (b) $z_{0.025} = 1.96$.

5.24 Given a random variable having the normal distribution with $\mu = 16.2$ and $\sigma^2 = 1.5625$, find the probabilities that it will take on a value

(a) greater than 16.8;   (b) less than 14.9;
(c) between 13.6 and 18.8;   (d) between 16.5 and 16.7.

5.25 A random variable has a normal distribution with $\mu = 62.4$. Find its standard deviation if the probability is 0.20 that it will take on a value greater than 79.2.

5.26 A random variable has a normal distribution with $\sigma = 10$. If the probability is 0.8212 that it will take on a value less than 82.5, what is the probability that it will take on a value greater than 58.3?

5.27 The time required to assemble a piece of machinery is a random variable having approximately a normal distribution with $\mu = 12.9$ minutes and $\sigma = 2.0$ minutes. What are the probabilities that the assembly of a piece of machinery of this kind will take

(a) at least 11.5 minutes;
(b) anywhere from 11.0 to 14.8 minutes?

5.28 Find the *quartiles*

$$-z_{0.25},\ z_{0.50},\ z_{0.25}$$

of the standard normal distribution.

5.29 In a photographic process, the developing time of prints may be looked upon as a random variable having the normal distribution with a mean of 16.28 seconds and a standard deviation of 0.12 second. Find the probability that it will take

(a) anywhere from 16.00 to 16.50 seconds to develop one of the prints;
(b) at least 16.20 seconds to develop one of the prints;
(c) at most 16.35 seconds to develop one of the prints.

5.30 With reference to the preceding exercise, for which value is the probability 0.95 that it will be exceeded by the time it takes to develop one of the prints?

5.31 Specifications for a certain job call for washers with an inside diameter of 0.300 ± 0.005 inch. If the inside diameters of the washers supplied by a given manufacturer may be looked upon as a random variable having the normal distribution with $\mu = 0.302$ inch and $\sigma = 0.003$ inch, what percentage of these washers will meet specifications?

5.32 With reference to the example on page 145, verify that if the variability of the filling machine is reduced to $\sigma = 0.025$ ounce, this will lower the required average amount of coffee to 4.05 ounces, yet keep 98% of the jars above 4 ounces.

5.33 A stamping machine produces can tops whose diameters are normally distributed with a standard deviation of 0.01 inch. At what "nominal" (mean) diameter should the machine be set so that no more than 5% of the can tops produced have diameters exceeding 3 inches?

5.34 Extruded plastic rods are automatically cut into nominal lengths of 6 inches. Actual lengths are normally distributed about a mean of 6 inches and their standard deviation is 0.06 inch.

(a) What proportion of the rods exceeds tolerance limits of 5.9 inches to 6.1 inches?

(b) To what value does the standard deviation need to be reduced if 99% of the rods must be within tolerance?

5.35 If a random variable has the binomial distribution with $n = 20$ and $p = 0.60$, use the normal approximation to determine the probabilities that it will take on

(a) the value 14; (b) a value less than 12.

5.36 A manufacturer knows that on the average 2% of the electric toasters that he makes will require repairs within 90 days after they are sold. Use the normal approximation to the binomial distribution to determine the probability that among 1,200 of these toasters at least 30 will require repairs within the first 90 days after they are sold.

5.37 The probability that an electronic component will fail in less than 1,000 hours of continuous use is 0.25. Use the normal approximation to find the probability that among 200 such components fewer than 45 will fail in less than 1,000 hours of continuous use.

5.38 A safety engineer feels that 30% of all industrial accidents in her plant are caused by failure of employees to follow instructions. If this figure is correct, find, approximately, the probability that among 84 industrial accidents in this plant anywhere from 20 to 30 (inclusive) will be due to failure of employees to follow instructions.

5.39 If 62% of all clouds seeded with silver iodide show spectacular growth, what is the probability that among 40 clouds seeded with silver iodide at most 20 will show spectacular growth?

5.40 To illustrate the law of large numbers mentioned on page 111, find the probabilities that the proportion of heads will be anywhere from 0.49 to 0.51 when a balanced coin is flipped

(a) 1,000 times; (b) 10,000 times.

5.41 Verify the identity $F(-z) = 1 - F(z)$ given on page 143.

5.42 Verify that the parameter μ in the expression for the normal density on page 143 is, in fact, its mean.

5.43 Verify that the parameter σ^2 in the expression for the normal density on page 143 is, in fact, its variance.

5.44 Show that the normal density has a relative maximum at $x = \mu$ and inflection points at $x = \mu - \sigma$ and $x = \mu + \sigma$.

5.4 OTHER PROBABILITY DENSITIES

In the application of statistics to problems in engineering and physical science, we shall encounter many probability densities other than the normal distribution. Among these are the t, F, and chi-square distributions, the fundamental sampling distributions that we introduce in Chapter 6, and the exponential and Weibull distributions, which we apply to problems of reliability and life testing in Chapter 15.

In the remainder of this chapter we shall discuss five continuous distributions, the **uniform distribution**, the **log-normal distribution**, the **gamma distribution**, the **beta distribution**, and the **Weibull distribution**, for the twofold purpose of giving further examples of probability densities and of laying the foundation for future applications.

5.5
THE UNIFORM DISTRIBUTION

The **uniform distribution** with the parameters α and β is defined by the equation

Uniform distribution

$$f(x) = \begin{cases} \dfrac{1}{\beta - \alpha} & \text{for } \alpha < x < \beta \\ 0 & \text{elsewhere} \end{cases}$$

and its graph is shown in Figure 5.4. Note that all values of x from α to β are "equally likely" in the sense that the probability that x lies in an interval of width Δx entirely contained in the interval from α to β is equal to $\Delta x/(\beta - \alpha)$, regardless of the exact location of the interval.

To illustrate how a physical situation might give rise to a uniform distribution, suppose that a wheel of a locomotive has the radius r and that x is the location of a point on its circumference measured along the circumference from some reference point 0. When the brakes are applied, some point will make sliding contact with the rail, and heavy wear will occur at that point. For repeated application of the

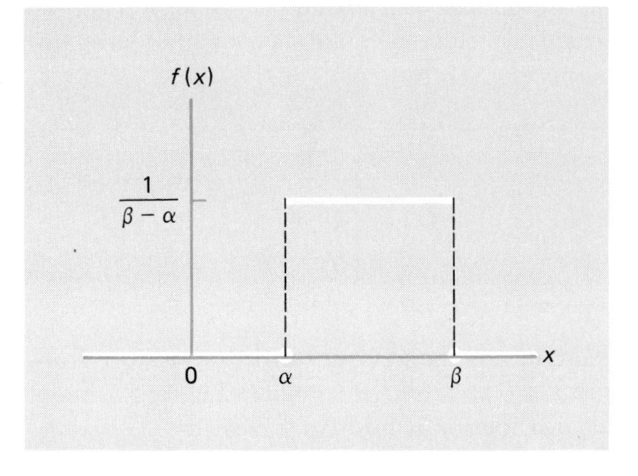

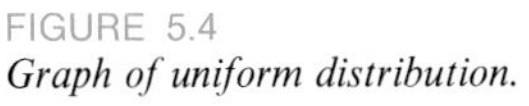
FIGURE 5.4
Graph of uniform distribution.

brakes, it would seem reasonable to assume that x is a value of a random variable having the uniform distribution with $\alpha = 0$ and $\beta = 2\pi r$. If this assumption were incorrect, that is, if some set of points on the wheel made contact more often than others, the wheel would eventually exhibit "flat spots" or wear out of round.

To determine the mean and the variance of the uniform distribution, we first evaluate the two integrals

$$\mu_1' = \int_\alpha^\beta x \cdot \frac{1}{\beta - \alpha}\, dx = \frac{\alpha + \beta}{2}$$

and

$$\mu_2' = \int_\alpha^\beta x^2 \cdot \frac{1}{\beta - \alpha}\, dx = \frac{\alpha^2 + \alpha\beta + \beta^2}{3}$$

Thus,

Mean of uniform distribution

$$\boldsymbol{\mu = \frac{\alpha + \beta}{2}}$$

and, making use of the formula $\sigma^2 = \mu_2' - \mu^2$, we find that

Variance of uniform distribution

$$\boldsymbol{\sigma^2 = \tfrac{1}{12}(\beta - \alpha)^2}$$

5.6

THE LOG-NORMAL DISTRIBUTION

The **log-normal distribution** occurs in practice whenever we encounter a random variable which is such that its logarithm has a normal distribution. Its probability density is given by

Log-normal distribution

$$f(x) = \begin{cases} \dfrac{1}{\sqrt{2\pi}\beta}\, x^{-1} e^{-(\ln x - \alpha)^2/2\beta^2} & \textit{for } x > 0,\ \beta > 0 \\ 0 & \textit{elsewhere} \end{cases}$$

where $\ln x$ is the natural logarithm of x. A graph of the log-normal distribution with $\alpha = 0$ and $\beta = 1$ is shown in Figure 5.5, and it can be seen from the figure that this distribution is positively skewed.

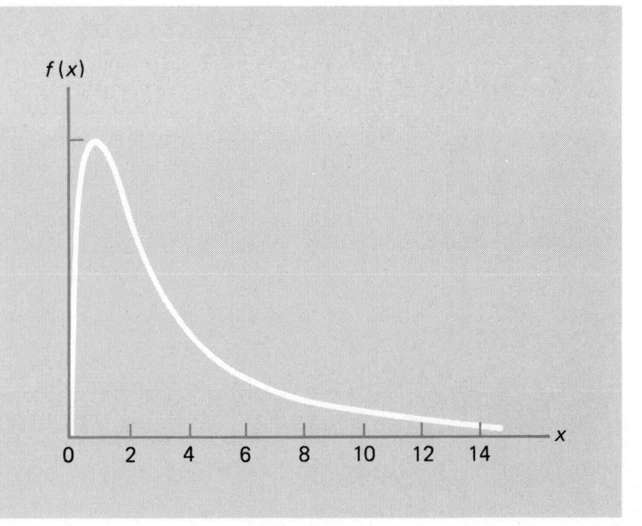

FIGURE 5.5
Graph of log-normal distribution.

To find the probability that a random variable having the log-normal distribution will take on a value between a and b $(0 < a < b)$, we must evaluate the integral

$$\int_a^b \frac{1}{\sqrt{2\pi}\beta} x^{-1} e^{-(\ln x - \alpha)^2/2\beta^2}\, dx$$

Changing variable by letting $y = \ln x$ and identifying the integrand as the normal density with $\mu = \alpha$ and $\sigma = \beta$, we find that the desired probability is given by

$$\int_{\ln a}^{\ln b} \frac{1}{\sqrt{2\pi}\beta} e^{-(y-\alpha)^2/2\beta^2}\, dy = F\left(\frac{\ln b - \alpha}{\beta}\right) - F\left(\frac{\ln a - \alpha}{\beta}\right)$$

where $\boldsymbol{F}$ is the distribution function of the standard normal distribution.

EXAMPLE The current gain of certain transistors is measured in units which make it equal to the logarithm of I_o/I_i, the ratio of the output to the input current. If it is normally distributed with $\mu = 2$ and $\sigma^2 = 0.01$, find the probability that I_o/I_i will take on a value between 6.1 and 8.2.

Solution Since $\alpha = 2$ and $\beta = 0.1$, we get

$$F\left(\frac{\ln 8.2 - 2}{0.1}\right) - F\left(\frac{\ln 6.1 - 2}{0.1}\right) = F(1.0) - F(-2.0)$$

$$= 0.8185$$

■

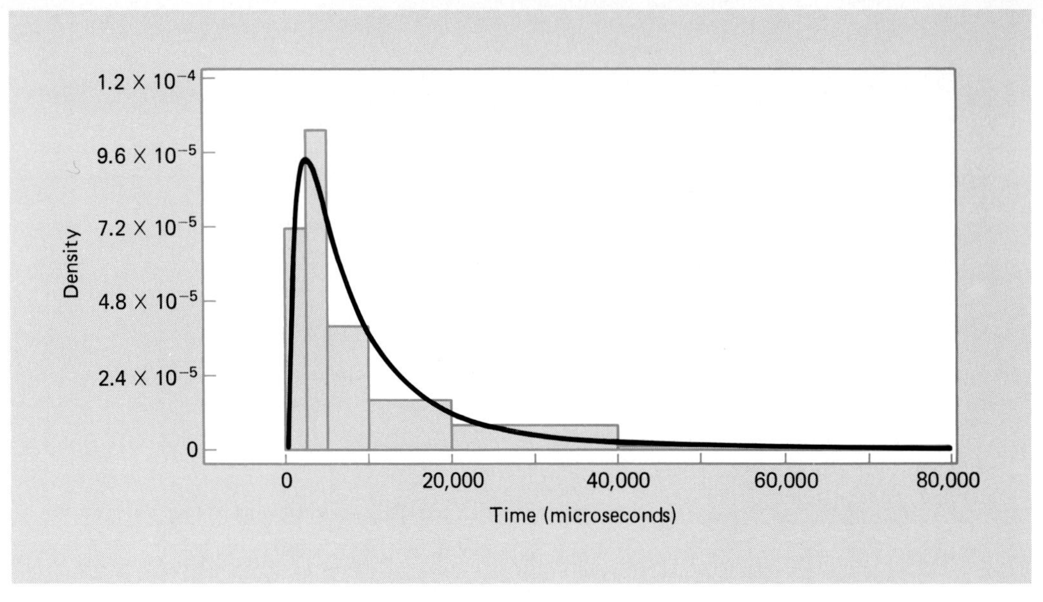

FIGURE 5.6
Histogram of Interrequest time.

EXAMPLE A histogram of the logarithm of the interrequest times on page 14 is shown in Figure 5.6. We have also plotted the log-normal density with $\alpha = 8.85$ and $\beta = 1.03$. The log-normal fit is explored further in Section 5.5 (see also Exercise 2.54).

EXAMPLE As part of a risk analysis concerning a nuclear power plant, engineers must model the strength of steam generator supports in terms of their ability to withstand the peak acceleration caused by earthquakes. Expert opinion suggests that the ln (strength) is normally distributed with $\mu = 4.0$ and $\sigma^2 = 0.09$. Find the probability that the supports will survive a peak acceleration of 33.

Solution Since $\alpha = 4.0$ and $\beta = 0.30$, we find

$$1 - F\left(\frac{\ln(33) - 4.0}{0.30}\right) = 1 - F(-1.68) = 0.9535$$

■

To find a formula for the mean of the log-normal distribution, we write

$$\mu = \frac{1}{\sqrt{2\pi}\beta} \int_0^\infty x \cdot x^{-1} e^{-(\ln x - \alpha)^2/2\beta^2} \, dx$$

and upon letting $y = \ln x$, this becomes

$$\mu = \frac{1}{\sqrt{2\pi}\beta} \int_{-\infty}^{\infty} e^{y} e^{-(y-\alpha)^2/2\beta^2}\, dy$$

This integral can be evaluated by completing the square on the exponent $y - (y - \alpha)^2/2\beta^2$, thus obtaining an integrand which has the form of a normal density. The final result, which the reader will be asked to verify in Exercise 5.48 on page 162, is

Mean of log-normal distribution

$$\mu = e^{\alpha + \beta^2/2}$$

Similar, but more lengthy, calculations yield

Variance of log-normal distribution

$$\sigma^2 = e^{2\alpha + \beta^2}(e^{\beta^2} - 1)$$

EXAMPLE With reference to the example on page 153, find the mean and the variance of the distribution of the ratio of the output to the input current.

Solution Substituting $\alpha = 2$ and $\beta = 0.1$ into the above formulas, we get

$$\mu = e^{2 + (0.1)^2/2} = 7.43$$

and

$$\sigma^2 = e^{4 + (0.1)^2}(e^{(0.1)^2} - 1) = 0.56$$

5.7 THE GAMMA DISTRIBUTION

Several important probability densities whose applications will be discussed later are special cases of the **gamma distribution**. This distribution is given by

Gamma distribution

$$f(x) = \begin{cases} \dfrac{1}{\beta^{\alpha}\Gamma(\alpha)} x^{\alpha - 1} e^{-x/\beta} & \text{for } x > 0, \alpha > 0, \beta > 0 \\ 0 & \text{elsewhere} \end{cases}$$

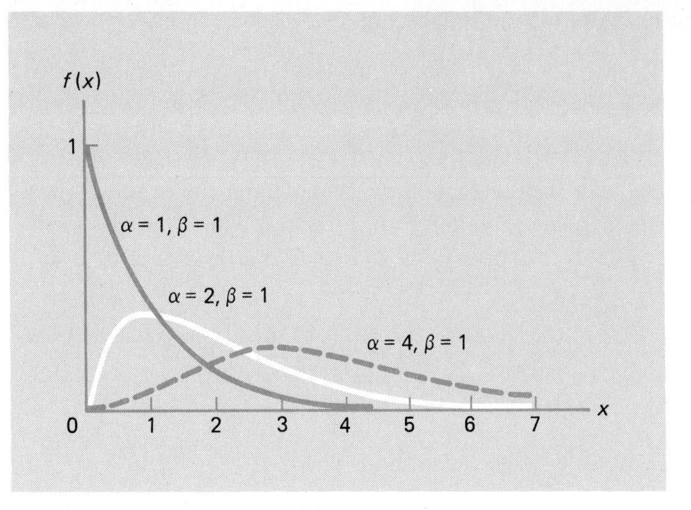

FIGURE 5.7
Graph of gamma distribution.

where $\Gamma(\alpha)$ is a value of the **gamma function**, defined by

$$\Gamma(\alpha) = \int_0^\infty x^{\alpha-1} e^{-x}\, dx$$

Integration by parts shows that

$$\Gamma(\alpha) = (\alpha - 1)\Gamma(\alpha - 1)$$

for any $\alpha > 1$ and, hence, that $\Gamma(\alpha) = (\alpha - 1)!$ when α is a positive integer. Graphs of several gamma distributions are shown in Figure 5.7 and they exhibit the fact that these distributions are positively skewed. In fact, the skewness decreases as α increases for any fixed value of β.

The mean and the variance of the gamma distribution may be obtained by making use of the gamma function and its special properties mentioned above. For the mean we have

$$\mu = \frac{1}{\beta^\alpha \Gamma(\alpha)} \int_0^\infty x \cdot x^{\alpha-1} e^{-x/\beta}\, dx$$

and, after letting $y = x/\beta$, we get

$$\mu = \frac{\beta}{\Gamma(\alpha)} \int_0^\infty y^\alpha e^{-y}\, dy = \frac{\beta\Gamma(\alpha + 1)}{\Gamma(\alpha)}$$

Then, making use of the identity $\Gamma(\alpha + 1) = \alpha \cdot \Gamma(\alpha)$, we arrive at the result that

Mean of gamma distribution

$$\boldsymbol{\mu = \alpha\beta}$$

Using similar methods, it can also be shown that the variance of the gamma distribution is given by

Variance of gamma distribution

$$\sigma^2 = \alpha\beta^2$$

In the special case where $\alpha = 1$, we get the **exponential distribution**, whose probability density is thus

Exponential distribution

$$f(x) = \begin{cases} \frac{1}{\beta} e^{-x/\beta} & \text{for } x > 0,\ \beta > 0 \\ 0 & \text{elsewhere} \end{cases}$$

and whose mean and variance are $\mu = \beta$ and $\sigma^2 = \beta^2$. Note that the distribution of the example on page 138 is an exponential distribution with $\beta = \frac{1}{2}$.

EXAMPLE A nuclear engineer observing a reaction measures the time intervals between the emissions of beta particles.

0.894	0.991	0.061	0.186	0.311	0.817	2.267	0.091	0.139	0.083
0.235	0.424	0.216	0.579	0.429	0.612	0.143	0.055	0.752	0.188
0.071	0.159	0.082	1.653	2.010	0.158	0.527	1.033	2.863	0.365
0.459	0.431	0.092	0.830	1.718	0.099	0.162	0.076	0.107	0.278
0.100	0.919	0.900	0.093	0.041	0.712	0.994	0.149	0.866	0.054

These decay times (in milliseconds) are presented as a histogram in Figure 5.8. The smooth curve is the exponential density with $\beta = 0.25$. Fit to an exponential density is further explored in Section 15.4.

The exponential distribution has many important applications; for instance, it can be shown that in connection with Poisson processes (see Section 4.7) the **waiting time** between successive arrivals (successes) has an exponential distribution. More specifically, it can be shown that if in a Poisson process the mean arrival rate (average number of arrivals per unit time) is α, the time until the first arrival, or the waiting time between successive arrivals, has an exponential distribution with $\beta = \frac{1}{\alpha}$ (see Exercise 5.62 on page 163).

EXAMPLE With reference to the example on page 119, where on the average three trucks arrived per hour to be unloaded at a warehouse, what are the probabilities that the time between the arrival of successive trucks will be

(a) less than 5 minutes;
(b) at least 45 minutes.

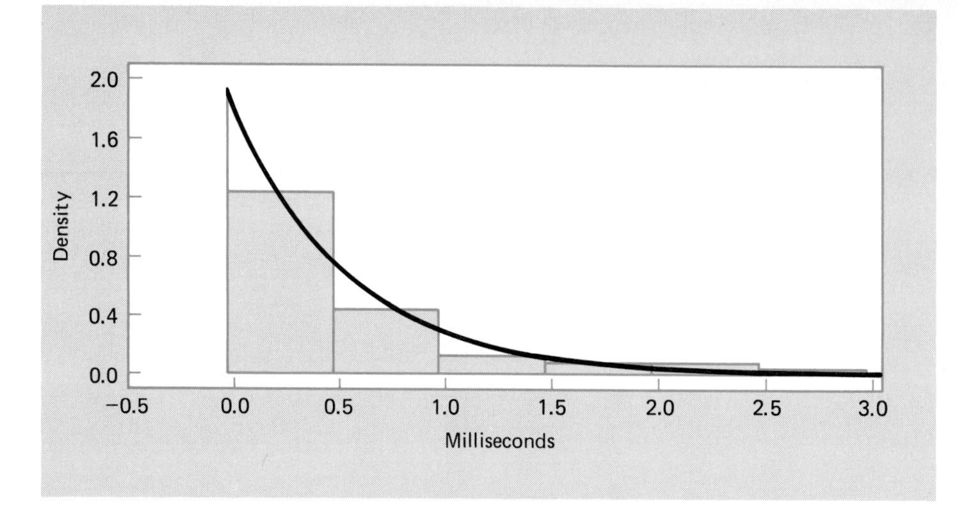

FIGURE 5.8
Histogram of decay times.

Solution Since $\beta = \frac{1}{3}$, we get

$$\int_0^{1/12} 3e^{-3x}\,dx = 1 - e^{-1/4} = 0.221$$

for part (a), and

$$\int_{3/4}^{\infty} 3e^{-3x}\,dx = e^{-9/4} = 0.105$$

for part (b).

■

5.8 THE BETA DISTRIBUTION

In Chapter 9 we shall need a probability density for a random variable which takes on values on the interval from 0 to 1, and most appropriate for this purpose is the **beta distribution**, whose probability density is

Beta distribution

$$f(x) = \begin{cases} \dfrac{\Gamma(\alpha+\beta)}{\Gamma(\alpha)\cdot\Gamma(\beta)}\, x^{\alpha-1}(1-x)^{\beta-1} & \text{for } 0 < x < 1, \alpha > 0, \beta > 0 \\ 0 & \text{elsewhere} \end{cases}$$

The mean and the variance of this distribution are given by

Mean and variance of beta distribution

$$\mu = \frac{\alpha}{\alpha + \beta} \quad \textit{and} \quad \sigma^2 = \frac{\alpha\beta}{(\alpha + \beta)^2(\alpha + \beta + 1)}$$

Note that for $\alpha = 1$ and $\beta = 1$ we obtain as a special case the uniform distribution of Section 5.5 defined on the interval from 0 to 1. The following example, pertaining to a proportion, illustrates a typical application of the beta distribution.

EXAMPLE

In a certain county, the proportion of highway sections requiring repairs in any given year is a random variable having the beta distribution with $\alpha = 3$ and $\beta = 2$ (shown in Figure 5.9). Find

(a) on the average what percentage of the highway sections require repairs in any given year;

(b) the probability that at most half of the highway sections will require repairs in any given year.

Solution

(a) $\mu = \dfrac{3}{3 + 2} = 0.60$, which means that on the average 60% of the highway sections require repairs in any given year. (b) Substituting $\alpha = 3$ and $\beta = 2$ into the

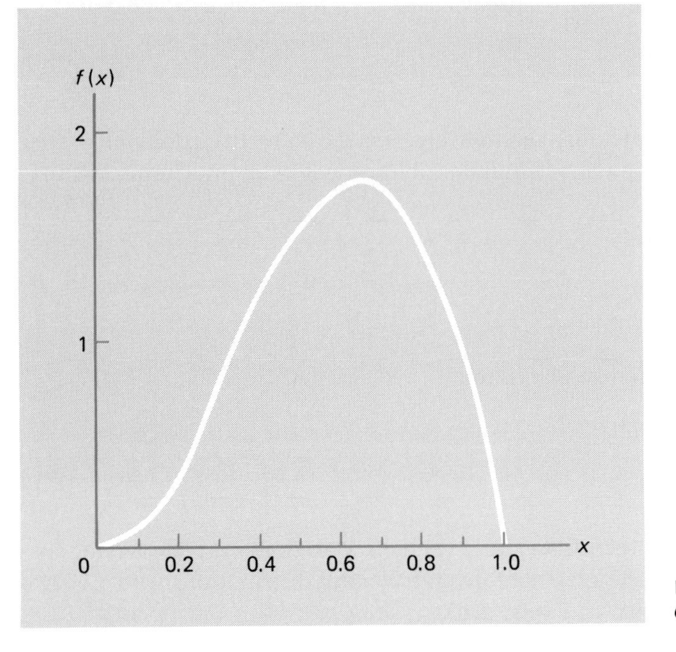

FIGURE 5.9
Graph of the beta distribution with $\alpha = 3$ and $\beta = 2$.

formula for the beta distribution and making use of the fact that $\Gamma(5) = 4! = 24$, $\Gamma(3) = 2! = 2$, and $\Gamma(2) = 1! = 1$, we get

$$f(x) = \begin{cases} 12x^2(1-x) & \text{for } 0 < x < 1 \\ 0 & \text{elsewhere} \end{cases}$$

Thus, the desired probability is given by

$$\int_0^{1/2} 12x^2(1-x)\,dx = \frac{5}{16}$$

■

In most realistically complex situations, probabilities relating to gamma and beta distributions are obtained from special tables or from computer packages.

5.9 THE WEIBULL DISTRIBUTION

Closely related to the exponential distribution is the **Weibull distribution**, whose probability density is given by

Weibull distribution

$$f(x) = \begin{cases} \alpha\beta x^{\beta-1}e^{-\alpha x^\beta} & \text{for } x > 0,\ \alpha > 0,\ \beta > 0 \\ 0 & \text{elsewhere} \end{cases}$$

To demonstrate this relationship, we evaluate the probability that a random variable having the Weibull distribution will take on a value less than a, namely, the integral

$$\int_0^a \alpha\beta x^{\beta-1}e^{-\alpha x^\beta}\,dx$$

Making the change of variable $y = x^\beta$, we get

$$\int_0^{a^\beta} \alpha e^{-\alpha y}\,dy = 1 - e^{-\alpha a^\beta}$$

and it can be seen that y is a value of a random variable having an exponential distribution. The graphs of several Weibull distributions with $\alpha = 1$ and $\beta = \frac{1}{2}$, 1, and 2 are shown in Figure 5.10.

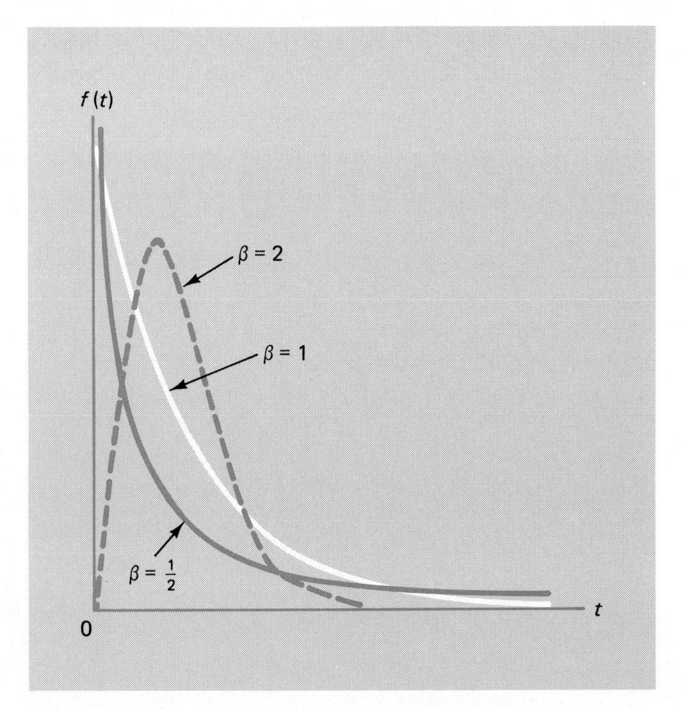

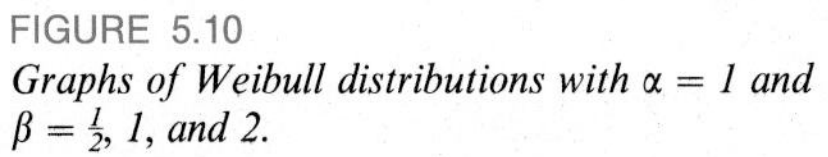
FIGURE 5.10
Graphs of Weibull distributions with $\alpha = 1$ *and* $\beta = \frac{1}{2}$, *1, and 2.*

The mean of the Weibull distribution having the parameters α and β may be obtained by evaluating the integral

$$\mu = \int_0^\infty x \cdot \alpha\beta x^{\beta-1} e^{-\alpha x^\beta}\, dx$$

Making the change of variable $u = \alpha x^\beta$, we get

$$\mu = \alpha^{-1/\beta} \int_0^\infty u^{1/\beta} e^{-u}\, du$$

and recognizing the integral as $\Gamma\left(1 + \frac{1}{\beta}\right)$, namely, as a value of the gamma function which we defined on page 156, we find that the mean of the Weibull distribution is given by

Mean of Weibull distribution

$$\boldsymbol{\mu = \alpha^{-1/\beta}\, \Gamma\left(1 + \frac{1}{\beta}\right)}$$

Using a similar method to determine first μ'_2, the reader will be asked to show in Exercise 5.70 on page 164 that the variance of this distribution is given by

Variance of Weibull distribution

$$\sigma^2 = \alpha^{-2/\beta}\left\{\Gamma\left(1 + \frac{2}{\beta}\right) - \left[\Gamma\left(1 + \frac{1}{\beta}\right)\right]^2\right\}$$

EXAMPLE Suppose that the lifetime of a certain kind of battery (in hours) is a random variable having the Weibull distribution with $\alpha = 0.1$ and $\beta = 0.5$. Find

(a) the mean lifetime of these batteries;
(b) the probability that such a battery will last more than 300 hours.

Solution (a) Substitution into the formula for the mean yields

$$\mu = (0.1)^{-2}\Gamma(3) = 200 \text{ hours}$$

(b) Performing the necessary integration, we get

$$\int_{300}^{\infty} (0.05)x^{-0.5}e^{-0.1x^{0.5}}\,dx = e^{-0.1(300)^{0.5}}$$
$$= 0.177$$

■

EXERCISES

5.45 Find the distribution function of a random variable which has the uniform distribution.

5.46 In certain experiments, the error made in determining the density of a substance is a random variable having the uniform density with $\alpha = -0.025$ and $\beta = 0.025$. What are the probabilities that such an error will be
(a) between 0.010 and 0.015;
(b) between -0.012 and 0.012?

5.47 From experience Mr. Harris has found that the low bid on a construction job can be regarded as a random variable having the uniform density

$$f(x) = \begin{cases} \dfrac{3}{4C} & \text{for } \dfrac{2C}{3} < x < 2C \\ 0 & \text{elsewhere} \end{cases}$$

where C is his own estimate of the cost of the job. What percentage should Mr. Harris add to his cost estimate when submitting bids to maximize his expected profit?

5.48 Verify the expression given on page 155 for the mean of the log-normal distribution.

5.49 With reference to the example on page 153, find the probability that I_o/I_i will take on a value between 7.0 and 7.5.

5.50 If a random variable has the log-normal distribution with $\alpha = -1$ and $\beta = 2$, find its mean and its standard deviation.

5.51 With reference to the preceding exercise, find the probabilities that the random variable will take on a value

(a) between 3.2 and 8.4; (b) greater than 5.0.

5.52 If a random variable has the gamma distribution with $\alpha = 2$ and $\beta = 2$, find the mean and the standard deviation of this distribution.

5.53 With reference to Exercise 5.52, find the probability that the random variable will take on a value less than 4.

5.54 In a certain city, the daily consumption of electric power (in millions of kilowatt-hours) can be treated as a random variable having a gamma distribution with $\alpha = 3$ and $\beta = 2$. If the power plant of this city has a daily capacity of 12 million kilowatt-hours, what is the probability that this power supply will be inadequate on any given day?

5.55 With reference to the example on page 154, suppose the expert opinion is in error. Calculate the probability that the supports will survive if

(a) $\mu = 3.0$ and $\sigma^2 = 0.09$; (b) $\mu = 4.0$ and $\sigma^2 = 0.36$.

5.56 Verify the expression for the variance of the gamma distribution given on page 157.

5.57 Show that when $\alpha > 1$, the graph of the gamma density has a relative maximum at $x = \beta(\alpha - 1)$. What happens when $0 < \alpha < 1$ and when $\alpha = 1$?

5.58 The amount of time that a watch will run without having to be reset is a random variable having the exponential distribution with $\beta = 50$ days. Find the probabilities that such a watch will

(a) have to be reset in less than 20 days;

(b) not have to be reset in at least 60 days.

5.59 With reference to Exercise 4.95, find the percent of the time that the interval between breakdowns of the computer will be

(a) less than 1 week; (b) at least 5 weeks.

5.60 With reference to Exercise 4.56, find the probabilities that the time between successive calls arriving at the switchboard of the consulting firm will be

(a) less than $\frac{1}{2}$ minute; (b) more than 3 minutes.

5.61 Given a Poisson process with on the average α arrivals per unit time, find the probability that there will be no arrivals during a time interval of length t, namely, the probability that the waiting times between successive arrivals will be at least of length t.

5.62 Use the result of Exercise 5.61 to find an expression for the probability density of the waiting time between successive arrivals.

5.63 Verify for $\alpha = 3$ and $\beta = 3$ that the integral of the beta density, from 0 to 1, is equal to 1.

5.64 If the annual proportion of erroneous income tax returns filed with the IRS can be looked upon as a random variable having a beta distribution with $\alpha = 2$ and $\beta = 9$, what is the probability that in any given year there will be fewer than 10% erroneous returns?

5.65 Suppose that the proportion of defectives shipped by a vendor, which varies somewhat from shipment to shipment, may be looked upon as a random variable having the beta distribution with $\alpha = 1$ and $\beta = 4$.

(a) Find the mean of this beta distribution, namely, the average proportion of defectives in a shipment from this vendor.

(b) Find the probability that a shipment from this vendor will contain 25% or more defectives.

5.66 Show that when $\alpha > 1$ and $\beta > 1$, the beta density has a relative maximum at

$$x = \frac{\alpha - 1}{\alpha + \beta - 2}$$

5.67 With reference to the example on page 162, find the probability that such a battery will not last 100 hours.

5.68 Suppose that the time to failure (in minutes) of certain electronic components subjected to continuous vibrations may be looked upon as a random variable having the Weibull distribution with $\alpha = \frac{1}{5}$ and $\beta = \frac{1}{3}$.

(a) How long can such a component be expected to last?

(b) What is the probability that such a component will fail in less than 5 hours?

5.69 Suppose that the service life (in hours) of a semiconductor is a random variable having the Weibull distribution with $\alpha = 0.025$ and $\beta = 0.500$. What is the probability that such a semiconductor will still be in operating condition after 4,000 hours?

5.70 Verify the formula for the variance of the Weibull distribution given on page 162.

5.10 JOINT PROBABILITY DENSITIES

There are many situations in which we describe an outcome by giving the values of several random variables. For instance, we may measure the weight and the hardness of a rock, the volume, pressure, and temperature of a gas, or the thickness, color, compressive strength, and potassium content of a piece of glass. If $x_1, x_2, \ldots, x_k$ are the values of k random variables, we shall refer to a function **f** with values $f(x_1, x_2, \ldots, x_k)$ as the **joint probability density** of these random variables, if the probability that $a_1 \le x_1 \le b_1$, $a_2 \le x_2 \le b_2, \ldots,$ and $a_k \le x_k \le b_k$ is given by the multiple integral

$$\int_{a_1}^{b_1} \int_{a_2}^{b_2} \cdots \int_{a_k}^{b_k} f(x_1, x_2, \ldots, x_k)\, dx_1\, dx_2 \cdots dx_k$$

Thus, not every function **f** with values $f(x_1, x_2, \ldots, x_k)$ can serve as a joint probability density, but if

$$f(x_1, x_2, \ldots, x_k) \ge 0$$

for all values of $x_1, x_2, \ldots, x_k$ for which the function is defined, and

$$\int_{-\infty}^{\infty} \int_{-\infty}^{\infty} \cdots \int_{-\infty}^{\infty} f(x_1, x_2, \ldots, x_k)\, dx_1\, dx_2 \cdots dx_k = 1$$

it can be shown that the axioms of probability (with the modification of the definition of "event" discussed in Section 5.1) are satisfied.

To extend the concept of a cumulative distribution function to the k-variable case, we write as $F(x_1, x_2, \ldots, x_k)$ the probability that the first random variable will take on a value less than or equal to x_1, the second random variable will take on a value less than or equal to $x_2, \ldots,$ and the kth random variable will take on a value less than or equal to x_k, and we refer to the corresponding function F as the **joint distribution function** of the k random variables.

EXAMPLE If the joint probability density of two random variables is given by

$$f(x_1, x_2) = \begin{cases} 6e^{-2x_1-3x_2} & \text{for } x_1 > 0, x_2 > 0 \\ 0 & \text{elsewhere} \end{cases}$$

find the probabilities that

(a) the first random variable will take on a value between 1 and 2 and the second random variable will take on a value between 2 and 3;

(b) the first random variable will take on a value less than 2 and the second random variable will take on a value greater than 2.

Solution Performing the necessary integrations, we get

$$\int_1^2 \int_2^3 6e^{-2x_1-3x_2}\, dx_1\, dx_2 = (e^{-2} - e^{-4})(e^{-6} - e^{-9})$$
$$= 0.0003$$

for part (a), and

$$\int_0^2 \int_2^\infty 6e^{-2x_1-3x_2}\, dx_1\, dx_2 = (1 - e^{-4})e^{-6}$$
$$= 0.0025$$

for part (b). ■

EXAMPLE Find the joint distribution function of the two random variables of the preceding exercise, and use it to find the probability that both random variables will take on values less than 1.

Solution By definition,

$$F(x_1, x_2) = \begin{cases} \int_0^{x_1} \int_0^{x_2} 6e^{-2u-3v}\, du\, dv & \text{for } x_1 > 0, x_2 > 0 \\ 0 & \text{elsewhere} \end{cases}$$

so that

$$F(x_1, x_2) = \begin{cases} (1 - e^{-2x_1})(1 - e^{-3x_2}) & \text{for } x_1 > 0, x_2 > 0 \\ 0 & \text{elsewhere} \end{cases}$$

and, hence,

$$F(1, 1) = (1 - e^{-2})(1 - e^{-3})$$
$$= 0.8216$$

■

Given the joint probability density of k random variables, the probability density of the ith random variable can be obtained by integrating out the other variables; symbolically,

Marginal density

$$f_i(x_i) = \int_{-\infty}^{\infty} \cdots \int_{-\infty}^{\infty} f(x_1, x_2, \ldots, x_k)\, dx_1 \cdots dx_{i-1}\, dx_{i+1} \cdots dx_k$$

and, in this context, the function $\mathbf{f}_i$ is called the **marginal density** of the ith random variable. Integrating out only some of the k random variables, we can similarly define **joint marginal densities** of any two, three, or more of the k random variables.

EXAMPLE With reference to the example on page 165, find the marginal density of the first random variable.

Solution Integrating out x_2, we get

$$f_1(x_1) = \begin{cases} \int_0^{\infty} 6e^{-2x_1-3x_2}\, dx_2 & \text{for } x_1 > 0 \\ 0 & \text{elsewhere} \end{cases}$$

or

$$f_1(x_1) = \begin{cases} 2e^{-2x_1} & \text{for } x_1 > 0 \\ 0 & \text{elsewhere} \end{cases}$$

■

To explain what we mean by the **independence** of continuous random variables, we could proceed as in Section 3.6 and define **conditional densities** first; however, it will be easier to say that

Independent random variables

> ***k random variables are independent if and only if***
>
> $$F(x_1, x_2, \ldots, x_k) = F_1(x_1) \cdot F_2(x_2) \cdots F_k(x_k)$$
>
> ***for all values of these random variables for which the respective functions are defined.***

In this notation, $F(x_1, x_2, \ldots, x_k)$ is, as before, a value of the joint distribution function of the k random variables, while $F_i(x_i)$ for $i = 1, 2, \ldots, k$ are the corresponding values of the individual distribution functions of the respective random variables.

EXAMPLE With reference to the example on page 165, check whether the two random variables are independent.

Solution As we already saw on page 166, the joint distribution function of the two random variables is given by

$$F(x_1, x_2) = \begin{cases} (1 - e^{-2x_1})(1 - e^{-3x_2}) & \text{for } x_1 > 0 \text{ and } x_2 > 0 \\ 0 & \text{elsewhere} \end{cases}$$

Now, since $F_1(x_1) = F(x_1, \infty)$ and $F_2(x_2) = F(\infty, x_2)$, it follows that

$$F_1(x_1) = \begin{cases} 1 - e^{-2x_1} & \text{for } x_1 > 0 \\ 0 & \text{elsewhere} \end{cases}$$

and

$$F_2(x_2) = \begin{cases} 1 - e^{-3x_2} & \text{for } x_2 > 0 \\ 0 & \text{elsewhere} \end{cases}$$

Thus, $F(x_1, x_2) = F_1(x_1) \cdot F_2(x_2)$ and the two random variables are independent. ■

Analogous to the special multiplication rule for independent events and its extension on page 71, it follows from our definition of independence that if k random variables are independent, any value of their joint probability density equals the product of the corresponding values of the marginal densities of the k random variables; symbolically,

$$f(x_1, x_2, \ldots, x_k) = f_1(x_1) \cdot f_2(x_2) \cdots f_k(x_k)$$

EXAMPLE With reference to the example on page 165, verify that $f(x_1, x_2) = f_1(x_1) \cdot f_2(x_2)$.

Solution On page 166 we showed that

$$f_1(x_1) = \begin{cases} 2e^{-2x_1} & \text{for } x_1 > 0 \\ 0 & \text{elsewhere} \end{cases}$$

and in the same way it can be shown that

$$f_2(x_2) = \begin{cases} 3e^{-3x_2} & \text{for } x_2 > 0 \\ 0 & \text{elsewhere} \end{cases}$$

Thus

$$f_1(x_1) \cdot f_2(x_2) = \begin{cases} 6e^{-2x_1 - 3x_2} & \text{for } x_1 > 0 \text{ and } x_2 > 0 \\ 0 & \text{elsewhere} \end{cases}$$

and it can be seen that $f_1(x_1) \cdot f_2(x_2) = f(x_1, x_2)$. ■

Given two random variables with values x_1 and x_2, we shall define the **conditional probability density** of the first given that the second takes on the value x_2 as

Conditional probability density

$$g_1(x_1 | x_2) = \frac{f(x_1, x_2)}{f_2(x_2)} \qquad f_2(x_2) \neq 0$$

where $f(x_1, x_2)$ and $f_2(x_2)$ are, as before, values of the joint density of the two random variables and the marginal density of the second. Note that this definition parallels that of conditional probability on page 70.

EXAMPLE If two random variables have the joint probability density

$$f(x_1, x_2) = \begin{cases} \frac{2}{3}(x_1 + 2x_2) & \text{for } 0 < x_1 < 1, 0 < x_2 < 1 \\ 0 & \text{elsewhere} \end{cases}$$

find the conditional density of the first given that the second takes on the value x_2.

Solution First we find the marginal density of the second random variable by integrating out x_1, and we get

$$f_2(x_2) = \int_0^1 \frac{2}{3}(x_1 + 2x_2)\, dx_1 = \frac{1}{3}(1 + 4x_2) \qquad \text{for } 0 < x_2 < 1$$

and $f_2(x_2) = 0$ elsewhere. Hence, by definition, the conditional density of the first random variable given that the second takes on the value x_2 is given by

$$g_1(x_1|x_2) = \frac{\frac{2}{3}(x_1 + 2x_2)}{\frac{1}{3}(1 + 4x_2)} = \frac{2x_1 + 4x_2}{1 + 4x_2} \qquad \text{for } 0 < x_1 < 1, 0 < x_2 < 1$$

and $g_1(x_1|x_2) = 0$ for $x_1 \leq 0$ or $x_1 \geq 1$ and $0 < x_2 < 1$.

■

The following is another concept which is of importance when we deal with k random variables. Suppose that a random variable takes on the value $g(x_1, x_2, \ldots, x_k)$ whenever k random variables take on the values $x_1, x_2, \ldots, x_k$. Then the expected value of this random variable (namely, the mean of its distribution) is given by

$$\int_{-\infty}^{\infty}\int_{-\infty}^{\infty}\cdots\int_{-\infty}^{\infty} g(x_1, \ldots, x_k)f(x_1, \ldots, x_k)\, dx_1, dx_2 \cdots dx_k$$

where $f(x_1, \ldots, x_k)$ is, as before, a value of the joint density of the k random variables.

EXAMPLE With reference to the example on page 165, find the expected value of the product of the two random variables.

Solution Substituting into the above formula, we get

$$\int_0^{\infty}\int_0^{\infty} x_1 x_2 \cdot 6e^{-2x_1 - 3x_2}\, dx_1\, dx_2 = \tfrac{1}{6}$$

■

EXERCISES

5.71 If two random variables have the joint density

$$f(x_1, x_2) = \begin{cases} x_1 x_2 & \text{for } 0 < x_1 < 1, 0 < x_2 < 2 \\ 0 & \text{elsewhere} \end{cases}$$

find the probabilities that

(a) both random variables will take on values less than 1;

(b) the sum of the values taken on by the two random variables will be less than 1.

5.72 With reference to the preceding exercise, find the marginal densities of the two random variables.

5.73 With reference to Exercise 5.71, find the joint distribution function of the two random variables, the distribution functions of the individual random variables, and check whether the two random variables are independent.

5.74 If two random variables have the joint density

$$f(x, y) = \begin{cases} \frac{6}{5}(x + y^2) & \text{for } 0 < x < 1, 0 < y < 1 \\ 0 & \text{elsewhere} \end{cases}$$

find the probability that $0.2 < x < 0.5$ and $0.4 < y < 0.6$.

5.75 With reference to the preceding exercise, find the joint distribution function of the two random variables and use it to verify the value obtained for the probability.

5.76 With reference to Exercise 5.74, find both marginal densities and use them to find the probabilities that

(a) $x > 0.8$; (b) $y < 0.5$.

5.77 With reference to Exercise 5.74, find

(a) an expression for $g_1(x|y)$ for $0 < y < 1$;

(b) an expression for $g_1(x|\frac{1}{2})$;

(c) the mean of the conditional density of the first random variable when the second takes on the value $\frac{1}{2}$.

5.78 With reference to the example on page 168, find expressions for

(a) the conditional density of the first random variable when the second takes on the value $x_2 = 0.25$;

(b) the conditional density of the second random variable when the first takes on the value x_1.

5.79 If three random variables have the joint density

$$f(x, y, z) = \begin{cases} k(x + y)e^{-z} & \text{for } 0 < x < 1, 0 < y < 2, z > 0 \\ 0 & \text{elsewhere} \end{cases}$$

find

(a) the value of k;

(b) the probability that $x < y$ and $z > 1$.

5.80 With reference to the preceding exercise, check whether

(a) the three random variables are independent;

(b) any two of the three random variables are pairwise independent.

5.81 A pair of random variables has the **circular normal distribution** if their joint density is given by

$$f(x_1, x_2) = \frac{1}{2\pi\sigma^2} e^{-[(x_1-\mu_1)^2+(x_2-\mu_2)^2]/2\sigma^2}$$

for $-\infty < x_1 < \infty$ and $-\infty < x_2 < \infty$.

(a) If $\mu_1 = 2$ and $\mu_2 = -2$, and $\sigma = 10$, use Table 3 to find the probability that $-8 < x_1 < 14$ and $-9 < x_2 < 3$.

(b) If $\mu_1 = \mu_2 = 0$ and $\sigma = 3$, find the probability that (x_1, x_2) is contained in the region between the two circles $x_1^2 + x_2^2 = 9$ and $x_1^2 + x_2^2 = 36$.

5.82 A precision drill, positioned over a target point, will make an acceptable hole if it is within 5 microns of the target. Using the target as the origin of a rectangular system of coordinates, assume that the coordinates (x, y) of the point of contact are values of a pair of random variables having the circular normal distribution (see Exercise 5.81) with $\mu_1 = \mu_2 = 0$ and $\sigma = 2$. What is the probability that the hole will be acceptable.

5.83 With reference to Exercise 5.71, find the expected value of the random variable whose values are given by $g(x_1, x_2) = x_1 + x_2$.

5.84 With reference to Exercise 5.74, find the expected value of the random variable whose values are given by $g(x, y) = x^2 y$.

5.85 If measurements of the length and the width of a rectangle have the joint density

$$f(x, y) = \begin{cases} \dfrac{1}{ab} & \text{for } L - \dfrac{a}{2} < x < L + \dfrac{a}{2},\ W - \dfrac{b}{2} < y < W + \dfrac{b}{2} \\ 0 & \text{elsewhere} \end{cases}$$

find the mean and the variance of the corresponding distribution of the area of the rectangle.

5.86 Establish a relationship between $g_1(x_1|x_2)$, $g_2(x_2|x_1)$, $f_1(x_1)$, and $f_2(x_2)$.

5.11 CHECKING IF THE DATA ARE NORMAL

In many instances, an experimenter needs to check whether a data set appears to be generated by a normally distributed random variable. As indicated in Figure 2.8 the normal distribution can serve to model variation in some quantities. Further, many commonly used statistical procedures, which we describe in later chapters, require that the probability distribution be nearly normal.

Although they involve an element of subjective judgment, graphical procedures are the most helpful for detecting serious departures from normality. Histograms can be checked for lack of symmetry. A single long tail certainly contradicts the assumption of a normal distribution. However, another special graph, called a **normal-scores plot**, is even more effective in detecting departures from normality. To introduce such a plot, we consider a sample of size 4. In practice, we need a minimum of 15-20 observations in order to evaluate the agreement with normality.

The term normal scores refers to an idealized sample from the standard normal distribution. It consists of the values of z that divide that axes into equal probability intervals. For sample size $n = 4$, the normal scores are

$$m_1 = -z_{0.20} = -0.84$$
$$m_2 = -z_{0.40} = -0.25$$
$$m_3 = z_{0.40} = 0.25$$
$$m_4 = z_{0.20} = 0.84$$

and these are illustrated in Figure 5.11.

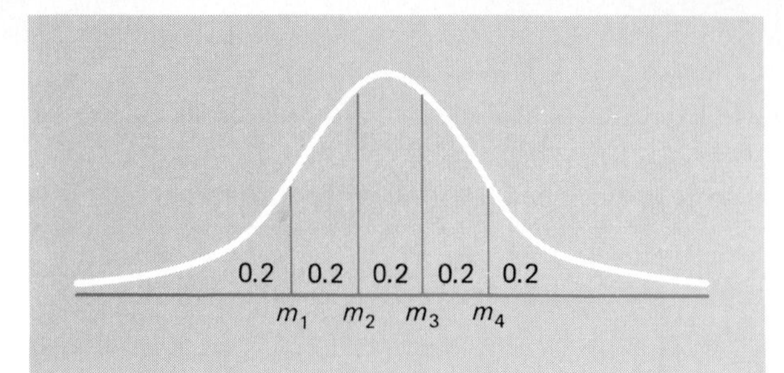

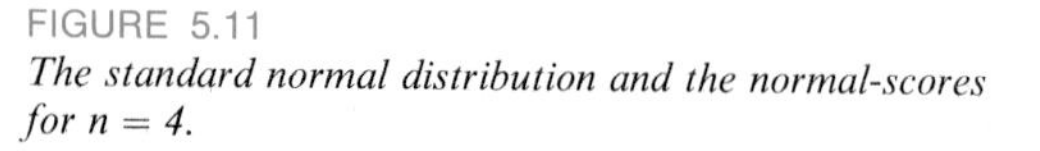
FIGURE 5.11
The standard normal distribution and the normal-scores for $n = 4$.

To construct a normal-scores plot

1. Order the data from smallest to largest;
2. Obtain the normal scores;
3. Plot the ith largest observation, versus the ith normal score m_i, for all i.

EXAMPLE Suppose the four observations are 67, 48, 76, 81. Construct a normal scores plot.

Solution The ordered observations are 48, 67, 76, 81. Earlier, we found that $m_1 = -z_{0.20} = -0.84$, so we plot the pair $(48, -0.84)$. Continuing, we obtain Figure 5.12. ■

If the data were from a standard normal distribution, we would expect the ith largest observation to approximate the ith normal score so that the normal-scores plot would resemble a 45^0 line through the origin. When the distribution is normal with an unspecified μ and σ, $z = \dfrac{(x - \mu)}{\sigma}$; so the idealized z values can be converted

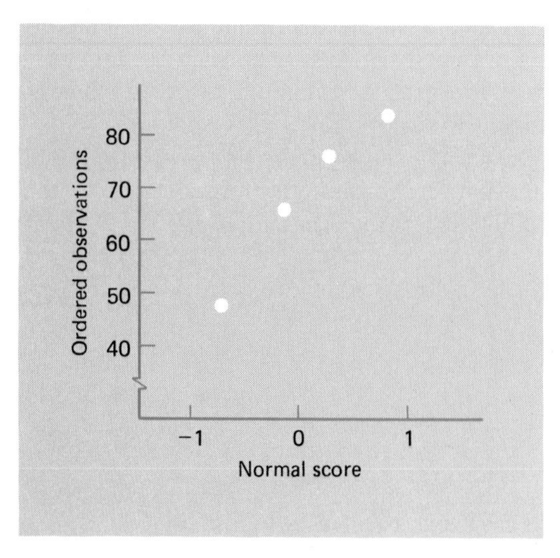

FIGURE 5.12
The normal-scores plot.

to idealized x values through the relation $x = \mu + \sigma z$. Because the idealized values have this linear relation, it is sufficient to plot the ordered observations versus the normal scores obtained from the standard normal distribution. If the normal distribution prevails, the pattern should still be a straight line. But, the line need not pass through the origin or have slope 1.

The construction of normal-scores plots by hand is a difficult task at best. Fortunately, they can be created easily with most statistical packages. (See Exercise 5.89, page 175.) Many slight variants are used in the calculation of the normal-scores but the plots are very similar if more than 20 observations are plotted. Whichever computer package you use, if a normal distribution is plausible, the plot will have a straight-line appearance.

Figure 5.13 shows the normal-scores plot for the interrequest times given on page 14. The bending shows that the largest values are larger than would be expected under a normal distribution. On the other hand, Figure 5.14 exhibits a plot of the sulfur emissions data (see the example on page 8) and a normal distribution appears to be plausible.

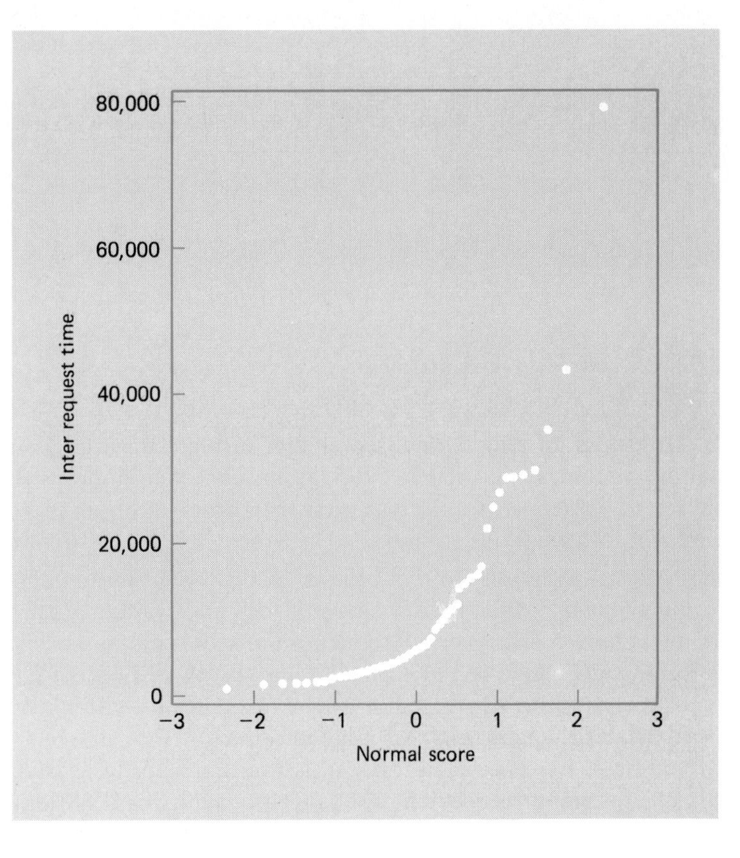

FIGURE 5.13
The normal-scores plot of the interrequest times.

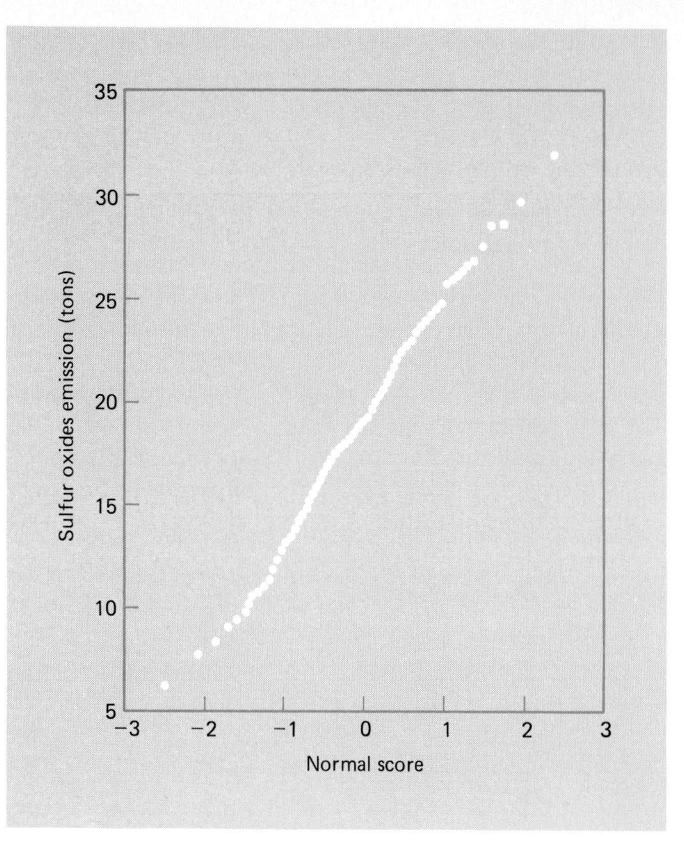

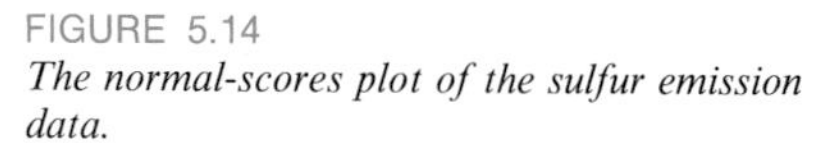

FIGURE 5.14
The normal-scores plot of the sulfur emission data.

5.12
TRANSFORMING OBSERVATIONS TO NEAR NORMALITY

When the histogram and normal-scores plot indicate that the assumption of a normal distribution is invalid, transformations can often improve the agreement with normality. Scientists regularly express their observations in logs. We consider a few other transformations, as indicated in Table 5.1. If the distribution has a long tail or a few stragglers on the right, the $\ln x$ or $\sqrt{x}$ transformations will pull the large values down further than the central or small values. If the transformed observations have a nearly straight line normal-scores plot, it is usually advantageous to take advantage of the normality and use this new scale to perform any statistical analysis. Further, the validity of many of the powerful statistical methods described in later chapters rests on the assumption that the probability distribution is nearly normal. By choosing a transformation that leads to nearly normal data, the investigator can greatly extend the range of validity to these techniques.

TABLE 5.1 *Some Useful Transformations*

Make large values smaller:	*Make large values larger:*
$\frac{-1}{x}$, $\ln x$, $x^{1/4}$, $\sqrt{x}$	x^2, x^3

EXAMPLE Transform the interrequest times in the example on page 14 to better approximate a normal distribution.

Solution On a computer, we calculate $\sqrt{x}$, take the square root again to obtain $x^{1/4}$, and take the natural logarithm $\ln x$ of all 50 values. The transformation $\log x$ appears to work best. The histogram and normal-scores plot are shown in Figure 5.15 for both the original and transformed data. The quality of the fit further confirms the lognormal model.

■

EXERCISES

5.87 For any seven observations,

(a) use Table 3 to verify that the normal scores are

$$-1.15, \ -0.67, \ -0.32, 0, 0.32, 0.67, 1.15$$

(b) construct a normal-scores plot using the observations

$$16, 10, 18, 27, 29, 19, 17$$

5.88 For any 11 observations,

(a) use Table 3 to verify that the normal scores are

$$-1.38, \ -0.97, \ -0.67, \ -0.43, \ -0.21, 0, 0.21, 0.43, 0.67, 0.97, 1.38$$

(b) construct a normal-scores plot using the observations on the times between neutrinos on page 7.

5.89 (Normal-scores plots) The *MINITAB* commands

```
NSCORE C1 PUT IN C2
PLOT C1 VS C2
```

will create a normal-scores plot from observations that were set in C1. (*MINITAB* uses a variant of the normal scores, m_i, that we defined.) Construct a normal-scores plot of

(a) the aluminum alloy strength data on page 15;

(b) the decay data on page 157.

FIGURE 5.15
The original and transformed data.

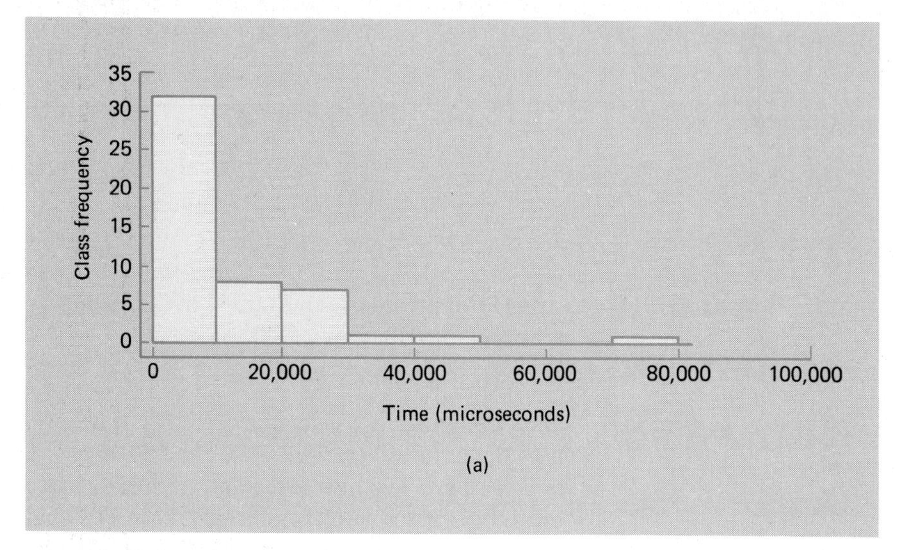

FIGURE 5.15(a)
Histogram of interrequest time.

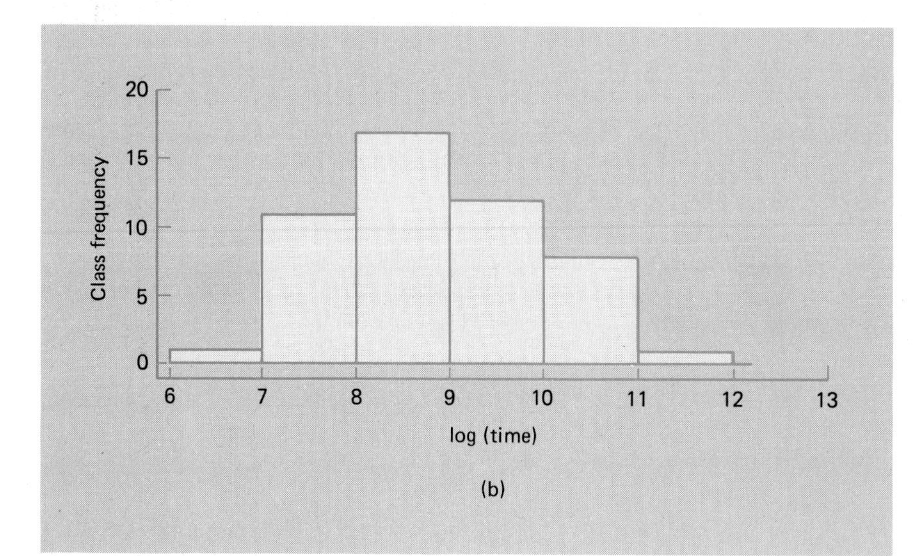

FIGURE 5.15(b)
Histogram of log(interrequest time).

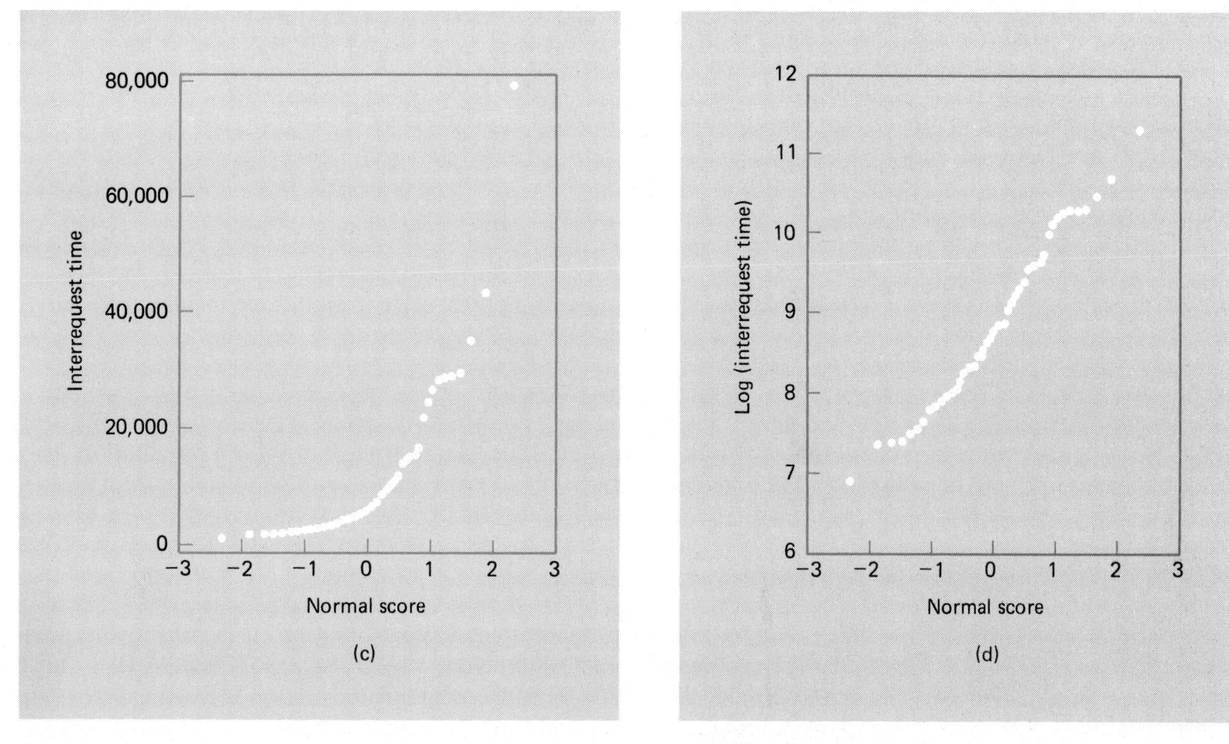

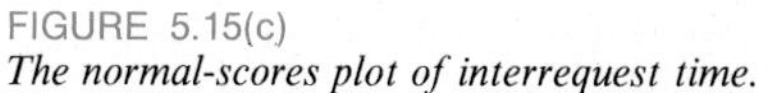
FIGURE 5.15(c)
The normal-scores plot of interrequest time.

FIGURE 5.15(d)
The normal-scores plot of log (interrequest time).

5.90 (Transformations). The *MINITAB* commands

```
LET C2= LOGE ( C1)
LET C3=  SQRT (C1)
LET C4=  SQRT (C3)
```

will place $\ln x$ in C2, $\sqrt{x}$ in C3, and $x^{1/4}$ in C4 for observations that are set in C1. Normal-scores plots can then be constructed as in Exercise 5.89. Try these three transformations and construct the corresponding normal-scores plots for

(a) the decay time data on page 157;
(b) the interrequest time data on page 14.

5.13 SIMULATION

Simulation techniques have grown up with computers. They are ideally suited for doing the repetitious calculations required. To **simulate** the observation of continuous random variables we usually start with uniform random numbers and relate these to the distribution function of interest. We could use two- or three-digit random integers, perhaps selected from Table 7, but most software packages have a continuous uniform random-number generator. That is, they produce approximations to random numbers from the uniform distribution

$$f(x) = \begin{cases} 1 & 0 < x < 1 \\ 0 & \text{elsewhere} \end{cases}$$

Suppose we wish to simulate an observation from the exponential distribution

$$F(x) = 1 - e^{-0.3x}, \qquad 0 < x < \infty$$

The computer would first produce the value u from the uniform distribution. Then we solve (see Exercise 5.94)

$$u = F(x) = 1 - e^{-0.3x}$$

so $x = [-\ln(1 - u)]/0.3$ is the corresponding value of an exponential random variable. For instance, if $u = 0.45$, then $x = [-\ln(1 - 0.45)]/0.3 = 1.993$. This is illustrated graphically in Figure 5.16, where u is located on the vertical scale and the corresponding x-value is read from the horizontal scale. (The theory on which this method is based involves the so-called probability integral transformation, which is treated in all the textbooks on mathematical statistics referred to in the bibliography). If we wish to simulate a sample from F, the preceding process is repeated with a different u for each new observation x.

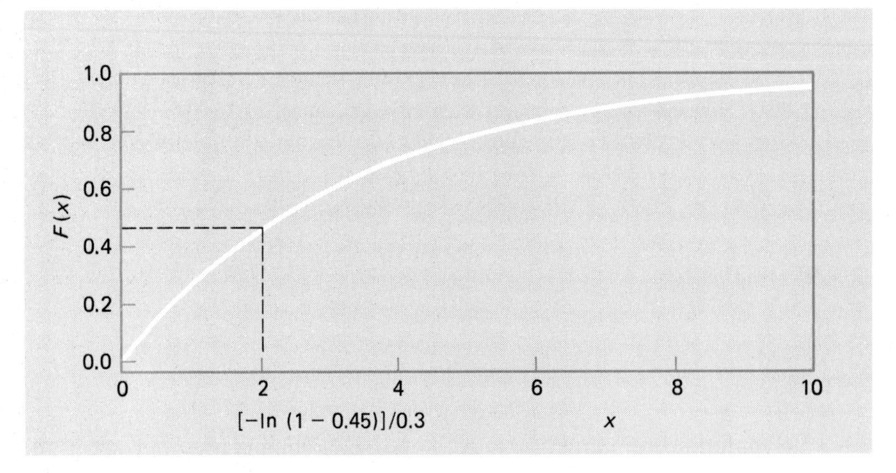

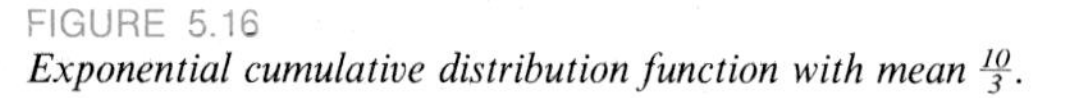
FIGURE 5.16
Exponential cumulative distribution function with mean $\frac{10}{3}$.

A similar procedure applies to the simulation of observations from a Weibull distribution. Starting with the value of a uniform variable u, we now solve (see Exercise 5.95)

$$u = F(x) = 1 - e^{-\alpha x^{\beta}}$$

for $x = \left[-\frac{1}{\alpha} \ln(1 - u) \right]^{1/\beta}$, which is the corresponding value of a Weibull random variable.

EXAMPLE Simulate five observations of a random variable having the Weibull distribution with $\alpha = 0.05$ and $\beta = 2.0$.

Solution A computer (see Exercise 5.97) generates the five values 0.57, 0.74, 0.26, 0.77, 0.12. (Alternatively, they could have been obtained by reading two digits at a time from Table 7.) We calculate

$$x = [-20.0 \ln(1 - 0.57)]^{1/2} = 4.108$$

$$x = [-20.0 \ln(1 - 0.74)]^{1/2} = 5.191$$

In Exercise 5.91 the reader will be asked to show that the last three uniform numbers yield $x = 2.454, 5.422, 1.599$.

■

Suppose we need to simulate values from the normal distribution with a specified μ and σ^2. By the relation $z = \frac{x - \mu}{\sigma}$ (page 144), it follows that $x = \mu + \sigma z$, so a value x can be calculated from the value of a standard normal variable z. Although z can be obtained from the value for a uniform variable u by numerically solving $u = F(z)$, another approach called the Box-Muller-Marsaglia method is almost universally preferred. It starts with a pair of independent uniform variables (u_1, u_2) and produces two standard normal variables

$$z_1 = \sqrt{-\ln(u_2)} \cos(2\pi u_1)$$

$$z_2 = \sqrt{-\ln(u_2)} \sin(2\pi u_1)$$

where the angle is expressed in radians. Then $x_1 = \mu + \sigma z_1$ and $x_2 = \mu + \sigma z_2$ are treated as two independent observations of normal random variables. (see Exercise 5.96) Most statistical packages include a normal generator. (see Exercise 5.99)

EXAMPLE Simulate two observations of a random variable having the normal distribution with $\mu = 50$ and $\sigma = 5$.

Solution A computer generates the two values 0.253 and 0.531 from a uniform distribution. (Alternatively, they could have been obtained by reading three digits at a time from Table 7.) We calculate

$$z_1 = \sqrt{-\ln(0.531)}\cos(2\pi \cdot 0.253) = 0.788$$

$$z_2 = \sqrt{-\ln(0.531)}\sin(2\pi \cdot 0.253) = 0.111$$

■

EXERCISES

5.91 Continue the example in the text, that is, find the simulated values of the Weibull random variable corresponding to the values 0.26, 0.77, 0.12 of a uniform random variable.

5.92 Suppose the number of hours it takes a person to learn how to operate a certain machine is a random variable having a normal distribution with $\mu = 5.8$ and $\sigma = 1.2$. Suppose it takes two persons to operate the machine. Simulate the time it takes four pairs of persons to learn how to operate the machine. That is, for each pair, calculate the maximum of the two learning times.

5.93 Suppose that the durability of paint (in years) is a random variable having an exponential distribution (see page 157) with mean $\beta = 2$. Simulating an experiment in which five houses are painted,

(a) find the time of the first failure;

(b) find the time of the fifth failure.

5.94 Verify that

(a) the exponential density $0.3e^{-0.3x}$, $x > 0$ corresponds to the distribution function $F(x) = 1 - e^{-0.3x}$, $x > 0$;

(b) the solution of $u = F(x)$ is given by $x = [-\ln(1-u)]/0.3$.

5.95 Verify that

(a) the Weibull density $\alpha\beta x^{\beta-1}e^{-\alpha x^\beta}$, $x > 0$, corresponds to the distribution function $F(x) = 1 - e^{-\alpha x^\beta}$, $x > 0$;

(b) the solution of $u = F(x)$ is given by $x = \left[-\frac{1}{\alpha}\ln(1-u)\right]^{1/\beta}$.

5.96 Consider two independent standard normal variables whose joint probability density is

$$\frac{1}{2\pi}\, e^{(z_1^2+z_2^2)/2}$$

Under a change to polar coordinates, $z_1 = r\cos(\theta)$, $z_2 = r\sin(\theta)$, we have $r^2 = z_1^2 + z_2^2$ and $dz_1\,dz_2 = r\,dr\,d\theta$, so the joint density of r and θ is

$$re^{-r^2/2}\frac{1}{2\pi}, \qquad 0 < \theta < 2\pi,\ r > 0$$

Show that

(a) r and θ are independent and that θ has a uniform distribution on the interval from 0 to 2π;

(b) $u_1 = \theta/2\pi$ and $u_2 = 1 - e^{-r^2/2}$ have independent uniform distributions;

(c) the relations between (u_1, u_2) and (z_1, z_2) on page 179 hold [note that $1 - u_2$ also has a uniform distribution, so $\ln(u_2)$ can be used in place of $\ln(1 - u_2)$].

5.97 The statistical package *MINITAB* has a continuous uniform random number generator. The commands

```
RANDOM 5 observations into C1;
 UNIFORM 0 to 1.
```

produces five continuous uniform random values and places them in column 1 of the MINITAB worksheet. In one case, this command produced the output

```
0.043281   0.418196   0.580963   0.797438   0.509641
```

(a) Generate eight uniform variates.
(b) Use the command

```
LET C2= -20*LOGE (1 - C1)
```

to place the values of $x = -20\ln(1-u)$ in column 2. As described on page 178, these are values of an exponential variable with $\alpha = 0.05$.

5.98 Repeat the simulation in Exercise 5.93 100 times and make a histogram of the values from parts (a) and (b).

5.99 The statistical package *MINITAB* has a normal random number generator. The command

```
RANDOM 5 observations into C1;
NORMAL  MU = 7  SIGMA = 4 .
```

produces five values of a normal variable with $\mu = 7$ and $\sigma = 4$ and places them in column 1 of the MINITAB worksheet. One call to this command produced the values

```
5.42137   6.98061   9.41352   7.05932   5.87297
```

Generate eight values for a normal variable with $\mu = 123$ and $\sigma^2 = 23.5$.

5.100 Repeat the simulation in Exercise 5.92 100 times and make a histogram of (a) the 400 learning times for the pairs of operators and (b) the 100 values representing the time to train four pairs of operators.

5.14 REVIEW EXERCISES

5.101 If the probability density of a random variable is given by

$$f(x) = \begin{cases} k(1-x^2) & \text{for } 0 < x < 1 \\ 0 & \text{elsewhere} \end{cases}$$

find the value of k and the probabilities that a random variable having this probability density will take on a value

(a) between 0.1 and 0.2;   (b) greater than 0.5.

5.102 With reference to the preceding exercise, find the corresponding distribution function and use it to determine the probabilities that a random variable having this distribution function will take on a value
(a) less than 0.3; (b) between 0.4 and 0.6.

5.103 In certain experiments, the error made in determining the density of a silicon compound is a random variable having the probability density

$$f(x) = \begin{cases} 25 & \text{for } -0.02 < x < 0.02 \\ 0 & \text{elsewhere} \end{cases}$$

Find the probabilities that such an error will be
(a) between -0.03 and 0.04; (b) between -0.005 and 0.005.

5.104 Find μ and σ^2 for the probability density of Exercise 5.101.

5.105 If a random variable has the standard normal distribution, find the probability that it will take on a value
(a) between 0 and 2.50; (b) between 1.22 and 2.35;
(c) between -1.33 and -0.33; (d) between -1.60 and 1.80.

5.106 The burning time of an experimental rocket is a random variable having the normal distribution with $\mu = 4.76$ seconds and $\sigma = 0.04$ second. What is the probability that this kind of rocket will burn
(a) less than 4.66 seconds;
(b) more than 4.80 seconds;
(c) anywhere from 4.70 to 4.82 seconds?

5.107 Verify that
(a) $z_{0.10} = 1.28$; (b) $z_{0.001} = 3.09$.

5.108 Referring to Exercise 5.28, find the *quartiles* of the normal distribution with $\mu = 102$ and $\sigma = 27$.

5.109 The probability density shown in Figure 5.6 is the log-normal distribution with $\alpha = 8.85$ and $\beta = 1.03$. Find the probability that
(a) the interrequest time is more than 200 microseconds;
(b) the interrequest time is less than 300 microseconds.

5.110 The probability density shown in Figure 5.8 is the exponential distribution

$$f(x) = \begin{cases} 0.25e^{-0.25x} & 0 < x \\ 0 & \text{elsewhere} \end{cases}$$

Find the probability that
(a) the time to observe a particle is more than 200 microseconds;
(b) the time to observe a particle is less than 10 microseconds.

5.111 Referring to the normal scores in Exercise 5.88, construct a normal-scores plot of the suspended solids data in Exercise 2.60.

5.112 Referring to the normal scores in Exercise 5.88, construct a normal-scores plot of the velocity of light data in Exercise 2.58.

5.113 If n salespeople are employed in a door-to-door selling campaign, the gross sales volume in thousands of dollars may be regarded as a random variable having the gamma distribution with $\alpha = 100\sqrt{n}$ and $\beta = \frac{1}{2}$. If the sales costs are \$5,000 per salesperson, how many salespeople should be employed to maximize the profit?

5.114 A mechanical engineer models the bending strength of a support beam in a transmission tower as a random variable having the Weibull distribution with $\alpha = 0.02$ and $\beta = 3.0$. What is the probability that the beam can support a load of 4.5?

5.115 Let the times to breakdown for the processors of a parallel processing machine have joint density

$$f(x, y) = \begin{cases} 0.04e^{-0.2x-0.2y} & \text{for } x > 0, y > 0 \\ 0 & \text{elsewhere} \end{cases}$$

where x is the time for the first processor and y is the time for the second. Find

(a) the marginal distributions and their means;

(b) the expected value of the random variable whose values are given by $g(x, y) = x + y$.

(c) Verify, in this example, that the mean of a sum is the sum of the means.

5.15

CHECK LIST OF KEY TERMS (with page references)

SAMPLING DISTRIBUTIONS

In most of the methods we shall study in this book, it will be assumed that we are dealing with a particular kind of sample called a random sample. This attention to random samples, which we discuss in Section 6.1, is due to their permitting valid, or logical, generalizations from sample data. Then, in Sections 6.2 through 6.4, we see how certain statistics (that is, certain quantities determined from samples) can be expected to vary from sample to sample. The concept of a sampling distribution, the distribution of a statistic calculated on the basis of a random sample, is basic to all of statistical inference.

6.1 POPULATIONS AND SAMPLES

Usage of the term **population** in statistics is a carryover from the days when statistics was applied mainly to sociological and economic phenomena. Today, it is applied to sets or collections of objects, actual or conceptual, and mainly to sets of numbers, measurements, or observations. For example, if we are interested in determining the average number of television sets per household in the United States, the totality of these figures, one for each household, constitutes the population for this study. Similarly, the population from which inspectors draw a sample to determine some quality characteristic of a manufactured product may be the corresponding measurements for all units in a given lot; depending on the objectives of the inspection, it may also consist of the corresponding measurements for all units that may conceivably be manufactured.

In some cases, such as the above example concerning the number of television sets per household, the population is **finite**; in other cases, such as the determination of some characteristic of all units past, present, and future, that might conceivably be manufactured by a given process, it is convenient to think of the population as **infinite**. Similarly, we look upon the results obtained in a series of flips of a coin as a sample from the hypothetically finite population consisting of all conceivably possible flips of the coin.

Populations are often described by the distributions of their values, and it is common practice to refer to a population in terms of this distribution. (For finite populations, we are referring here to the actual distribution of its values; for infinite poulations, we are referring to the corresponding probability distribution or probability density.) For example, we may refer to a number of flips of a coin as a sample from a "binomial population" or to certain measurements as a sample from a "normal population." Hereafter, when referring to a "population $f(x)$" we shall mean a population such that its elements have a frequency distribution, a probability distribution, or a density with values given by $f(x)$.

If a population is infinite it is impossible to observe all its values, and even if it is finite it may be impractical or uneconomical to observe it in its entirety. Thus, it is usually necessary to use a **sample**, a part of a population, and infer from it results pertaining to the entire population. Clearly, such results can be useful only if the sample is in some way "representative" of the population. It would be unreasonable, for instance, to expect useful generalizations about the population of 1989 family incomes in the United States on the basis of data pertaining to home owners only. Similarly, we can hardly expect reasonable generalizations about the performance of a tire if it is tested only on smooth roads. To assure that a sample is representative of the population from which it is obtained, and to provide a framework for the application of probability theory to problems of sampling, we shall limit our discussion to **random samples**. For sampling from finite populations, they are defined as follows:

Random sample (finite population)

> ***A set of observations $x_1, x_2, \ldots, x_n$ constitutes a random sample of size n from a finite population of size N, if it is chosen so that each subset of n of the N elements of the population has the same probability of being selected.***

Note that this definition of randomness pertains essentially to the manner in which the sample values are selected. This holds also for the following definition of a random sample from an infinite population:

Random sample (infinite population)

> ***A set of observations $x_1, x_2, \ldots, x_n$ constitutes a random sample of size n from the infinite population $f(x)$ if:***
>
> **1. *Each x_i is a value of a random variable whose distribution has the values $f(x)$.***
> **2. *These n random variables are independent.***

(In recent years, it has become the custom to apply the term random sample to the random variables instead of their values.)

There are several ways of assuring the selection of a sample that is at least approximately random. When dealing with a finite population, we can serially number the elements of the population and then select a sample with the aid of a table of random digits (see discussion on page 128). For instance, if a population has $N = 500$ elements and we wish to select a random sample of size $n = 10$, we can use three arbitrarily selected columns of Table 7 to obtain 10 different three-digit numbers less than or equal to 500, which will then serve as the serial numbers of the elements to be included in the sample.

When the population size is large, the use of random numbers can become very laborious and at times practically impossible. For instance, if a sample of five cartons of canned peaches is to be chosen for inspection from among the many thousands stored in a warehouse, one can hardly expect to number all the cartons, make a selection with the use of random numbers, and then pull out the ones that were chosen. In a situation like this, one really has very little choice but to make the selection relatively haphazard, hoping that this will not seriously violate the assumption of randomness which is basic to most statistical theory.

When dealing with infinite populations, the situation is somewhat different since we cannot physically number the elements of the population; but efforts should be made to approach conditions of randomness by the use of artificial devices. For example, in selecting a sample from a production line we may be able to approximate conditions of randomness by choosing one unit each half hour; when tossing a coin we can try to flip it in such a way that neither side is intentionally favored, and so forth. The proper use of artificial or mechanical devices for selecting random samples is always preferable to human judgment, as it is extremely difficult to avoid unconscious biases when making almost any kind of selection.

Even with the careful choice of artificial devices, it is all too easy to commit gross errors in the selection of a random sample. To illustrate some of these pitfalls, suppose we have the task of selecting logs being fed into a sawmill by a constant-speed conveyer belt, for the purpose of obtaining a random sample of their lengths. One sampling device, which at first sight would seem to assure randomness, consists of measuring the logs which pass a given point at the end of a certain number of 10-minute intervals. However, further thought reveals that this method of selection favors the longer logs, since they require more time to pass the given point. Thus, the sample is not random since the longer logs have a better chance of being included. Another common mistake in selecting a sample is that of sampling from the wrong population or from a poorly specified population. As we have pointed out earlier, we would hardly get a sample from which we could generalize about family incomes in the United States if we limited our sample to home owners. Similarly, if we wanted to determine the effect of vibrations on a structural member, we should be careful to delineate the frequency band of vibrations that is of relevance, and to vibrate test specimens only at frequencies selected randomly from this band.

The purpose of most statistical investigations is to generalize from information contained in random samples about the population from which the samples were obtained. In particular, we are usually concerned with the problem of making inferences about the **parameters** of populations, such as the mean μ or the standard deviation σ. In making such inferences, we use **statistics** such as $\bar{x}$ and s, namely, quantities calculated on the basis of sample observations. In practice, the term statistic is also applied to the corresponding random variables.

Since the selection of a random sample is controlled largely by chance, so are the values we obtain for statistics, and the remainder of this chapter will be devoted to sampling distributions, namely, to distributions which describe the chance fluctuations of statistics calculated on the basis of random samples.

6.2 THE SAMPLING DISTRIBUTION OF THE MEAN (σ KNOWN)

Suppose that a random sample of n observations has been taken from some population and that $\bar{x}$ has been computed, say, to estimate the mean of the population. It should be clear that if we took a second random sample of size n from this population, it would be quite unreasonable to expect the identical value for $\bar{x}$, and if we took several more samples, probably no two of the $\bar{x}$'s would be alike. The differences among such $\bar{x}$'s are generally attributed to chance, and this raises important questions concerning their distribution, specifically concerning the extent of their chance fluctuations.

To approach this question experimentally, suppose that 50 random samples of size $n = 10$ are to be taken from a population having the **discrete uniform distribution**

$$f(x) = \begin{cases} \frac{1}{10} & \text{for } x = 0, 1, 2, \ldots, 9 \\ 0 & \text{elsewhere} \end{cases}$$

Sampling is **with replacement**, so to speak, so that we are sampling from an infinite population. A convenient way of obtaining these samples is to use a table of random digits, like Table 7, letting each sample consist of 10 consecutive digits in arbitrarily chosen rows or columns. Actually proceeding in this way, we got 50 samples whose means are

4.4	3.2	5.0	3.5	4.1	4.4	3.6	6.5	5.3	4.4
3.1	5.3	3.8	4.3	3.3	5.0	4.9	4.8	3.1	5.3
3.0	3.0	4.6	5.8	4.6	4.0	3.7	5.2	3.7	3.8
5.3	5.5	4.8	6.4	4.9	6.5	3.5	4.5	4.9	5.3
3.6	2.7	4.0	5.0	2.6	4.2	4.4	5.6	4.7	4.3

Grouping these means into a distribution with the classes 2.0–2.9, 3.0–3.9, . . . , and 6.0–6.9, we get

$\bar{x}$	*Frequency*
2.0–2.9	2
3.0–3.9	14
4.0–4.9	19
5.0–5.9	12
6.0–6.9	3
	50

and it is apparent from this distribution as well as its histogram shown in Figure 6.1 that the distribution of the means is fairly **bell-shaped**, even though the population itself has a uniform distribution. This raises the question whether our result is typical of what we might expect; that is, whether we would get similar distributions if we repeated the experiment again and again.

To answer this kind of question, we shall have to investigate the **theoretical sampling distribution** of the mean which, for the given example, provides us with the probabilities of getting means from 2.0 to 2.9, from 3.0 to 3.9, . . . , from 6.0 to 6.9, and perhaps values less than 2.0 or greater than 6.9. Although we could evaluate these probabilities for this particular example, it is usually sufficient to refer to some general theorems concerning sampling distributions. The first of these, stated

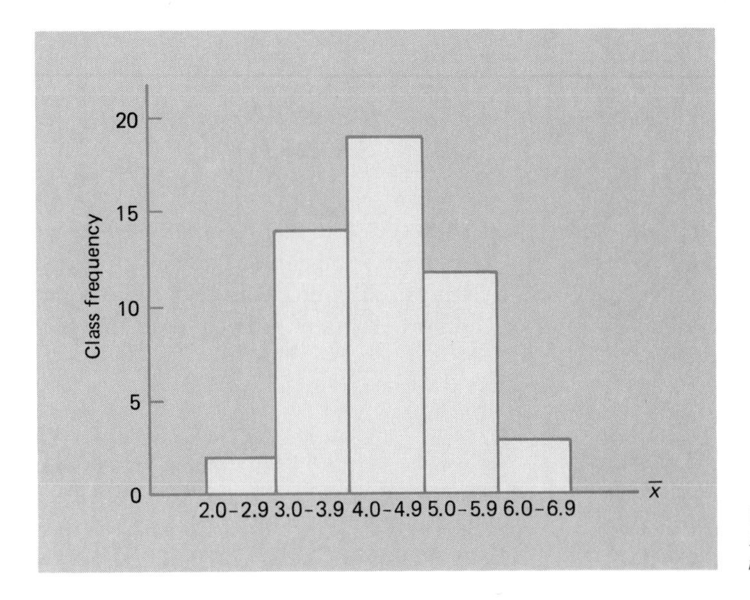

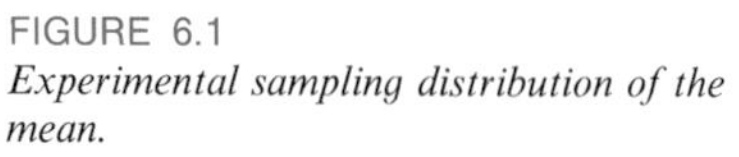
FIGURE 6.1
Experimental sampling distribution of the mean.

as follows, gives expressions for the mean $\mu_{\bar{x}}$ and the variance $\sigma_{\bar{x}}^2$ of sampling distributions of the mean:

Formulas for $\mu_{\bar{x}}$ and $\sigma_{\bar{x}}^2$

> ***Theorem 6.1*** ***If a random sample of size n is taken from a population having the mean μ and the variance σ^2, then $\bar{x}$ is a value of a random variable whose distribution has the mean μ.***
>
> ***For samples from infinite populations the variance of this distribution is $\frac{\sigma^2}{n}$;***
>
> ***For samples from a finite population of size N the variance is $\frac{\sigma^2}{n}\frac{N-n}{N-1}$.***

To prove that $\mu_x = \mu$ for the continuous case, we use the definition on page 169 and write

$$\mu_{\bar{x}} = \int_{-\infty}^{\infty}\int_{-\infty}^{\infty}\cdots\int_{-\infty}^{\infty}\sum_{i=1}^{n}\frac{x_i}{n}f(x_1, x_2, \ldots, x_n)\,dx_1\,dx_2\cdots dx_n$$

$$= \frac{1}{n}\sum_{i=1}^{n}\int_{-\infty}^{\infty}\int_{-\infty}^{\infty}\cdots\int_{-\infty}^{\infty}x_i f(x_1, x_2, \ldots, x_n)\,dx_1\,dx_2\cdots dx_n$$

where $f(x_1, x_2, \ldots, x_n)$ is a value of the joint density of the random variables whose values constitute the random sample. Using the assumption of randomness, we can write

$$f(x_1, x_2, \ldots, x_n) = f(x_1)f(x_2)\cdots f(x_n)$$

and we now have

$$\mu_{\bar{x}} = \frac{1}{n}\sum_{i=1}^{n}\int_{-\infty}^{\infty}f(x_1)\,dx_1\cdots\int_{-\infty}^{\infty}x_i f(x_i)\,dx_i\cdots\int_{-\infty}^{\infty}f(x_n)\,dx_n$$

Since each integral except the one with the integrand $x_i f(x_i)$ equals 1, and the one with the integrand $x_i f(x_i)$ equals μ, we finally obtain

$$\mu_{\bar{x}} = \frac{1}{n}\sum_{i=1}^{n}\mu = \mu$$

and this completes the proof. (For the discrete case the proof follows the same steps, with integral signs replaced by $\sum$'s.)

To prove that $\sigma_{\bar{x}}^2 = \sigma^2/n$ for the continuous case, we shall make the simplifying assumption that $\mu = 0$, which does not involve any loss of generality, as the reader will be asked to show in Exercise 6.18 on page 197. Using the definition on page 169, we thus have

$$\sigma_{\bar{x}}^2 = \int_{-\infty}^{\infty} \int_{-\infty}^{\infty} \cdots \int_{-\infty}^{\infty} \bar{x}^2 f(x_1, x_2, \ldots, x_n)\, dx_1\, dx_2 \cdots dx_n$$

and making use of the fact that

$$\bar{x}^2 = \frac{1}{n^2}\left(\sum_{i=1}^{n} x_i\right)^2 = \frac{1}{n^2}\left(\sum_{i=1}^{n} x_i^2 + \sum_{i \neq j}\sum x_i x_j\right)$$

we obtain

$$\sigma_{\bar{x}}^2 = \frac{1}{n^2}\sum_{i=1}^{n} \int_{-\infty}^{\infty} \int_{-\infty}^{\infty} \cdots \int_{-\infty}^{\infty} x_i^2 f(x_1, x_2, \ldots, x_n)\, dx_1\, dx_2 \cdots dx_n$$

$$+ \frac{1}{n^2}\sum_{i \neq j}\sum \int_{-\infty}^{\infty} \int_{-\infty}^{\infty} \cdots \int_{-\infty}^{\infty} x_i x_j f(x_1, x_2, \ldots, x_n)\, dx_1\, dx_2 \cdots dx_n$$

where $\sum_{i \neq j}\sum$ extends over all i and j from 1 to n, not including the terms where $i = j$. Again using the fact that

$$f(x_1, x_2, \ldots, x_n) = f(x_1)f(x_2) \cdots f(x_n)$$

we can write each of the preceding multiple integrals as a product of simple integrals, where each integral with integrand $f(x)$ equals 1. We thus obtain

$$\sigma_{\bar{x}}^2 = \frac{1}{n^2}\sum_{i=1}^{n} \int_{-\infty}^{\infty} x_i^2 f(x_i)\, dx_i + \frac{1}{n^2}\sum_{i \neq j}\sum \int_{-\infty}^{\infty} x_i f(x_i)\, dx_i \cdot \int_{-\infty}^{\infty} x_j f(x_j)\, dx_j$$

and since each integral in the first sum equals σ^2 while each integral in the second sum equals 0, we finally have

$$\sigma_{\bar{x}}^2 = \frac{1}{n^2}\sum_{i=1}^{n} \sigma^2 = \frac{\sigma^2}{n}$$

This completes the proof of the second part of the theorem. We shall not prove the corresponding result for random samples from finite populations, but it should be noted that in the resulting formula for $\sigma_{\bar{x}}^2$ the factor $\frac{N-n}{N-1}$, often called the **finite population correction factor**, is close to 1 (and can be omitted for most practical purposes) unless the sample constitutes a substantial portion of the population.

EXAMPLE Find the value of the finite population correction factor for $n = 10$ and $N = 1{,}000$.

Solution

$$\frac{1{,}000 - 10}{1{,}000 - 1} = 0.991$$

■

Although it should not come as a surprise that $\mu_{\bar{x}} = \mu$, the fact that $\sigma_{\bar{x}}^2 = \sigma^2/n$ for random samples from infinite populations is interesting and important. To point out its implications, let us apply Chebyshev's theorem to the sampling distribution of the mean, substituting $\bar{x}$ for x and $\sigma/\sqrt{n}$ for σ in the formula for the alternate form of the theorem (see page 110). We thus obtain

$$P\left(|\bar{x} - \mu| < \frac{k\sigma}{\sqrt{n}}\right) \geq 1 - \frac{1}{k^2}$$

and letting $k\sigma/\sqrt{n} = \varepsilon$, we get

$$P(|\bar{x} - \mu| < \varepsilon) \geq 1 - \frac{\sigma^2}{n\varepsilon^2}$$

Thus, for any given $\varepsilon > 0$, the probability that $\bar{x}$ differs from μ by less than ε can be made arbitrarily close to 1 by choosing n sufficiently large. In less rigorous language, the larger the sample size, the closer we can expect $\bar{x}$ to be to the mean of the population. In this sense we can say that the mean becomes more and more reliable as an estimate of μ as the sample size is increased.

The reliability of the mean as an estimate of μ is often measured by $\sigma_{\bar{x}}^2 = \sigma/\sqrt{n}$, also called the **standard error of the mean**. Note that this measure of the reliability of the mean decreases in proportion to the square root of n; for instance, it is necessary to quadruple the size of the sample in order to halve the standard deviation of the sampling distribution of the mean. This also indicates what might be called a "law of diminishing returns" so far as increasing the sample size is concerned. Usually it does not pay to take excessively large samples since the extra labor and expense is not accompanied by a proportional gain in reliability. For instance, if we increase the size of a sample from 25 to 2,500, the errors to which we are exposed are reduced only by a factor of 10.

Let us now return to the experimental sampling distribution on page 187, and let us check how closely its mean and variance correspond to the values we should expect in accordance with Theorem 6.1. Since the population from which the 50 samples of size $n = 10$ were obtained has the mean

$$\mu = \sum_{x=0}^{9} x \cdot \tfrac{1}{10} = 4.5$$

and the variance

$$\sigma^2 = \sum_{x=0}^{9} (x - 4.5)^2 \tfrac{1}{10} = 8.25$$

Theorem 6.1 leads us to expect a mean of $\mu_{\bar{x}} = 4.5$ and a variance of $\sigma_{\bar{x}}^2 = 8.25/10 = 0.825$. Calculating the mean and the variance from the frequency distribution on page 188, we get $\bar{x}_x = 4.45$ and $s_{\bar{x}}^2 = 0.939$, which are reasonably close to the theoretical values.

Theorem 6.1 provides only partial information about the theoretical sampling distribution of the mean. In general, it is impossible to determine such a distribution exactly without knowledge of the actual form of the population, but it is possible to find the limiting distribution as $n \to \infty$ of a random variable whose values are closely related to $\bar{x}$, assuming only that the population has a finite variance σ^2. The random variable we are referring to here is the **standardized mean**; its values are given by

$$z = \frac{\bar{x} - \mu}{\sigma/\sqrt{n}}$$

namely, by the difference between $\bar{x}$ and μ divided by the standard error of the mean. With reference to this random variable, we can now state the following theorem, called the **central limit theorem**:

Central limit theorem

***Theorem 6.2** If $\bar{x}$ is the mean of a random sample of size n taken from a population having the mean μ and the finite variance σ^2, then*

$$z = \frac{\bar{x} - \mu}{\sigma/\sqrt{n}}$$

is the value of a random variable whose distribution function approaches that of the standard normal distribution as $n \to \infty$.

The central limit theorem provides a normal distribution that allows us to assign probabilities to intervals of values for $\bar{x}$. Regardless of the form of the population distribution, the distribution of $\bar{x}$ is approximately normal with mean μ and variance σ^2/n whenever n is large. This tendency towards normality is illustrated in Figure 6.2 for a uniform population distribution and a exponential population distribution.

Although proving the central limit theorem is beyond the scope of this text, we can obtain experimental verification by constructing a normal-scores plot of the 50 sample means on page 187, which were obtained by sampling with replacement

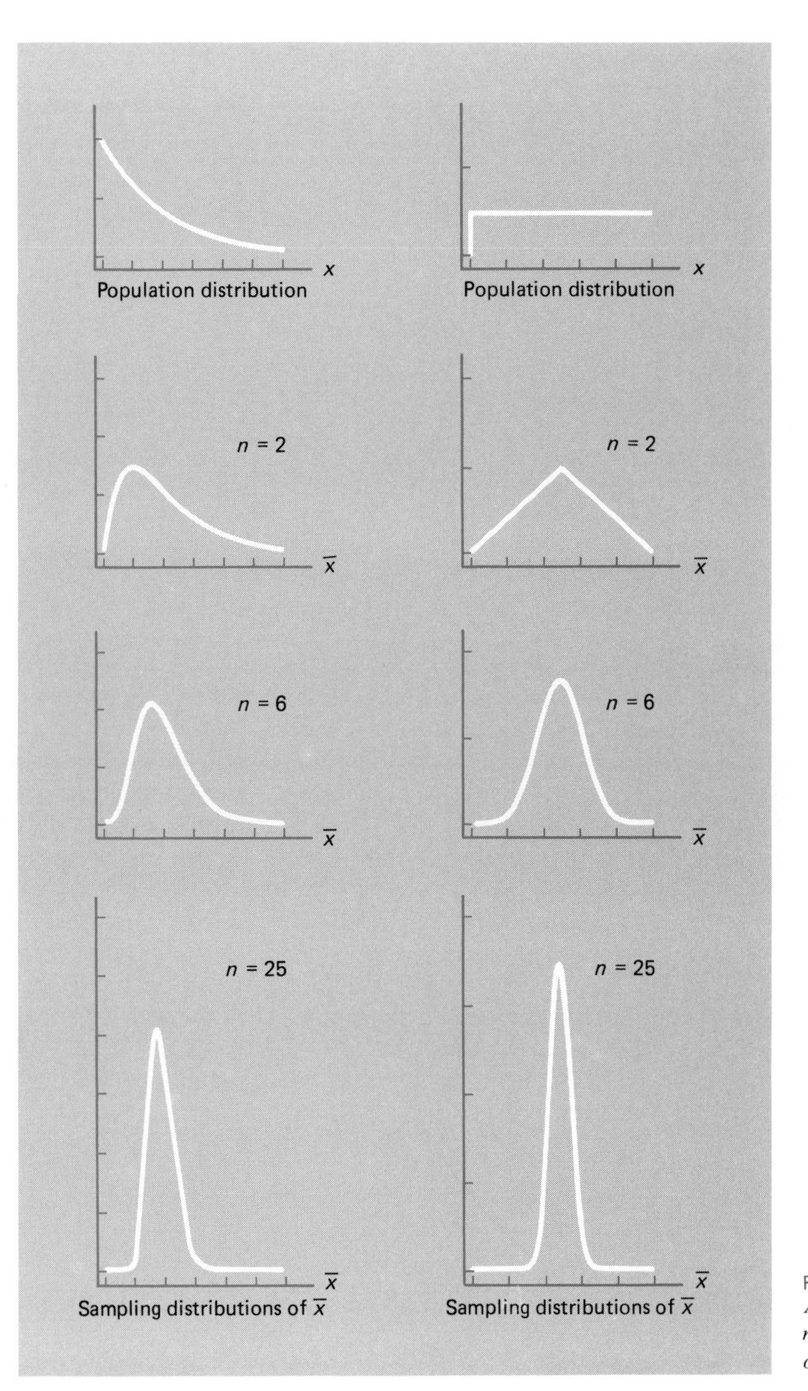

FIGURE 6.2
An illustration of the tendency toward normality for the sampling distribution of $\bar{x}$ as sample size increases.

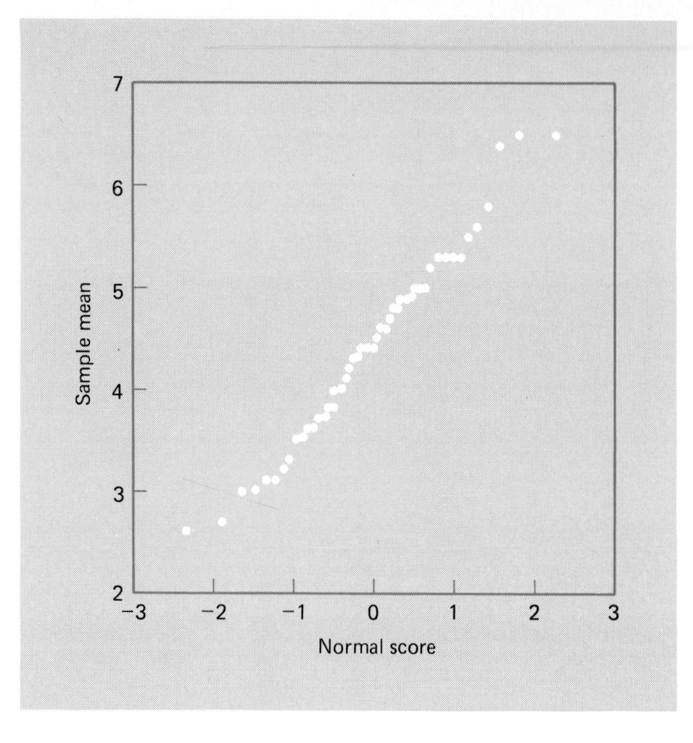

FIGURE 6.3
Experimental verification of central limit theorem.

from a discrete uniform population (Figure 6.3). As can be seen, the points fall close to a straight line, and it seems that even for $n = 10$ the sampling distribution of the mean for this example follows the overall pattern of a normal distribution. In practice, the normal distribution provides an excellent approximation to the sampling distribution of the mean for n as small as 25 or 30, with hardly any restrictions on the shape of the population. As we saw in our example, the sampling distribution of the mean has the general shape of a normal distribution even for samples of size $n = 10$ from a discrete uniform distribution. Note that for random samples from a normal population the sampling distribution of the mean is normal regardless of the size of the sample.

EXAMPLE If a 1-gallon can of a certain kind of paint covers on the average 513.3 square feet with a standard deviation of 31.5 square feet, what is the probability that the mean area covered by a sample of 40 of these 1-gallon cans will be anywhere from 510.0 to 520.0 square feet?

Solution By Theorem 6.2 we shall have to find the normal curve area between

$$z = \frac{510.0 - 513.3}{31.5/\sqrt{40}} = -0.66 \quad \text{and} \quad z = \frac{520.0 - 513.3}{31.5/\sqrt{40}} = 1.34$$

and checking these values in Table 3 we obtain a probability of 0.6553. Note that if $\bar{x}$ turned out to be much less than 513.3, say, less than 500.0, this might cause serious doubt whether the sample actually came from a population having $\mu = 513.3$ and $\sigma = 31.5$; the probability of obtaining such a small value (a z-value less than -2.67) is only 0.0038.

■

EXERCISES

6.1 An inspector examines every twentieth piece coming off an assembly line. List some of the conditions under which this method of sampling might not yield a random sample.

6.2 In 1932 the *Literary Digest* predicted the presidential election by random sampling from telephone directories and from its list of subscribers. The prediction was grossly incorrect; explain why.

6.3 Explain why the following will not lead to random samples of the desired populations.

(a) To determine what the average person spends on a vacation, a researcher interviews passengers on a luxury cruise.

(b) To determine the average income of its graduates 10 years after graduation, the alumni office of a university sent questionnaires in 1988 to all the members of the class of 1978 and bases its estimate on the questionnaires returned.

(c) To determine public sentiment about certain import restrictions, an interviewer asks voters: "Do you feel that this unfair practice should be stopped?"

6.4 A market research organization wants to try a new product in 8 of the 50 states. Use Table 7 to make this selection.

6.5 How many different samples of size $n = 2$ can be chosen from a finite population of size

(a) $N = 6$;

(b) $N = 25$?

6.6 With reference to Exercise 6.5, what is the probability of each sample in part (a) and the probability of each sample in part (b), if the samples are to be random?

6.7 Take 30 slips of paper and label five each -4 and 4, four each -3 and 3, three each -2 and 2, and two each -1, 0, and 1.

(a) If each slip of paper has the same probability of being drawn, find the probability of getting $-4, -3, -2, -1, 0, 1, 2, 3, 4$, and find the mean and the variance of this distribution.

(b) Draw 50 samples of size 10 from this population, each sample being drawn without replacement, and calculate their means.

(c) Calculate the mean and the variance of the 50 means obtained in part (b).

(d) Compare the results obtained in part (c) with the corresponding values expected according to Theorem 6.1. [Note that μ and σ^2 were obtained in part (a).]

6.8 Repeat Exercise 6.7, but select each sample with replacement; that is, replace each slip of paper and reshuffle before the next one is drawn.

6.9 Given the infinite population whose distribution is given by

x	$f(x)$
1	0.25
2	0.25
3	0.25
4	0.25

list the 16 possible samples of size 2 and use this list to construct the distribution of $\bar{x}$ for random samples of size 2 from the given population. Verify that the mean and the variance of this sampling distribution are identical with the corresponding values expected according to Theorem 6.1.

6.10 Suppose that we convert the 50 samples referred to on page 187 into 25 samples of size $n = 20$ by combining the first two, the next two, and so on. Find the means of these samples and calculate their mean and their standard deviation. Compare this mean and this standard deviation with the corresponding values expected in accordance with Theorem 6.1.

6.11 When we sample from an infinite population, what happens to the standard error of the mean if the sample size is
(a) increased from 50 to 200; (b) increased from 400 to 900;
(c) decreased from 225 to 25; (d) decreased from 640 to 40?

6.12 What is the value of the finite population correction factor in the formula for $\sigma_{\bar{x}}^2$ when
(a) $n = 5$ and $N = 200$; (b) $n = 10$ and $N = 400$;
(c) $n = 100$ and $N = 5{,}000$?

6.13 For large sample size n, verify that there is a fifty-fifty chance that the mean of a random sample from an infinite population with the standard deviation σ will differ from μ by less than $0.6745 \cdot \frac{\sigma}{\sqrt{n}}$. It has been the custom to refer to this quantity as the **probable error of the mean.**

6.14 The mean of a random sample of size $n = 25$ is used to estimate the mean of an infinite population with the standard deviation $\sigma = 2.4$. What can we assert about the probability that the error will be less than 1.2, if we use
(a) Chebyshev's theorem; (b) the central limit theorem?

6.15 A random sample of size 100 is taken from an infinite population having the mean $\mu = 76$ and the variance $\sigma^2 = 256$. What is the probability of getting an $\bar{x}$ between 75 and 78?

6.16 A wire-bonding process is said to be in control if the mean pull strength is 10 pounds. It is known that the pull-strength measurements are normally distributed with a standard deviation of 1.5 pounds. Periodic random samples of size 4 are taken from this process and the process is said to be "out of control" if a sample mean is less than 7.75 pounds. Comment.

6.17 If the distribution of the weights of all men traveling by air between Dallas and El Paso has a mean of 163 pounds and a standard deviation of 18 pounds, what is the probability that the combined gross weight of 36 men traveling on a plane between these two cities is more than 6,000 pounds?

6.18 If x is a value of a continuous random variable and $y = ax + b$, show that

(a) $\mu_y = a\mu_x + b$; (b) $\sigma_y^2 = a^2\sigma_x^2$.

6.19 Prove that $\mu_{\bar{x}} = \mu$ for random samples from discrete (finite or countably infinite) populations.

6.3
THE SAMPLING DISTRIBUTION OF THE MEAN (σ UNKNOWN)

Application of the theory of the preceding section requires knowledge of the population standard deviation σ. If n is large, this does not pose any problems even when σ is unknown, as it is reasonable in that case to substitute for it the sample standard deviation s. However, when it comes to the random variable whose values are given by $\frac{\bar{x} - \mu}{s/\sqrt{n}}$, very little is known about its exact sampling distribution for small values of n unless we make the assumption that the sample comes from a normal population. Under this assumption, one can prove the following.

Value of random variable having the t distribution

***Theorem 6.3** If $\bar{x}$ is the mean of a random sample of size n taken from a normal population having the mean μ and the variance σ^2, then*

$$t = \frac{\bar{x} - \mu}{s/\sqrt{n}}$$

is the value of a random variable having the t distribution with the parameter $\nu = n - 1$.

This theorem is more general than Theorem 6.2 in the sense that it does not require knowledge of σ; on the other hand, it is less general than Theorem 6.2 in the sense that it requires the assumption of a normal population.

As can be seen from Figure 6.4, the overall shape of a ***t* distribution** is similar to that of a normal distribution—both are bell-shaped and symmetrical about the mean. Like the standard normal distribution, the t distribution has the mean 0, but its variance depends on the parameter ν (*nu*), called the number of **degrees of freedom**. The variance of the t distribution exceeds 1, but it approaches 1 as $n \to \infty$. In fact, it can be shown that the t distribution with ν degrees of freedom approaches the standard normal distribution as $\nu \to \infty$.

Table 4 at the end of the book contains selected values of t_α for various values of ν, where t_α is such that the area under the t distribution to its right is equal to α. In this table the left-hand column contains values of ν, the column headings are areas α in the right-hand tail of the t distribution, and the entries are values of t_α (see

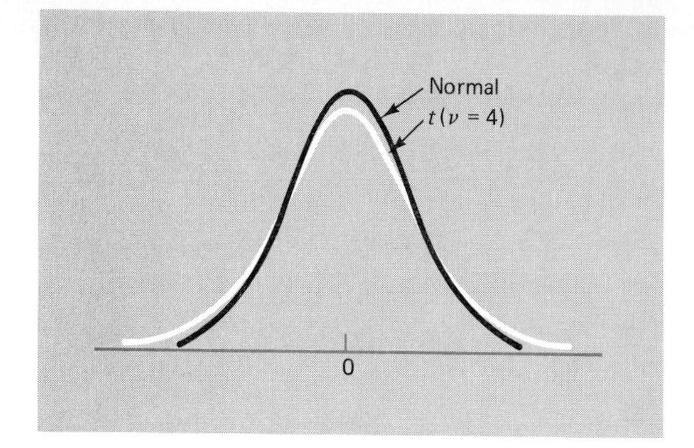

FIGURE 6.4
t distribution and standard normal distribution.

also Figure 6.5). It is not necessary to tabulate values of t_α for $\alpha > 0.50$, as it follows from the symmetry of the t distribution that $t_{1-\alpha} = -t_\alpha$; thus, the value of t that corresponds to a left-hand tail area of α is $-t_\alpha$.

Note that in the bottom row of Table 4 the entries correspond to the values of z that cut off right-hand tails of area α under the standard normal curve. Using the notation z_α for such a value of z, it can be seen, for example, that $z_{0.025} = 1.96 = t_{0.025}$ for $\nu = \infty$. This result should really have been expected, since the t distribution approaches the standard normal distribution as $\nu \to \infty$. In fact, observing that the values of t_α for 29 or more degrees of freedom are close to the corresponding values of z_α, we conclude that **the standard normal distribution provides a good approximation to the *t* distribution for samples of size 30 or more.**

EXAMPLE A manufacturer of fuses claims that with a 20% overload, the fuses will blow in 12.40 minutes on the average. To test this claim, a sample of 20 of the fuses was subjected to a 20% overload, and the times it took them to blow had a mean of

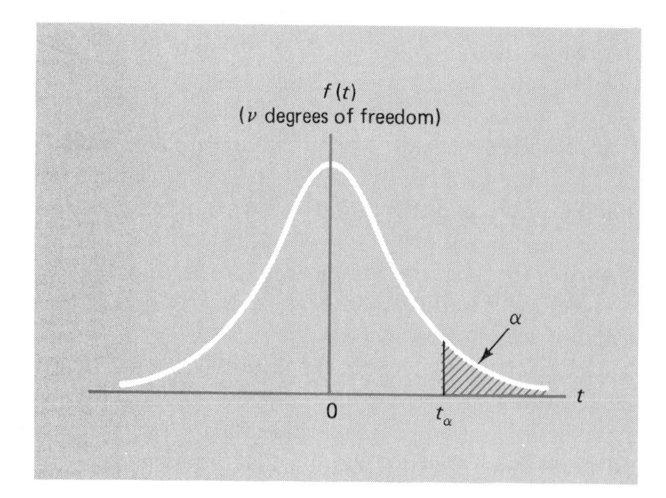

FIGURE 6.5
Tabulated values of t.

10.63 minutes and a standard deviation of 2.48 minutes. If it can be assumed that the data constitute a random sample from a normal population, do they tend to support or refute the manufacturer's claim?

Solution First we calculate

$$t = \frac{10.63 - 12.40}{2.48/\sqrt{20}} = -3.19$$

which is a value of a random variable having the t distribution with $\nu = 20 - 1 = 19$ degrees of freedom. Now, from Table 4 we find that for $\nu = 19$ the probability that t will exceed 2.861 is 0.005, and hence that the probability that t will be less than -2.861 is also 0.005. Since $t = -3.19$ is less than -2.861 and 0.005 is a very small probability, we conclude that the data tend to refute the manufacturer's claim. In all probability, the mean blowing time of his fuses with a 20% overload is less than 12.40 minutes.

■

The assumption that the sample must come from a normal population is not so severe a restriction as it may seem. Studies have shown that the distribution of random variables with the values $t = \frac{\bar{x} - \mu}{s/\sqrt{n}}$ is fairly close to a t distribution even for samples from certain nonnormal populations. In practice, it is necessary to make sure primarily that the population from which we are sampling is approximately bell-shaped and not too skewed. A practical way of checking this assumption is to construct a normal-scores plot, as described on page 172. (If such a plot shows a distinct curve rather than a straight line, it may be possible to "straighten it out" by transforming the data—say, by taking their logarithms or their square roots, as discussed in Section 5.12.)

6.4
THE SAMPLING DISTRIBUTION OF THE VARIANCE

So far we have discussed only the sampling distribution of the mean, but if we had taken the medians or the standard deviations of the 50 samples mentioned in the example on page 187, we would similarly have obtained experimental sampling distributions of these statistics. In this section we shall be concerned with the theoretical sampling distribution of the sample variance for random samples from normal populations. Since s^2 cannot be negative, we should suspect that this sampling distribution is not a normal curve; in fact, it is related to the gamma distribution (see page 155) with $\alpha = v/2$ and $\beta = 2$, called the **chi-square distribution**. Specifically, we have the following theorem.

Value of random variable having the chi-square distribution

> ***Theorem 6.4*** *If s^2 is the variance of a random sample of size n taken from a normal population having the variance σ^2, then*
>
> $$\chi^2 = \frac{(n-1)s^2}{\sigma^2}$$
>
> *is a value of a random variable having the chi-square distribution with the parameter $v = n - 1$.*

Table 5 at the end of the book contains selected values of χ^2_α for various values of v, again called the **number of degrees of freedom**, where χ^2_α is such that the area under the chi-square distribution to its right is equal to α. In this table the left-hand-column contains values of v, the column headings are areas α in the right-hand tail of the chi-square distribution, and the entries are values of χ^2_α (see also Figure 6.6). Unlike the t distribution, it is necessary to tabulate values of χ^2_α for $\alpha > 0.50$, because the chi-square distribution is not symmetrical.

EXAMPLE An optical firm purchases glass to be ground into lenses, and it is known from past experience that the variance of the refractive index of this kind of glass is $1.26 \cdot 10^{-4}$. As it is important that the various pieces of glass have nearly the same index of refraction, the firm rejects such a shipment if the sample variance of 20 pieces selected at random exceeds $2.00 \cdot 10^{-4}$. Assuming that the sample values may be looked upon as a random sample from a normal population, what is the probability that a shipment will be rejected even though $\sigma^2 = 1.26 \cdot 10^{-4}$?

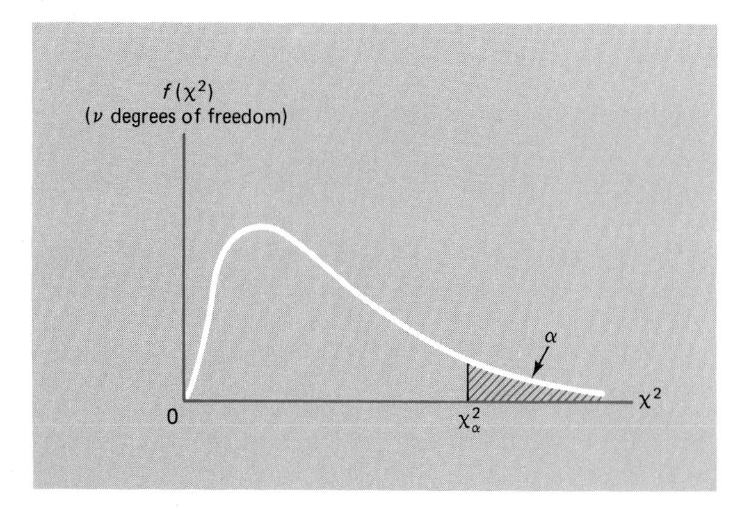

FIGURE 6.6
Tabulated values of chi-square.

Solution Substituting into the formula for the chi-square statistic, we get

$$\chi^2 = \frac{19(2.00 \cdot 10^{-4})}{1.26 \cdot 10^{-4}} = 30.2$$

and then we find from Table 5 that for 19 degrees of freedom $\chi^2_{0.05} = 30.1$. Thus, the probability that a good shipment will erroneously be rejected is less than 0.05.

■

A problem closely related to that of finding the distribution of the sample variance is that of finding the distribution of the ratio of the variances of two independent random samples. This problem is important because it arises in tests in which we want to determine whether two samples come from populations having equal variances. If they do, the two sample variances should be nearly the same; that is, their ratio should be close to 1. To determine whether the ratio of two sample variances is too small or too large, we use the theory given in the following theorem.

Value of random variable having the F distribution

> ***Theorem 6.5*** ***If s_1^2 and s_2^2 are the variances of independent random samples of size n_1 and n_2, respectively, taken from two normal populations having the same variance, then***
>
> $$F = \frac{s_1^2}{s_2^2}$$
>
> ***is a value of a random variable having the F distribution with the parameters $\nu_1 = n_1 - 1$ and $\nu_2 = n_2 - 1$.***

The ***F*** **distribution** is related to the beta distribution (page 158), and its two parameters, ν_1 and ν_2, are called the **numerator** and **denominator degrees of freedom**. As it would require too large a table to give values of F_α corresponding to many different right-hand tail probabilities α, and since $\alpha = 0.05$ and $\alpha = 0.01$ are most commonly used in practice, Table 6 contains only values of $F_{0.05}$ and $F_{0.01}$ for various combinations of values of ν_1 and ν_2 (see also Figure 6.7).

EXAMPLE If two independent random samples of size $n_1 = 7$ and $n_2 = 13$ are taken from a normal population, what is the probability that the variance of the first sample will be at least three times as large as that of the second sample?

Solution From Table 6 we find that $F_{0.05} = 3.00$ for $\nu_1 = 7 - 1 = 6$ and $\nu_2 = 13 - 1 = 12$; thus, the desired probability is 0.05.

■

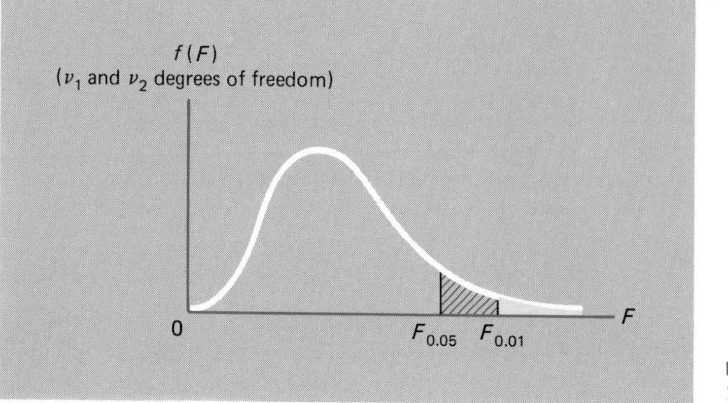

FIGURE 6.7
Tabulated values of F.

It is possible to use Table 6 also to find values of F corresponding to left-hand tail probabilities of 0.05 or 0.01. Writing $F_\alpha(\nu_1, \nu_2)$ for F_α with ν_1 and ν_2 degrees of freedom, we simply use the identity

$$F_{1-\alpha}(\nu_1, \nu_2) = \frac{1}{F_\alpha(\nu_2, \nu_1)}$$

EXAMPLE Find the value of $F_{0.95}$ (corresponding to a left-hand tail probability of 0.05) for $\nu_1 = 10$ and $\nu_2 = 20$ degrees of freedom.

Solution Making use of the identity and Table 6, we get

$$F_{0.95}(10, 20) = \frac{1}{F_{0.05}(20, 10)} = \frac{1}{2.77} = 0.36$$

■

Note that Theorems 6.4 and 6.5 require the assumption that we are sampling from normal populations. Unlike the situation with the t distribution, deviations from an underlying normal distribution, such as a long tail, may have a serious effect on these sampling distributions. Consequently, it is best to transform to near normality using the approach in Section 5.12 before invoking the sampling distributions in this section.

EXERCISES

6.20 A random sample of size 25 from a normal population has the mean $\bar{x} = 47.5$ and the standard deviation $s = 8.4$. Does this information tend to support or refute the claim that the mean of the population is $\mu = 42.1$?

6.21 The following are the times between six calls for an ambulance (in a certain city) and the patient's arrival at the hospital: 27, 15, 20, 32, 18, and 26 minutes. Use these figures to judge the reasonableness of the ambulance service's claim that it takes on the average 20 minutes between the call for an ambulance and the patient's arrival at the hospital.

6.22 A process for making certain bearings is under control if the diameters of the bearings have a mean of 0.5000 cm. What can we say about this process if a sample of 10 of these bearings has a mean diameter of 0.5060 cm and a standard deviation of 0.0040 cm?

6.23 The claim that the variance of a normal population is $\sigma^2 = 21.3$ is rejected if the variance of a random sample of size 15 exceeds 39.74. What is the probability that the claim will be rejected even though $\sigma^2 = 21.3$?

6.24 A random sample of 10 observations is taken from a normal population having the variance $\sigma^2 = 42.5$. Find the approximate probability of obtaining a sample standard deviation between 3.14 and 8.94.

6.25 If independent random samples of size $n_1 = n_2 = 8$ come from normal populations having the same variance, what is the probability that either sample variance will be at least seven times as large as the other?

6.26 Find the values of

(a) $F_{0.95}$ for 12 and 15 degrees of freedom;
(b) $F_{0.99}$ for 6 and 20 degrees of freedom.

6.27 The chi-square distribution with 4 degrees of freedom is given by

$$f(x) = \begin{cases} \frac{1}{4} \cdot x \cdot e^{-x/2} & x > 0 \\ 0 & x \leq 0 \end{cases}$$

Find the probability that the variance of a random sample of size 5 from a normal population with $\sigma = 12$ will exceed 180.

6.28 The t distribution with 1 degree of freedom is given by

$$f(t) = \frac{1}{\pi}(1 + t^2)^{-1} \qquad -\infty < t < \infty$$

Verify the value given for $t_{0.05}$ for $v = 1$ in Table 4.

6.29 The F distribution with 4 and 4 degrees of freedom is given by

$$f(F) = \begin{cases} 6F(1 + F)^{-4} & F > 0 \\ 0 & F \leq 0 \end{cases}$$

If random samples of size 5 are taken from two normal populations having the same variance, find the probability that the ratio of the larger to the smaller sample variance will exceed 3.

6.5 REVIEW EXERCISES

6.30 The panel for a national science fair wishes to select 10 states to send a student representative chosen at random from the students participating in the state science fair.

(a) Use Table 7 to select the 10 states.
(b) Does the total selection process give each student that participates in a state science fair an equal chance of being selected to be a representative at the national science fair?

6.31 How many different samples of size $n = 2$ can be chosen from a finite population of size

(a) $N = 8$; (b) $N = 20$?

6.32 With reference to Exercise 6.31, what is the probability of each sample in part (a) and the probability of each sample in part (b), if the samples are to be random?

6.33 Referring to Exercise 6.31, find the value of the finite population correction factor in the formula for $\sigma_{\bar{x}}^2$ for part (a) and part (b).

6.34 A random sample of size $n = 36$ is taken from an infinite population with the mean $\mu = 63$ and the variance $\sigma^2 = 81$. What can we assert about the probability of getting a sample mean greater than 66.75, if we use

(a) Chebyshev's theorem; (b) the central limit theorem?

6.35 If measurements of the specific gravity of a metal can be looked upon as a sample from a normal population having a standard deviation of 0.04, what is the probability that the mean of a random sample of size 25 will be "off" by at most 0.02?

6.36 If two independent random samples of size $n_1 = 9$ and $n_2 = 16$ are taken from a normal population, what is the probability that the variance of the first sample will be at least four times as large as the variance of the second sample?

6.37 When we sample from an infinite population, what happens to the standard error of the mean if the sample size is

(a) increased from 100 to 200; (b) increased from 200 to 300;
(c) decreased from 360 to 90?

6.38 Explain why the following may not lead to random samples from the desired populations:

(a) To determine the smoothness of shafts, a manufacturer measures the roughness of the first piece made each morning.
(b) To determine the mix of cars, trucks, and buses in the rush hour, an engineer records the type of vehicle passing a fixed point at 1-minute intervals.

6.6
CHECK LIST OF KEY TERMS (with page references)

INFERENCES CONCERNING MEANS

In the beginning of Chapter 6, we stated that the purpose of most statistical investigations is to generalize from information contained in random samples about the populations from which the samples were obtained. In the classical approach the methods of statistical inference are divided into two major areas—estimation and tests of hypotheses. In Sections 7.1 through 7.3 we shall present some theory and some methods which pertain to the estimation of means. Sections 7.4 and 7.5 deal with the basic concepts of hypothesis testing, and Sections 7.6 through 7.7 and 7.9 deal with tests of hypotheses concerning means. Randomization, a procedure to help meet the assumptions, is presented in Section 7.10.

7.1 POINT ESTIMATION

Basically, **point estimation** concerns the choosing of a statistic, that is, a single number calculated from sample data (and perhaps other information) for which we have some expectation, or assurance, that it is "reasonably close" to the parameter it is supposed to estimate. To explain what we mean here by reasonably close is not an easy task; first, the value of the parameter is unknown, and second, the value of the statistic is unknown until after the sample has been obtained. Thus, we can only ask whether, upon repeated sampling, the distribution of the statistic has certain desirable properties akin to "closeness." For instance, we know from Theorem 6.1 that the sampling distribution of the mean has the same mean as the population from which the sample is obtained; hence, we can expect that the means of repeated random samples from a given population will center on the mean of this population and not about some other value. To formulate this property more generally, let us now make the following definition:

Unbiased estimate

A statistic $\hat{\theta}$ is said to be an unbiased estimate, or the value of an unbiased estimator, if and only if the mean of the sampling distribution of the estimator equals θ.

Thus, we call a statistic unbiased if "on the average" its values will equal the parameter it is supposed to estimate. Note that we have distinguished here between an estimator, a random variable, and an estimate, which is one of its values; also, it is customary to apply the term "statistic" to both, estimates and estimators (see page 187).

Generally speaking, the property of unbiasedness is one of the more desirable properties in point estimation, although it is by no means essential and it is sometimes outweighed by other factors. One shortcoming of the criterion of unbiasedness is that it will generally not provide a unique statistic for a given problem of estimation. For instance, it can be shown that for a random sample of size $n = 2$ with the values x_1 and x_2, the mean $\frac{x_1 + x_2}{2}$ as well as the weighted mean $\frac{ax_1 + bx_2}{a + b}$, where a and b are positive constants, are unbiased estimates of the mean of the population. If we assume, furthermore, that the population is symmetric, so are the median and the midrange (the mean of the largest value and the smallest) for random samples of any size.

This suggests that we must seek a further criterion for deciding which of several unbiased estimators is "best" for estimating a given parameter. Such a criterion becomes evident when we compare the sampling distributions of the median and the mean for random samples of size n from the same normal population. Although these two sampling distributions have the same mean, the population mean μ, and although they are both symmetrical and bell-shaped, their variances differ. From Theorem 6.1, the variance of the sampling distribution of the mean for random samples from infinite populations is $\frac{\sigma^2}{n}$, and it can be shown that for random samples of the same size from normal populations the variance of the sampling distribution of the median is approximately $1.5708 \cdot \frac{\sigma^2}{n}$. Thus, it is more likely that the mean will be closer to μ than the median. Despite this long-run average property, given a particular sample we have no way of knowing which of the two is closest.

We formalize this important comparison of sampling distributions of statistics on the basis of their variances.

More efficient unbiased estimator

> ***A statistic $\hat{\theta}_1$ is said to be a more efficient unbiased estimate of the parameter θ than the statistic $\hat{\theta}_2$ if***
>
> 1. ***$\hat{\theta}_1$ and $\hat{\theta}_2$ are both unbiased estimates of θ;***
> 2. ***the variance of the sampling distribution of the first estimator is less than that of the second.***

We have thus seen that for random samples from normal populations the mean is more efficient than the median as an estimate of μ; in fact, it can be shown that in most practical situations where we estimate a population mean μ, the variance of the sampling distribution of no other statistic is less than that of the sampling distribution of the mean. In other words, in most practical situations the sample mean is an acceptable statistic for estimating a population mean μ. (There exist several other criteria for assessing the "goodness" of methods of point estimation, but we shall not discuss them in this book.)

When we use a sample mean to estimate the mean of a population, we know that although we are using a method of estimation which has certain desirable properties, the chances are slim, virtually nonexistent, that the estimate will actually equal μ. Hence, it would seem desirable to accompany such a point estimate of μ with some statement as to how close we might reasonably expect the estimate to be. The error, $\bar{x} - \mu$, is the difference between the estimate and the quantity it is supposed to estimate. To examine this error, let us make use of the fact that for large n

$$\frac{\bar{x} - \mu}{\sigma/\sqrt{n}}$$

is a value of a random variable having approximately the standard normal distribution. Consequently, we can assert with probability $1 - \alpha$ that the inequality

$$-z_{\alpha/2} \leq \frac{\bar{x} - \mu}{\sigma/\sqrt{n}} \leq z_{\alpha/2}$$

will be satisfied or that

$$\frac{|\bar{x} - \mu|}{\sigma/\sqrt{n}} \leq z_{\alpha/2}$$

where $z_{\alpha/2}$ is such that the normal curve area to its right equals $\alpha/2$. If we now let E stand for the maximum value of $|\bar{x} - \mu|$, the maximum error of estimate, we have

Maximum error of estimate

$$E = z_{\alpha/2} \cdot \frac{\sigma}{\sqrt{n}}$$

with probability $1 - \alpha$. In other words, if we intend to estimate μ with the mean of a large ($n \geq 30$) random sample, we can assert with probability $1 - \alpha$ that the error, $|\bar{x} - \mu|$, will be at most $z_{\alpha/2} \cdot \frac{\sigma}{\sqrt{n}}$. The most widely used values for $1 - \alpha$ are 0.95 and 0.99, and the corresponding values of $z_{\alpha/2}$ are $z_{0.025} = 1.96$ and $z_{0.005} = 2.575$ (see Exercise 5.23 on page 149).

EXAMPLE An industrial engineer intends to use the mean of a random sample of size $n = 150$ to estimate the average mechanical aptitude (as measured by a certain test) of assembly line workers in a large industry. If, on the basis of experience, the engineer can assume that $\sigma = 6.2$ for such data, what can he assert with probability 0.99 about the maximum size of his error?

Solution Substituting $n = 150$, $\sigma = 6.2$, and $z_{0.005} = 2.575$ into the preceding formula for E, we get

$$E = 2.575 \cdot \frac{6.2}{\sqrt{150}} = 1.30$$

Thus, the engineer can assert with probability 0.99 that his error will be at most 1.30.

■

Suppose now that the engineer of this example collects his data and gets $\bar{x} = 69.5$. Can he still assert with probability 0.99 that the error is at most 1.30? After all, $\bar{x} = 69.5$ differs from the true average by at most 1.30 or it does not, and he does not know which. Well, he can, but it must be understood that the 0.99 probability applies to the method he used to determine the maximum error (getting the sample data and using the formula for E) and not directly to the parameter he is trying to estimate. To make this distinction, it has become the custom to use the word **confidence** here instead of *probability*. **In general, we make probability statements about future values of random variables (say, the potential error of an estimate) and confidence statements once the data have been obtained**. Accordingly, we would say in our example that the engineer can be 99% confident that the error of his estimate, $\bar{x} = 69.5$, is at most 1.30.

The formula for E (see above) can also be used to determine the sample size that is needed to attain a desired degree of precision. Suppose that we want to use the mean of a large random sample to estimate the mean of a population, and we want to be able to assert with probability $1 - \alpha$ that the error will be at most some prescribed quantity E [or assert later with $(1 - \alpha)100\%$ confidence that the error is at most E]. As before, we write

$$E = z_{\alpha/2} \cdot \frac{\sigma}{\sqrt{n}}$$

and upon solving this equation for n we get

Sample size

$$\mathbf{n = \left[\frac{z_{\alpha/2} \cdot \sigma}{E}\right]^2}$$

To be able to use this formula we must know $1 - \alpha$, E, and σ, and for the latter we often substitute an estimate based on prior data of a similar kind (or, if necessary, a good guess).

EXAMPLE A research worker wants to determine the average time it takes a mechanic to rotate the tires of a car, and she wants to be able to assert with 95% confidence that the mean of her sample is off by at most 0.50 minute. If she can presume from past experience that $\sigma = 1.6$ minutes, how large a sample will she have to take?

Solution Substituting $E = 0.50$, $\sigma = 1.6$, and $z_{0.025} = 1.96$ into the formula for n, we get

$$n = \left[\frac{1.96 \cdot 1.6}{0.50}\right]^2 = 39.3$$

or 40 rounded up to the nearest integer. Thus, the research worker will have to time 40 mechanics performing the task of rotating the tires of a car.

■

The methods discussed so far in this section require that σ be known or that it can be approximated with the sample standard deviation s, thus requiring that n be large. However, if it is reasonable to assume that we are sampling from a normal population, we can base our argument on Theorem 6.3 instead of Theorem 6.2, namely, on the fact that

$$t = \frac{\bar{x} - \mu}{s/\sqrt{n}}$$

is a value of a random variable having the t distribution with $n - 1$ degrees of freedom. Duplicating the steps on page 207, we thus arrive at the result that we can assert with probability $1 - \alpha$ that the error we will make in using $\bar{x}$ to estimate μ will be at most

Maximum error of estimate

$$E = t_{\alpha/2} \cdot \frac{s}{\sqrt{n}}$$

or with $(1 - \alpha)100\%$ confidence that the error is less than this quantity.

EXAMPLE In six determinations of the melting point of tin, a chemist obtained a mean of 232.26 degrees Celsius with a standard deviation of 0.14 degree. If he uses this mean as the actual melting point of tin, what can the chemist assert with 98% confidence about the maximum error?

Solution Substituting $n = 6$, $s = 0.14$, and $t_{0.01} = 3.365$ (for $n - 1 = 5$ degrees of freedom) into the formula for E, we get

$$E = 3.365 \cdot \frac{0.14}{\sqrt{6}} = 0.19$$

Thus, the chemist can assert with 98% confidence that his figure for the melting point of tin is off by at most 0.19 degree.

■

7.2 INTERVAL ESTIMATION

Since point estimates cannot really be expected to coincide with the quantities they are intended to estimate, it is sometimes preferable to replace them with **interval estimates**, that is, with intervals for which we can assert with a reasonable degree of certainty that they will contain the parameter under consideration. To illustrate the construction of such an interval, suppose that we have a large ($n \geq 30$) random sample from a population with the unknown mean μ and the known variance σ^2. Referring to the inequality

$$-z_{\alpha/2} < \frac{\bar{x} - \mu}{\sigma/\sqrt{n}} < z_{\alpha/2}$$

which, as shown on page 207, will be satisfied with probability $1 - \alpha$, we now apply simple algebra and rewrite it as

Large-sample confidence interval for μ, σ known

$$\bar{x} - z_{\alpha/2} \cdot \frac{\sigma}{\sqrt{n}} < \mu < \bar{x} + z_{\alpha/2} \cdot \frac{\sigma}{\sqrt{n}}$$

Thus, when a sample has been obtained and the value of $\bar{x}$ has been calculated, we can claim with $(1 - \alpha)100\%$ confidence that the interval from $\bar{x} - z_{\alpha/2} \cdot \frac{\sigma}{\sqrt{n}}$ to $\bar{x} + z_{\alpha/2} \cdot \frac{\sigma}{\sqrt{n}}$ contains μ. It is customary to refer to an interval of this kind as a **confidence interval** for μ having the **degree of confidence** $1 - \alpha$ or $(1 - \alpha)100\%$ and to its endpoints as the **confidence limits**.

EXAMPLE A random sample of size $n = 100$ is taken from a population with $\sigma = 5.1$. Given that the sample mean is $\bar{x} = 21.6$, construct a 95% confidence interval for the population mean μ.

Solution Substituting the given values of n, $\bar{x}$, σ, and $z_{0.025} = 1.96$ into the confidence-interval formula, we get

$$21.6 - 1.96 \cdot \frac{5.1}{\sqrt{100}} < \mu < 21.6 + 1.96 \cdot \frac{5.1}{\sqrt{100}}$$

or $20.6 < \mu < 22.6$. Of course, the interval from 20.6 to 22.6 contains the population mean μ, or it does not, but we are 95% confident that it does. As was explained on page 208, this means that the method by which the interval was obtained "works" 95% of the time. In other words, in repeated applications of the confidence-interval formula, 95% of the intervals can be expected to contain the means of the respective populations.

■

The preceding confidence-interval formula is exact only for random samples from normal populations, but for large samples it will generally provide good approximations. Since σ is unknown in many practical applications, we may have to make the further approximation of substituting for σ the sample standard deviation s.

EXAMPLE With reference to the sulfur oxides emission data on page 8, for which we had $n = 80$, $\bar{x} = 18.85$, and $s^2 = 30.77$ (hence, $s = 5.55$), construct a 99% confidence interval for the plant's true average daily emission of sulfur oxides.

Solution Substituting into the confidence-interval formula with $s = 5.55$ in place of σ, we get

$$18.85 - 2.575 \cdot \frac{5.55}{\sqrt{80}} < \mu < 18.85 + 2.575 \cdot \frac{5.55}{\sqrt{80}}$$

or $17.25 < \mu < 20.45$. We are 99% confident that the interval from 17.25 tons to 20.45 tons contains the true average daily emission.

■

For small samples ($n < 30$), we proceed as on page 210, provided it is reasonable to assume that we are sampling from a normal population. Thus, with $t_{\alpha/2}$ defined as on page 197, we get the $(1 - \alpha)100\%$ confidence-interval formula

Small-sample confidence interval for μ

$$\bar{x} - t_{\alpha/2} \cdot \frac{s}{\sqrt{n}} < \mu < \bar{x} + t_{\alpha/2} \cdot \frac{s}{\sqrt{n}}$$

This formula applies to samples from normal populations, but in accordance with the discussion on page 199, it may be used as long as the sample does not exhibit any pronounced departures from normality.

EXAMPLE The mean weight loss of $n = 16$ grinding balls after a certain length of time in mill slurry is 3.42 grams with a standard deviation of 0.68 gram. Construct a 99% confidence interval for the true mean weight loss of such grinding balls under the stated conditions.

Solution Substituting $n = 16$, $\bar{x} = 3.42$, $s = 0.68$, and $t_{0.005} = 2.947$ for $n - 1 = 15$ degrees of freedom into the small-sample confidence-interval formula for μ, we get

$$3.42 - 2.947 \cdot \frac{0.68}{\sqrt{16}} < \mu < 3.42 + 2.947 \cdot \frac{0.68}{\sqrt{16}}$$

or $2.92 < \mu < 3.92$. We are 99% confident that the interval from 2.92 grams to 3.92 grams contains the mean weight loss.

■

Because confidence intervals are an important way of making inferences, we review their interpretation in the context of 95% confidence intervals for μ.

Before the observations are made, $\bar{x}$ and s are random variables, so we have the following conclusions:

1. The interval from $\bar{x} - t_{0.025}\dfrac{s}{\sqrt{n}}$ to $\bar{x} + t_{0.025}\dfrac{s}{\sqrt{n}}$ is a random interval. It is centered at $\bar{x}$ and its length is proportional to s.

2. The interval from $\bar{x} - t_{0.025}\frac{s}{\sqrt{n}}$ to $\bar{x} + t_{0.025}\frac{s}{\sqrt{n}}$ will cover the true (fixed) μ with probability 0.95.

Once the observations are made and we have numerical values for $\bar{x}$ and s.

3. The calculated interval from $\bar{x} - t_{0.025}\frac{s}{\sqrt{n}}$ to $\bar{x} + t_{0.025}\frac{s}{\sqrt{n}}$ is fixed. It is no longer possible to talk about the probability of covering μ. The interval either covers μ or it does not. Further, in any particular application, we have no way of knowing if μ is covered or not.

However, because 0.95 is the probability that we cover μ in each application, the long run relative frequency interpretation of probability promises that

$$\frac{\text{number of intervals that cover the true mean}}{\text{number of intervals calculated}} \to 0.95$$

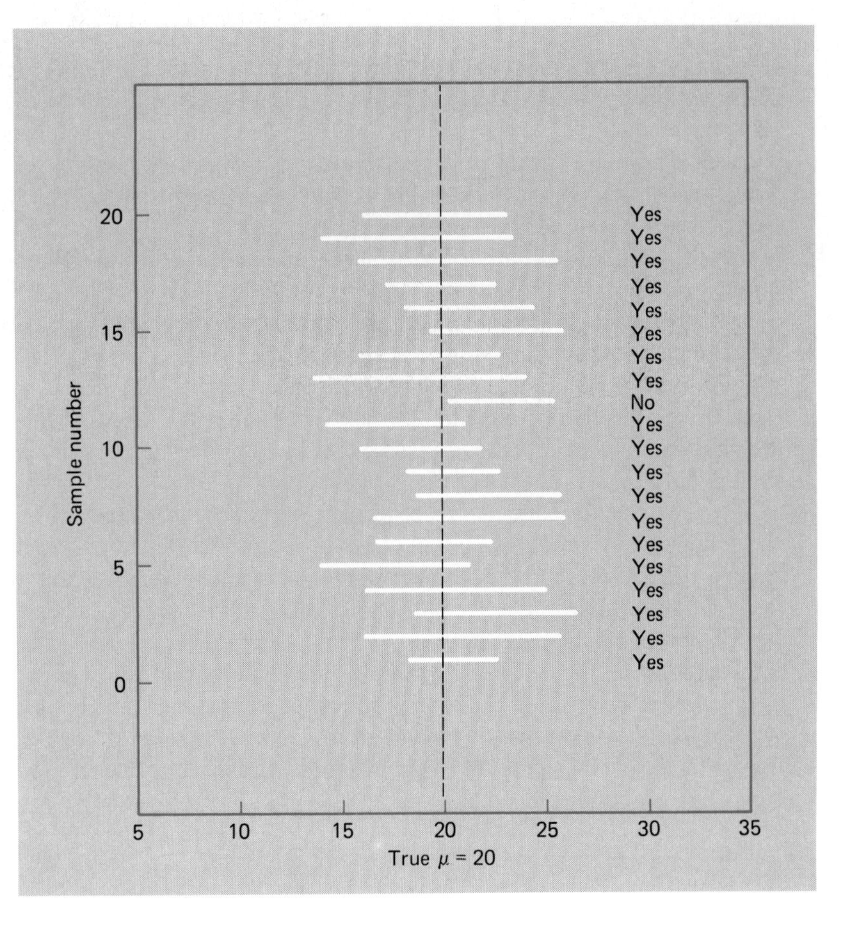

FIGURE 7.1
Interpretation of the confidence interval for population mean, true mean = 20.

when the intervals are calculated for a large number of different problems. This is what gives us 95% confidence! The proportion of intervals that cover μ should be nearly 0.95.

To emphasize these points, we simulated a sample of size $n = 10$ from a normal distribution with $\mu = 20$ and $\sigma = 5$. The 95% confidence interval was then calculated and graphed in Figure 7.1. This procedure was repeated 20 times. The different samples produce different values for $\bar{x}$ and, consequently, the intervals are centered at different points. The different values of s gave rise to intervals of different lengths. Unlike a real application, here we know that the true fixed mean is $\mu = 20$. The proportion of intervals that cover the true value $\mu = 20$ should be near 0.95 and, in this instance, we happen to have exactly that proportion $19/20 = 0.95$.

7.3 BAYESIAN ESTIMATION†

In recent years there has been mounting interest in methods of inference that look upon parameters as random variables. This idea is not really new, but these **Bayesian methods**, as they are called, have received considerable impetus and much wider applicability through the concept of personal, or subjective, probability. In fact, this is why supporters of the subjective concept of probability refer to themselves as **Bayesians**, or **Bayesian statisticians**.

In this section we shall present a Bayesian method of estimating the mean of a population, looking upon μ as a random variable whose distribution is indicative of how strongly a person feels about the various values which μ can take on. He or she will feel most strongly about some particular value of μ, but this enthusiasm will diminish for values of μ which are farther and farther away from the value liked most. Like any distribution that we use in practice, this kind of subjective **prior distribution** has a mean that we shall denote by μ_0 and a standard deviation that we shall denote by σ_0.

To illustrate the concept of a prior distribution of the mean, let us refer again to the sulfur oxides emission example on page 8, and let us suppose that before the sample data were actually obtained the chief engineer of the plant said that his feelings about the true average daily emission of sulfur oxides can best be described by a normal distribution with $\mu_0 = 17.50$ tons and $\sigma_0 = 2.50$ tons. With this information, we might ask, for example, how he felt about the possibility that the plant's true average daily emission of sulfur oxides is somewhere between 18.00 and 19.00 tons. The answer to this question is given by the area of the ruled region of the upper diagram of Figure 7.2, and—as can easily be verified—it represents a subjective probability of 0.15.

† This section may be omitted without loss of continuity.

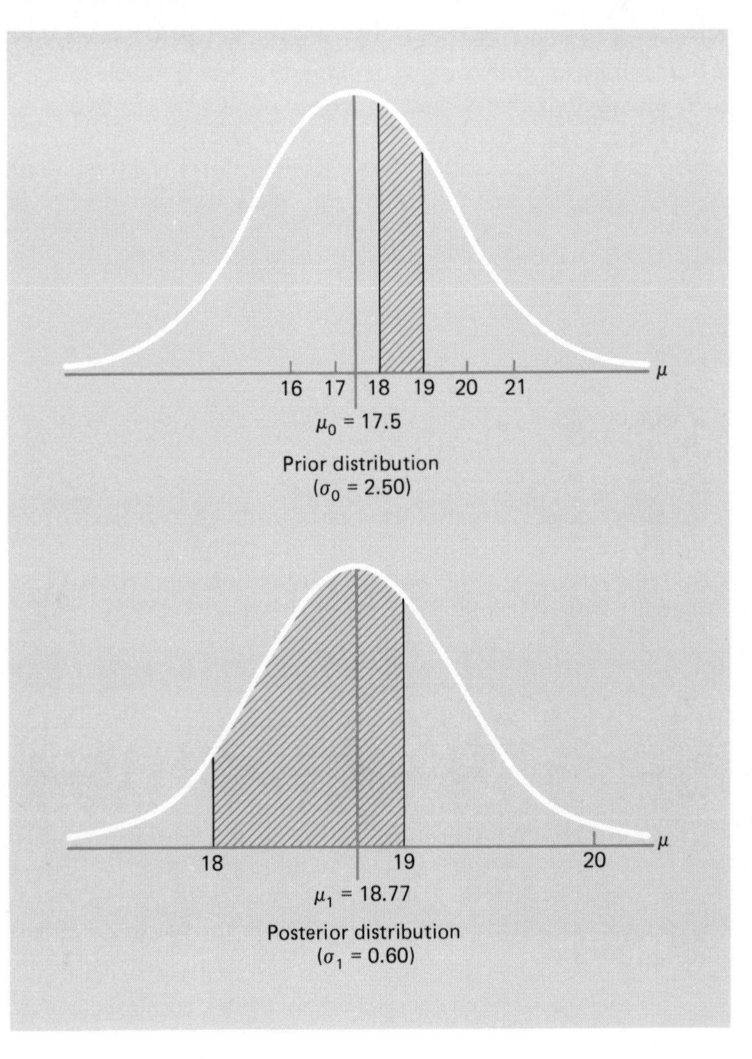

FIGURE 7.2
Prior and posterior distribution of μ.

In **Bayesian estimation**, prior feelings about the possible values of μ are combined with direct sample evidence. Without going into any detail, let us merely point out that this leads to a **posterior distribution** of μ, which, under fairly general conditions, can be approximated by a normal distribution with

Mean and standard deviation of posterior distribution

$$\mu_1 = \frac{n\bar{x}\sigma_0^2 + \mu_0\sigma^2}{n\sigma_0^2 + \sigma^2} \quad \textit{and} \quad \sigma_1 = \sqrt{\frac{\sigma^2\sigma_0^2}{n\sigma_0^2 + \sigma^2}}$$

Details about these results may be found in the texts on the theory of statistics listed in the bibliography.

To illustrate, let us refer again to the sulfur oxides emission example and combine the prior information given above, that based on the feelings of the chief engineer, with the fact that a random sample of size $n = 80$ yielded $\bar{x} = 18.85$ and $s = 5.55$. Using s as an estimate of σ and substituting these values together with $\mu_0 = 17.50$ and $\sigma_0 = 2.5$ into the above formulas for μ_1 and σ_1, we obtain

$$\mu_1 = \frac{80(18.85)(2.5)^2 + 17.50(5.55)^2}{80(2.5)^2 + (5.55)^2} = 18.77$$

and

$$\sigma_1 = \sqrt{\frac{(5.55)^2(2.5)^2}{80(2.5)^2 + (5.55)^2}} = 0.60$$

Now, the posterior probability that μ lies on the interval from 18.00 to 19.00 is given by the area of the ruled region of the lower diagram of Figure 7.2, and as can easily be verified, it is 0.55. Note that the information supplied by the sample raised the probability from 0.15 to 0.55.

EXERCISES

7.1 To illustrate that the mean of a random sample is an unbiased estimate of the mean of the population, consider five slips of paper numbered 3, 6, 9, 15, and 27.

(a) List all possible samples of size 3 that can be taken *without* replacement from this finite population.

(b) Calculate the mean of each of the samples listed in (a), and, assigning each sample a probability of $\frac{1}{10}$, verify that the mean of these $\bar{x}$'s equals 12, namely, the mean of the population.

7.2 Suppose that we observe a random variable having the binomial distribution and get x successes in n trials.

(a) Show that $\frac{x}{n}$ is an unbiased estimate of the binomial parameter p.

(b) Show that $\frac{x+1}{n+2}$ is not an unbiased estimate of the binomial parameter p.

7.3 To verify the claim that the mean is generally more efficient than the median (that it is subject to smaller chance fluctuations), a student conducted an experiment consisting of 12 tosses of three dice. The following are his results: 2, 4, and 6; 5, 3, and 5; 4, 5, and 3; 5, 2, and 3; 6, 1, and 5; 2, 3, and 1; 3, 1, and 4; 5, 5, and 2; 3, 3, and 4; 1, 6, and 2; 3, 3, and 3; and 4, 5, and 3.

(a) Calculate the 12 medians and the 12 means.

(b) Plot the medians and the means obtained in part (a) into separate Stem-and-leaf display.

(c) Draw dot diagrams of the two distributions obtained in part (b) and explain how they illustrate the claim that the mean is generally more efficient than the median.

7.4 With reference to the $n = 50$ interrequest time observations on page 14, which have mean 11,795 and standard deviation 14,054, what can one assert with 95% confidence about the maximum error if $\bar{x} = 11{,}795$ is used as a point estimate of the true population mean interrequest time?

7.5 With reference to the previous exercise, construct a 95% confidence interval for the true mean interrequest time.

7.6 To estimate the average time it takes to assemble a certain computer component, the industrial engineer at an electronics firm timed 40 technicians in the performance of this task, getting a mean of 12.73 minutes and a standard deviation of 2.06 minutes.

(a) What can we say with 99% confidence about the maximum error if $\bar{x} = 12.73$ is used as a point estimate of the actual average time required to do the job?

(b) Use the given data to construct a 98% confidence interval for the true average time it takes to assemble the computer component.

7.7 With reference to Exercise 7.6, with what confidence can we assert that the sample mean does not differ from the true mean by more than 30 seconds?

7.8 What is the maximum error one can expect to make with probability 0.90 when using the mean of a random sample of size $n = 64$ to estimate the mean of a population with $\sigma^2 = 2.56$?

7.9 In a study of automobile collision insurance costs, a random sample of 80 body repair costs for a particular kind of damage had a mean of \$472.36 and a standard deviation of \$62.35. If $\bar{x} = \$472.36$ is used as a point estimate of the true average repair cost of this kind of damage, with what confidence can one assert that the error does not exceed \$10?

7.10 If we want to determine the average mechanical aptitude of a large group of workers, how large a random sample will we need to be able to assert with probability 0.95 that the sample mean will not differ from the true mean by more than 3.0 points? Assume that it is known from past experience that $\sigma = 20.0$.

7.11 The dean of a college wants to use the mean of a random sample to estimate the average amount of time students take to get from one class to the next, and she wants to be able to assert with 99% confidence that the error is at most 0.25 minute. If it can be presumed from experience that $\sigma = 1.40$ minutes, how large a sample will she have to take?

7.12 It is desired to estimate the mean time of continuous use until an answering machine will first require service. If it can be assumed that $\sigma = 60$ days, how large a sample is needed so that one will be able to assert with 90% confidence that the sample mean is off by at most 10 days?

7.13 Modify the formula for E on page 208 so that it applies to large samples which constitute substantial portions of finite populations, and use the resulting formula for the following problems:

(a) A sample of 50 scores on the admission test for a school of engineering is drawn at random from the scores of the 420 persons who applied to the school in 1988. If the sample mean and standard deviation are $\bar{x} = 546$ and $s = 85$, what can we assert with 95% confidence about the maximum error if $\bar{x} = 546$ is used as an estimate of the mean score of all the applicants?

(b) A random sample of 40 drums of a chemical, drawn from among 200 such drums whose weights can be expected to have the standard deviation $\sigma = 12.2$ pounds, has a mean weight of 240.8 pounds. If we estimate the mean weight of all 200 drums as 240.8 pounds, what can we assert with 99% confidence about the maximum error?

7.14 Modify the large-sample confidence-interval formula for μ on page 211 so that it applies to large samples which constitute substantial portions of finite populations, and use the resulting formula for the following problems:

(a) With reference to part (a) of Exercise 7.13, construct a 99% confidence interval for the mean score of the 420 persons who applied to the school of engineering in 1988.

(b) In a random sample of 100 of 500 experimental batteries produced by a firm, the lifetimes have a mean of 148.2 hours with a standard deviation of 24.9 hours. Find a 76.60% confidence interval for the mean life of the 500 batteries.

7.15 A random sample of 100 teachers in a large metropolitan area revealed a mean weekly salary of \$487 with a standard deviation of \$48. With what degree of confidence can we assert that the average weekly salary of all teachers in the metropolitan area is between \$472 and \$502?

7.16 Instead of the large-sample confidence-interval formula for μ on page 211, we could have given the alternative formula

$$\bar{x} - z_{\alpha/3} \cdot \frac{\sigma}{\sqrt{n}} < \mu < \bar{x} + z_{2\alpha/3} \cdot \frac{\sigma}{\sqrt{n}}$$

Explain why the one on page 211 is narrower, and hence preferable, to the one given here.

7.17 In the example on page 198, 20 fuses were subjected to a 20% overload, and the times it took them to blow had a mean of 10.63 minutes and a standard deviation of 2.48 minutes. If we use $\bar{x} = 10.63$ minutes as a point estimate of the true average it takes such fuses to blow with a 20% overload, what can we assert with 95% confidence about the maximum error?

7.18 A major truck stop has kept extensive records on various transactions with its customers. If a random sample of $n = 18$ of these records shows average sales of 63.84 gallons of diesel fuel with a standard deviation of 2.75 gallons, and we use $\bar{x} = 63.84$ as an estimate of the truck stop's mean sales of diesel fuel per customer, what can we say with 99% confidence about the maximum error?

7.19 Ten bearings made by a certain process have a mean diameter of 0.5060 cm with a standard deviation of 0.0040 cm. Assuming that the data may be looked upon as a random sample from a normal population, construct a 95% confidence interval for the actual average diameter of bearings made by this process.

7.20 Inspecting ceramic tiles prior to their shipment, a quality-control engineer detects 2, 3, 6, 0, 4, and 9 defectives in six cartons, each containing 144 tiles. Assuming that the data can be looked upon as a random sample from a population which can be approximated closely by a normal distribution, what can he assert with 99% confidence about the maximum error, if he uses the mean of the sample as a point estimate of the true average number of defectives per carton?

7.21 With reference to Exercise 7.20, construct a 98% confidence interval for the true average number of defectives per carton.

7.22 In an air-pollution study performed at an experiment station, the following amounts of suspended benzene-soluble organic matter (in micrograms per cubic meter) were obtained for eight different samples of air: 2.2, 1.8, 3.1, 2.0, 2.4, 2.0, 2.1, and 1.2. Assuming that the population sampled is normal, construct a 95% confidence interval for the corresponding true mean.

7.23 (Maximum likelihood estimation) Consider a random sample of size n from a discrete population $f(x; \theta)$ that depends on a parameter θ. The joint distribution is

$$f(x_1; \theta)f(x_2; \theta)\cdots f(x_n; \theta)$$

Once the observations become available, we could substitute their actual values x_1, $x_2, \ldots, x_n$ into the joint density. After the substitution, the resulting function of θ

$$L(\theta) = f(x_1; \theta)f(x_2; \theta)\cdots f(x_n; \theta)$$

is called the **likelihood function**. The **maximum likelihood estimator** of θ is the value for θ that maximizes the probability of the observed sample. Because this is an after-the-fact calculation, we say it maximizes the likelihood.

Find the maximum likelihood estimator

(a) for λ when $f(x; \lambda)$ is the Poisson distribution;

(b) for p when $f(x; p) = p^x(1 - p)^{1-x}$ for $x = 0, 1$.

7.24 Refer to Exercise 7.23. The same procedure for obtaining the maximum likelihood estimator applies also in the continuous case.

Find the maximum likelihood estimator

(a) for α when $f(x; \alpha)$ is the exponential distribution;

(b) for μ when $f(x; \mu)$ is a normal distribution with $\sigma^2 = 1$.

7.25 A sales manager's feelings about the average monthly demand for one of his company's products may be described by means of a normal distribution with $\mu_0 = 3{,}800$ units and $\sigma_0 = 260$ units.

(a) What probability is he, thus, assigning to the true average monthly demand being on the interval from 3,500 to 4,000 units?

(b) If data for 9 months show an average demand for 3,702 units with a standard deviation of 390 units, how does this information affect the probability determined in part (a)? Use $s = 390$ as an estimate of σ.

7.26 A professor is making up a final examination for a course in computer-aided manufacturing, which is to be given to a large group of students. Her feelings about the average grade they should get is expressed subjectively by a normal distribution with $\mu_0 = 67.2$ and $\sigma_0 = 1.5$.

(a) What prior probability does she thus assign to the average grade being on the interval from 65.0 to 70.0?

(b) If the examination is tried on a random sample of 40 of the students and the mean and the standard deviation of their grades are 74.9 and 7.4, calculate the values of μ_1 and σ_1. Use $s = 7.4$ as an estimate of σ.

(c) Use the results of part (b) to calculate the posterior probability which the professor will thus assign to the average grade being on the interval from 65.0 to 70.0.

7.4 TESTS OF HYPOTHESES

There are many problems in which, rather than estimate the value of a parameter, we must decide whether a statement concerning a parameter is true or false; that is, we must test a hypothesis about a parameter. For example, in quality control work a random sample may serve to determine whether the "process mean" (for a given kind of measurement) has remained unchanged or whether it has changed to such an extent that the process has gone "out of control" and adjustments have to be made.

To illustrate the general concepts involved in this kind of decision problems, suppose that a consumer protection agency wants to test a paint manufacturer's claim that the average drying time of his new "fast-drying" paint is 20 minutes. So it instructs a member of its research staff to take 36 boards and paint them with paint from 36 different 1-gallon cans of the paint, with the intention of rejecting the claim if the mean of the drying times exceeds 20.75 minutes; otherwise, it will accept the claim and in either case it will take whatever action is called for in its plans.

This provides a clear-cut criterion for accepting or rejecting the claim, but unfortunately it is not infallible. Since the decision is based on a sample, there is the possibility that the sample mean may exceed 20.75 minutes even though the true mean drying time is $\mu = 20$ minutes, and there is also the possibility that the sample mean may be 20.75 minutes or less even though the true mean drying time is, say, $\mu = 21$ minutes. Thus, before adopting the criterion, it would seem wise to investigate the chances that the criterion may lead to a wrong decision.

Assuming that it is known from past experience that the standard deviation of such drying times can be expected to equal $\sigma = 2.4$ minutes, let us first investigate the possibility that the sample mean may exceed 20.75 minutes even though the true mean drying time is $\mu = 20$. The probability that this will happen purely due to chance is given by the area of the ruled region of Figure 7.3, and it can easily be determined by approximating the sampling distribution of the mean with a normal curve. Assuming that the population is large enough to be treated as infinite, we have $\sigma_{\bar{x}} = \frac{2.4}{\sqrt{36}} = 0.4$, and the dividing line of the criterion, in standard units is

$$z = \frac{20.75 - 20}{0.4} = 1.875$$

It follows from Table 3 that the area of the ruled region of Figure 7.3 is $1 - 0.9696 = 0.0304$ (by interpolation) and hence that the probability of erroneously rejecting the hypothesis $\mu = 20$ minutes is approximately 0.03.

Let us now consider the other possibility, where the procedure fails to detect that $\mu \neq 20$ minutes. Suppose again, for the sake of argument, that the true mean drying time is $\mu = 21$ minutes, so that the probability of getting a sample mean less than or equal to 20.75 minutes (and, hence, erroneously accepting the claim that $\mu = 20$ minutes) is given by the area of the ruled region of Figure 7.4. As before, $\sigma_{\bar{x}} = 0.4$, so that the dividing line of the criterion, in standard units, is now

$$z = \frac{20.75 - 21}{0.4} = -0.625$$

It follows from Table 3 that the area of the ruled region of Figure 7.4 is $1 - 0.7340 = 0.2660$ (by interpolation), and hence that the probability of erroneously accepting the hypothesis $\mu = 20$ is approximately 0.27.

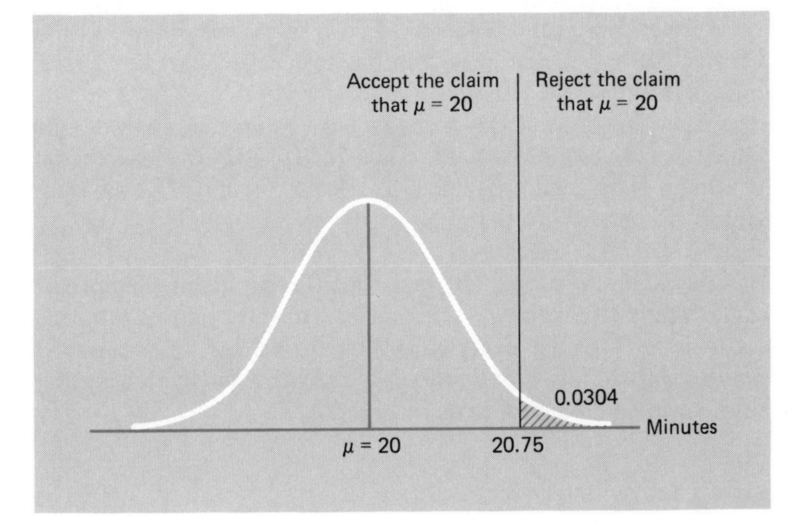

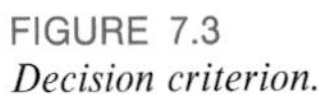
FIGURE 7.3
Decision criterion.

The situation described in this example is typical of testing a statistical hypothesis, and it may be summarized in the following table, where we refer to the hypothesis being tested as hypothesis H:

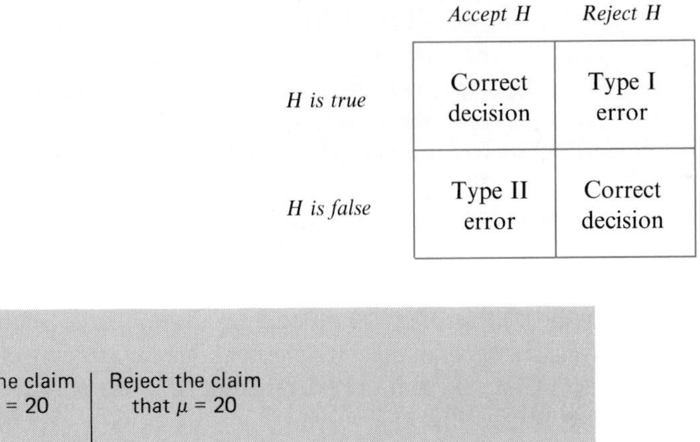

	Accept H	*Reject H*
H is true	Correct decision	Type I error
H is false	Type II error	Correct decision

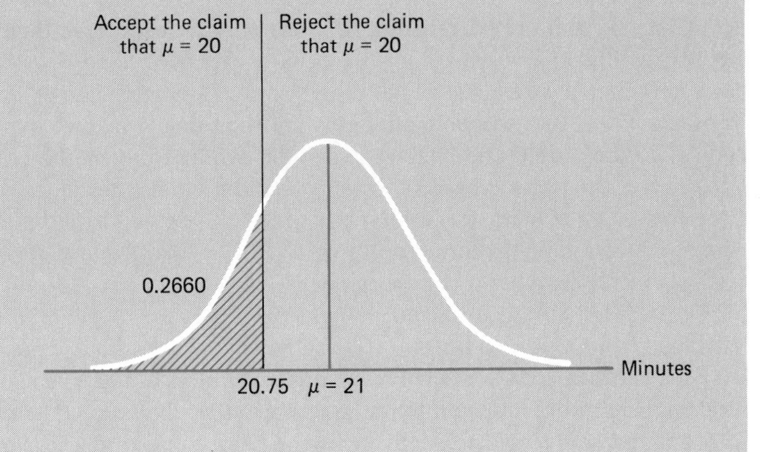

FIGURE 7.4
Decision criterion.

If hypothesis H is true and accepted or false and rejected, the decision is in either case correct. If hypothesis H is true but rejected, it is rejected in error, and if hypothesis H is false but accepted, it is accepted in error. The first of these errors is called a **Type I error** and the probability of committing it is designated by the Greek letter α (*alpha*); the second is called a **Type II error** and the probability of committing it is designated by the Greek letter β (*beta*). Thus, in our example we showed that for the given test criterion $\alpha = 0.03$ and $\beta = 0.27$ when $\mu = 21$ minutes.

In calculating the probability of a Type II error in our example, we arbitrarily chose the alternative value $\mu = 21$ minutes. However, in this problem, as in most others, there are infinitely many other alternatives, and for each one of them there is a positive probability β of erroneously accepting the hypothesis H. What to do about this will be discussed further in Section 7.8.

7.5 NULL HYPOTHESES AND SIGNIFICANCE TESTS

In the drying-time example of the preceding section we were able to calculate the probability of a Type I error because we formulated the hypothesis H as a single value for the parameter μ; that is, we formulated the hypothesis H so that μ was completely specified. Had we formulated instead, $\mu \leq 20$ minutes, where μ can take on more than one possible value, we would not have been able to calculate the probability of a Type I error without specifying by how much μ is less than 20 minutes.

We often formulate hypotheses to be tested as a single value for a parameter; at least, we do this whenever possible. This usually requires that we hypothesize the opposite of what we hope to prove. For instance, if we want to show that one method of teaching computer programming is more efficient than another, we hypothesize that the two methods are equally effective. Similarly, if we want to show that one method of irrigating the soil is more expensive than another, we hypothesize that the two methods are equally expensive; and if we want to show that a new copper-bearing steel has a higher yield strength than ordinary steel, we hypothesize that the two yield strengths are the same. Since we hypothesize that there is no difference in the effectiveness of the two teaching methods, no difference in the cost of the two methods of irrigation, and no difference in the yield strength of the two kinds of steel, we call hypotheses like these **null hypotheses** and denote them H_0. Nowadays, the term "null hypothesis" is used for any hypothesis set up primarily to see whether it can be rejected.

The idea of setting up a null hypothesis is not an uncommon one, even in nonstatistical thinking. In fact, this is exactly what is done in an American court of law, where an accused is assumed to be innocent unless he is proven guilty "beyond a reasonable doubt." The null hypothesis states that the accused is not guilty, and the probability expressed subjectively by the phrase "beyond a reasonable doubt" reflects the probability α of risking a Type I error. Note that the "burden of proof" is always on the prosecution in the sense that the accused is found not guilty unless the null hypothesis of innocence is clearly disproved. This does not imply that the defendant has been proved innocent if found not guilty; it implies only that he has not been proved guilty. Of course, since we cannot legally "reserve judgment" if proof of guilt is not established, the accused is freed and we act as if the null hypothesis of innocence were accepted. Note that this is precisely what we may have to do in tests of statistical hypotheses, when we cannot afford the luxury of reserving judgment.

When reserving judgment is a viable alternative to the rejection of a null hypothesis, Type II errors can be sidestepped altogether. To illustrate, let us return to the drying time example of Section 7.4 and restate the criterion as follows:

> *Reject the hypothesis* $\mu = 20$ *minutes (and accept the alternative* $\mu > 20$ *minutes) if the mean of the* 36 *sample values exceeds* 20.75 *minutes; otherwise, reserve judgment.*

If judgment is reserved as in this criterion, there is no possibility of a Type II error; no matter what happens, the null hypothesis is never accepted.

The above criterion may well be described as a test of whether $\bar{x}$ is significantly greater than $\mu = 20$ minutes, where "significantly greater" means that the discrepancy between $\bar{x}$ and $\mu = 20$ is too large to be reasonably attributed to chance. Thus, we refer to this kind of test as a **significance test**.

To approach problems of hypothesis testing systematically, it will help to proceed as outlined in the following five steps:

1. **We formulate a null hypothesis and an appropriate alternative hypothesis which we accept when the null hypothesis must be rejected.**†

In the drying-time example the null hypothesis is $\mu = 20$ minutes and the alternative hypothesis is $\mu > 20$ minutes (since the paint was referred to as "fastdrying"). This kind of alternative, which is called a **one-sided alternative**, may also have the inequality going the other way. For instance, if we hope to be able to show that the average time required to do a certain job is less than 15 minutes, we would test the null hypothesis $\mu = 15$ against the alternative hypothesis $\mu < 15$.

The following is an example in which we would use the **two-sided alternative** $\mu \neq \mu_0$, where μ_0 is the value assumed under the null hypothesis: A food processor wants to check whether the average amount of coffee that goes into his 4-ounce jars is indeed 4 ounces. Since the food processor cannot afford to put much less than 4 ounces into each jar for fear of losing customer acceptance, nor can he afford to put much more than 4 ounces into each jar for fear of losing part of his profit, the appropriate alternative hypothesis is $\mu \neq 4$.

---

† See also the discussion on page 224.

As in the examples of the two preceding paragraphs, alternative hypotheses usually specify that the population mean (or whatever other parameter may be of concern) is not equal to, greater than, or less than the value assumed under the null hypothesis. For any given problem, the choice of an appropriate alternative depends mostly on what we hope to be able to show, or better, where we want to put the burden of proof.

EXAMPLE An appliance manufacturer is considering the purchase of a new machine for stamping out sheet metal parts. If μ_0 is the average number of good parts stamped out per hour by her old machine and μ is the corresponding average for the new machine, the manufacturer wants to test the null hypothesis $\mu = \mu_0$ against a suitable alternative. What should the alternative be if

(a) she does not want to buy the new machine unless it is more productive than the old one;

(b) she wants to buy the new machine (which has some other nice features) unless it is less productive than the old one?

Solution (a) The manufacturer should use the alternative hypothesis $\mu > \mu_0$ and purchase the new machine only if the null hypothesis can be rejected. (b) The manufacturer should use the alternative hypothesis $\mu < \mu_0$ and purchase the new machine unless the null hypothesis can be rejected.

■

Having formulated the null hypothesis and an alternative hypothesis, we proceed with the following step:

2. **We specify the probability of a Type I error; if possible, desired, or necessary, we may also specify the probabilities of Type II errors for particular alternatives.**

The probability of a Type I error is also called the **level of significance**, and it is usually set at $\alpha = 0.05$ or $\alpha = 0.01$. Which value we choose in any given problem will have to depend on the risks, or consequences, of committing a Type I error. Observe, however, that we should not make the probability of a Type I error too small, because this will have the tendency to make the probabilities of serious Type II errors too large.

Step 2 cannot be performed unless the null hypothesis is a simple hypothesis, but this is not as restrictive as it may seem. To illustrate, let us investigate briefly what might be done in the drying time example if we wanted to allow for the possibility that the paint may be better (faster drying) than claimed, and hence test the null hypothesis $\mu \leq 20$ minutes against the alternative hypothesis $\mu > 20$ minutes. Now, the probability of a Type I error cannot be calculated, but observe that if μ is less than 20 minutes, the normal curve of Figure 7.3 on page 221 is shifted to the left, and the area under the curve to the right of 20.75 becomes less than 0.0304. Thus, if the null hypothesis is $\mu \leq 20$ minutes, we can say that the probability of a Type I error is at most 0.0304, and we write $\alpha \leq 0.0304$. In general, if the null hypothesis is of the form $\mu \leq \mu_0$ or $\mu \geq \mu_0$, we can only specify the maximum probability of a Type I error, and by performing the test as if the null hypothesis were $\mu = \mu_0$, we protect ourselves against the worst possibility (see the example on page 230).

After the null hypothesis, the alternative hypothesis, and the level of significance have been specified, the remaining steps are:

3. **Based on the sampling distribution of an appropriate statistic, we construct a criterion for testing the null hypothesis against the given alternative.**
4. **We calculate from the data the value of the statistic on which the decision is to be based.**
5. **We decide whether to reject the null hypothesis, whether to accept it, or whether to reserve judgment.**

In the drying-time example we studied the criterion using the normal-curve approximation to the sampling distribution of the mean; in general, step 3 depends not only on the statistic on which we want to base the decision and on its sampling distribution, but also on the alternative hypothesis we happen to choose. In the drying-time example we used a **one-sided criterion (one-sided test** or **one-tailed test)** with the one-sided alternative $\mu > 20$ minutes, rejecting the null hypothesis only for large values of the mean. In the example dealing with the food processor and his 4-ounce jars of coffee, we would use a **two-sided criterion (two-sided test** or **two-tailed test)** to go with the two-sided alternative $\mu \neq 4$ ounces. In general, a test is said to be two-sided if the null hypothesis is rejected for values of the test statistic falling into either tail of its sampling distribution.

The purpose of the discussion of this and the preceding section has been to introduce some of the basic problems connected with the testing of statistical hypotheses. Although the methods we have presented are objective—that is, two experimenters analyzing the same data under the same conditions would arrive at the identical results—their use does entail some arbitrary, or subjective, considerations. For instance, in the example on page 220 it was partially a subjective decision to "draw the line" between satisfactory and unsatisfactory values of μ at 21 minutes. It was also partially a subjective decision to use a sample of 36 one-gallon cans of the paint, and to reject the manufacturer's claim for values of $\bar{x}$ exceeding 20.75 minutes. Approaching the problem differently, the government agency investigating the paint manufacturer's claim could have specified values of α and β, thus controlling the risks to which they are willing to be exposed. The choice of α, the probability of a Type I error, could have been based on the consequences of making that kind of error, namely, the manufacturer's cost of having a good product condemned, the possible cost of subsequent litigation, the manufacturer's cost of unnecessarily adjusting his machinery, the cost to the public of not having the product available when needed, and so forth. The choice of β, the probability of a Type II error, could similarly have been based on the consequences of making that kind of error, namely, the cost to the public of buying an inferior product, the manufacturer's savings in using inferior ingredients but loss in good will, again the cost of possible litigation, and so forth. It should be obvious that it would be extremely difficult to put "cash values" on all these eventualities, but they must nevertheless be considered, at least indirectly, in choosing suitable criteria for testing statistical hypotheses.

In recent years, attempts have been made to incorporate all these matters within a formal theory called **decision theory**. Although many important advances have been made, it should be recognized that such a theory does not eliminate the arbitrariness, or subjectiveness, discussed above; it merely incorporates these matters within the theory. This means that the use of decision theory requires that we actually put cash values on all possible consequences of our decisions. Although this has the advantage that it makes the experimenter (the engineer) more "cost conscious," it also has the disadvantage of requiring information which very often cannot be obtained.

In this text we shall discuss mainly the **Neyman-Pearson theory**, also called the **classical theory** of testing hypotheses. This means that we shall consider cost factors and other considerations that are partly arbitrary and partly subjective only insofar as they will affect the choice of a sample size, the choice of an alternative hypothesis, the choice of α and β, and so forth.

In this approach, the maximum value of α is controlled over the null hypothesis. For instance, if the null hypothesis is $\mu < \mu_0$, then μ as well as σ is unspecified. This is a **composite** hypothesis. Otherwise, if all the parameters are completely specified, the hypothesis is **simple**. Even with a composite null hypothesis, we set the critical region so that the error probability is α on the boundary $\mu = \mu_0$. Then, the error probabilities will be even smaller under values of μ that are less than μ_0.

Because the probability of falsely rejecting the null hypothesis is controlled, the null hypothesis is retained unless the observations strongly contradict it. Consequently, if the goal of an experiment is to establish an assertion or hypothesis, that hypothesis must be taken as the alternative hypothesis.

Guideline for selecting the null hypothesis

When the goal of an experiment is to establish an assertion, the negation of the assertion should be taken as the null hypothesis. The assertion becomes the alternative hypothesis.

Similar reasoning suggests that even when costs are difficult to determine but the consequences of one error are much more serious than for the other, the hypotheses should be labeled so that the most serious error is the Type I error.

Before we discuss various special tests about means in the remainder of this chapter, let us point out that the concepts we have introduced here apply equally well to hypotheses concerning proportions, standard deviations, the randomness of samples, and relationships among several variables.

EXERCISES

7.27 Suppose that an engineering firm is asked to check the safety of a dam. What type of error would it commit if it erroneously rejects the null hypothesis that the dam is safe? What type of error would it commit if it erroneously accepts the null hypothesis that the dam is safe?

7.28 Suppose that we want to test the null hypothesis that an antipollution device for cars is effective. Explain under what conditions we would commit a Type I error and under what conditions we would commit a Type II error.

7.29 If the criterion on page 220 is modified so that the paint manufacturer's claim is rejected for $\bar{x} > 20.50$ minutes, find
(a) the probability of a Type I error;
(b) the probability of a Type II error when $\mu = 21$ minutes.

7.30 A process for making steel pipe is under control if the diameter of the pipe has a mean of 3.0000 inches with a standard deviation of 0.0250 inch. To check whether the process is under control, a random sample of size $n = 30$ is taken each day and the null hypothesis $\mu = 3.0000$ is rejected if $\bar{x}$ is less than 2.9960 or greater than 3.0040. Find
(a) the probability of a Type I error;
(b) the probability of a Type II error when $\mu = 3.0050$ inches.

7.31 Suppose that in the drying-time example on page 220, n is changed from 36 to 50 while everything else remains the same. Find
(a) the probability of a Type I error;
(b) the probability of a Type II error when $\mu = 21$ minutes.

7.32 It is desired to test the null hypothesis $\mu = 100$ pounds against the alternative hypothesis $\mu < 100$ pounds on the basis of a random sample of size $n = 40$ from a population with $\sigma = 12$. For what values of $\bar{x}$ must the null hypothesis be rejected if the probability of a Type I error is to be $\alpha = 0.01$?

7.33 Suppose that for a given population with $\sigma = 8.4$ in^2 we want to test the null hypothesis $\mu = 80.0$ in^2 against the alternative hypothesis $\mu < 80.0$ square inches on the basis of a random sample of size $n = 100$.
(a) If the null hypothesis is rejected for $\bar{x} < 78.0$ in^2 and otherwise it is accepted, what is the probability of a Type I error?
(b) What is the answer to part (a) if the null hypothesis is $\mu \geq 80.0$ square inches instead of $\mu = 80.0$ square inches?

7.34 If the null hypothesis $\mu = \mu_0$ is to be tested on the basis of a large random sample against the alternative hypothesis $\mu < \mu_0$ (or $\mu > \mu_0$), the probability of a Type I error is to be α and the probability of a Type II error is to be β for $\mu = \mu_1$, it can be shown that the required sample size is

$$n = \frac{\sigma^2(z_\alpha + z_\beta)^2}{(\mu_1 - \mu_0)^2}$$

where σ^2 is the variance of the population.
(a) It is desired to test the null hypothesis $\mu = 40$ against the alternative hypothesis $\mu < 40$ on the basis of a large random sample from a population with $\sigma = 4$. If the probability of a Type I error is to be 0.05 and the probability of a Type II error is to be 0.12 for $\mu = 38$, find the required size of the sample.
(b) Suppose that we want to test the null hypothesis $\mu = 64$ against the alternative hypothesis $\mu < 64$ for a population whose standard deviation is $\sigma = 7.2$. How large a sample must we take if α is to be 0.05 and β is to be 0.01 for $\mu = 61$? Also, for what values of $\bar{x}$ will the null hypothesis have to be rejected?

7.35 If the alternative hypothesis is two-sided, the formula of Exercise 7.34 must be modified by substituting $z_{\alpha/2}$ for z_α. How large a sample must we take if it is desired to test the null hypothesis $\mu = 100$ against the alternative hypothesis $\mu \neq 100$ on the basis of a random sample from a population with $\sigma = 16$, the probability of a Type I error is to be 0.01, and the probability of a Type II error is to be 0.20 for $\mu = 92$?

7.36 A city's police department is considering replacing the tires on its cars with radial tires. If μ_1 is the average number of miles they get out of their old tires and μ_2 is the average number of miles they will get out of the new tires, the null hypothesis they shall want to test is $\mu_1 - \mu_2 = 0$.

(a) What alternative hypothesis should they use if they do not want to buy the radial tires unless they are definitely proven to give a better mileage? In other words, the burden of proof is put on the radial tires and the old tires are to be kept unless the null hypothesis can be rejected.

(b) What alternative hypothesis should they use if they are anxious to get the new tires (which have some other nice features) unless they actually give a poorer mileage than the old tires? Note that now the burden of proof is on the old tires, which will be kept only if the null hypothesis can be rejected.

7.37 A producer of extruded plastic products finds that his mean daily inventory is 1,250 pieces. A new marketing policy has been put into effect and it is desired to test the null hypothesis that the mean daily inventory is still the same. What alternative hypothesis should be used if

(a) it is desired to know whether or not the new policy changes the mean daily inventory;

(b) it is desired to demonstrate that the new policy actually reduces the mean daily inventory;

(c) the new policy will be retained so long as it cannot be shown that it actually increases the mean daily inventory?

7.38 Specify the null hypothesis and the alternative hypothesis in each of the following cases.

(a) An automobile manufacturer wants to check a supplier's claim that the maximum resistance in a wiring harness is less than 50 ohms.

(b) An investigator wants to check the research department's claim that a new filament will increase mean bulb life to more than 3,000 hours.

7.6 HYPOTHESES CONCERNING ONE MEAN

Having used tests concerning means to illustrate the basic principles of hypothesis testing, let us now see how we proceed in actual practice. Suppose, for instance, that we want to test on the basis of $n = 35$ determinations and at the 0.05 level of significance whether the thermal conductivity of a certain kind of cement brick is 0.340, as has been claimed. From information gathered in similar studies, we can expect that the variability of such determinations is given by $\sigma = 0.010$.

Following the outline of the preceding section, we begin with Steps 1 and 2 by writing

1. *Null hypothesis*: $\mu = 0.340$
   Alternative hypothesis: $\mu \neq 0.340$
2. *Level of significance*: $\alpha = 0.05$

The alternative hypothesis is two-sided because we shall want to reject the null hypothesis if the mean of the determinations is significantly less than or significantly greater than 0.340.

Next, in Step 3, we depart from the procedure used in the example of the preceding section and base the test on the statistic

Statistic for test concerning mean

$$z = \frac{\bar{x} - \mu_0}{\sigma/\sqrt{n}}$$

instead of $\bar{x}$. The reason for working with standard units, or z values, is that it enables us to formulate criteria which are applicable to a great variety of problems, not just one.

If z_α is, as before, such that the area under the standard normal curve to its right equals α, the **critical regions**, namely, the sets of values of z for which we reject the null hypothesis $\mu = \mu_0$, can be expressed as in the following table:

Critical Regions for Testing $\mu = \mu_0$
(Normal population and σ known, or large sample)

Alternative hypothesis	*Reject null hypothesis if:*
$\mu < \mu_0$	$z < -z_\alpha$
$\mu > \mu_0$	$z > z_\alpha$
$\mu \neq \mu_0$	$z < -z_{\alpha/2}$ or $z > z_{\alpha/2}$

If $\alpha = 0.05$, the dividing lines, or **critical values**, of the criteria are -1.645 and 1.645 for the one-sided alternatives, and -1.96 and 1.96 for the two-sided alternative. If $\alpha = 0.01$, the dividing lines of the criteria are -2.33 and 2.33 for the one-sided alternatives, and -2.575 and 2.575 for the two-sided alternative. These results come from the example on page 144 and Exercise 5.23.

Returning now to the example dealing with the thermal conductivity of the cement bricks, suppose that the mean of the 35 determinations is 0.343. So we continue by writing

3. *Criterion*: Reject the null hypothesis if $z < -1.96$ or $z > 1.96$, where

$$z = \frac{\bar{x} - \mu_0}{\sigma/\sqrt{n}}$$

4. *Calculations*:

$$z = \frac{0.343 - 0.340}{0.010/\sqrt{35}} = 1.77$$

5. *Decision*: Since $z = 1.77$ falls on the interval from -1.96 to 1.96, the null hypothesis cannot be rejected; to put it another way, the difference between $\bar{x} = 0.343$ and $\mu = 0.340$ can be attributed to chance. Whether we actually accept the null hypothesis will have to depend on whether the situations call for a decision one way or the other.

In problems like this, some research workers accompany the calculated value of z with a corresponding **tail probability**, or ***P* value**, with the probability of getting a difference between $\bar{x}$ and μ_0 greater than or equal to that actually observed. For instance, in the above example, it is given by the total area under the standard normal curve to the left of -1.77 and to the right of 1.77, and it equals $2(1 - 0.9616) = 0.0768$. This value exceeds 0.05, which agrees with our earlier result, but observe that giving a tail probability does not relieve us of the responsibility of specifying the level of significance before the test is actually performed.

The test we have described in this section is essentially an approximate large-sample test; it is exact only when the population we are sampling is normal and σ is known. In many practical situations where σ is unknown, we must make the further approximation of substituting for it the sample standard deviation s.

EXAMPLE A trucking firm suspects the claim that the average lifetime of certain tires is at least 28,000 miles. To check the claim, the firm puts 40 of these tires on its trucks and gets a mean lifetime of 27,463 miles with a standard deviation of 1,348 miles. What can it conclude if the probability of a Type I error is to be at most 0.01?

Solution

1. *Null hypothesis*: $\mu \geq 28{,}000$ miles
   Alternative hypothesis: $\mu < 28{,}000$ miles
2. *Level of significance*: $\alpha \leq 0.01$
3. *Criterion*: Since the probability of a Type I error is greatest when $\mu = 28{,}000$ miles, we proceed as if we were testing the null hypothesis $\mu = 28{,}000$ miles against the alternative hypothesis $\mu < 28{,}000$ miles at the

0.01 level of significance. Thus, the null hypothesis must be rejected if $z < -2.33$, where

$$z = \frac{\bar{x} - \mu_0}{\sigma/\sqrt{n}}$$

with σ replaced by s.

4. *Calculations*:

$$z = \frac{27{,}463 - 28{,}000}{1{,}348/\sqrt{40}} = -2.52$$

5. *Decision*: Since $z = -2.52$ is less than -2.33, the null hypothesis must be rejected at level of significance 0.05. In other words, the trucking firm's suspicion that $\mu < 28{,}000$ miles is confirmed.

■

If the sample size is small and σ is unknown, the tests just described cannot be used. However, if the sample comes from a normal population (to within a reasonable degree of approximation), we can make use of the theory discussed in Section 6.3 and base the test of the null hypothesis $\mu = \mu_0$ on the statistic

Statistic for small-sample test concerning mean

$$\boldsymbol{t = \frac{\bar{x} - \mu_0}{s/\sqrt{n}}}$$

which is a value of a random variable having the t distribution with $n - 1$ degrees of freedom. The criteria for the **one-sample *t* test** based on this statistic are like those given in the table on page 229 with z replaced by t and z_α and $z_{\alpha/2}$ replaced by t_α and $t_{\alpha/2}$.

EXAMPLE

The specifications for a certain kind of ribbon call for a mean breaking strength of 180 pounds. If five pieces of the ribbon (randomly selected from different rolls) have a mean breaking strength of 169.5 pounds with a standard deviation of 5.7 pounds, test the null hypothesis $\mu = 180$ pounds against the alternative hypothesis $\mu < 180$ pounds at the 0.01 level of significance.

Solution

1. *Null hypothesis*: $\mu = 180$ pounds
   Alternative hypothesis: $\mu < 180$ pounds
2. *Level of significance*: $\alpha = 0.01$
3. *Criterion*: Reject the null hypothesis if $t < -3.747$, where 3.747 is the value of $t_{0.01}$ for $5 - 1 = 4$ degrees of freedom and

$$t = \frac{\bar{x} - \mu_0}{s/\sqrt{n}}$$

4. *Calculations*:

$$t = \frac{169.5 - 180}{5.7/\sqrt{5}} = -4.12$$

5. *Decision*: Since $t = -4.12$ is less than -3.747, the null hypothesis must be rejected at level $\alpha = 0.01$. In other words, the breaking strength is below specifications. The exact tail probability, or P value, cannot be determined from Table 4, but it is 0.0073.

■

7.7
THE RELATION BETWEEN TESTS AND CONFIDENCE INTERVALS

We now describe an important connection between tests for two-sided alternatives and confidence intervals. This relation provides the reason that most statisticians prefer the information available in a confidence interval statement as opposed to the information that the null hypothesis $\mu = \mu_0$ was or was not rejected.

To develop the relation, we consider the $100(1 - \alpha)\%$ confidence interval for μ given on page 212

$$\bar{x} - t_{\alpha/2}\frac{s}{\sqrt{n}} < \mu < \bar{x} + t_{\alpha/2}\frac{s}{\sqrt{n}}$$

This interval is closely connected to the level α test of H_0: $\mu = \mu_0$ versus the two-sided alternative H_1: $\mu \neq \mu_0$. This test has critical region

$$\left|\frac{\bar{x} - \mu_0}{s/\sqrt{n}}\right| = |t| \geq t_{\alpha/2}$$

The acceptance region of this test is obtained by reversing the inequality to obtain all the values of $\bar{x}$ that do not lead to the rejection of μ_0.

$$\textit{acceptance region:} \quad \left|\frac{\bar{x} - \mu_0}{s/\sqrt{n}}\right| < t_{\alpha/2}$$

The acceptance region can also be expressed as

$$\textit{acceptance region:} \quad \bar{x} - t_{\alpha/2}\frac{s}{\sqrt{n}} < \mu_0 < \bar{x} + t_{\alpha/2}\frac{s}{\sqrt{n}}$$

where the limits of the interval are identical to the preceding confidence interval. That is, the null hypothesis μ_0 will not be rejected at level α if μ_0 lies within the 100 $(1 - \alpha)\%$ confidence interval for μ.

The set of plausible values for μ, as determined by the confidence interval, tells us at once about the outcome of all possible two-sided tests of hypothesis that specify a single value for μ.

EXAMPLE Referring to the example on page 212, $n = 16$ measurements of weight loss yielded $\bar{x} = 3.42$ and $s = 0.68$. Since $t_{0.025} = 2.131$ with 15 degrees of freedom, the 95% confidence interval becomes

$$\bar{x} - t_{\alpha/2}\frac{s}{\sqrt{n}} < \mu < \bar{x} + t_{\alpha/2}\frac{s}{\sqrt{n}}, \quad \text{or} \quad 3.06 < \mu < 3.78$$

Use the relation between 95% confidence intervals and $\alpha = 0.05$ level tests to test the null hypothesis $\mu = 3.7$ versus the alternative hypothesis $\mu \neq 3.7$. Also test the null hypothesis $\mu = 3.0$ versus the alternative hypothesis $\mu \neq 3.0$.

Solution In view of the relation just established, a test of $\mu = 3.7$ versus $\mu \neq 3.7$ would not be rejected at the 5% level, since $\mu = 3.7$ falls within the confidence interval.

On the other hand $\mu = 3.0$ does not fall within the 95% confidence interval and hence the null hypothesis would be rejected at level $\alpha = 0.05$.

■

The relation holds quite generally. Suppose, for any value θ_0 of a parameter θ, we have a level α test of the null hypothesis $\theta = \theta_0$ versus the alternative $\theta \neq \theta_0$. Collect all the values θ_0 that would not be rejected. These form a $100(1 - \alpha)\%$ confidence interval for θ.

A confidence interval statement provides a more comprehensive inference than a statement concerning a two-sided test of a single null hypothesis. Consequently, we favor a confidence-interval approach when one is available.

7.8 OPERATING CHARACTERISTIC CURVES

So far we have not paid much attention to Type II errors; in the drying-time example of Section 7.4 we calculated one probability of a Type II error, and all the tests of the previous sections were looked upon as tests of significance. Since the choice of the alternative hypothesis $\mu = 21$ minutes in the drying-time example was essentially arbitrary, it may be of interest to see how the testing procedure will perform when other values of μ prevail. The actual mean drying time depends on the chemical reactions governing the paint and the environmental conditions.

Lacking a total scientific explanation, we do not know precisely what value to expect for this population mean. Consequently, we must investigate the probability of accepting the null hypothesis under a range of possible values for μ. To this end, let

$$L(\mu) = \text{probability of accepting the null hypothesis when } \mu \text{ prevails}$$

The function $L(\mu)$ completely characterizes the testing procedure whatever the value of population mean μ. If μ equals a value where the null hypothesis is true, then $1 - L(\mu)$ is the probability of the Type I error. When μ has a value where the alternative hypothesis is true, then $L(\mu)$ is the probability of a Type II error. That is, the function $L(\mu)$ carries complete information about the probabilities of both types of error.

To illustrate the calculation of $L(\mu)$, we continue with the drying-time example on page 220, where we had $\mu_0 = 20$, $\sigma = 2.4$, $n = 36$, and the dividing line of the criterion was $\bar{x} = 20.75$. If the prevailing population mean is $\mu = 20.5$, then $\sqrt{n}(\bar{x} - 20.5)/\sigma = 6(\bar{x} - 20.5)/2.4$ is a standard normal variable. We reason that $L(20.5)$ is the probability of observing

$$\bar{x} < 20.75, \quad \text{or} \quad \frac{6(\bar{x} - 20.5)}{2.4} < \frac{6(20.75 - 20.5)}{2.4}$$

or $z < 0.625$. Therefore, from Table 3, $L(20.5) = 0.73$. Continuing with other possible values for μ, we obtain the results shown in the following table and also in Figure 7.5 (see Exercise 7.50):

Value of μ	*Probability of accepting null hypothesis*
19.50	0.999
19.75	0.99
20.00	0.97
20.25	0.89
20.50	0.73
20.75	0.50
21.00	0.27
21.25	0.11
21.50	0.03
21.75	0.01
22.00	0.001

Note that the probability of committing a Type II error diminishes when μ is increased, and that the probability of not committing a Type I error approaches 1 when μ becomes much smaller than 20 (and the paint is even better than claimed).

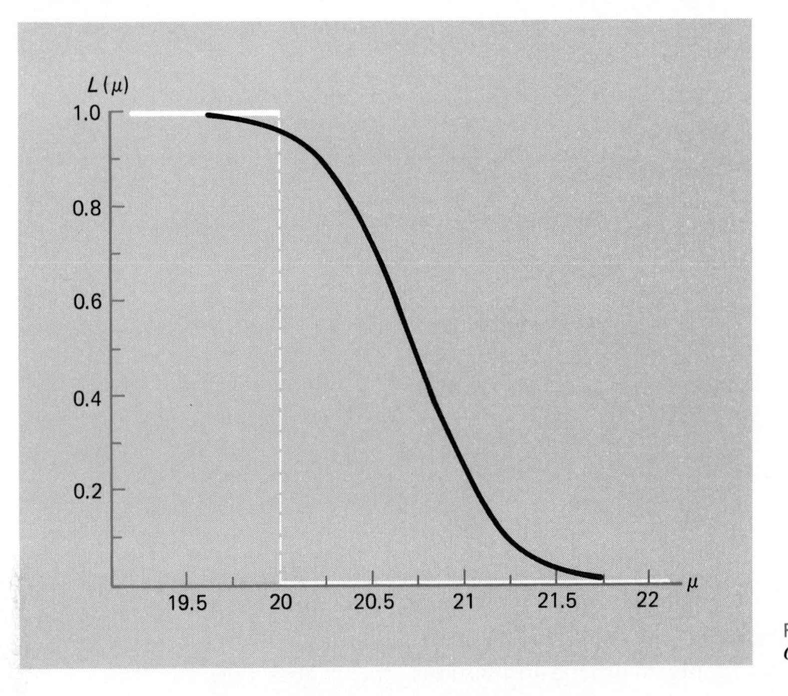

FIGURE 7.5
Operating characteristic curve.

The graph of $L(\mu)$ for various values of μ, shown in Figure 7.5, is called the **operating characteristic curve**, or simply the **OC curve**, of the test.

In the context of sampling in order to decide whether or not to accept a shipment of paint on the basis of drying time, we would like to accept the shipment if the mean drying time is low and reject it if it is high. Based on the operating characteristic curve in Figure 7.5, the engineer can decide if the proposed procedure has small enough error probabilities at values of μ she deems important. Ideally, we should want to reject the null hypothesis $\mu = \mu_0$ when actually μ exceeds μ_0, and to accept it when μ is less than or equal to μ_0. Thus, the ideal *OC* curve for our example would be given by the white horizontal lines of Figure 7.5. In actual practice, *OC* curves can only approximate such ideal curves, with the approximation becoming better as the sample size is increased. To illustrate this point, the reader will be asked in Exercise 7.91 to recalculate the probabilities in the table on page 234 when the sample size is increased from 36 to 50.

Figure 7.5 presents the picture of a typical *OC* curve for the case where the alternative hypothesis is $\mu > \mu_0$. When the alternative hypothesis is $\mu < \mu_0$, the *OC* curve becomes the mirror image of that of Figure 7.5, reflected about the dashed vertical line through μ_0.

When the alternative hypothesis is two-sided, H_1: $\mu < \mu_0$ or $\mu > \mu_0$, then a typical *OC* curve will have a maximum at $\mu = \mu_0$ and it will fall off as the value for μ moves away from μ_0 in either direction. An *OC* curve for a two-tailed test is shown in Figure 7.6. It should be observed that $L(\mu)$ now is the probability of a Type II error for all values of μ except μ_0; at μ_0 it is still the probability of not committing a Type I error. Due to the symmetry of this kind of *OC* curve about the dashed

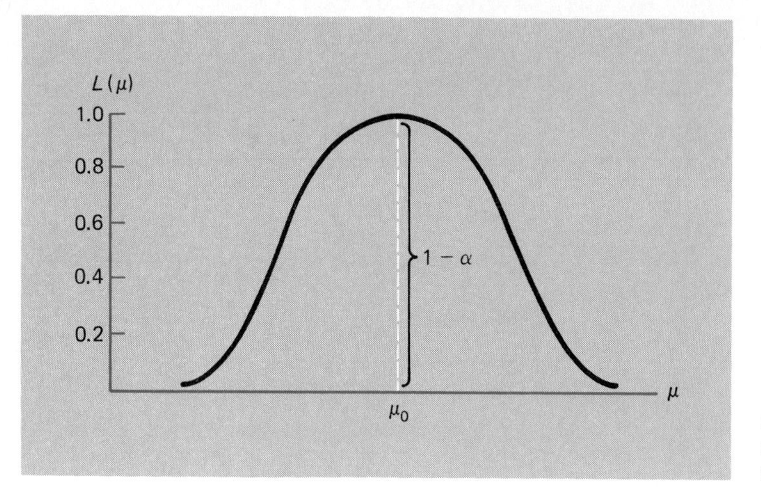

FIGURE 7.6
Operating characteristic curve for two-tailed test.

vertical line through μ_0, all problems relating to Type II errors can be answered on the basis of the values of $L(\mu)$ for $\mu > \mu_0$.

In many practical situations, the probabilities of Type II errors can be determined directly from charts like those given in Table 8 at the end of the book. They enable us to read the values of β that correspond to given values of μ_0, μ_1, σ, n, and α, and they are based on the assumption that the sampling distribution of the mean is a normal distribution; hence, they can be used when n is large or the population we are sampling has roughly the shape of a normal distribution. Charts (a) and (b) apply to one-tailed tests with $\alpha = 0.05$ or $\alpha = 0.01$, and they apply regardless of whether the alternative hypothesis is $\mu > \mu_0$ or $\mu < \mu_0$, since the quantity that we plot on the horizontal scale is not μ but

Quantity needed for use of Table 8

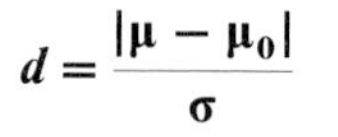

$$d = \frac{|\mu - \mu_0|}{\sigma}$$

namely, the absolute value of the difference between μ and μ_0 divided by σ. The following examples illustrate this technique.

EXAMPLE Suppose we want to investigate the claim that the sound intensity of certain vacuum cleaners is a random variable having a normal distribution with a mean of 75.20 dB and a standard deviation of 3.6 dB. Specifically, we shall want to test the null hypothesis $\mu = 75.20$ against the alternative hypothesis $\mu > 75.20$ on the basis of measurements of the sound intensity of $n = 15$ of these machines. If the probability of a Type I error is to be $\alpha = 0.05$, what is the probability of a Type II error for $\mu = 77.00$ dB?

Solution Since $d = \dfrac{|77.00 - 75.20|}{3.6} = 0.50$, we mark this point on the horizontal scale of chart (a) of Table 8, and, going up vertically until we come to the point where the line is crossed by the OC curve for $n = 15$, we find that the corresponding probability of a Type II error is about 0.38.

■

EXAMPLE Suppose we want to check the claim that in city traffic the mileage a person will get per tankful with a certain kind of car may be looked upon as a random variable having a normal distribution with a mean of 25.0 miles per gallon and a standard deviation of 0.8 mile per gallon. Specifically, we shall want to test the null hypothesis $\mu = 25.0$ against the alternative hypothesis $\mu < 25.0$ on the basis of the mileage obtained in city traffic with a tankful of gas by 10 of these cars. If the probability of a Type I error is to be $\alpha = 0.01$, what is the probability of a Type II error for $\mu = 24.0$ miles per gallon?

Solution Since $d = \dfrac{|24.0 - 25.0|}{0.8} = 1.25$, we mark this point on the horizontal scale of chart (b) of Table 8, and, going up vertically until we come to the point where the line is crossed by the OC curve for $n = 10$, we find that the corresponding probability of a Type II error is about 0.06.

■

For two-tailed tests, that is, for tests of the null hypothesis $\mu = \mu_0$ against the alternative hypothesis $\mu \neq \mu_0$, the procedure is the same as before, with the exception that we must now refer to charts (c) and (d) of Table 8. Note that these charts contain only the right-hand halves of the respective OC curves (as shown in Figure 7.6), and that is all we need, since these OC curves are symmetrical about μ_0.

EXAMPLE Suppose that the length of certain machine parts may be looked upon as a random variable having a normal distribution with a mean of 2.000 cm and a standard deviation of 0.050 cm. Specifically, we shall want to test the null hypothesis $\mu = 2.000$ against the alternative hypothesis $\mu \neq 2.000$ on the basis of the mean of a random sample of size $n = 30$. If the probability of a Type I error is to be $\alpha = 0.05$, what is the probability of a Type II error for $\mu = 2.010$?

Solution Since $d = \dfrac{|2.010 - 2.000|}{0.050} = 0.20$, we mark this point on the horizontal scale of chart (c) of Table 8, and, going up vertically until we come to the point where the line is crossed by the OC curve for $n = 30$, we find that the corresponding probability of a Type II error is about 0.80.

■

Table 8 can also be used to solve problems similar to the one of Exercise 7.34 on page 227, in which we are asked to determine the sample size that is required by given values of μ_0, μ_1, α, and β.

EXAMPLE With reference to the drying time example on page 220, where we had $\mu_0 = 20$, $\sigma = 2.4$, and $\alpha = 0.05$, how large a sample would we need so that $\beta = 0.10$ for $\mu = 21$?

Solution Since $d = \dfrac{|21 - 20|}{2.4} = 0.42$, we locate the point corresponding to $d = 0.42$ and $\beta = 0.10$ on chart (c) of Table 8, and we find that it lies barely above the OC curve for $n = 50$. Thus, a sample of size $n = 50$ will serve the stated purpose.

■

EXERCISES

7.39 According to the norms established for a mechanical aptitude test, persons who are 18 years old should average 73.2 with a standard deviation of 8.6. If 45 randomly selected persons of that age averaged 76.7, test the null hypothesis $\mu = 73.2$ against the alternative hypothesis $\mu > 73.2$ at the 0.01 level of significance.

7.40 Tests performed with a random sample of 40 diesel engines produced by a large manufacturer show that they have a mean thermal efficiency of 31.4% with a standard deviation of 1.6%. At the 0.01 level of significance, test the null hypothesis $\mu = 32.3\%$ against the alternative hypothesis $\mu \neq 32.3\%$.

7.41 In 64 randomly selected hours of production, the mean and the standard deviation of the number of acceptable pieces produced by an automatic stamping machine are $\bar{x} = 1{,}038$ and $s = 146$. At the 0.05 level of significance does this enable us to reject the null hypothesis $\mu = 1{,}000$ against the alternative hypothesis $\mu > 1{,}000$?

7.42 An oceanographer wants to check whether the average depth of the ocean in a certain region is 67.4 fathoms, as has previously been recorded. What can he conclude at the 0.01 level of significance if soundings taken at 40 random locations in the given region yielded a mean of 69.3 fathoms with a standard deviation of 5.4 fathoms?

7.43 In a labor-management discussion it was brought up that workers at a certain large plant take on the average 32.6 minutes to get to work. If a random sample of 60 workers took on the average 33.8 minutes with a standard deviation of 6.1 minutes, can we reject the null hypothesis $\mu = 32.6$ against the alternative hypothesis $\mu > 32.6$ at the 0.05 level of significance?

7.44 A random sample of six steel beams has a mean compressive strength of 58,392 psi (pounds per square inch) with a standard deviation of 648 psi. Use this information and the level of significance $\alpha = 0.05$ to test whether the true average compressive strength of the steel from which this sample came is 58,000 psi.

7.45 Given a random sample of 5 pints from different production lots, we want to test whether the fat content of a certain kind of ice cream exceeds 14%. What can we conclude at the 0.01 level of significance about the null hypothesis $\mu = 14\%$ if the sample has the mean $\bar{x} = 14.9\%$ and the standard deviation $s = 0.42\%$?

7.46 Test runs with six models of an experimental engine showed that they operated for 24, 28, 21, 23, 32, and 22 minutes with a gallon of a certain kind of fuel. If the probability of a Type I error is to be at most 0.01, is this evidence against a hypothesis that on the average this kind of engine will operate for at least 29 minutes per gallon with this kind of fuel?

7.47 A random sample from a company's very extensive files shows that orders for a certain piece of machinery were filled, respectively, in 10, 12, 19, 14, 15, 18, 11, and 13 days. Use the level of significance $\alpha = 0.01$ to test the claim that on the average such orders are filled in 10.5 days. Choose the alternative hypothesis so that rejection of the null hypothesis $\mu = 10.5$ implies that it takes longer than indicated.

7.48 Five measurements of the tar content of a certain kind of cigarette yielded 14.5, 14.2, 14.4, 14.3, and 14.6 mg per cigarette. Show that the difference between the mean of this sample, $\bar{x} = 14.4$, and the average tar claimed by the manufacturer, $\mu = 14.0$, is significant at $\alpha = 0.05$.

7.49 Suppose that in the preceding exercise the first measurement is recorded incorrectly as 16.0 instead of 14.5. Show that now the difference between the mean of the sample, $\bar{x} = 14.7$, and the average tar content claimed by the cigarette manufacturer, $\mu = 14.0$, is not significant at $\alpha = 0.05$. Explain the apparent paradox that even though the difference between $\bar{x}$ and μ has increased, it is no longer significant.

7.50 Use the method by which we calculated the probability of a Type II error on page 220, to verify the probabilities given in the table on page 234.

7.51 With reference to the vacuum cleaner example on page 236, use Table 8 to find the probabilities of Type II errors for

(a) $\mu = 76.00$; (b) $\mu = 78.00$.

7.52 With reference to the car mileage example on page 237, use Table 8 to find the probability of a Type II error for $\mu = 24.5$.

7.53 With reference to the machine parts example on page 237, use Table 8 to find the probabilities of Type II errors for

(a) $\mu = 2.020$; (b) $\mu = 2.030$;
(c) $\mu = 2.040$.

7.54 If a random sample of size $n = 6$ is used to test the hypothesis that the mean of a normal population with $\sigma = 4$ is $\mu_0 = 50$ against the alternative hypothesis that $\mu > 50$, and $\alpha = 0.01$, use Table 8 to determine the probability of committing a Type II error when

(a) $\mu = 52$; (b) $\mu = 53$;
(c) $\mu = 54$; (d) $\mu = 55$;
(e) $\mu = 56$; (f) $\mu = 57$.

7.55 Rework Exercise 7.54, using the alternative hypothesis $\mu \neq 50$ instead of the alternative hypothesis $\mu > 50$.

7.56 If a random sample of size $n = 8$ is used to test the hypothesis that the mean of a normal population with $\sigma = 20$ is $\mu_0 = 200$ against the alternative hypothesis $\mu \neq 200$, and $\alpha = 0.05$, use Table 8 to determine the probability of committing a Type II error when

(a) $\mu = 190$; (b) $\mu = 185$;
(c) $\mu = 180$; (d) $\mu = 175$;
(e) $\mu = 170$.

7.57 Rework Exercise 7.56, using the alternative hypothesis $\mu < 200$ instead of the alternative hypothesis $\mu \neq 200$.

7.58 Use Table 8 to rework part (b) of Exercise 7.31 on page 227.

7.59 Use Table 8 to rework Exercise 7.35 on page 228.

7.60 The statistical package *MINITAB* will calculate t tests. With the sulfur emission data in C1, the command

```
TTEST MU=20.0 , C1
```

produces the output

```
TEST OF MU = 20.000 VS MU N.E. 20.000

              N       MEAN     STDEV   SE MEAN        T    P VALUE
C1           80     18.896     5.656     0.632    -1.75      0.085
```

You must compare your preselected α with the printed P value in order to obtain the conclusion of your test. Here we would reject H_0: $\mu = 20.0$ at level 0.10.

(a) Test H_0: $\mu = 21.5$ with $\alpha = 0.05$.

(b) Test H_0: $\mu = 16.0$ for the data in Exercise 2.58 with $\alpha = 0.05$.

7.61 The statistical package *MINITAB* will calculate the small sample confidence interval for μ. With the sulfur emissions data in C1, the command

```
TINTERVAL 95 PERCENT    C1.
```

produces the output

```
              N       MEAN     STDEV   SE MEAN    95.0 PERCENT C.I.
C1           80     18.896     5.656     0.632   ( 17.637,  20.155)
```

(a) Is the test in Exercise 7.60 consistent with this interval?

(b) Obtain a 90% confidence interval for μ.

(c) Obtain a 95% confidence interval for μ with the aluminum alloy data on page 15.

7.62 You can simulate the coverage of the small sample confidence intervals for μ by generating 20 samples of size 10 from a normal distribution with $\mu = 20$ and $\sigma = 5$ and computing the 95% confidence intervals according to the formula on page 212. The *MINITAB* commands are

```
RANDOM 10 C1-C20;
NORMAL MU = 20 SIGMA = 5.
TINTERVAL 95 PERCENT C1-C20
```

(a) From your output, determine the proportion of the 20 intervals that cover the true mean $\mu = 20$.

(b) Repeat with 20 samples of size 5.

7.9 INFERENCES CONCERNING TWO MEANS

There are many statistical problems in which we are faced with decisions about the relative size of the means of two or more populations. Leaving the general problem until Chapter 12, we shall devote this section to tests concerning the difference between two means. For example, if two methods of welding are being considered for use with railroad rails, we may take samples and decide which is better by comparing their mean strengths; also, if a licensing examination is given to engineers who graduated from two different colleges, we may want to decide whether any observed difference between the means of the scores of the students from the two colleges is significant or whether it may be attributed to chance.

Formulating the problem more generally, we shall consider two populations having the means μ_1 and μ_2 and the variances σ_1^2 and σ_2^2, and we shall want to test the null hypothesis $\mu_1 - \mu_2 = \delta$, where δ is a specified constant, on the basis of independent random samples of size n_1 and n_2. Analogous to the tests concerning one mean, we shall consider tests of this null hypothesis against each of the alternatives $\mu_1 - \mu_2 < \delta$, $\mu_1 - \mu_2 > \delta$, and $\mu_1 - \mu_2 \neq \delta$. The test, itself, will depend on the difference between the sample means, $\bar{x}_1 - \bar{x}_2$, and if both samples come from normal populations with known variances, it can be based on the statistic

$$z = \frac{(\bar{x}_1 - \bar{x}_2) - \delta}{\sigma_{\bar{x}_1 - \bar{x}_2}}$$

which is a value of a random variable having the standard normal distribution. Here $\sigma_{\bar{x}_1 - \bar{x}_2}$ is the standard deviation of the sampling distribution of the difference between the sample means, and its value for random samples from infinite populations may be obtained with the use of the following theorem, which we shall state without proof.

Mean and variance of sum (or difference) of two random variables

Theorem 7.1 ***If the distributions of two independent random variables have the means μ_1 and μ_2 and the variances σ_1^2 and σ_2^2, then the distribution of their sum (or difference) has the mean $\mu_1 + \mu_2$ (or $\mu_1 - \mu_2$) and the variance $\sigma_1^2 + \sigma_2^2$.***

To find the variance of the difference between the means of two independent random samples of size n_1 and n_2 from infinite populations, note first that the variances of the two means, themselves, are

$$\sigma_{\bar{x}_1}^2 = \frac{\sigma_1^2}{n_1} \quad \text{and} \quad \sigma_{\bar{x}_2}^2 = \frac{\sigma_2^2}{n_2}$$

where σ_1^2 and σ_2^2 are the variances of the respective populations. Thus, by Theorem 7.1 we have

$$\sigma_{\bar{x}_1 - \bar{x}_2}^2 = \frac{\sigma_1^2}{n_1} + \frac{\sigma_2^2}{n_2}$$

and the test statistic can be written as

Statistic for test concerning difference between two means

$$z = \frac{(\bar{x}_1 - \bar{x}_2) - \delta}{\sqrt{\dfrac{\sigma_1^2}{n_1} + \dfrac{\sigma_2^2}{n_2}}}$$

which is a value of a random variable having the standard normal distribution. In deriving this result we assumed that we are sampling normal populations. However, the above statistic can also be used when our samples are large enough so that we can apply the central limit theorem and approximate σ_1 and σ_2 with s_1 and s_2, namely, when n_1 and n_2 are both greater than or equal to 30.

Analogous to the table on page 229, the critical regions for testing the null hypothesis $\mu_1 - \mu_2 = \delta$ are as follows:

Critical Regions for Testing $\mu_1 - \mu_2 = \delta$
(Normal populations with σ_1 and σ_2 known, or large samples)

Alternative hypothesis	*Reject null hypothesis if:*
$\mu_1 - \mu_2 < \delta$	$z < -z_\alpha$
$\mu_1 - \mu_2 > \delta$	$z > z_\alpha$
$\mu_1 - \mu_2 \neq \delta$	$z < -z_{\alpha/2}$ or $z > z_{\alpha/2}$

Although δ can be any constant, it is worth noting that in the great majority of problems its value is zero and we test the null hypothesis of "no difference," namely, the null hypothesis $\mu_1 = \mu_2$.

EXAMPLE To test the claim that the resistance of electric wire can be reduced by more than 0.050 ohm by alloying, 32 values obtained for standard wire yielded $\bar{x}_1 = 0.136$ ohm and $s_1 = 0.004$ ohm, and 32 values obtained for alloyed wire yielded $\bar{x}_2 = 0.083$ ohm and $s_2 = 0.005$ ohm. At the 0.05 level of significance, does this support the claim?

Solution

1. *Null hypothesis*: $\mu_1 - \mu_2 = 0.050$
   Alternative hypothesis: $\mu_1 - \mu_2 > 0.050$
2. *Level of significance*: $\alpha = 0.05$
3. *Criterion*: Reject the null hypothesis if $z > 1.645$, where z is given by the formula on page 241.
4. *Calculations*:

$$z = \frac{0.136 - 0.083 - 0.050}{\sqrt{\frac{(0.004)^2}{32} + \frac{(0.005)^2}{32}}} = 2.65$$

5. *Decision*: Since $z = 2.65$ exceeds 1.645, the null hypothesis must be rejected; that is, the data substantiate the claim. From Table 3, the P value is 0.004.

■

EXAMPLE A company claims that its light bulbs are superior to those of its main competitor. If a study showed that a sample of $n_1 = 40$ of its bulbs had a mean lifetime of 647 hours of continuous use with a standard deviation of 27 hours, while a sample of $n_2 = 40$ bulbs made by its main competitor had a mean lifetime of 638 hours of continuous use with a standard deviation of 31 hours, does this substantiate the claim at the 0.05 level of significance?

Solution

1. *Null hypothesis*: $\mu_1 - \mu_2 = 0$
   Alternative hypothesis: $\mu_1 - \mu_2 > 0$
2. *Level of significance*: $\alpha = 0.05$
3. *Criterion*: Reject the null hypothesis if $z > 1.645$, where z is given by the formula on page 241.
4. *Calculations*:

$$z = \frac{647 - 638}{\sqrt{\dfrac{27^2}{40} + \dfrac{31^2}{40}}} = 1.38$$

5. *Decision*: Since $z = 1.38$ does not exceed 1.645, the null hypothesis cannot be rejected; that is, the observed difference between the two sample means is not significant.

■

To judge the strength of support for the null hypothesis when it is not rejected, we consider Type II errors, for which the probabilities depend on the actual alternative differences $\delta' = \mu_1 - \mu_2$. Fortunately, these can be determined with the use of Table 8 (so long as we are sampling from normal populations with known standard deviations or both samples are large). The quantity that we mark on the horizontal scale is

Quantity needed for use of Table 8

$$d = \frac{|\delta - \delta'|}{\sqrt{\sigma_1^2 + \sigma_2^2}}$$

where δ is the value of $\mu_1 - \mu_2$ assumed under the null hypothesis and δ' is the alternative value of $\mu_1 - \mu_2$ with which we are concerned. If $n_1 = n_2 = n$, the probability of a Type II error is read off the OC curve corresponding to this value of n; if $n_1 \neq n_2$, it is read off the OC curve corresponding to

$$n = \frac{\sigma_1^2 + \sigma_2^2}{\dfrac{\sigma_1^2}{n_1} + \dfrac{\sigma_2^2}{n_2}}$$

EXAMPLE With reference to the preceding example, what is the probability of a Type II error for $\delta' = 16$ hours?

Solution Since $d = \dfrac{|0 - 16|}{\sqrt{27^2 + 31^2}} = 0.39$, we mark this point on the horizontal scale of chart (a) of Table 8, and, going up vertically until we come to the OC curve for $n = 40$, we find that the probability of a Type II error is approximately 0.20.

■

When n_1, n_2, or both are small and the population variances are unknown, we can base tests of the null hypothesis $\mu_1 - \mu_2 = \delta$ on a suitable t statistic, provided we can assume that both populations are normal with $\sigma_1 = \sigma_2 (= \sigma)$. Under these conditions, it can be shown that

$$t = \frac{(\bar{x}_1 - \bar{x}_2) - \delta}{\hat{\sigma}_{\bar{x}_1 - \bar{x}_2}}$$

is a value of a random variable having the t distribution with $n_1 + n_2 - 2$ degrees of freedom, where $\hat{\sigma}_{\bar{x}_1 - \bar{x}_2}$ is the square root of an estimate of

$$\sigma_{\bar{x}_1 - \bar{x}_2}^2 = \frac{\sigma_1^2}{n_1} + \frac{\sigma_2^2}{n_2} = \sigma^2\left(\frac{1}{n_1} + \frac{1}{n_2}\right)$$

with σ^2 estimated by **pooling** the sums of the squared deviations from the respective sample means. That is, we estimate σ^2 as

$$\frac{\sum (x_1 - \bar{x}_1)^2 + \sum (x_2 - \bar{x}_2)^2}{n_1 + n_2 - 2} = \frac{(n_1 - 1)s_1^2 + (n_2 - 1)s_2^2}{n_1 + n_2 - 2}$$

where $\sum (x_1 - \bar{x}_1)^2$ is the sum of the squared deviations from the mean for the first sample, while $\sum (x_2 - \bar{x}_2)^2$ is the sum of the squared deviations from the mean for the second sample. We divide by $n_1 + n_2 - 2$, since there are $n_1 - 1$ independent deviations from the mean in the first sample, $n_2 - 1$ in the second, and we thus have $n_1 + n_2 - 2$ independent deviations from the mean to estimate the population variance. Substituting this estimate of σ^2 into the above expression for $\sigma_{\bar{x}_1 - \bar{x}_2}^2$ and then substituting the square root of the result into the denominator of the formula for t above, we finally obtain

Statistic for small-sample test concerning difference between two means

$$t = \frac{(\bar{x}_1 - \bar{x}_2) - \delta}{\sqrt{(n_1 - 1)s_1^2 + (n_2 - 1)s_2^2}} \sqrt{\frac{n_1 n_2 (n_1 + n_2 - 2)}{n_1 + n_2}}$$

which is a value of a random variable having the t distribution with $n_1 + n_2 - 2$ degrees of freedom. The criteria for the **two-sample t test** based on this statistic are like those given in the table on page 242 with z replaced by t and z_α and $z_{\alpha/2}$ replaced by t_α and $t_{\alpha/2}$. In the application of this test, n_1 and n_2 may be small, yet $n_1 + n_2 - 2$ may be 30 or more; in that case we use the critical values given in the bottom line of Table 4.

EXAMPLE The following random samples are measurements of the heat-producing capacity (in millions of calories per ton) of specimens of coal from two mines:

Mine 1:	8,260,	8,130,	8,350,	8,070,	8,340	
Mine 2:	7,950,	7,890,	7,900,	8,140,	7,920,	7,840

Use the 0.01 level of significance to test whether the difference between the means of these two samples is significant.

Solution

1. *Null hypothesis*: $\mu_1 - \mu_2 = 0$
   Alternative hypothesis: $\mu_1 - \mu_2 \neq 0$
2. *Level of significance*: $\alpha = 0.01$
3. *Criterion*: Reject the null hypothesis if $t < -3.250$ or $t > 3.250$, where 3.250 is the value of $t_{0.005}$ for $5 + 6 - 2 = 9$ degrees of freedom and t is given by the formula above.
4. *Calculations*: The means and the variances of the two samples are $\bar{x}_1 = 8{,}230$, $\bar{x}_2 = 7{,}940$, $s_1^2 = \frac{63{,}000}{4} = 15{,}750$, and $s_2^2 = \frac{54{,}600}{5} = 10{,}920$, so that

$$t = \frac{(8{,}230 - 7{,}940)}{\sqrt{63{,}000 + 54{,}600}} \sqrt{\frac{5 \cdot 6 \cdot 9}{11}} = 4.19$$

5. *Decision*: Since $t = 4.19$ exceeds 3.250, the null hypothesis must be rejected at level $\alpha = 0.01$. We conclude that the average heat-producing capacity of the coal from the two mines is not the same.

■

In the preceding example we went ahead and performed the two-sample t test, tacitly assuming that the population variances are equal. Fortunately, the test is not overly sensitive to small differences between the population variances, and the procedure used in this instance is justifiable. As a rule of thumb, if one variance is four times the other, we should be concerned. A transformation will often improve the situation. As another alternative there is the Smith-Satterthwaithe test mentioned in Exercise 7.70.

Confidence intervals follow directly from the acceptance region for the tests. For two normal populations with equal variances,

Small-sample confidence interval

> ***The*** **100(1 − α)%** ***confidence interval for δ is***
>
> $$\bar{x}_1 - \bar{x}_2 \pm t_{\alpha/2}\sqrt{\frac{(n_1-1)s_1^2+(n_2-1)s_2^2}{n_1+n_2-2}}\sqrt{\frac{n_1+n_2}{n_1 n_2}}$$
>
> ***where*** $t_{\alpha/2}$ **is based on** $\nu = n_1 - n_2 - 2$ ***degrees of freedom.***

EXAMPLE Strength measurements on an aluminum alloy were introduced on page 15. A second alloy yielded measurements given in the following stem-and-leaf display. Find a 95% confidence interval for the difference in mean strength δ.

Solution We first place the observations on the two alloys in stem-and-leaf displays. Note that the observations from the first alloy appear normal, but those on the second alloy may deviate. Since the sample sizes are relatively large, this will not cause any difficulty.

Alloy 1, $N = 58$
Leaf unit = 0.10

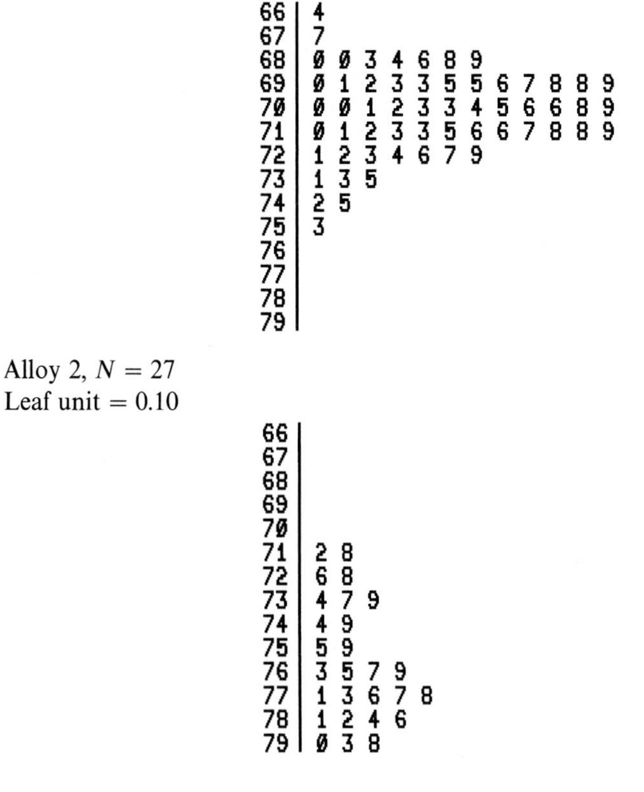

```
66 | 4
67 | 7
68 | Ø Ø 3 4 6 8 9
69 | Ø 1 2 3 3 5 5 6 7 8 8 9
7Ø | Ø Ø 1 2 3 3 4 5 6 6 8 9
71 | Ø 1 2 3 3 5 6 6 7 8 8 9
72 | 1 2 3 4 6 7 9
73 | 1 3 5
74 | 2 5
75 | 3
76 |
77 |
78 |
79 |
```

Alloy 2, $N = 27$
Leaf unit = 0.10

```
66 |
67 |
68 |
69 |
7Ø |
71 | 2 8
72 | 6 8
73 | 4 7 9
74 | 4 9
75 | 5 9
76 | 3 5 7 9
77 | 1 3 6 7 8
78 | 1 2 4 6
79 | Ø 3 8
```

A computer calculation gives the sample means and standard deviations

```
            N      MEAN       STDEV
ALLOY 1    58      70.70       1.80
ALLOY 2    27      76.13       2.42
```

From another computer calculation (or by interpolation in Table 4), we find $t_{0.025} = 1.99$, so the 95% confidence interval is

$$\bar{x}_1 - \bar{x}_2 \pm t_{\alpha/2}\sqrt{\frac{(n_1-1)s_1^2+(n_2-1)s_2^2}{n_1+n_2-2}}\sqrt{\frac{n_1+n_2}{n_1 n_2}}$$

$$= 70.70 - 76.13 \pm 1.99\sqrt{\frac{57(1.80)^2+26(2.42)^2}{83}}\sqrt{\frac{85}{58 \times 27}}$$

or

$$-6.4 < \mu_1 - \mu_2 < -4.5$$

We are 95% confident that the mean strength of alloy 2 is 4.5 to 6.4 thousand pounds per square inch higher than the mean of alloy 1.

■

It is good practice to show stem-and-leaf displays, boxplots, or histograms. Often they reveal more than a mean difference.

The large sample confidence interval follows from the test on page 241.

Large-sample confidence interval

$$\bar{x}_1 - \bar{x}_2 \pm z_{\alpha/2}\sqrt{\frac{s_1^2}{n_1}+\frac{s_2^2}{n_2}}$$

EXAMPLE Referring to the previous example, find the 95% large sample confidence interval.

Solution

$$\bar{x}_1 - \bar{x}_2 \pm z_{\alpha/2}\sqrt{\frac{s_1^2}{n_1}+\frac{s_2^2}{n_2}} = 70.70 - 76.13 \pm 1.96\sqrt{\frac{(1.80)^2}{58}+\frac{(2.42)^2}{27}}$$

or

$$-6.5 < \mu_1 - \mu_2 < -4.4$$

There is not much difference between this 95% confidence interval and the one in the previous example, where the variances were pooled.

■

In the application of the two-sample t test we shall also have to watch that the samples are independent. For instance, the test cannot be used when we deal with "before and after" kind of data, the I.Q.'s of husbands and wives, and numerous other kinds of situations where the data are naturally paired. In that case we work with the (signed) differences of the paired data and test whether these differences may be looked upon as a random sample from a population for which $\mu = \delta$, usually $\mu = 0$. If the sample is small, we use the one-sample t test on page 231; otherwise, we use the corresponding large-sample test on page 229.

EXAMPLE The following are the average weekly losses of worker-hours due to accidents in 10 industrial plants before and after a certain safety program was put into operation:

45 and 36, 73 and 60, 46 and 44, 124 and 119, 33 and 35,
57 and 51, 83 and 77, 34 and 29, 26 and 24, and 17 and 11

Use the 0.05 level of significance to test whether the safety program is effective.

Solution

1. *Null hypothesis*: $\mu = 0$ (where μ is the mean of the population of differences sampled)
   Alternative hypothesis: $\mu > 0$
2. *Level of significance*: $\alpha = 0.05$
3. *Criterion*: Reject the null hypothesis if $t > 1.833$, the value of $t_{0.05}$ for $10 - 1 = 9$ degrees of freedom, where

$$t = \frac{\bar{x} - \mu_0}{s/\sqrt{n}}$$

   and $\bar{x}$ and s are the mean and the standard deviation of the differences.
4. *Calculations*: The differences are 9, 13, 2, 5, -2, 6, 6, 5, 2, and 6, their mean is $\bar{x} = 5.2$, their standard deviation is $s = 4.08$, so that

$$t = \frac{5.2 - 0}{4.08/\sqrt{10}} = 4.03$$

5. *Decision*: Since $t = 4.03$ exceeds 1.833, the null hypothesis must be rejected at level $\alpha = 0.05$. We conclude that the industrial safety program is effective. The evidence is very strong, since the P value is less than 0.005.

■

In connection with this kind of problem, the one-sample t test is referred to as the **paired-sample t test**.

EXAMPLE Referring to the previous example, find a 90% confidence interval for the mean improvement in lost worker-hours.

Solution The $n = 10$ differences have $\bar{x} = 5.2$ and $s = 4.08$. Since $t_{0.05} = 1.833$, the 90% confidence interval for μ, the improvement, is

$$5.2 - 1.833 \frac{4.08}{\sqrt{10}} < \mu < 5.2 + 1.833 \frac{4.08}{\sqrt{10}}, \quad \text{or} \quad 4.0 < \mu < 6.4$$

worker-hours per week. We are 90% confident that the between 4.0 and 6.4 fewer worker-hours per week are lost, on average.

■

EXERCISES

7.63 The diameters of rotor shafts in a lot have a mean of 0.249 inch and a standard deviation of 0.003 inch. The inner diameters of bearings in another lot have a mean of 0.255 inch and a standard deviation of 0.002 inch.

(a) What are the mean and the standard deviation of the clearances between shafts and bearings selected from these lots?

(b) If a shaft and a bearing are selected at random, what is the probability that the shaft will not fit inside the bearing? (Assume that both dimensions are normally distributed.)

7.64 An investigation of two kinds of photocopying equipment showed that 71 failures of the first kind of equipment took on the average 83.2 minutes to repair with a standard deviation of 19.3 minutes, while 75 failures of the second kind of equipment took on the average 90.8 minutes to repair with a standard deviation of 21.4 minutes.

(a) Test the null hypothesis $\mu_1 - \mu_2 = 0$ (namely, the hypothesis that on the average it takes an equal amount of time to repair either kind of equipment) against the alternative hypothesis $\mu_1 - \mu_2 \neq 0$ at the level of significance $\alpha = 0.05$.

(b) Using 19.3 and 21.4 as estimates of σ_1 and σ_2 and referring to Table 8, find the probability of accepting the null hypothesis $\mu_1 - \mu_2 = 0$ with the criterion of part (a) when actually $\mu_1 - \mu_2 = -12$.

7.65 Suppose that we want to investigate whether on the average men earn more than $20 per week more than women in a certain industry. If sample data show that 60 men earn on the average $\bar{x}_1 = \$292.50$ per week with a standard deviation of $s_1 = \$15.60$, while 60 women earn on the average $\bar{x}_2 = \$266.10$ per week with a standard deviation of $s_2 = \$18.20$, what can we conclude at the 0.01 level of significance?

7.66 Studying the flow of traffic at two busy intersections between 4 P.M. and 6 P.M. (to determine the possible need for turn signals), it was found that on 40 weekdays there were on the average 247.3 cars approaching the first intersection from the south which made left turns, while on 30 weekdays there were on the average 254.1 cars approaching the second intersection from the south which made left turns. The corresponding sample standard deviations are $s_1 = 15.2$ and $s_2 = 18.7$.

(a) Test the null hypothesis $\mu_1 - \mu_2 = 0$ against the alternative hypothesis $\mu_1 - \mu_2 \neq 0$ at the level of significance $\alpha = 0.01$.

(b) Using 15.2 and 18.7 as estimates of σ_1 and σ_2 and referring to Table 8, find the probability of accepting the null hypothesis $\mu_1 - \mu_2 = 0$ when actually $|\mu_1 - \mu_2| = 15.6$.

7.67 Measuring specimens of nylon yarn taken from two spinning machines, it was found that eight specimens from the first machine had a mean denier of 9.67 with a standard deviation of 1.81 while 10 specimens from the second machine had a mean denier of 7.43 with a standard deviation of 1.48. Assuming that the populations sampled are normal and have the same variance, test the null hypothesis $\mu_1 - \mu_2 = 1.5$ against the alternative hypothesis $\mu_1 - \mu_2 > 1.5$ at the 0.05 level of significance.

7.68 As part of an industrial training program, some trainees are instructed by Method A, which is straight teaching-machine instruction, and some are instructed by Method B, which also involves the personal attention of an instructor. If random samples of size 10 are taken from large groups of trainees instructed by each of these two methods, and the scores which they obtained in an appropriate achievement test are

Method A:	71,	75,	65,	69,	73,	66,	68,	71,	74,	68
Method B:	72,	77,	84,	78,	69,	70,	77,	73,	65,	75

use the 0.05 level of significance to test the claim that Method B is more effective. Assume that the populations sampled can be approximated closely with normal distributions having the same variance.

7.69 The following are the number of sales which a sample of nine salespeople of industrial chemicals in California and a sample of six salespeople of industrial chemicals in Oregon made over a certain fixed period of time:

California:	59,	68,	44,	71,	63,	46,	69,	54,	48
Oregon:	50,	36,	62,	52,	70,	41			

Assuming that the populations sampled can be approximated closely with normal distributions having the same variance, test the null hypothesis $\mu_1 - \mu_2 = 0$ against the alternative hypothesis $\mu_1 - \mu_2 \neq 0$ at the 0.01 level of significance.

7.70 When we deal with two independent random samples from normal populations whose variances seem to be unequal, the following test, called the **Smith–Satterthwaite test**, can be used instead of the two-sample t test. The test statistic

$$t' = \frac{(\bar{x}_1 - \bar{x}_2) - \delta}{\sqrt{\dfrac{s_1^2}{n_1} + \dfrac{s_2^2}{n_2}}}$$

is a value of a random variable having approximately the t distribution with

$$\frac{\left(\dfrac{s_1^2}{n_1} + \dfrac{s_2^2}{n_2}\right)^2}{\dfrac{(s_1^2/n_1)^2}{n_1 - 1} + \dfrac{(s_2^2/n_2)^2}{n_2 - 1}}$$

degrees of freedom. Use this test in the following problems:

(a) The following are the Brinell hardness values obtained for samples of two magnesium alloys:

Alloy 1:	66.3,	63.5,	64.9,	61.8,	64.3,	64.7,	65.1,
	64.5,	68.4,	63.2				
Alloy 2:	71.3,	60.4,	62.6,	63.9,	68.8,	70.1,	64.8,
	68.9,	65.8,	66.2				

Use the 0.05 level of significance to test the null hypothesis $\mu_1 - \mu_2 = 0$ against the alternative hypothesis $\mu_1 - \mu_2 < 0$.

(b) To compare two kinds of bumper guards, six of each kind were mounted on a certain kind of compact car. Then each car was run into a concrete wall at 5 miles per hour, and the following are the costs of the repairs (in dollars):

Bumper guard 1:	107,	148,	123,	165,	102,	119
Bumper guard 2:	134,	115,	112,	151,	133,	129

Use the 0.01 level of significance to test whether the difference between the two sample means is significant.

7.71 The following data were obtained in an experiment designed to check whether there is a systematic difference in the weights obtained with two different scales:

	Weight in grams	
	Scale I	*Scale II*
Rock specimen 1	11.23	11.27
Rock specimen 2	14.36	14.41
Rock specimen 3	8.33	8.35
Rock specimen 4	10.50	10.52
Rock specimen 5	23.42	23.41
Rock specimen 6	9.15	9.17
Rock specimen 7	13.47	13.52
Rock specimen 8	6.47	6.46
Rock specimen 9	12.40	12.45
Rock specimen 10	19.38	19.35

Use the paired-sample t test at the 0.05 level of significance to test whether the difference of the means of the weights obtained with the two scales is significant.

7.72 In a study of the effectiveness of physical exercise in weight reduction, a group of 16 persons engaged in a prescribed program of physical exercise for one month showed the following results:

Weight before (pounds)	*Weight after (pounds)*	*Weight before (pounds)*	*Weight after (pounds)*
209	196	170	164
178	171	153	152
169	170	183	179
212	207	165	162
180	177	201	199
192	190	179	173
158	159	243	231
180	180	144	140

Use the 0.01 level of significance to test whether the prescribed program of exercise is effective.

7.73 With reference to the example on page 242, construct a 95% confidence interval for the true difference between the average resistance of the two kinds of wire.

7.10
RANDOMIZATION AND PAIRING

In many comparative studies, the investigator applies one or the other treatment to an object we call an experimental unit. The assignment of experimental units to groups or pairs can be crucial to the validity of the statistical procedures for comparing the means of two populations. Suppose a chemist has a new formula for waterproofing that she applies to several persons shoes that are almost like new. She also applies the old formula to several pairs of scuffed shoes. At the end of a month, she will measure the ability of each pair of shoes to withstand water. It doesn't take a statistician to see that this is not a good experimental design. The persons with scuffed shoes probably walk a lot more and do so in all kinds of weather. These sources of variation could very well lead to systematic biases that make the new formula seem better than the old even when this was not the case. The pairs of shoes need to be assigned to the treatments, old and new formula waterproofing, in a random manner.

When possible, the $n = n_1 + n_2$ experimental units should be assigned at random to the two treatments. This means that all $\binom{n}{n_1}$ possible selections of n_1 units to receive the first treatment are equally likely. Practically, the assignment is accomplished by selecting n_1 random integers between 1 and n. The corresponding experimental units are assigned to the first treatment. Generally, a test will have more power if the two sample sizes are equal.

In summary,

> ***Randomization prevents uncontrolled sources of variation from exerting a systematic influence on the responses.***

The object of **pairing** experimental units, according to a characteristic that is likely to influence the response, is to eliminate this source of variation from the comparison. In the context of waterproofing for shoes, each person could have the old formula on one shoe and the new formula on the other. Since the paired t analysis only uses differences from the same pair, this experimental strategy should eliminate most of the variation in response due to different terrain, distance covered and weather conditions.

Even after units are paired, there is a need for randomization. For each pair, a fair coin should be flipped to assign the treatments. In a context of the waterproofing example, the old formula could be applied to the right shoe if heads and the left shoe if tails. The new formula is applied to the other shoe. This randomization, restricted to be within pairs, would prevent systematic influences caused by the fact that a majority of persons would tend to kick things with their right shoe.

Notice that in the example on page 248, the experimenter had no control over the before and after. Many uncontrolled variables may also have changed over the course of the experiment, fewer working hours due to strikes, phasing out of an old type of equipment etc. One of these could have been the cause for the improvement rather than the safety program.

We pursue the ideas of randomization and blocking in Chapter 12. Our purpose here was to show what practical steps can be taken to meet the idealistic assumptions of random samples.

EXERCISES

7.74 An investigator wants to compare two busy network protocols by recording the number of messages that are successfully passed by the network in a day. Describe how to select five of the next ten working days for trying Protocol 1. Protocol 2 would be tried on the other five days.

7.75 An electrical engineer has developed a modified power source for reducing noise in either one of the two channels of a stereo amplifier. If six modified power sources and six stereo amplifiers are available for a comparative test of the old versus the modified power source,

(a) Describe how you would select the three amplifiers in which to insert two modified power sources in each.

(b) Describe how you would conduct a paired comparison and then randomize within the pair.

7.76 It takes an average of 10 weeks to train a typical employee to run a computer-aided machine. The instructor has a new approach that she feels will lead to faster learning. She intends to teach 5 persons by the new method and then compare the results with 10 week average for the old method. In order to obtain 5 students from the 25 available candidates, she asks for volunteers. Why is this a bad idea?

7.77 How would you randomize, for a two-sample test, if fifty cars are available for an emissions study and you want to compare a modified air pollution device with the current production?

7.11 REVIEW EXERCISES

7.78 Specify the null hypothesis and the alternative hypothesis in each of the following cases.

(a) An engineer hopes to establish that an additive will increase the viscosity of an oil.

(b) An electrical engineer hopes to establish that an modified circuit board will give a computer a higher average operating speed.

7.79 With reference to the example on page 246, find a 95% confidence interval for the mean strength of aluminum alloy 1.

7.80 While performing a certain task under simulated weightlessness, the pulse rate of 32 astronaut trainees increased on the average by 26.4 beats per minute with a standard deviation of 4.28 beats per minute. What can one assert with 95% confidence about the maximum error if $\bar{x} = 26.4$ is used as a point estimate of the true average increase in the pulse rate of astronaut trainees performing the given task?

7.81 With reference to the preceding exercise, construct a 95% confidence interval for the true average increase in the pulse rate of astronaut trainees performing the given task.

7.82 It is desired to estimate the mean number of hours of continuous use until a certain kind of computer will first require repairs. If it can be assumed that $\sigma = 48$ hours, how large a sample is needed so that one will be able to assert with 90% confidence that the sample mean is off by at most 10 hours?

7.83 A sample of 12 cam shafts intended for use in gasoline engines has an average eccentricity of 1.02 and a standard deviation of 0.044 inch. Assuming the data may be treated as a random sample from a normal population, determine a 95% confidence interval for the actual mean eccentricity of the cam shafts.

7.84 In order to test the durability of a new paint, a highway department had test strips painted across heavily traveled roads in 15 different locations. If on the average the test strips disappeared after they had been crossed by 146,692 cars with a standard deviation of 14,380 cars, construct a 99% confidence interval for the true average number of cars it takes to wear off the paint. Assume normal population.

7.85 Referring to Exercise 7.84 and using 14,380 as an estimate of σ, find the sample size that would have been needed to be able to assert with 95% confidence that the sample mean is off by at most 10,000. [*Hint*: First estimate n_1 by using $z = 1.96$, then use $t_{0.025}$ for $n_1 - 1$ degrees of freedom to obtain a second estimate n_2, and repeat this procedure until the last two values of n thus obtained are equal.]

7.86 With reference to Exercise 2.56, test that the mean charge of the electron is the same for both tubes. Use $\alpha = 0.05$.

7.87 With reference to the previous exercise, find a 90% confidence interval for the difference of the two means.

7.88 With reference to the example on page 7, test that the mean copper content is the same for both heats.

7.89 With reference to the previous exercise, find a 90% confidence interval for the difference of the two means.

7.90 A laboratory technician is timed 20 times in the performance of a task, getting $\bar{x} = 7.9$ and $s = 1.2$. If the probability of a Type I error is to be at most 0.05, does this constitute evidence against the hypothesis that the average time is less than 7.5 minutes?

7.91 Suppose that in the drying-time example on page 220, n is changed from 36 to 50, while the other quantities remain $\mu_0 = 20$, $\sigma = 2.4$, and $\alpha = 0.03$. Find

(a) the new dividing line of the test criterion;

(b) the probability of Type II errors for the same values of μ as shown in the table on page 234.

Also plot the *OC* curve and compare it with the one shown in Figure 7.5.

7.92 Use Table 8 to rework part (a) of Exercise 7.34 on page 227.

7.93 Use Table 8 to rework part (b) of Exercise 7.34 on page 227.

7.94 In an air-pollution study, ozone measurements were taken in a large California city at 5:00 P.M. The eight readings (in parts per million) were: 7.9, 11.3, 6.9, 12.7, 13.2, 8.8, 9.3, 10.6. Assuming the population sampled is normal, construct a 95% confidence interval for the corresponding true mean.

7.95 Random samples are taken from two normal populations with $\sigma_1 = 10.8$ and $\sigma_2 = 14.4$ to test the null hypothesis $\mu_1 - \mu_2 = 53.2$ against the alternative hypothesis $\mu_1 - \mu_2 > 53.2$ at the level of significance $\alpha = 0.01$. Use Table 8 to determine the common sample size $n = n_1 = n_2$ that is required if the probability of accepting the null hypothesis is to be 0.09 when $\mu_1 - \mu_2 = 66.7$.

7.96 With reference to the example on page 246, find a 90% confidence interval for the difference of mean strengths of the alloys

(a) using the pooled procedure;
(b) using the large sample procedure.

7.97 How would you randomize, for a two-sample test, in each of the following cases?

(a) Twenty cars are available for a mileage study and you want to compare a modified spark plug with the regular.
(b) A new oven will be compared with the old. Fifteen ceramic specimens are available for baking.

7.98 With reference to part (a) of Exercise 7.97, how would you pair and then randomize for a paired test?

7.99 Two samples in C1 and C2 can be analyzed using the *MINITAB* command

```
TWOSAMPLE C1 C2;
POOLED.
```

The output relating to the example on page 246 is

```
TWOSAMPLE T FOR ALLOY1 VS ALLOY2
            N    MEAN   STDEV   SE MEAN
ALLOY1     58   70.70    1.80      0.24
ALLOY2     27   76.13    2.42      0.47

95 PCT CI FOR MU ALLOY1 - MU ALLOY2: (-6.36, -4.50)

TTEST MU ALLOY1 = MU ALLOY2 (VS NE): T=-11.58 P=0.0000 DF=83.0
```

Perform the test for the data in Exercise 7.68.

7.12
CHECK LIST OF KEY TERMS (with page references)

Alternative hypothesis 223
Bayesian estimation 215
Classical theory 226
Composite hypothesis 226
Confidence 208
Confidence interval 211
Confidence limits 211
Critical regions 229
Critical values 229
Decision theory 226
Degree of confidence 211
Interval estimate 210
Level of significance 224
Likelihood function 219
Maximum likelihood estimator 219
Neyman-Pearson theory 226
Null hypothesis 222
One-sample t test 231
One-sided alternative 223
One-sided criterion 225

INFERENCES CONCERNING VARIANCES

In Chapter 7 we learned how to judge the size of the error in estimating a population mean, how to construct confidence intervals for means, and how to perform tests of hypotheses about the means of one and of two populations. As we shall see in this and in subsequent chapters, very similar methods apply to inferences about other population parameters.

In this chapter we shall concentrate on population variances, or standard deviations, which are not only important in their own right, but which must sometimes be estimated before inferences about other parameters can be made. Section 8.1 is devoted to the estimation of σ^2 and σ, and Sections 8.2 and 8.3 deal with tests of hypotheses about these parameters.

8.1 THE ESTIMATION OF VARIANCES

In the preceding chapter, there were several instances where we estimated a population standard deviation by means of a sample standard deviation—we substituted s for σ in the large-sample confidence interval for μ on page 211, in the large-sample test concerning μ on page 230, and in the large-sample test concerning the difference between two means on page 242. Since there are many statistical procedures in which s is thus substituted for σ, or s^2 for σ^2, let us show first that the sample variance is, in fact, an unbiased estimate of the population variance.

If $f(x_1, x_2, \ldots, x_n)$ is the joint density of the sample values $x_1, x_2, \ldots,$ and x_n, it follows from the discussion on page 169 that the mean of the sampling distribution of s^2 is given by

$$\int_{-\infty}^{\infty} \int_{-\infty}^{\infty} \cdots \int_{-\infty}^{\infty} s^2 f(x_1, x_2, \ldots, x_n)\, dx_1\, dx_2 \cdots dx_n$$
$$= \int_{-\infty}^{\infty} \int_{-\infty}^{\infty} \cdots \int_{-\infty}^{\infty} \sum_{i=1}^{n} \frac{(x_i - \bar{x})^2}{n-1} f(x_1, x_2, \ldots, x_n)\, dx_1\, dx_2 \cdots dx_n$$

If we now write

$$\sum_{i=1}^{n} (x_i - \bar{x})^2 = \sum_{i=1}^{n} x_i^2 - n\bar{x}^2$$

and interchange the operations of summation and integration, the expression for the mean of the distribution of s^2 becomes

$$\frac{1}{n-1} \sum_{i=1}^{n} \int_{-\infty}^{\infty} \int_{-\infty}^{\infty} \cdots \int_{-\infty}^{\infty} x_i^2 f(x_1, x_2, \ldots, x_n)\, dx_1\, dx_2 \cdots dx_n$$
$$- \frac{n}{n-1} \int_{-\infty}^{\infty} \int_{-\infty}^{\infty} \cdots \int_{-\infty}^{\infty} \bar{x}^2 f(x_1, x_2, \ldots, x_n)\, dx_1\, dx_2 \cdots dx_n$$

Assuming without loss of generality that the population mean μ is equal to zero, we find that these last two integrals have already been evaluated on page 190, where it was shown that their respective values are σ^2 and σ^2/n. Thus, the mean of the sampling distribution of s^2 is given by

$$\frac{1}{n-1} \sum_{i=1}^{n} \sigma^2 - \frac{n}{n-1} \cdot \frac{\sigma^2}{n} = \frac{n\sigma^2}{n-1} - \frac{\sigma^2}{n-1} = \sigma^2$$

and this completes the proof of the unbiasedness of s^2 as an estimate of σ^2. (Note that, had we divided by n instead of $n - 1$ in defining s^2, the resulting estimator would have been biased; the mean of its sampling distribution would have been $\frac{n-1}{n} \cdot \sigma^2$.)

Although the sample variance is an unbiased estimator of σ^2, it does not follow that the sample standard deviation is also an unbiased estimator of σ; in fact, it is not. However, for large samples the bias is small and it is common practice to estimate σ with s.

Besides s, population standard deviations are sometimes estimated in terms of the **sample range** R, which we defined in Section 2.6 as the largest value of a sample minus the smallest. Given a random sample of size n from a normal population, it can be shown that the sampling distribution of the range has the mean $d_2\sigma$ and the standard deviation $d_3\sigma$, where d_2 and d_3 are constants which depend on the size of the sample. For $n = 1, 2, \ldots,$ and 10, their values are as shown in the following table:

n	2	3	4	5	6	7	8	9	10
d_2	1.128	1.693	2.059	2.326	2.534	2.704	2.847	2.970	3.078
d_3	0.853	0.888	0.880	0.864	0.848	0.833	0.820	0.808	0.797

Thus, R/d_2 is an unbiased estimate of σ, and for very small samples, $n \leq 5$, it provides nearly as good an estimate of σ as does s; as the sample size increases, it becomes more efficient to use s instead of R/d_2. Nowadays, the range is used to estimate σ primarily in problems of industrial quality control, where sample sizes are usually small and computational ease is of prime concern. This application will be discussed in Chapter 14, where we shall need the above values of the constant d_3.

EXAMPLE With reference to the example on page 245, use the range of the first sample to estimate σ for the heat-producing capacity of coal from the first mine.

Solution Since the smallest value is 8,070, the largest value is 8,350, and $n = 5$ so that $d_2 = 2.326$, we get

$$\frac{R}{d_2} = \frac{8{,}350 - 8{,}070}{2.326} = 120.4$$

Note that this is fairly close to the sample standard deviation $s = 125.5$.

■

In most practical applications, interval estimates of σ or σ^2 are based on the sample standard deviation or the sample variance. For random samples from normal populations, we make use of Theorem 6.4, according to which

$$\frac{(n-1)s^2}{\sigma^2}$$

is a value of a random variable having the chi-square distribution with $n - 1$ degrees of freedom. Thus, with χ^2_α defined as on page 200 for a chi-square distribution with $n - 1$ degrees of freedom, we can assert with probability $1 - \alpha$ that the inequality

$$\chi^2_{1-\alpha/2} < \frac{(n-1)s^2}{\sigma^2} < \chi^2_{\alpha/2}$$

will be satisfied; once the data have been obtained, we make the same assertion with $(1 - \alpha)100\%$ confidence. Solving this inequality for σ^2, we obtain the following result:

Confidence interval for σ^2

$$\frac{(n-1)s^2}{\chi^2_{\alpha/2}} < \sigma^2 < \frac{(n-1)s^2}{\chi^2_{1-\alpha/2}}$$

If we take the square root of each member of this inequality, we obtain a corresponding $(1 - \alpha)100\%$ confidence interval for σ.

Note that confidence intervals for σ or σ^2 obtained by taking "equal tails," as in the above formula, do not actually give the narrowest confidence intervals, because the chi-square distribution is not symmetrical (see Exercise 7.16). Nevertheless, they are used in most applications in order to avoid fairly complicated calculations.

EXAMPLE Returning to the example on page 200, suppose that the refractive indices of 20 pieces of glass (randomly selected from a large shipment purchased by the optical firm) have a variance of $1.20 \cdot 10^{-4}$. Construct a 95% confidence interval for σ, the standard deviation of the population sampled.

Solution For $20 - 1 = 19$ degrees of freedom, $\chi^2_{0.975} = 8.907$ and $\chi^2_{0.025} = 32.852$ according to Table 5, so that substitution into the formula yields

$$\frac{(19)(1.20 \cdot 10^{-4})}{32.852} < \sigma^2 < \frac{(19)(1.20 \cdot 10^{-4})}{8.907}$$

$$0.000069 < \sigma^2 < 0.000256$$

and, hence,

$$0.0083 < \sigma < 0.0160$$

This means we are 95% confident that the interval from 0.0083 to 0.0160 contains σ, the true standard deviation of the refractive index.

■

The method which we have discussed applies only to random samples from normal populations (or at least to random samples from populations which can be approximated closely with normal distributions).

EXERCISES

8.1 Use the data of Exercise 7.46 on page 238 to estimate σ for the length of time the experimental engine will operate with the given fuel in terms of

(a) the sample standard deviation;

(b) the sample range.

Compare the two estimates by expressing their difference as a percentage of the first.

8.2 With reference to the example on page 245, use the range of the second sample to estimate σ for the heat-producing capacity of coal from the second mine, and compare the result with the standard deviation of the second sample.

8.3 Use the data of part (a) of Exercise 7.70 to estimate σ for the Brinell hardness of Alloy 1 in terms of

(a) the sample standard deviation;

(b) the sample range.

Compare the two estimates by expressing their difference as a percentage of the first.

8.4 With reference to Exercise 7.47, construct a 99% confidence interval for the variance of the amount of time it takes the company to fill an order for a piece of the given kind of machinery.

8.5 With reference to Exercise 7.48, construct a 99% confidence interval for the variance of the population sampled.

8.6 Use the value of s obtained in Exercise 8.3 to construct a 98% confidence interval for σ, measuring the actual variability in the hardness of Alloy 1.

8.2
HYPOTHESES CONCERNING ONE VARIANCE

In this section we shall consider the problem of testing the null hypothesis that a population variance equals a specified constant against a suitable one-sided or two-sided alternative; that is, we shall test the null hypothesis $\sigma^2 = \sigma_0^2$ against one of the alternatives $\sigma^2 < \sigma_0^2$, $\sigma^2 > \sigma_0^2$, or $\sigma^2 \neq \sigma_0^2$. Tests like these are important whenever it is desired to control the uniformity of a product or an operation. For example, suppose that a silicon disc, or "wafer," is to be cut into small squares, or "dice," to be used in the manufacture of a semiconductor device. Since certain electrical characteristics of the finished device may depend on the thickness of the die, it is important that all dice cut from a wafer have approximately the same thickness. Thus, not only must the mean thickness of a wafer be kept within specifications, but also the variation in thickness from location to location on the wafer.

Using the same sampling theory as on page 259, we base such tests on the fact that for random samples from a normal population with the variance σ_0^2

Statistic for test concerning variance

$$\chi^2 = \frac{(n-1)s^2}{\sigma_0^2}$$

is a value of a random variable having the chi-square distribution with $n - 1$ degrees of freedom. The critical regions for such tests are as shown in the following table:

Critical Regions for Testing $\sigma^2 = \sigma_0^2$
(Normal population)

Alternative hypothesis	*Reject null hypothesis if:*
$\sigma^2 < \sigma_0^2$	$\chi^2 < \chi_{1-\alpha}^2$
$\sigma^2 > \sigma_0^2$	$\chi^2 > \chi_{\alpha}^2$
$\sigma^2 \neq \sigma_0^2$	$\chi^2 < \chi_{1-\alpha/2}^2$ or $\chi^2 > \chi_{\alpha/2}^2$

In this table χ_α^2 is as defined on page 200. Note that "equal tails" are used for the two-sided alternative, and this is actually not the best procedure since the chi-square distribution is not symmetrical.

EXAMPLE The lapping process which is used to grind certain silicon wafers to the proper thickness is acceptable only if σ, the population standard deviation of the thickness of dice cut from the wafers, is at most 0.50 mil. Use the 0.05 level of significance to test the null hypothesis $\sigma = 0.50$ against the alternative hypothesis $\sigma > 0.50$, if the thicknesses of 15 dice cut from such wafers have a standard deviation of 0.64 mil.

Solution

1. *Null hypothesis*: $\sigma = 0.50$
   Alternative hypothesis: $\sigma > 0.50$
2. *Level of significance*: $\alpha = 0.05$
3. *Criterion*: Reject the null hypothesis if $\chi^2 > 23.685$, the value of $\chi_{0.05}^2$ for 14 degrees of freedom, where

$$\chi^2 = \frac{(n-1)s^2}{\sigma_0^2}$$

4. *Calculations*:

$$\chi^2 = \frac{(15-1)(0.64)^2}{(0.50)^2} = 22.94$$

5. *Decision*: Since $\chi^2 = 22.94$ does not exceed 23.685, the null hypothesis cannot be rejected; even though the sample standard deviation exceeds 0.50, there is not sufficient evidence to conclude that the lapping process is unsatisfactory.

■

There exist tables such as Table 8, which enable us to read the probabilities of Type II errors connected with this kind of test. As given in the *National Bureau of Standards Handbook 91* (see the bibliography), they contain the *OC* curves for the different one-sided and two-sided alternatives, for $\alpha = 0.05$ and $\alpha = 0.01$, and for various values of n. The quantity that we mark on the horizontal scale is the ratio $\frac{\sigma_1}{\sigma_0}$, where σ_1 is the alternative value of σ for which we want to determine the probability of erroneously accepting the null hypothesis $\sigma = \sigma_0$.

8.3
HYPOTHESES CONCERNING TWO VARIANCES

The two-sample t test, described in Section 7.9, requires that the variances of the two populations sampled are equal. In this section we describe a test of the null hypothesis $\sigma_1^2 = \sigma_2^2$, which applies to independent random samples from two normal populations; it must be used with some discretion as it is very sensitive to departures from this assumption.

If independent random samples of size n_1 and n_2 are taken from normal populations having the same variance, it follows from Theorem 6.5 that

Statistic for test of equality of two variances

$$F = \frac{s_1^2}{s_2^2}$$

is a value of a random variable having the F distribution with $n_1 - 1$ and $n_2 - 1$ degrees of freedom. Thus, if the null hypothesis $\sigma_1^2 = \sigma_2^2$ is true, the ratio of the sample variances s_1^2 and s_2^2 provides a statistic on which tests of the null hypothesis can be based.

The critical region for testing the null hypothesis $\sigma_1^2 = \sigma_2^2$ against the alternative hypothesis $\sigma_1^2 > \sigma_2^2$ is $F > F_\alpha$, where F_α is as defined on page 201. Similarly, the critical region for testing the null hypothesis against the alternative hypothesis $\sigma_1^2 < \sigma_2^2$ is $F < F_{1-\alpha}$, and this causes some difficulties since Table 6 only contains values corresponding to right-hand tails of $\alpha = 0.05$ and $\alpha = 0.01$. As a result, we use the reciprocal of the original test statistic and make use of the relation

$$F_{1-\alpha}(\nu_1, \nu_2) = \frac{1}{F_\alpha(\nu_2, \nu_1)}$$

first given on page 202. Thus, we base the test on the statistic $F = s_2^2/s_1^2$ and the critical region for testing the null hypothesis $\sigma_1^2 = \sigma_2^2$ against the alternative hypothesis $\sigma_1^2 < \sigma_2^2$ becomes $F > F_\alpha$, where F_α is the appropriate critical value of F for $n_2 - 1$ and $n_1 - 1$ degrees of freedom.

For the two-sided alternative $\sigma_1^2 \neq \sigma_2^2$ the critical region is $F < F_{1-\alpha/2}$ or $F > F_{\alpha/2}$, where $F = s_1^2/s_2^2$ and the degrees of freedom are $n_1 - 1$ and $n_2 - 1$. In practice, we modify this test as in the preceding paragraph, so that we can again use the table of F values corresponding to right-hand tails of $\alpha = 0.05$ and $\alpha = 0.01$. To this end we let s_M^2 represent the larger of the two sample variances, s_m^2 the smaller, and we write the corresponding sample sizes as n_M and n_m. Thus, the test statistic becomes $F = s_M^2/s_m^2$ and the critical region is as shown in the following table:

Critical Regions for Testing $\sigma_1^2 = \sigma_2^2$
(Normal populations)

Alternative hypothesis	*Test statistic*	*Reject null hypothesis if:*
$\sigma_1^2 < \sigma_2^2$	$F = \dfrac{s_2^2}{s_1^2}$	$F > F_\alpha(n_2 - 1, n_1 - 1)$
$\sigma_1^2 > \sigma_2^2$	$F = \dfrac{s_1^2}{s_2^2}$	$F > F_\alpha(n_1 - 1, n_2 - 1)$
$\sigma_1^2 \neq \sigma_2^2$	$F = \dfrac{s_M^2}{s_m^2}$	$F > F_{\alpha/2}(n_M - 1, n_m - 1)$

The level of significance of these tests is α and the figures indicated in parentheses are the respective degrees of freedom. Note that, as in the chi-square test, "equal tails" are used in the two-tailed test as a matter of mathematical convenience, even though the F distribution is not symmetrical.

EXAMPLE It is desired to determine whether there is less variability in the silver plating done by Company 1 than in that done by Company 2. If independent random samples of size 12 of the two companies' work yield $s_1 = 0.035$ mil and $s_2 = 0.062$ mil, test the null hypothesis $\sigma_1^2 = \sigma_2^2$ against the alternative hypothesis $\sigma_1^2 < \sigma_2^2$ at the 0.05 level of significance.

Solution

1. *Null hypothesis*: $\sigma_1^2 = \sigma_2^2$
   Alternative hypothesis: $\sigma_1^2 < \sigma_2^2$
2. *Level of significance*: $\alpha = 0.05$

3. *Criterion*: Reject the null hypothesis if $F > 2.82$, the value of $F_{0.05}$ for 11 and 11 degrees of freedom, where

$$F = \frac{s_2^2}{s_1^2}$$

4. *Calculations*:

$$F = \frac{(0.062)^2}{(0.035)^2} = 3.14$$

5. *Decision*: Since $F = 3.14$ exceeds 2.82, the null hypothesis must be rejected; in other words, the data support the contention that the plating done by Company 1 is less variable than that done by Company 2.

■

EXAMPLE With reference to the example dealing with the heat-producing capacity of coal from two mines on page 245, use the 0.02 level of significance to test whether it is reasonable to assume that the variances of the two populations sampled are equal.

Solution

1. *Null hypothesis*: $\sigma_1^2 = \sigma_2^2$
   Alternative hypothesis: $\sigma_1^2 \neq \sigma_2^2$
2. *Level of significance*: $\alpha = 0.02$
3. *Criterion*: Reject the null hypothesis if $F > 11.4$, the value of $F_{0.01}$ for 4 and 5 degrees of freedom, where

$$F = \frac{s_1^2}{s_2^2}$$

   since $s_1^2 = 15{,}750$ is greater than $s_2^2 = 10{,}920$.

4. *Calculations*:

$$F = \frac{15{,}750}{10{,}920} = 1.44$$

5. *Decision*: Since $F = 1.44$ does not exceed 11.4, the null hypothesis cannot be rejected; there is no real reason to doubt the equality of the variances of the two populations.

■

Had we wanted to use the level of significance $\alpha = 0.05$ or $\alpha = 0.01$ in this example, we would have required tables of the values of $F_{0.025}(\nu_1, \nu_2)$ or

$F_{0.005}(\nu_1, \nu_2)$; such tables may be found in the *Biometrika Tables for Statisticians* listed in the Bibliography under Pearson and Hartley. Also, *OC* curves for the one-tailed F tests may be found in the *National Bureau of Standards Handbook 91*.

In marked contrast to the procedures for making inferences about μ, the validity of the procedures in this chapter depend rather strongly on the assumption that the underlying population is normal. The sampling variance of s^2 can change when the population departs from normality by having, for instance, a single long tail. It can be shown that, when the underlying population is normal, the sampling variance of s^2 is $2\sigma^4/(n-1)$. However, for nonnormal distributions, the sampling variance of s^2 depends not only on σ^2 but also on the population third and fourth moments, μ_3 and μ_4 (see page 108). Consequently, it could be much larger than $2\sigma^4/(n-1)$. This behavior completely invalidates any tests of hypothesis or confidence intervals for σ^2. We say that these procedures for making inferences about σ^2 are not **robust** with respect to deviations from normality.

EXERCISES

8.7 With reference to Exercise 7.44 on page 238, test the null hypothesis $\sigma = 600$ psi for the compressive strength of the given kind of steel against the alternative hypothesis $\sigma > 600$ psi. Use the 0.05 level of significance.

8.8 If 12 determinations of the specific heat of iron have a standard deviation of 0.0086, test the null hypothesis that $\sigma = 0.010$ for such determinations. Use the alternative hypothesis $\sigma \neq 0.010$ and the level of significance $\alpha = 0.01$.

8.9 With reference to Exercise 7.64, test the null hypothesis that $\sigma = 15.0$ minutes for the time that is required for repairs of the first kind of photocopying equipment against the alternative hypothesis that $\sigma > 15.0$ minutes. Use the 0.05 level of significance and assume normality.

8.10 Use the 0.01 level of significance to test the null hypothesis that $\sigma = 0.015$ inch for the diameters of certain bolts against the alternative hypothesis that $\sigma \neq 0.015$ inch, given that a random sample of size 15 yielded $s^2 = 0.00011$.

8.11 Playing 10 rounds of golf on his home course, a golf professional averaged 71.3 with a standard deviation of 1.32. Test the null hypothesis that the consistency of his game on his home course is actually measured by $\sigma = 1.20$, against the alternative hypothesis that he is less consistent. Use the level of significance $\alpha = 0.05$.

8.12 The security department of a large office building wants to test the null hypothesis that $\sigma = 2.0$ minutes for the time it takes a guard to walk his round against the alternative hypothesis that $\sigma \neq 2.0$ minutes. What can it conclude at the 0.01 level of significance if a random sample of size $n = 31$ yields $s = 1.8$ minutes?

8.13 Justify the use of the two-sample t test in Exercise 7.67 on page 250 by testing the null hypothesis that the two populations have equal variance. Use the 0.02 level of significance.

8.14 With reference to Exercise 7.68 on page 250, use the 0.02 level of significance to test the assumption that the two populations have equal variances.

8.15 Two different lighting techniques are compared by measuring the intensity of light at selected locations in areas lighted by the two methods. If 15 measurements in the first area had a standard deviation of 2.7 foot-candles and 21 measurements in the second area had a standard deviation of 4.2 foot-candles, can it be concluded that the lighting in the second area is less uniform? Use a 0.01 level of significance. What assumptions must be made as to how the two samples are obtained?

8.16 With reference to Exercise 7.66, where we had $n_1 = 40$, $n_2 = 30$, $s_1 = 15.2$, and $s_2 = 18.7$, use the 0.05 level of significance to test the claim that there is a greater variability in the number of cars which make left turns approaching from the south between 4 P.M. and 6 P.M. at the second intersection. Assume the distributions are normal.

8.17 Random samples of size n_1 and n_2, respectively, are taken from two log-normal populations, and the resulting sample means are $\bar{x}_1 = 3.74$ and $\bar{x}_2 = 13.91$. You wish to test whether the second population has a mean value four times as large as the first.

(a) Can you directly use a two-sample test? Why?

(b) Is there a transformation that can be made on the data that could conceivably allow the use of a two-sample test?

8.4
REVIEW EXERCISES

8.18 With reference to the example on page 231, construct a 95% confidence interval for the true standard deviation of the breaking strength of the given kind of ribbon.

8.19 With reference to the example on page 246, find separate 95% confidence intervals for the standard deviations of the two aluminum alloys.

8.20 While performing a strenuous task, the pulse rate of 25 workers increased on the average by 18.4 beats per minute with a standard deviation of 4.9 beats per minute. Find a 95% confidence interval for the corresponding population standard deviation.

8.21 With reference to Exercise 8.20, use the 0.05 level of significance to test the null hypothesis that $\sigma^2 = 30.0$ for such increases in the pulse rate (while performing the given task) against the alternative hypothesis that $\sigma^2 < 30.0$.

8.22 If 31 measurements of the boiling point of sulfur have a standard deviation of 0.83 degree Celsius, construct a 98% confidence interval for the true standard deviation of such measurements.

8.23 Past data indicate that the variance of measurements made on sheet metal stampings by experienced quality control inspectors is 0.18 square inch. Such measurements made by an inexperienced inspector could have too large a variance (perhaps because of inability to read instruments properly) or too small a variance (perhaps because unusually high or low measurements are discarded). If a new inspector measures 101 stampings with a variance of 0.13 square inch, test at the 0.05 level of significance whether the inspector is making satisfactory measurements.

8.24 With reference to Exercise 7.69 on page 250, use the 0.02 level of significance to test the assumption that the two populations have equal variances.

8.25 Pull-strength tests on 10 soldered leads for a semiconductor device yield the following results in pounds force required to rupture the bond:

15.8, 12.7, 13.2, 16.9, 10.6, 18.8, 11.1, 14.3, 17.0, 12.5

Another set of eight leads was tested after encapsulation to determine whether the pull strength has been increased by encapsulation of the device, with the following results:

24.9, 23.6, 19.8, 22.1, 20.4, 21.6, 21.8, 22.5

As a preliminary to the two-sample t test, use the 0.02 level of significance to test whether it is reasonable to assume that the two samples come from populations with equal variances.

8.26 With reference to the example on page 246, test the equality of the variances for the two aluminum alloys. Use the 0.02 level of significance.

8.5
CHECK LIST OF KEY TERMS (with page reference)

9 INFERENCES CONCERNING PROPORTIONS

Many engineering problems deal with proportions, percentages, or probabilities. In acceptance sampling we are concerned with the proportion of defectives in a lot, and in life testing we are concerned with the percentage of certain components which will perform satisfactorily during a stated period of time, or the probability that a given component will last at least a given number of hours. It should be clear from these examples that problems concerning proportions, percentages, or probabilities are really equivalent; a percentage is merely a proportion multiplied by 100, and a probability may be interpreted as a proportion in the long run.

Sections 9.1 and 9.2 deal with the estimation of proportions; Section 9.3 deals with tests concerning proportions; Section 9.4 deals with tests concerning two or more proportions; in Section 9.5 we shall learn how to analyze data tallied into a two-way classification; and in Section 9.6 we shall learn how to judge whether differences between an observed frequency distribution and corresponding expectations can be attributed to chance.

9.1

ESTIMATION OF PROPORTIONS

The information that is usually available for the estimation of a proportion is the number of times, x, that an appropriate event has occurred in n trials, occasions, or observations. The point estimate, itself, is usually the **sample proportion** $\frac{x}{n}$, namely, the proportion of the time that the event has actually occurred. If the n trials satisfy the assumptions underlying the binomial distribution listed on page 92, we know that the mean and the standard deviation of the number of successes are given by np and $\sqrt{np(1-p)}$. If we divide both of these quantities by n, we find that the mean and the standard deviation of the proportion of successes (namely, of the sample proportion) are given by

$$\frac{np}{n} = p \quad \text{and} \quad \frac{\sqrt{np(1-p)}}{n} = \sqrt{\frac{p(1-p)}{n}}$$

The first of these results shows that the sample proportion is an unbiased estimator of the binomial parameter p, namely, of the true proportion we are trying to estimate on the basis of a sample.

In the construction of confidence intervals for the binomial parameter p, we meet several obstacles. First, since x and $\frac{x}{n}$ are values of discrete random variables, it may be impossible to get an interval for which the degree of confidence is exactly $(1-\alpha)100\%$. Second, the standard deviation of the sampling distribution of the number of successes, as well as that of the proportion of successes, involves the parameter p that we are trying to estimate.

To construct a confidence interval for p having approximately the degree of confidence $(1-\alpha)100\%$, we first determine for a given set of values of p the corresponding quantities x_0 and x_1, where x_0 is the largest integer for which

$$\sum_{k=0}^{x_0} b(k; n, p) \leq \frac{\alpha}{2}$$

while x_1 is the smallest integer for which

$$\sum_{k=x_1}^{n} b(k; n, p) \leq \frac{\alpha}{2}$$

To emphasize the point that x_0 and x_1 depend on the value of p, we shall write these quantities as $x_0(p)$ and $x_1(p)$. Thus, we can assert with a probability of approximately $1-\alpha$, and at least $1-\alpha$, that the inequality

$$x_0(p) < x < x_1(p)$$

will be satisfied; here x is a value of a random variable and p is a fixed constant. To change inequalities like these into confidence intervals for p, we can use a simple graphical method which is illustrated by the following example: Suppose, for instance, that we want to find approximate 95% confidence intervals for p for samples of size $n = 20$. Using Table 1 at the end of the book, we first determine x_0 and x_1 for selected values of p such that x_0 is the largest integer for which

$$B(x_0; 20, p) \leq 0.025$$

while x_1 is the smallest integer for which

$$1 - B(x_1 - 1; 20, p) \leq 0.025$$

Letting p equal 0.1, 0.2, ..., and 0.9, we thus obtain the values shown in the following table:

p	0.1	0.2	0.3	0.4	0.5	0.6	0.7	0.8	0.9
x_0	—	0	1	3	5	7	9	11	14
x_1	6	9	11	13	15	17	19	20	—

Plotting the points with coordinates p and $x(p)$ as in Figure 9.1, and drawing smooth curves, one through the x_0 points and one through the x_1 points, we can now "solve" for p. For any given value of x we can obtain approximate 95% confidence limits for p by going horizontally to the two curves and marking off the corresponding values of p (see Figure 9.1). Thus, for $x = 4$ we obtain the approximate 95% confidence interval

$$0.06 < p < 0.45$$

Graphs similar to the one shown in Figure 9.1 are given in Tables 9(a) and 9(b) at the end of the book for various values of n and for the 95% and 99% degrees of confidence. These tables differ from the one of Figure 9.1 in that the sample proportion $\frac{x}{n}$ is used instead of x, thus making it possible to graph curves corresponding to various values of n on the same diagram. Also, for increased accuracy, Tables 9(a) and 9(b) are arranged so that values of $\frac{x}{n}$ from 0.00 to 0.50 are marked on the bottom scale while those from 0.50 to 1.00 are marked on the top scale of the diagram. For values of $\frac{x}{n}$ from 0.00 to 0.50 the confidence limits for

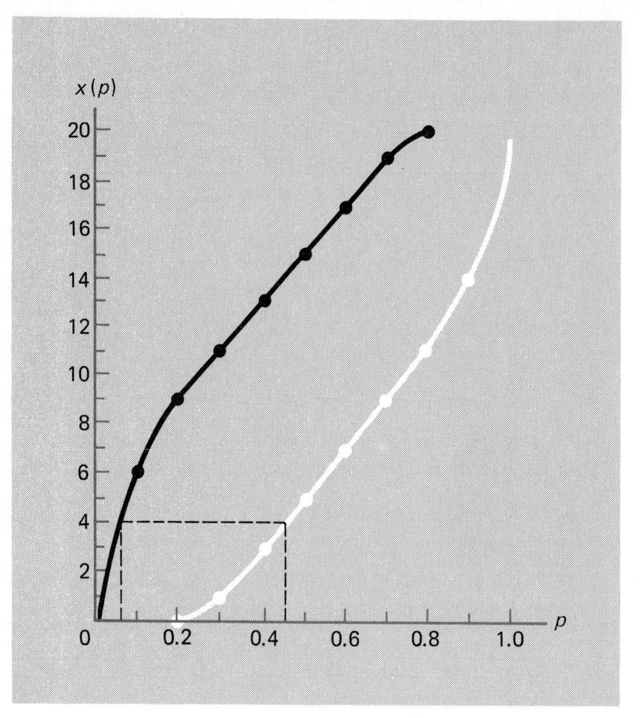

FIGURE 9.1
95% confidence intervals for proportions ($n = 20$).

p are read off the left-hand scale of the diagram, while for values of $\frac{x}{n}$ from 0.50 to 1.00 they are read off the right-hand scale. Note that for $n = 20$ and $x = 4$, Table 9(a) yields the 95% confidence interval $0.06 < p < 0.44$, which is very close, indeed, to the results obtained with Figure 9.1.

On page 148 we gave the general rule of thumb that the normal distribution provides a good approximation to the binomial distribution when np and $n(1 - p)$ are both greater than 5. Thus, for $n = 50$ the normal curve approximation may be used if it can be assumed that p lies between 0.10 and 0.90; for $n = 100$ it may be used if it can be assumed that p lies between 0.05 and 0.95; for $n = 200$ it may be used if it can be assumed that p lies between 0.025 and 0.975; and so forth. This is what we shall mean here, and later in this chapter, by "n being large."

When n is large, we can construct approximate confidence intervals for the binomial parameter p by using the normal approximation to the binomial distribution. Accordingly, we can assert with probability $1 - \alpha$ that the inequality

$$-z_{\alpha/2} < \frac{x - np}{\sqrt{np(1 - p)}} < z_{\alpha/2}$$

will be satisfied. Solving this quadratic inequality for p, we can obtain a corresponding set of approximate confidence limits for p (see Exercise 9.14 on page 280), but since the necessary calculations are involved, we shall make the further approximation of substituting $\frac{x}{n}$ for p in $\sqrt{np(1-p)}$. This yields

Large-sample confidence interval for p

$$\frac{x}{n} - z_{\alpha/2}\sqrt{\frac{\frac{x}{n}\left(1-\frac{x}{n}\right)}{n}} < p < \frac{x}{n} + z_{\alpha/2}\sqrt{\frac{\frac{x}{n}\left(1-\frac{x}{n}\right)}{n}}$$

where the degree of confidence is $(1-\alpha)100\%$.

EXAMPLE If $x = 36$ of $n = 100$ persons interviewed are familiar with the tax incentives for installing certain energy-saving devices, construct a 95% confidence interval for the corresponding true proportion.

Solution Substituting $\frac{x}{n} = \frac{36}{100} = 0.36$ and $z_{\alpha/2} = 1.96$ into the above formula, we get

$$0.36 - 1.96\sqrt{\frac{(0.36)(0.64)}{100}} < p < 0.36 + 1.96\sqrt{\frac{(0.36)(0.64)}{100}}$$

or

$$0.266 < p < 0.454$$

We are 95% confident that p is contained in the interval from 0.266 to 0.454.

Note that if we had used Table 9(a), we would have obtained

$$0.27 < p < 0.46$$

■

The magnitude of the error we make when we use $\frac{x}{n}$ as an estimate of p is given by $\left|\frac{x}{n} - p\right|$. Again using the normal approximation, we can thus assert with probability $1-\alpha$ that the inequality

$$\left|\frac{x}{n} - p\right| \le z_{\alpha/2}\sqrt{\frac{p(1-p)}{n}}$$

will be satisfied, namely, that the error will be at most $z_{\alpha/2}\sqrt{\frac{p(1-p)}{n}}$. With $\frac{x}{n}$ substituted for p this yields

Maximum error of estimate

$$\mathbf{E = z_{\alpha/2}\sqrt{\frac{\frac{x}{n}\left(1-\frac{x}{n}\right)}{n}}}$$

EXAMPLE In a sample survey conducted in a large city, 136 of 400 persons answered Yes to the question whether their city's public transportation is adequate. With 99% confidence, what can we say about the maximum error, if $\frac{x}{n} = \frac{136}{400} = 0.34$ is used as an estimate of the corresponding true proportion?

Solution Substituting $\frac{x}{n} = 0.34$ and $z_{\alpha/2} = 2.575$ into the above formula, we find that the error is at most

$$E = 2.575\sqrt{\frac{(0.34)(0.66)}{400}} = 0.061$$

■

The preceding formula for E can also be used to determine the sample size that is needed to attain a desired degree of precision. Solving for n, we get

Sample size

$$\mathbf{n = p(1-p)\left[\frac{z_{\alpha/2}}{E}\right]^2}$$

but this formula cannot be used as it stands unless we have some information about the possible size of p (on the basis of collateral data, say, a pilot sample). If no such information is available, we can make use of the fact that $p(1-p)$ is at most $\frac{1}{4}$, corresponding to $p = \frac{1}{2}$, as can be shown by the methods of elementary calculus. Thus, if

Sample size

$$\mathbf{n = \frac{1}{4}\left[\frac{z_{\alpha/2}}{E}\right]^2}$$

we can assert with a probability of at least $1 - \alpha$ that the error in using $\frac{x}{n}$ as an estimate of p will not exceed E; once the data have been obtained, we will be able to assert with at least $(1 - \alpha)100\%$ confidence that the error does not exceed E.

EXAMPLE Suppose that we want to estimate the true proportion of defectives in a very large shipment of adobe bricks, and that we want to be at least 95% confident that the error is at most 0.04. How large a sample will we need if

(a) we have no idea what the true proportion might be;

(b) we know that the true proportion does not exceed 0.12?

Solution (a) Using the second of the two formulas for the sample size, we get

$$n = \frac{1}{4}\left[\frac{1.96}{0.04}\right]^2 = 600.25$$

or $n = 601$ rounded up to the nearest integer. (b) Using the first of the two formulas for the sample size with $p = 0.12$ (the possible value closest to $p = \frac{1}{2}$), we get

$$n = (0.12)(0.88)\left[\frac{1.96}{0.04}\right]^2 = 253.55$$

or $n = 254$ rounded up to the nearest integer. This serves to illustrate how some collateral information about the possible size of p can substantially reduce the size of the required sample. ■

When p is very close to 0, as is the case in problems of high reliability and p is the probability of failure, none of the confidence intervals we have discussed provides a satisfactory solution. What we really need here are **one-sided confidence intervals** of the form $p < C$, where C is a constant depending on the degree of confidence and the size of the sample. As we already pointed out on page 113, the binomial distribution is best approximated with a Poisson distribution with $\lambda = np$ when p is small and n is large. Based on this approximation, it can be shown that

One-sided confidence interval for p

$$p < \frac{1}{2n} \cdot \chi^2_\alpha$$

is a one-sided confidence interval for p, where χ^2_α is as defined on page 200 and the number of degrees of freedom equals $2(x + 1)$. A discussion of this result may be found in the book by A. Hald mentioned in the Bibliography.

EXAMPLE If there are $x = 4$ failures among $n = 2{,}000$ parts used continuously for a month, construct a one-sided 99% confidence interval for the probability that one such part will fail under the stated conditions.

Solution Since $\chi^2_{0.01} = 23.209$ for $2(4 + 1) = 10$ degrees of freedom, substitution into the formula yields

$$p < \frac{1}{2(2{,}000)} \cdot 23.209$$

and, hence,

$$p < 0.0058$$

That is, 0.0058 is an approximate 95% upper confidence bound for p. ■

9.2 BAYESIAN ESTIMATION†

In the preceding section we looked upon the true proportions we tried to estimate as unknown constants; in Bayesian estimation these parameters are looked upon as random variables having prior distributions which reflect the strength of one's belief about the possible values they can take on, or other indirect information. As in Section 7.3, we are thus faced with the problem of combining prior information with direct sample evidence.

To illustrate how this might be done, suppose that a manufacturer, who regularly receives large shipments of electronic components from a vendor, knows that about 25% of the time 0.005 (half of 1%) of the components are defective, about 25% of the time 0.01 of the components are defective, and about 50% of the time 0.02 of the components are defective. Thus, before a shipment from this vendor is inspected, we have the following prior distribution for the proportion of defectives:

Value of p	*Prior probability*
0.005	0.25
0.01	0.25
0.02	0.50

† This section may be omitted without loss of continuity.

Now suppose that 200 of these components, randomly selected from the shipment, are inspected, and only one of them is found to be defective. The probability of this happening when $p = 0.005$, $p = 0.01$, or $p = 0.02$ are, respectively,

$$\binom{200}{1}(0.005)^1(0.995)^{199} = 0.37$$

$$\binom{200}{1}(0.01)^1(0.99)^{199} = 0.27, \quad \text{and} \quad \binom{200}{1}(0.02)^1(0.98)^{199} = 0.07$$

where we used the formula for the binomial distribution on page 93 and logarithms to simplify the calculations. Combining these probabilities by means of the formula for Bayes' theorem (Theorem 3.11 on page 75), we find that the posterior probability for $p = 0.005$ is

$$\frac{(0.25)(0.37)}{(0.25)(0.37) + (0.25)(0.27) + (0.50)(0.07)} = 0.47$$

and that the corresponding posterior probabilities for $p = 0.01$ and $p = 0.02$ are 0.35 and 0.18. We have thus arrived at the following posterior distribution for the proportion of defective components:

Value of p	*Posterior probability*
0.005	0.47
0.01	0.35
0.02	0.18

Note that whereas the odds were originally 3 to 1 against $p = 0.005$, it is now almost an even bet; of course, this shift is accounted for by the fact that in the sample only $\frac{1}{200} = 0.005$ of the components inspected were defective.

In the preceding example we assumed that p had to be 0.005, 0.01, or 0.02, and this restriction was imposed mainly to simplify the calculations; the method would have been the same if we had considered 10 different values of p, or even 100. It would be more logical, perhaps, to let p take on any value on the continuous interval from 0 to 1, and in that case it is customary to use as the prior distribution the beta distribution of Section 5.8. The parameters of this distribution are α and β, and its mean and variance can be expressed in terms of α and β in accordance with the formulas on page 159. It can then be shown that the posterior distribution of p, namely, the conditional distribution of p for a given (observed) value of x, is also a beta distribution, and that its parameters are $x + \alpha$ and $n - x + \beta$ instead of α and

β.† Thus, the mean and the variance of the posterior distribution may be obtained by substituting $x + \alpha$ for α and $n - x + \beta$ for β in the formulas on page 159.

EXAMPLE A person doing research for a large oil company feels that the proportion of persons requiring oil as well as gasoline at one of the oil company's service stations is a random variable having the beta distribution with $\alpha = 10$ and $\beta = 400$. In a random sample of size $n = 800$, she finds that only $x = 3$ persons required oil as well as gasoline. Find the mean and the variance of

(a) the prior distribution of p;
(b) the posterior distribution of p.

Solution (a) For the prior distribution we get

$$\mu_0 = \frac{10}{10 + 400} = 0.024$$

and

$$\sigma_0^2 = \frac{10 \cdot 400}{410^2 \cdot 411} = 0.000058$$

(b) For the posterior distribution we get

$$\mu_1 = \frac{3 + 10}{10 + 400 + 800} = 0.011$$

and

$$\sigma_1^2 = \frac{(3 + 10)(800 - 3 + 400)}{(10 + 400 + 800)^2(10 + 400 + 800 + 1)} = 0.0000088$$

■

If we have mathematical tables giving the values of beta integrals, we can continue with an example like this and calculate prior as well as posterior probabilities associated with various intervals of the values of p.

---

† Proofs of these results may be found in the book by John E. Freund and Ronald E. Walpole listed in the bibliography.

EXERCISES

9.1 In a random sample of 200 claims filed against an insurance company writing collision insurance on cars, 84 exceeded \$1,200. Construct a 95% confidence interval for the true proportion of claims filed against this insurance company that exceed \$1,200, using

(a) Table 9;

(b) the large-sample confidence-interval formula.

9.2 With reference to Exercise 9.1, what can we say with 99% confidence about the maximum error, if we use the sample proportion as an estimate of the true proportion of claims filed against this insurance company that exceed \$1,200?

9.3 In a random sample of 400 industrial accidents, it was found that 231 were due at least partially to unsafe working conditions. Construct a 99% confidence interval for the corresponding true proportion using

(a) Table 9;

(b) the large-sample confidence-interval formula.

9.4 With reference to Exercise 9.3, what can we say with 95% confidence about the maximum error if we use the sample proportion to estimate the corresponding true proportion?

9.5 In a sample survey of the "safety explosives" used in certain mining operations, explosives containing potassium nitrate were found to be used in 95 of 250 cases.

(a) Use Table 9 to construct a 95% confidence interval for the corresponding true proportion.

(b) If $\frac{95}{250} = 0.38$ is used as an estimate of the corresponding true proportion, what can we say with 95% confidence about the maximum error?

9.6 In a random sample of 60 sections of pipe in a chemical plant, 8 showed signs of serious corrosion. Construct a 95% confidence interval for the true proportion of pipe sections showing signs of serious corrosion, using

(a) Table 9;

(b) the large-sample confidence-interval formula.

9.7 In a recent study, 69 of 120 meteorites were observed to enter the earth's atmosphere with a velocity of less than 26 miles per second. If we estimate the corresponding true proportion as $\frac{69}{120} = 0.575$, what can we say with 95% confidence about the maximum error?

9.8 Among 100 fish caught in a large lake, 18 were inedible due to the pollution of the environment. If we use $\frac{18}{100} = 0.18$ as an estimate of the corresponding true proportion, with what confidence can we assert that the error of this estimate is at most 0.065?

9.9 A random sample of 300 shoppers at a supermarket includes 204 who regularly use cents-off coupons. Construct a 98% confidence interval for the probability that any one shopper at the supermarket, selected at random, will regularly use cents-off coupons.

9.10 What is the size of the smallest sample required to estimate an unknown proportion to within a maximum error of 0.06 with at least 95% confidence?

9.11 With reference to Exercise 9.10, how would the required sample size be affected if it is known that the proportion to be estimated is at least 0.75?

9.12 Suppose that we want to estimate what percentage of all drivers exceed the 55-mile per hour speed limit on a certain stretch of road. How large a sample will we need to be at least 99% confident that the error of our estimate, the sample percentage, is at most 3.5%?

9.13 With reference to Exercise 9.12, how would the required sample size be affected if it is known that the percentage to be estimated is at most 40%?

9.14 Show that the inequality on page 272 leads to the following $(1-\alpha)100\%$ confidence limits:

$$\frac{x + \frac{1}{2}z_{\alpha/2}^2 \pm z_{\alpha/2}\sqrt{\frac{x(n-x)}{n} + \frac{1}{4}z_{\alpha/2}^2}}{n + z_{\alpha/2}^2}$$

9.15 Use the formula of Exercise 9.14 to rework Exercise 9.3.

9.16 Use the formula of Exercise 9.14 to rework Exercise 9.6.

9.17 In a random sample of 500 remote controls for home entertainment centers, 7 failed during the 90-day warranty period. Construct an upper 95% confidence limit for the true probability of failure during warranty.

9.18 Observing the amount of pollutants in the air in a western city on 500 days, it was found that it exceeded 200 micrograms per cubic meter only four times. Construct an upper 99% confidence limit for the probability that the air pollution in this city will exceed 200 micrograms per cubic meter on any one day.

9.19 The head of a highway department feels that four out of five road building jobs stay within cost estimates, while his assistant feels that it should be only three out of five.

(a) If the head of the highway department is regarded to be "three times as good" as his assistant in determining figures like these, what prior probabilities should we assign to their claims?

(b) What posterior probabilities should we assign to their claims if it is found that among 12 road building jobs (randomly selected from the department's files) only two stayed within cost estimates?

9.20 The purchasing agent of a firm feels that the probability is 0.80 that any one of several shipments of steel recently received will meet specifications. The head of the firm's quality control department feels that this probability is 0.90, and the chief engineer feels (somewhat more pessimistically) that it is 0.60.

(a) If the managing director of the firm feels that in this matter the purchasing agent is 10 times as reliable as the chief engineer while the head of the quality control department is 14 times as reliable as the chief engineer, what prior probabilities would she assign to their claims?

(b) If five of the shipments are inspected and only two meet specifications, what posterior probabilities should the managing director of the firm assign to the respective claims?

9.21 The output of a certain transistor production line is checked daily by inspecting a sample of 200 units. Over a long period of time, the process has maintained a yield of 80%, that is, a proportion defective of 0.20, and the variation of the proportion defective (from day to day) is measured by a standard deviation of 0.0125. If on a certain day the sample contains 86 defectives, find the mean of the posterior distribution of the proportion defective as an estimate of that day's proportion defective. Assume that the prior distribution of the proportion defective can be approximated closely with a beta distribution.

9.22 Records of the dean of an engineering school (collected over many years) show that on the average 75% of all applicants have an IQ of at least 115. Of course, the percentage varies somewhat from year to year and this variation is measured by a standard deviation of 2.15%.

(a) Verify that if the prior distribution of the proportion of applicants with an IQ of at least 115 can be approximated closely with a beta distribution, we can use the beta distribution with $\alpha = 300$ and $\beta = 100$.

(b) If a sample check of 25 of this year's applicants shows that only 16 of them have an IQ of at least 115, use the results and the assumptions of part (a) to find the mean and the standard deviation of the posterior distribution of the proportion of this year's applicants who have an IQ of at least 115.

9.3 HYPOTHESES CONCERNING ONE PROPORTION

Many of the methods used in sampling inspection, quality control, and reliability verification are based on tests of the null hypothesis that a proportion (percentage, or probability) equals some specified constant. The details of the application of such tests to quality control will be discussed in Chapter 14, where we shall also go into some problems of sampling inspection; applications to reliability and life testing will be taken up in Chapter 15.

Although there are exact tests based on the binomial distribution that can be performed with the use of Table 1, we shall consider here only approximate large-sample tests based on the normal approximation to the binomial distribution. In other words, we shall test the null hypothesis $p = p_0$ against one of the alternatives $p < p_0$, $p > p_0$, or $p \neq p_0$ with the use of the statistic

Statistic for large-sample test concerning p

$$z = \frac{x - np_0}{\sqrt{np_0(1 - p_0)}}$$

which is a value of a random variable having approximately the standard normal distribution.† The critical regions are like those shown in the table on page 229 with p and p_0 substituted for μ and μ_0.

EXAMPLE In a study designed to investigate whether certain detonators used with explosives in coal mining meet the requirement that at least 90% will ignite the explosive when charged, it is found that 174 of 200 detonators function properly. Test the null hypothesis $p = 0.90$ against the alternative hypothesis $p < 0.90$ at the 0.05 level of significance.

Solution

1. *Null hypothesis*: $p = 0.90$
   Alternative hypothesis: $p < 0.90$
2. *Level of significance*: $\alpha = 0.05$

† Some authors write the numerator of this formula for z as $x \pm \frac{1}{2} - np_0$, whichever is numerically smaller, but there is generally no need for this continuity correction so long as n is large.

3. *Criterion*: Reject the null hypothesis if $z < -1.645$, where

$$z = \frac{x - np_0}{\sqrt{np_0(1 - p_0)}}$$

4. *Calculations*: Substituting $x = 174$, $n = 200$, and $p_0 = 0.90$ into the formula for z, we get

$$z = \frac{174 - 200(0.90)}{\sqrt{200(0.90)(0.10)}} = -1.41$$

5. *Decision*: Since $z = -1.41$ is not less than -1.645, the null hypothesis cannot be rejected; in other words, there is not sufficient evidence to say that the given kind of detonator fails to meet the required standard.

■

9.4 HYPOTHESES CONCERNING SEVERAL PROPORTIONS

When we compare the consumer response (percentage favorable and percentage unfavorable) to two different products, when we decide whether the proportion of defectives of a given process remains constant from day to day, when we judge whether there is a difference in political persuasion among several nationality groups, and in many similar situations, we are interested in testing whether two or more binomial populations have the same parameter p. Referring to these parameters as $p_1, p_2, \ldots,$ and p_k, we are, in fact, interested in testing the null hypothesis

$$p_1 = p_2 = \cdots = p_k = p$$

against the alternative hypothesis that these population proportions are not all equal. To perform a suitable large-sample test of this hypothesis, we require independent random samples of size $n_1, n_2, \ldots,$ and n_k from the k populations; then, if the corresponding numbers of "successes" are $x_1, x_2, \ldots,$ and x_k, the test we shall use is based on the fact that (1) for large samples the sampling distribution of

$$z_i = \frac{x_i - n_i p_i}{\sqrt{n_i p_i(1 - p_i)}}$$

is approximately the standard normal distribution, (2) the square of a random variable having the standard normal distribution is a random variable having the chi-square distribution with 1 degree of freedom, and (3) the sum of k independent random variables having chi-square distributions with 1 degree of freedom is a random variable having the chi-square distribution with k degrees of freedom. (Proofs of these last two results may be found in the book by John E. Freund and Ronald E. Walpole mentioned in the bibliography.) Thus,

$$\chi^2 = \sum_{i=1}^{k} \frac{(x_i - n_i p_i)^2}{n_i p_i (1 - p_i)}$$

is a value of a random variable having approximately the chi-square distribution with k degrees of freedom, and in practice we substitute for the p_i, which under the null hypothesis are all equal, the pooled estimate

$$\hat{p} = \frac{x_1 + x_2 + \cdots + x_k}{n_1 + n_2 + \cdots + n_k}$$

Since the null hypothesis should be rejected if the differences between the x_i and the $n_i \hat{p}$ are large, the critical region is $\chi^2 > \chi^2_\alpha$, where χ^2_α is as defined on page 200 and the number of degrees of freedom is $k - 1$. The loss of one degree of freedom results from substituting for p the estimate $\hat{p}$.

In actual practice, when we compare two or more sample proportions it is convenient to determine the value of the χ^2 statistic by looking at the data as arranged in the following way:

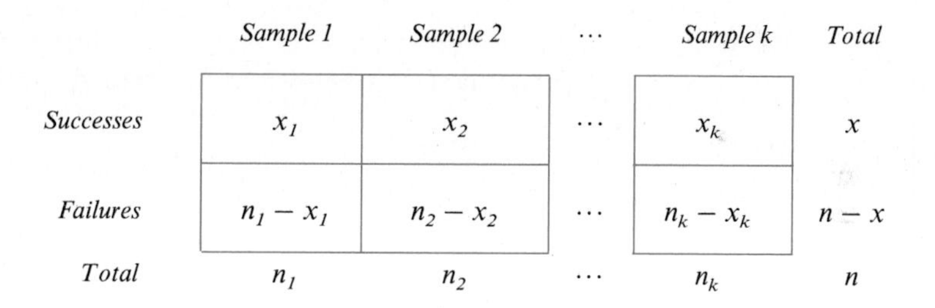

	Sample 1	*Sample 2*	$\cdots$	*Sample k*	*Total*
Successes	x_1	x_2	$\cdots$	x_k	x
Failures	$n_1 - x_1$	$n_2 - x_2$	$\cdots$	$n_k - x_k$	$n - x$
Total	n_1	n_2	$\cdots$	n_k	n

The notation is the same as before, except for x and n, which represent, respectively, the total number of successes and the total number of trials for all samples combined. With reference to this table, the entry in the cell belonging to the ith row and jth column is called the **observed cell frequency** o_{ij} with $i = 1, 2$ and $j = 1, 2, \ldots, k$.

Under the null hypothesis $p_1 = p_2 = \cdots = p_k = p$, we estimate p, as before, as the total number of successes divided by the total number of trials, which we now write as $\hat{p} = \frac{x}{n}$. Hence, the expected number of successes and failures for the jth sample are estimated by

$$e_{1j} = n_j \cdot \hat{p} = \frac{n_j \cdot x}{n}$$

and

$$e_{2j} = n_j(1 - \hat{p}) = \frac{n_j \cdot (n - x)}{n}$$

The quantities e_{1j} and e_{2j} are called the **expected cell frequencies** for $j = 1, 2, \ldots, k$. Note that **the expected frequency for any given cell may be obtained by multiplying the totals of the column and the row to which it belongs and then dividing by the grand total *n*.**

In this notation, the χ^2 statistic on page 283, with $\hat{p}$ substituted for the p_i, can be written in the form

Statistic for test concerning difference among proportions

$$\chi^2 = \sum_{i=1}^{2} \sum_{j=1}^{k} \frac{(o_{ij} - e_{ij})^2}{e_{ij}}$$

as the reader will be asked to verify in Exercise 9.40 on page 291. This formula has the advantage that it can easily be extended to the more general case, to be treated in Section 9.5, where each trial permits more than two possible outcomes, and there are, thus, more than two rows in the tabular presentation of the various frequencies.

EXAMPLE Samples of three kinds of materials, subjected to extreme temperature changes, produced the results shown in the following table:

	Material A	*Material B*	*Material C*	*Total*
Crumbled	41	27	22	90
Remained intact	79	53	78	210
Total	120	80	100	300

Use the 0.05 level of significance to test whether, under the stated conditions, the probability of crumbling is the same for the three kinds of materials.

Solution

1. *Null hypothesis*: $p_1 = p_2 = p_3$
   Alternative hypothesis: p_1, p_2, and p_3 are not all equal.
2. *Level of significance*: $\alpha = 0.05$
3. *Criterion*: Reject the null hypothesis if $\chi^2 > 5.991$, the value of $\chi^2_{0.05}$ for $3 - 1 = 2$ degrees of freedom, where χ^2 is given by the formula above.

4. *Calculations*: The expected frequencies for the first two cells of the first row are

$$e_{11} = \frac{90 \cdot 120}{300} = 36 \quad \text{and} \quad e_{12} = \frac{90 \cdot 80}{300} = 24$$

and, as it can be shown that **the sum of the expected frequencies for any row or column equals that of the corresponding observed frequencies** (see Exercise 9.41 on page 291), we find by subtraction that $e_{13} = 90 - (36 + 24) = 30$, and that the expected frequencies for the second row are $e_{21} = 120 - 36 = 84$, $e_{22} = 80 - 24 = 56$, and $e_{23} = 100 - 30 = 70$. Then, substituting these values together with the observed frequencies into the formula for χ^2, we get

$$\begin{aligned} \chi^2 &= \frac{(41-36)^2}{36} + \frac{(27-24)^2}{24} + \frac{(22-30)^2}{30} \\ &\quad + \frac{(79-84)^2}{84} + \frac{(53-56)^2}{56} + \frac{(78-70)^2}{70} \\ &= 4.575 \end{aligned}$$

5. *Decision*: Since $\chi^2 = 4.575$ does not exceed 5.991, the null hypothesis cannot be rejected; in other words, the data do not refute the hypothesis that, under the stated conditions, the probability of crumbling is the same for the three kinds of material.

■

It is customary in problems of this kind to round the expected cell frequencies to the nearest integer or to one decimal. Most of the entries of Table 5 are given to three decimal places, but, since rounding errors tend to average out, there is seldom any need to carry more than two decimal places when calculating the value of the χ^2 statistic. Also, the test we have been discussing here is only an approximate test since the sampling distribution of the χ^2 statistic is only approximately the chi-square distribution, and it should not be used when one or more of the expected frequencies is less than 5. If this is the case, we can sometimes combine two or more of the samples in such a way that none of the e's is less than 5.

If the null hypothesis of equal proportions is rejected, it is a good practice to graph the confidence intervals (see page 272) for the individual proportions p_i. The graph helps illuminate differences between the proportions.

EXAMPLE Four methods are under development for making discs of a super conducting material. Fifty discs are made by each method and they are checked for superconductivity when cooled with liquid nitrogen.

	Method 1	*Method 2*	*Method 3*	*Method 4*	*Total*
Super conductors	31	42	22	25	120
Failures	19	8	28	25	80
Total	50	50	50	50	200

Perform a chi-square test with $\alpha = 0.05$. If there is a significant difference between the proportions of super conductors produced, plot the individual confidence intervals.

Solution

1. *Null hypothesis*: $p_1 = p_2 = p_3 = p_4$
   Alternative hypothesis: p_1, p_2, p_3, and p_4 are not all equal.
2. *Level of significance*: $\alpha = 0.05$
3. *Criterion*: Reject the null hypothesis if $\chi^2 > 7.815$, the value of $\chi^2_{0.05}$ for $4 - 1 = 3$ degrees of freedom.
4. *Calculations*: Each cell in the first row has expected frequency $120 \cdot \frac{50}{200} = 30$, and each cell in the second row has expected frequency $80 \cdot \frac{50}{200} = 20$.

   The chi-square statistic is

$$\chi^2 = \frac{1}{30} + \frac{144}{30} + \frac{64}{30} + \frac{25}{30} + \frac{1}{20} + \frac{144}{20} + \frac{64}{20} + \frac{25}{20} = 19.50$$

5. *Decision*: Since 19.50 greatly exceeds 7.815, we reject the null hypothesis of equal proportions at the 5% level of significance.

The confidence intervals, obtained from the large-sample formula on page 273 are

$$0.62 \pm 0.13,\ 0.84 \pm 0.14,\ 0.44 \pm 0.14,\ 0.50 \pm 0.14$$

These are plotted in Figure 9.2. Note how Method 2 stands out as being better. ■

So far, the alternative hypothesis has been that $p_1, p_2, \ldots$, and p_k are not all equal, and for $k = 2$ this reduces to the alternative hypothesis $p_1 \neq p_2$. In problems

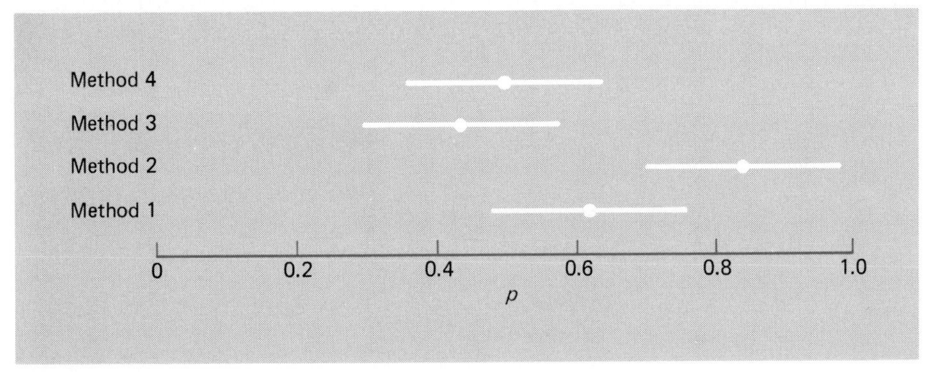

FIGURE 9.2
Confidence intervals for several proportions.

where the alternative hypothesis may also be $p_1 < p_2$ or $p_1 > p_2$, we can base the test on the statistic

Statistic for test concerning difference between two proportions

$$z = \frac{\dfrac{x_1}{n_1} - \dfrac{x_2}{n_2}}{\sqrt{\hat{p}(1-\hat{p})\left(\dfrac{1}{n_1} + \dfrac{1}{n_2}\right)}} \quad \text{with} \quad \hat{p} = \frac{x_1 + x_2}{n_1 + n_2}$$

which, for large samples, is a value of a random variable having approximately the standard normal distribution. The test based on this statistic is equivalent to the one based on the χ^2 statistic on page 284 with $k = 2$, in the sense that the square of this z statistic actually equals the χ^2 statistic (see Exercise 9.42 on page 291). The critical regions for this alternative test of the null hypothesis $p_1 = p_2$ are like those shown in the table on page 229 with p_1 and p_2 substituted for μ and μ_0.

EXAMPLE A study shows that 16 of 200 tractors produced on one assembly line required extensive adjustments before they could be shipped, while the same was true for 14 of 400 tractors produced on another assembly line. At the 0.01 level of significance, does this support the claim that the second production line does superior work?

Solution

1. *Null hypothesis*: $p_1 = p_2$
   Alternative hypothesis: $p_1 > p_2$
2. *Level of significance*: $\alpha = 0.01$
3. *Criterion*: Reject the null hypothesis if $z > 2.33$, where z is given by the above formula.

4. *Calculations*: Substituting $x_1 = 16$, $n_1 = 20$, $x_2 = 14$, $n_2 = 400$, and

$$\hat{p} = \frac{16 + 14}{200 + 400} = 0.05$$

into the formula for z, we get

$$z = \frac{\frac{16}{200} - \frac{14}{400}}{\sqrt{(0.05)(0.95)(\frac{1}{200} + \frac{1}{400})}} = 2.38$$

5. *Decision*: Since $z = 2.38$ exceeds 2.33, the null hypothesis must be rejected; we conclude that the true proportion of tractors requiring extensive adjustments is greater for the first assembly line than for the second.

■

The test we have described here applies to the null hypothesis $p_1 = p_2$, but it can easily be modified (see Exercise 9.38 on page 290) so that it applies also to the null hypothesis $p_1 - p_2 = \delta$.

The statistic for testing $p_1 = p_2$ leads to a confidence interval which provides the set of plausible values for $p_1 - p_2$.

Large-sample confidence interval for the difference of two proportions

$$\frac{x_1}{n_1} - \frac{x_2}{n_2} \pm z_{\alpha/2}\sqrt{\frac{\frac{x_1}{n_1}\left(1 - \frac{x_1}{n_1}\right)}{n_1} + \frac{\frac{x_2}{n_2}\left(1 - \frac{x_2}{n_2}\right)}{n_2}}$$

EXAMPLE With reference to the preceding example, find the large-sample 95% confidence interval for $p_1 - p_2$.

Solution Since $x_1/n_1 = \hat{p}_1 = \frac{16}{200} = 0.08$ and $x_2/n_2 = \hat{p}_2 = \frac{14}{400} = 0.035$

$$\frac{x_1}{n_1} - \frac{x_2}{n_2} \pm z_{\alpha/2}\sqrt{\frac{\frac{x_1}{n_1}\left(1 - \frac{x_1}{n_1}\right)}{n_1} + \frac{\frac{x_2}{n_2}\left(1 - \frac{x_2}{n_2}\right)}{n_2}}$$

$$= 0.08 - 0.035 \pm 1.96\sqrt{\frac{(0.08)(0.92)}{200} + \frac{(0.035)(0.965)}{400}},$$

or $0.003 < p_1 - p_2 < 0.087$

■

EXERCISES

9.23 A manufacturer of submersible pumps claims that at most 30% of the pumps require repairs within the first 5 years of operation. If a random sample of 120 of these pumps includes 47 which required repairs within the first 5 years, test the null hypothesis $p = 0.30$ against the alternative hypothesis $p > 0.30$ at the 0.05 level of significance.

9.24 The performance of a computer is observed over a period of 2 years to check the claim that the probability is 0.20 that its down time will exceed 5 hours in any given week. Testing the null hypothesis $p = 0.20$ against the alternative hypothesis $p \neq 0.20$, what can we conclude at the level of significance $\alpha = 0.05$, if there were only 11 weeks in which the downtime of the computer exceeded 5 hours?

9.25 To check on an ambulance service's claim that at least 40% of its calls are life-threatening emergencies, a random sample was taken from its files, and it was found that only 49 of 150 calls were life-threatening emergencies. Can the null hypothesis $p \geq 0.40$ be rejected against the alternative hypothesis $p < 0.40$ if the probability of a Type I error is to be at most 0.01?

9.26 In a random sample of 600 cars making a right turn at a certain intersection, 157 pulled into the wrong lane. Test the null hypothesis that actually 30% of all drivers make this mistake at the given intersection, using the alternative hypothesis $p \neq 0.30$ and the level of significance

(a) $\alpha = 0.05$; (b) $\alpha = 0.01$.

9.27 An airline claims that only 6% of all lost luggage is never found. If, in a random sample, 17 of 200 pieces of lost luggage are not found, test the null hypothesis $p = 0.06$ against the alternative hypothesis $p > 0.06$ at the 0.05 level of significance.

9.28 Suppose that 4 of 13 undergraduate engineering students state that they will go on to graduate school. Test the dean's claim that 60% of the undergraduate students will go on to graduate school, using the alternative hypothesis $p < 0.60$ and the level of significance $\alpha = 0.05$. [*Hint*: Use Table 1 to determine the probability of getting "at most 4 successes in 13 trials" when $p = 0.60$.]

9.29 Suppose that we want to test the "honesty" of a coin on the basis of the number of heads we will get in 15 flips. Using Table 1, determine how few or how many heads we would have to get so that we could reject the null hypothesis $p = 0.50$ against the alternative hypothesis $p \neq 0.50$ at the level of significance $\alpha = 0.05$. What is the *actual* level of significance we would be using with this criterion?

9.30 It costs more to test a certain type of ammunition than to manufacture it, and, hence, only three rounds are tested from each large lot. If the lot is rejected unless all three rounds function according to specifications,

(a) sketch the *OC* curve for this test;

(b) find the actual proportion of defectives for which the test procedure will cause a lot to be rejected with a probability of 0.10.

9.31 Tests are made on the proportion of defective castings produced by five different molds. If there were 14 defectives among 100 castings made with Mold I, 33 defectives among 200 castings made with Mold II, 21 defectives among 180 castings made with Mold III, 17 defectives among 120 castings made with Mold IV, and 25 defectives among 150 castings made with Mold V, use the 0.01 level of significance to test whether the true proportion of defectives is the same for each mold.

9.32 A study showed that 64 of 180 persons who saw a photocopying machine advertised during the telecast of a baseball game and 75 of 180 other persons who saw it advertised on a variety show remembered 2 hours later the brand name. Use the χ^2 statistic to test at the 0.05 level of significance whether the difference between the corresponding sample proportions is significant.

9.33 The following data come from a study in which random samples of the employees of three government agencies were asked questions about their pension plan:

	Agency 1	*Agency 2*	*Agency 3*
For the pension plan	67	84	109
Against the pension plan	33	66	41

Use the 0.01 level of significance to test the null hypothesis that the actual proportions of employees favoring the pension plan are the same.

9.34 The owner of a machine shop must decide which of two cigarette-vending machines to install in his shop. If each machine is tested 250 times, the first machine fails to work (neither delivers the cigarettes nor returns the money) 13 times, and the second machine fails to work 7 times, test at the 0.05 level of significance whether the difference between the corresponding sample proportions is significant, using

(a) the χ^2 statistic on page 284; (b) the z statistic on page 287.

9.35 With reference to the preceding exercise, verify that the square of the value obtained for z in part (b) equals the value obtained for χ^2 in part (a).

9.36 Two groups of 80 patients each took part in an experiment in which one group received pills containing an antiallergy drug, while the other group received a placebo, a pill containing no drug. If in the group given the drug 23 exhibited allergic symptoms while in the group given the placebo 41 exhibited such symptoms, is this sufficient evidence to conclude at the 0.01 level of significance that the drug is effective in reducing these symptoms?

9.37 With reference to Exercise 9.36, find a large-sample 99% confidence interval for the true difference of the proportions.

9.38 To test the null hypothesis that the difference between two population proportions equals some constant δ, not necessarily 0, we can use the statistic

$$z = \frac{\dfrac{x_1}{n_1} - \dfrac{x_2}{n_2} - \delta}{\sqrt{\dfrac{\dfrac{x_1}{n_1}\left(1 - \dfrac{x_1}{n_1}\right)}{n_1} + \dfrac{\dfrac{x_2}{n_2}\left(1 - \dfrac{x_2}{n_2}\right)}{n_2}}}$$

which, for large samples, is a value of a random variable having the standard normal distribution.

(a) With reference to Exercise 9.36, use this statistic to test at the 0.05 level of significance whether the true percentage of patients exhibiting allergic symptoms is at least 8% less for those who actually receive the drug.

(b) In a true-false test, a test item is considered to be good if it discriminates between well-prepared students and poorly prepared students. If 205 of 250 well-prepared students and 137 of 250 poorly prepared students answer a certain item correctly, test at the 0.01 level of significance whether for the given item the proportion of correct answers can be expected to be at least 15% higher among well-prepared students than among poorly prepared students.

9.39 With reference to part (b) of Exercise 9.38, find a large-sample 99% confidence interval for the true difference of the proportions.

9.40 Verify that the formulas for the χ^2 statistic on page 283 (with $\hat{p}$ substituted for the p_i) and on page 284 are equivalent.

9.41 Verify that if the expected frequencies are determined in accordance with the rule on page 284, the sum of the expected frequencies for each row and column equals the sum of the corresponding observed frequencies.

9.42 Verify that the square of the z statistic on page 287 equals the χ^2 statistic on page 284 for $k = 2$.

9.5 THE ANALYSIS OF $r \times c$ TABLES

As we suggested earlier, the method by which we analyzed the example on page 284 lends itself also to the analysis of ***r* × *c* tables**, or r-by-c tables, that is, tables in which data are tallied into a two-way classification having r rows and c columns. Such tables arise in essentially two kinds of problems. First, we might again have samples from several populations, with the distinction that now each trial permits more than two possible outcomes. This might happen, for example, if persons belonging to different income groups are asked whether they favor a certain political candidate, whether they are against him, or whether they are indifferent or undecided. The other situation giving rise to an $r \times c$ table is one in which we sample from one population but classify each item with respect to two (usually qualitative) categories. This might happen, for example, if a consumer testing service rates cars as excellent, superior, average, or poor with regard to performance and also with regard to appearance. Each car tested would then fall into one of the 16 cells of a 4×4 table, and it is mainly in connection with problems of this kind that $r \times c$ tables are referred to as **contingency tables**.

The essential difference between the two kinds of situations giving rise to $r \times c$ tables is that in the first case the column totals (the sample sizes) are fixed, while in the second case only the **grand total** (the total for the entire table) is fixed. As a result, there are also differences in the null hypotheses we shall want to test. In the first case we want to test whether the probability of obtaining an observation in the ith row is the same for each column; symbolically, we shall want to test the null hypothesis

$$p_{i1} = p_{i2} = \cdots = p_{ic} \qquad \text{for } i = 1, 2, \ldots, r$$

where p_{ij} is the probability of obtaining an observation belonging to the ith row and the jth column, and $\sum_{i=1}^{r} p_{ij} = 1$ for each column. The alternative hypothesis is that the p's are not all equal for at least one row. In the second case we shall want to test the null hypothesis that the random variables represented by the two classifications are independent, so that p_{ij} is the product of the probability of getting a value belonging to the ith row and the probability of getting a value belonging to the jth column. The alternative hypothesis is that the two random variables are not independent.

In spite of the differences we have described, the analysis of an $r \times c$ table is the same for both cases. First we calculate the expected cell frequencies e_{ij} as on page 284, namely, by multiplying the totals of the respective rows and columns and then dividing by the grand total. In practice, we make use of the fact that the observed frequencies and the expected frequencies total the same for each row and column, so that only $(r-1)(c-1)$ of the e_{ij} have to be calculated directly, while the others can be obtained by subtraction from appropriate row or column totals. We then substitute into the formula

Statistic for analysis of $r \times c$ table

$$\chi^2 = \sum_{i=1}^{r} \sum_{j=1}^{c} \frac{(o_{ij} - e_{ij})^2}{e_{ij}}$$

and we reject the null hypothesis if the value of this statistic exceeds χ^2_α for $(r-1)(c-1)$ degrees of freedom. This expression for the number of degrees of freedom is justified by the above observation that after we determine $(r-1)(c-1)$ of the expected cell frequencies, the others are automatically determined, that is, they may be obtained by subtraction from appropriate row or column totals.

EXAMPLE

To determine whether there really is a relationship between an employee's performance in the company's training program and his or her ultimate success in the job, it takes a sample of 400 cases from its very extensive files and obtains the results shown in the following table:

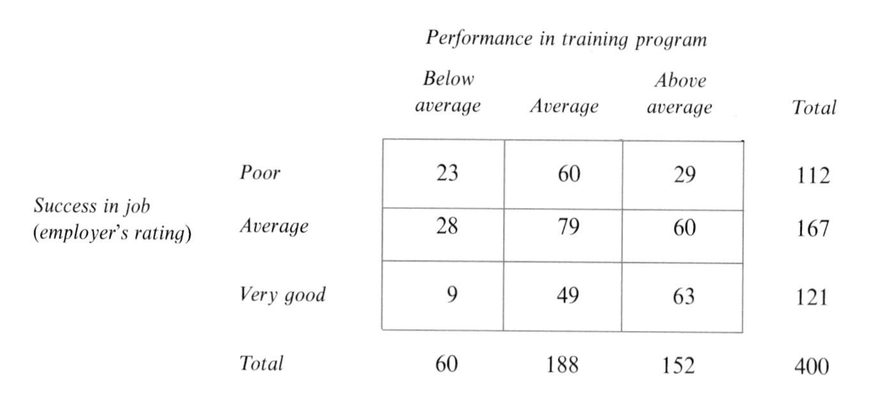

		Performance in training program			
		Below average	*Average*	*Above average*	*Total*
Success in job (employer's rating)	*Poor*	23	60	29	112
	Average	28	79	60	167
	Very good	9	49	63	121
	Total	60	188	152	400

Use the 0.01 level of significance to test the null hypothesis that performance in the training program and success in the job are independent.

Solution

1. *Null hypothesis*: Performance in training program and success in job are independent.
   Alternative hypothesis: Performance in training program and success in job are not independent.
2. *Level of significance*: $\alpha = 0.01$
3. *Criterion*: Reject the null hypothesis if $\chi^2 > 13.277$, the value of $\chi^2_{0.01}$ for $(3-1)(3-1) = 4$ degrees of freedom, where χ^2 is given by the formula on page 292.
4. *Calculations*: Calculating first the expected cell frequencies for the first two cells of the first two rows, we get

$$e_{11} = \frac{112 \cdot 60}{400} = 16.8, \qquad e_{12} = \frac{112 \cdot 188}{400} = 52.6,$$

$$e_{21} = \frac{167 \cdot 60}{400} = 25.0, \qquad e_{22} = \frac{167 \cdot 188}{400} = 78.5$$

   Then, by subtraction, we find that the expected frequencies for the third cell of the first two rows are 42.6 and 63.5, and those for the third row are 18.2, 56.9, and 45.9. Thus,

$$\chi^2 = \frac{(23-16.8)^2}{16.8} + \frac{(60-52.6)^2}{52.6} + \frac{(29-42.6)^2}{42.6}$$
$$+ \frac{(28-25.0)^2}{25.0} + \frac{(79-78.5)^2}{78.5} + \frac{(60-63.5)^2}{63.5}$$
$$+ \frac{(9-18.2)^2}{18.2} + \frac{(49-56.9)^2}{56.9} + \frac{(63-45.9)^2}{45.9}$$
$$= 20.34$$

5. *Decision*: Since $\chi^2 = 20.34$ exceeds 13.277, the null hypothesis must be rejected; we conclude that there is a dependence between an employee's performance in the training program and his or her success in the job.

■

We pursue this example further in order to determine the form of the dependence.

EXAMPLE With reference to the preceding example, find the individual contributions to the chi-square.

Solution We display the contingency table, but this time we include the expected frequencies just below the observed frequencies.

		Performance in training program			
		Below average	*Average*	*Above average*	*Total*
Success in job (employer's rating)	*Poor*	23 **16.8**	60 **52.6**	29 **42.6**	112
	Average	28 **25.0**	79 **78.5**	60 **63.5**	167
	Very good	9 **18.1**	49 **56.9**	63 **46.0**	121
	Total	60	188	152	400

Also, we write

$$\begin{aligned}\chi^2 &= 2.29 + 1.04 + 4.34 + 0.36 + 0.00 + 0.19 + 4.65 + 1.10 + 6.37 \\ &= 20.34\end{aligned}$$

From these two displays, it is clear that there is a positive dependence between performance in training and job success. For the three individual cells with the largest contributions to χ^2, the *above average–very good* cell frequency is high, whereas the *above average–poor* and *below average–very good* cell frequencies are low.

■

9.6 GOODNESS OF FIT

We speak of **goodness of fit** when we try to compare an observed frequency distribution with the corresponding values of an expected, or theoretical, distribution. To illustrate, suppose that during 400 five-minute intervals the air-traffic control of an airport received 0, 1, 2, . . . , or 13 radio messages with respective frequencies of 3, 15, 47, 76, 68, 74, 46, 39, 15, 9, 5, 2, 0, and 1. Suppose, furthermore, that we want to check whether these data substantiate the claim that the number of radio messages which they receive during a 5-minute interval may be looked upon as a random variable having the Poisson distribution with $\lambda = 4.6$. Looking up the corresponding Poisson probabilities in Table 2 and multiplying them by 400 to get the expected frequencies, we arrive at the result shown in the following table together with the original data:

Number of radio messages	*Observed frequencies*	*Poisson probabilities*	*Expected frequencies*
0	3 } 18	0.010	4.0 } 22.4
1	15	0.046	18.4
2	47	0.107	42.8
3	76	0.163	65.2
4	68	0.187	74.8
5	74	0.173	69.2
6	46	0.132	52.8
7	39	0.087	34.8
8	15	0.050	20.0
9	9	0.025	10.0
10	5 } 8	0.012	4.8 } 8.0
11	2	0.005	2.0
12	0	0.002	0.8
13	1	0.001	0.4
	400		400.0

Note that we combined some of the data so that none of the expected frequencies is less than 5.

To test whether the discrepancies between the observed and expected frequencies can be attributed to chance, we use the statistic

Statistic for test of goodness of fit

$$\chi^2 = \sum_{i=1}^{k} \frac{(o_i - e_i)^2}{e_i}$$

where the o_i and e_i are the observed and expected frequencies. The sampling distribution of this statistic is approximately the chi-square distribution with $k - m$ degrees of freedom, where k is the number of terms in the formula for χ^2 and m is the number of quantities, obtained from the observed data, that are needed to calculate the expected frequencies.

EXAMPLE With reference to the illustration on page 294, test at the 0.01 level of significance whether the data can be looked upon as values of a random variable having the Poisson distribution with $\lambda = 4.6$.

Solution

1. *Null hypothesis*: Random variable has Poisson distribution with $\lambda = 4.6$.
   Alternative hypothesis: Random variable does not have Poisson distribution with $\lambda = 4.6$.
2. *Level of significance*: $\alpha = 0.01$
3. *Criterion*: Reject the null hypothesis if $\chi^2 > 16.919$, the value of $\chi^2_{0.01}$ for $k - m = 10 - 1 = 9$ degrees of freedom, where χ^2 is given by the formula above. (The number of degrees of freedom is $10 - 1 = 9$, since only one quantity, the total frequency of 400, is needed from the observed data to calculate the expected frequencies.)
4. *Calculations*: Substitution into the formula for χ^2 yields

$$\chi^2 = \frac{(18 - 22.4)^2}{22.4} + \frac{(47 - 42.8)^2}{42.8} + \cdots + \frac{(9 - 10.0)^2}{10.0} + \frac{(8 - 8.0)^2}{8.0}$$

$$= 6.749$$

5. *Decision*: Since $\chi^2 = 6.749$ does not exceed 16.919, the null hypothesis cannot be rejected; we conclude that the Poisson distribution with $\lambda = 4.6$ provides a good fit.

■

EXERCISES

9.43 The results of polls conducted two weeks and four weeks before a gubernatorial election are shown in the following table:

	Two weeks before election	*Four weeks before election*
For Republican candidate	79	91
For Democratic candidate	84	66
Undecided	37	43

Use the 0.05 level of significance to test whether there has been a change in opinion during the 2 weeks between the two polls.

9.44 A large electronics firm that hires many handicapped workers wants to determine whether their handicaps affect such workers' performance. Use the level of significance $\alpha = 0.05$ to decide on the basis of the sample data shown in the following table whether it is reasonable to maintain that the handicaps have no effect on the workers' performance:

	Performance		
	Above average	*Average*	*Below average*
Blind	21	64	17
Deaf	16	49	14
No handicap	29	93	28

9.45 Tests of the fidelity and the selectivity of 190 radio receivers produced the results shown in the following table:

		Fidelity		
		Low	*Average*	*High*
	Low	6	12	32
Selectivity	*Average*	33	61	18
	High	13	15	0

Use the 0.01 level of significance to test whether there is a relationship (dependence) between fidelity and selectivity.

9.46 A quality-control engineer takes daily samples of $n = 4$ tractors coming off an assembly line and on 200 consecutive working days the data summarized in the following table are obtained:

Number requiring adjustments	*Number of days*
0	101
1	79
2	19
3	1

To test the claim that 10% of all the tractors coming off this assembly line require adjustments, look up the corresponding probabilities in Table 1, calculate the expected frequencies, and perform the chi-square test at the 0.01 level of significance.

9.47 With reference to Exercise 9.46, verify that the mean of the observed distribution is 0.60, corresponding to 15% of the tractors requiring adjustments. Then look up the probabilities for $n = 4$ and $p = 0.15$ in Table 1, calculate the expected frequencies, and test at the 0.01 level of significance whether the binomial distribution with $n = 4$ and $p = 0.15$ provides a suitable model for this situation.

9.48 Suppose that in the example on page 294 we had shown first that the mean of the distribution, rounded to one decimal, is 4.5, and then tested whether the Poisson distribution with $\lambda = 4.5$ provides a good fit. What would have been the number of degrees of freedom for the appropriate chi-square criterion?

9.49 The following is the distribution of the hourly number of trucks arriving at a company's warehouse:

Trucks arriving per hour	*Frequency*
0	52
1	151
2	130
3	102
4	45
5	12
6	5
7	1
8	2

Find the mean of this distribution, and using it (rounded to one decimal place) as the parameter λ, fit a Poisson distribution. Test for goodness of fit at the 0.05 level of significance.

9.50 Using any four columns of Table 7 (that is, a total of 200 random digits), construct a table showing how many times each of the digits 0, 1, ..., and 9 occurred. Comparing the observed frequencies with the corresponding expected frequencies (based on the assumption that the digits are randomly generated), test at the 0.05 level of significance whether the assumption of randomness is tenable.

9.51 The following is the distribution of the sulfur oxides emission data on page 8, for which we showed that $\bar{x} = 18.85$ and $s = \sqrt{30.77} = 5.55$:

Class limits (tons)	*Frequency*
5.0– 8.9	3
9.0–12.9	10
13.0–16.9	14
17.0–20.9	25
21.0–24.9	17
25.0–28.9	9
29.0–32.9	2
	80

(a) Find the probabilities that a random variable having a normal distribution with $\mu = 18.85$ and $\sigma = 5.55$ takes on a value less than 8.95, between 8.95 and 12.95, between 12.95 and 16.95, between 16.95 and 20.95, between 20.95 and 24.95, between 24.95 and 28.95, and greater than 28.95.

(b) Multiply the probabilities obtained in part (a) by the total frequency, $n = 80$, thus getting the expected normal curve frequencies corresponding to the seven classes of the given distribution (with the first one changed to "8.9 or less" and the last one changed to "29.0 or more").

(c) Use the 0.05 level of significance to test the null hypothesis that the given data may be looked upon as a random sample from a normal population. Explain why the number of degrees of freedom for this χ^2 test is $k - 3$, where k is the number of terms in the χ^2 statistic.

9.52 Among 100 vacuum tubes used in an experiment, 46 had a service life of less than 20 hours, 19 had a service life of 20 or more but less than 40 hours, 17 had a service life of 40 or more but less than 60 hours, 12 had a service life of 60 or more but less than 80 hours, and 6 had a service life of 80 hours or more. Using steps similar to those outlined in the preceding exercise, test at the 0.01 level of significance whether the lifetimes may be regarded as a sample from an exponential population with $\mu = 40$ hours.

9.53 A chi-square test is easily implemented on a computer. The *MINITAB* commands

```
READ INTO C1 C2 C3 C4
31 42 25 22
19  8 25 28
```

place the table from the example on page 285 into columns 1–4. Then,

```
CHISQUARE C1-C4
```

produces the output

```
Expected counts are printed below observed counts

         Method 1 Method 2 Method 3 Method 4     Total
     1      31       42       25       22         120
          30.0     30.0     30.0     30.0

     2      19        8       25       28          80
          20.0     20.0     20.0     20.0

 Total      50       50       50       50         200

ChiSq =   0.03 +   4.80 +   0.83 +   2.13  +
          0.05 +   7.20 +   1.25 +   3.20 = 19.50
  df = 3
```

Repeat the analysis using only the data from the first three methods.

9.54 The procedure in Exercise 9.53 also calculates the chi-square test for independence. Do Exercise 9.44 using the computer.

9.7

REVIEW EXERCISES

9.55 In a sample of 100 ceramic pistons made for an experimental diesel engine, 18 were cracked. Construct a 95% confidence interval for the true proportion of cracked pistons, using

(a) Table 9;

(b) the large-sample confidence-interval formula.

9.56 With reference to Exercise 9.55, test the null hypothesis $p = 0.20$ versus the alternative hypothesis $p < 0.20$ at the 0.05 level.

9.57 In a random sample of 160 workers exposed to a certain amount of radiation, 24 experienced some ill effects. Construct a 99% confidence interval for the corresponding true percentage, using

(a) Table 9;

(b) the large-sample confidence-interval formula.

9.58 With reference to Exercise 9.57, test the null hypothesis $p = 0.15$ versus the alternative hypothesis $p \neq 0.15$ at the 0.01 level.

9.59 In a random sample of 100 packages shipped by air freight, 13 had some damage. Construct a 95% confidence interval for the true proportion of damaged packages, using

(a) Table 9;

(b) the large-sample confidence-interval formula.

9.60 With reference to Exercise 9.59, test the null hypothesis $p = 0.10$ versus the alternative hypothesis $p > 0.10$ at the 0.01 level.

9.61 In 4,000 firings of a certain kind of rocket there were 10 instances in which a rocket exploded upon ignition. Construct an upper 95% confidence limit for the probability that such a rocket will explode upon ignition.

9.62 If 26 of 200 Brand *A* tires fail to last 20,000 miles, whereas the corresponding figures for 200 tires each of Brands *B*, *C*, and *D* are 23, 15, and 32, use the 0.05 level of significance to test the null hypothesis that there is no difference in the quality of the four kinds of tires with regard to their durability.

9.63 One method of seeding clouds was successful in 57 of 150 attempts while another method was successful in 33 of 100 attempts. At the 0.05 level of significance, can we conclude that the first method is better than the second?

9.64 With reference to Exercise 9.63, find a large-sample 95% confidence interval for the true difference of probabilities.

9.65 Two bonding agents, A and B, are available for making a laminated beam. Of 50 beams made with Agent A, 11 failed a stress test, whereas 19 of the 50 beams made with Agent B failed. At the 0.05 level, can we conclude that Agent A is better than Agent B?

9.66 With reference to Exercise 9.65, find a large-sample 95% confidence interval for the true difference of the probabilities of failure.

9.67 Cooling pipes at three nuclear power plants are investigated for deposits that would inhibit the flow of water. From 30 randomly selected spots at each plant, 13 from the first plant, 8 from the second plant, and 19 from the third were clogged.

(a) Use the 0.05 level to test the null hypothesis of equality.

(b) Plot the confidence intervals for the three probabilities of being clogged.

9.68 Suppose that in Exercise 9.62 we had been interested also in how many of the tires lasted more than 30,000 miles and obtained the results shown in the following table:

	Brand A	*Brand B*	*Brand C*	*Brand D*
Failed to last 20,000 miles	26	23	15	32
Lasted from 20,000 to 30,000	118	93	116	121
Lasted more than 30,000 miles	56	84	69	47

Use the 0.01 level of significance to test the null hypothesis that there is no difference in the quality of the four kinds of tires with regard to their durability.

9.69 The following is the distribution of the daily number of power failures reported in a western city on 300 days:

Number of power failures	*Number of days*
0	9
1	43
2	64
3	62
4	42
5	36
6	22
7	14
8	6
9	2

Test at the 0.05 level of significance whether the daily number of power failures in this city is a random variable having the Poisson distribution with $\lambda = 3.2$.

9.70 With reference to the example on page 292, repeat the analysis after combining the categories *below average* and *average* in the training program and the categories *poor* and *average* in success. Comment on the form of the dependence.

9.71 Mechanical engineers, testing a new arc welding technique, classified welds both with respect to appearance and an X-ray inspection.

		Appearance			
		Bad	*Normal*	*Good*	*Total*
	Bad	20	7	3	30
X ray	*Normal*	13	51	16	80
	Good	7	12	21	40
	Total	40	70	40	150

Test for independence using $\alpha = 0.05$.

9.8
CHECK LIST OF KEY TERMS (with page references)

Contingency table 291
Expected cell frequency 284
Goodness of fit 294
Observed cell frequency 283
One-sided confidence interval 275
$r \times c$ table 291
Sample proportion 270

NONPARAMETRIC TESTS

Most of the methods of inference that we have studied are based on the assumption that the observations come from normal populations. If this is the case, these methods extract all the information that is available in a sample, and they usually attain the best possible precision. However, since there are many situations where it is doubtful whether the assumption of normality can be met, statisticians have developed alternative techniques based on less stringent assumptions, which have become known as nonparametric tests.

Of the many nonparametric tests that have been developed in recent years, the sign test is presented in Section 10.2; tests based on rank sums are given in Section 10.3; a test of randomness is introduced in Section 10.4; and a goodness of fit test is given in Section 10.5.

10.1 INTRODUCTION

In this chapter, we expand the choice of statistical methods available for inferences concerning one or more populations. The assumption of a normal population underlies most of the "standard methods" discussed in the previous chapters, but it is often difficult to verify this tentative assumption. Here we introduce tests that depend only on order relationships among the observations. Consequently, much less has to be assumed about the form of the underlying populations. The main advantage of these **nonparametric tests** is that exact inferences can be made when the assumptions underlying the so-called standard methods cannot be met. When the normal assumption is met, the standard tests will have more power. However, asymmetry or other departures from normality will have no effect on the nonparametric methods. Moreover, their power is usually satisfactory even when the populations deviate from normality.

Also, nonparametric tests apply even when the choice of a particular numerical scale of measurement is arbitrary. Still, their strongest virtue is the fact that the level of significance is exact even when the populations are quite non-normal.

10.2 THE SIGN TEST

In this section we shall describe a nonparametric alternative to the one-sample t test, the paired-sample t test, and corresponding large-sample tests. As an alternative to the one-sample t test or the corresponding large-sample test, the **sign test** applies when we sample a continuous symmetrical population, so that the probability of getting a sample value less than the mean and the probability of getting a sample value greater than the mean are both $\frac{1}{2}$. More generally, because symmetry is often difficult to verify with small or moderate sample sizes, we can formulate the hypotheses in terms of the population median $\tilde{\mu}$. To test the null hypothesis $\tilde{\mu} = \tilde{\mu}_0$ against an appropriate alternative on the basis of a random sample of size n, we replace each sample value greater than $\tilde{\mu}_0$ with a plus sign and each sample value less than $\tilde{\mu}_0$ with a minus sign, and then we test the null hypothesis that these plus and minus signs are the outcomes of binomial trials with $p = \frac{1}{2}$. If a sample value equals $\tilde{\mu}_0$, which may well happen since the values of continuous random valuables are virtually always rounded, it is discarded.

To perform this kind of test when the sample is small, we refer directly to a table of binomial probabilities such as Table 1 at the end of the book; when the sample is large we use the test described in Section 9.3.

EXAMPLE The following data constitute a random sample of 15 measurements of the octane rating of a certain kind of gasoline:

99.0	102.3	99.8	100.5	99.7	96.2	99.1	102.5
103.3	97.4	100.4	98.9	98.3	98.0	101.6	

Test the null hypothesis $\tilde{\mu} = 98.0$ against the alternative hypothesis $\tilde{\mu} > 98.0$ at the 0.01 level of significance.

Solution Since one of the sample values equals 98.0 and must be discarded, the sample size for the sign test is only $n = 14$.

1. *Null hypothesis*: $\tilde{\mu} = 98.0$ $(p = \frac{1}{2})$
   Alternative hypothesis: $\tilde{\mu} > 98.0$ $(p > \frac{1}{2})$
2. *Level of significance*: $\alpha = 0.01$
3. *Criterion*: The criterion may be based on the number of plus signs or the number of minus signs. Using the number of plus signs, denoted by x, reject the null hypothesis if the probability of getting x or more plus signs is less than or equal to 0.05.
4. *Calculations*: Replacing each value greater than 98.0 with a plus sign and each value less than 98.0 with a minus sign, the 14 sample values yield

   $$+ + + + + - + + + - + + + +$$

   Thus, $x = 12$ and Table 1 shows that for $n = 14$ and $p = 0.50$ the probability of $x \geq 12$ is $1 - 0.9935 = 0.0065$.
5. *Decision*: Since 0.0065 is less than 0.01, the null hypothesis must be rejected; we conclude that the median octane rating of the given kind of gasoline exceeds 98.0.

■

The sign test can also be used as a nonparametric alternative to the paired-sample t test or the corresponding large-sample test. In such problems, each pair of sample values is replaced with a plus sign if the first value is greater than the second, with a minus sign if the first value is smaller than the second, or it is discarded if the two values are equal. The procedure is the same as before. Let $\tilde{\mu}_D$ denote the median of the differences.

EXAMPLE With reference to the example on page 248, which dealt with the effectiveness of an industrial safety program, use the sign test at the 0.05 level of significance to test whether the safety program is effective.

Solution

1. *Null hypothesis*: $\tilde{\mu}_D = 0$ $(p = \frac{1}{2})$
   Alternative hypothesis: $\tilde{\mu}_D > 0$ $(p > \frac{1}{2})$
2. *Level of significance*: $\alpha = 0.05$
3. *Criterion*: If x is the number of plus signs, reject the null hypothesis if the probability of getting x or more plus signs is less than or equal to 0.05.
4. *Calculations*: Replacing each pair of values with a plus sign if the first value is greater than the second or with a minus sign if the first value is smaller than the second, the 10 sample pairs yield

   $$+ + + + - + + + + +$$

Thus, $x = 9$ and Table 1 shows that for $n = 10$ and $p = 0.50$ the probability of $x \geq 9$ is $1 - 0.9990 = 0.0010$.

5. *Decision:* Since 0.0010 is less than 0.05, the null hypothesis must be rejected; we conclude that the safety program is effective.

■

10.3 RANK-SUM TESTS

In this section we shall introduce two tests based on **rank sums**—the ***U* test** will be presented as a nonparametric alternative to the two-sample t test, and the ***H* test** will be presented as a nonparametric alternative to the one-way analysis of variance, which we shall study in Chapter 12. In other words, the H test serves to test the null hypothesis that k samples come from identical populations against the alternative that the populations are not identical.

To illustrate how the U test (also called the **Wilcoxon test** or the **Mann–Whitney test**, named after the statisticians who contributed to its development) is performed, suppose that in a study of sedimentary rocks, the following diameters (in millimeters) were obtained for two kinds of sand:

Sand I: 0.63, 0.17, 0.35, 0.49, 0.18, 0.43, 0.12, 0.20, 0.47, 1.36, 0.51, 0.45, 0.84, 0.32, 0.40

Sand II: 1.13, 0.54, 0.96, 0.26, 0.39, 0.88, 0.92, 0.53, 1.01, 0.48, 0.89, 1.07, 1.11, 0.58

The problem is to decide whether the two populations are the same or if one is more likely to produce larger observations than the other. Let x_1 be the value of a random variable having the first distribution and x_2 be the value of a random variable having the second distribution. If $P(a < x_1) \leq P(a < x_2)$ for all a, with strict inequality for some a, we say that the second population (distribution) is **stochastically larger** than the first population (distribution). We formulate one-sided hypotheses in terms of this stochastic order relation.

We begin the U test by ranking the data jointly, as if they comprise one sample, in an increasing order of magnitude, and for our data we get

0.12	0.17	0.18	0.20	0.26	0.32	0.35	0.39	0.40	0.43
I	I	I	I	II	I	I	II	I	I
0.45	0.47	0.48	0.49	0.51	0.53	0.54	0.58	0.63	0.84
I	I	II	I	I	II	II	II	I	I
0.88	0.89	0.92	0.96	1.01	1.07	1.11	1.13	1.36	
II	II	II	II	II	II	II	II	I	

Note that we indicated for each value whether it is a measurement of Sand I or Sand II. Assigning the data in this order the ranks 1, 2, 3, ..., and 29, we find that the values of the first sample (Sand I) occupy ranks 1, 2, 3, 4, 6, 7, 9, 10, 11, 12, 14, 15, 19, 20, and 29, while those of the second sample (Sand II) occupy ranks 5, 8, 13, 16, 17, 18, 21, 22, 23, 24, 25, 26, 27, and 28. There are no ties here among values belonging to different samples, but if there were, we would assign to each of the tied observations the mean of the ranks which they jointly occupy. (For instance, if the third and fourth values are identical we would assign each the rank $\frac{3+4}{2} = 3.5$, and if the ninth, tenth, and eleventh values are identical we would assign each the rank $\frac{9+10+11}{3} = 10$.)

The null hypothesis we want to test is that the two samples come from identical populations, and it stands to reason that in that case the means of the ranks assigned to the values of the two samples should be more or less the same. Instead of the means, we can also compare the sums of the ranks assigned to the values of the two samples, suitably accounting for a possible difference in their size. For our two samples, the sums of the ranks are $W_1 = 162$ and $W_2 = 273$, and it remains to be seen whether their difference is large enough to reject the null hypothesis.

When the use of rank sums was first proposed as a nonparametric alternative to the two-sample t test, the decision was based on W_1 or W_2, but now the decision is based on either of the related statistics

U_1 and U_2 statistics

$$U_1 = W_1 - \frac{n_1(n_1+1)}{2}$$

or

$$U_2 = W_2 - \frac{n_2(n_2+1)}{2}$$

or on the statistic U, which always equals the smaller of the two. The sizes of the two samples are n_1 and n_2, and as it does not matter how we number the samples, we shall use here the statistic U_1.†

† The tests based on U_1 or U_2 are equivalent to those based on W_1 or W_2, but they have the advantage that they lend themselves more readily to the construction of tables of critical values. Not only do U_1 and U_2 take on values on the interval from 0 to n_1n_2—indeed, their sum is always equal to n_1n_2— but their sampling distributions are symmetrical about $\frac{n_1n_2}{2}$. The use of U, which always equals the smaller of the values of U_1 and U_2, has the added advantage that the resulting test is one-tailed regardless of the alternative hypothesis, and hence easier to tabulate.

Under the null hypothesis that the two samples come from identical populations, it can be shown that the mean and the variance of the sampling distribution of U_1 are

Mean and variance of U_1 statistic

$$\mu_{U_1} = \frac{n_1 n_2}{2}$$

and

$$\sigma^2_{U_1} = \frac{n_1 n_2 (n_1 + n_2 + 1)}{12}$$

If there are ties in rank, these formulas provide only approximations, but if the number of ties is small, these approximations will generally be good.

Since numerical studies have shown that the sampling distribution of U_1 can be approximated closely by a normal distribution when n_1 and n_2 are both greater than 8, the test of the null hypothesis that the two samples come from identical populations can be based on

Statistic for large-sample U test

$$z = \frac{U_1 - \mu_{U_1}}{\sigma_{U_1}}$$

which is a value of a random variable having approximately the standard normal distribution. For small samples, we can base the test on special tables; for instance, on those in the book by Johnson and Bhattacharyya listed in the Bibliography.

Note that when we test the null hypothesis that the two samples come from identical populations against the alternative hypothesis:

population 2 is stochastically larger than population 1.

we reject the null hypothesis if $z < -z_\alpha$ since small values of U_1 correspond to small values of W_1; correspondingly, if the alternative hypothesis is

population 1 is stochastically larger than population 2.

we reject the null hypothesis if $z > z_\alpha$ since large values of U_1 correspond to large values of W_1.

EXAMPLE With reference to the grain-size data on page 306, use the U test at the 0.01 level of significance to test the null hypothesis that the two samples come from identical populations against the alternative hypothesis that the populations are not identical.

Solution

1. *Null hypothesis*: Populations are identical.
   Alternative hypothesis: The populations are not identical.
2. *Level of significance*: $\alpha = 0.01$
3. *Criterion*: Reject the null hypothesis if $z < -2.575$ or $z > 2.575$, where z is given by the above formula.
4. *Calculations*: Since $n_1 = 15$, $n_2 = 14$, and we have already shown that $W_1 = 162$, we find that

$$U_1 = 162 - \frac{15 \cdot 16}{2} = 42$$

$$\mu_{U_1} = \frac{15 \cdot 14}{2} = 105$$

and

$$\sigma^2_{U_1} = \frac{15 \cdot 14 \cdot 30}{12} = 525$$

and it follows that

$$z = \frac{42 - 105}{\sqrt{525}} = -2.75$$

5. *Decision*: Since $z = -2.75$ is less than -2.575, the null hypothesis must be rejected; we conclude that there is a difference in the populations of grain size.

■

The **H test**, or **Kruskal–Wallis test**, is a generalization of the U test in that it enables us to test the null hypothesis that k independent random samples come from identical populations. As in the U test, all the observations are ranked jointly, and if R_i is the sum of the ranks occupied by the n_i observations of the ith sample and $n_1 + n_2 + \cdots + n_k = n$, the test is based on the statistic

Statistic for H test

$$H = \frac{12}{n(n+1)} \sum_{i=1}^{k} \frac{R_i^2}{n_i} - 3(n+1)$$

When $n_i > 5$ for all i and the null hypothesis is true, the sampling distribution of the H statistic is well approximated by the chi-square distribution with $k - 1$ degrees of freedom. There exist special tables of critical values for the H test for selected small values of the n_i and k.

EXAMPLE An experiment designed to compare three preventive methods against corrosion yielded the following maximum depths of pits (in thousandths of an inch) in pieces of wire subjected to the respective treatments:

Method A:	77	54	67	74	71	66	
Method B:	60	41	59	65	62	64	52
Method C:	49	52	69	47	56		

Use the 0.05 level of significance to test the null hypothesis that the three samples come from identical populations.

Solution

1. *Null hypothesis*: Populations are identical.
   Alternative hypothesis: Populations are not all equal.
2. *Level of significance*: $\alpha = 0.05$
3. *Criterion*: Reject the null hypothesis if $H > 5.991$, the value of $\chi^2_{0.05}$ for 2 degrees of freedom, where H is given by the formula on page 309.
4. *Calculations*: Ranking these measurements jointly from smallest to largest, we find that those of the first sample occupy ranks 6, 13, 14, 16, 17, and 18; those of the second sample occupy ranks 1, 4.5, 8, 9, 10, 11, and 12; and those of the third sample occupy ranks 2, 3, 4.5, 5, 7, and 15. Thus, $R_1 = 84$, $R_2 = 55.5$, $R_3 = 31.5$, and substitution into the formula for H yields

$$H = \frac{12}{18 \cdot 19}\left(\frac{84^2}{6} + \frac{55.5^2}{7} + \frac{31.5^2}{5}\right) - 3 \cdot 19$$
$$= 6.7$$

5. Since $H = 6.7$ exceeds 5.991, the null hypothesis must be rejected; we conclude that the three preventive methods against corrosion are not equally effective. ■

EXERCISES

10.1 In a laboratory experiment, 18 determinations of the coefficient of friction between leather and metal yielded the following results: 0.59, 0.56, 0.49, 0.55, 0.65, 0.55, 0.51, 0.60, 0.56, 0.47, 0.58, 0.61, 0.54, 0.68, 0.56, 0.50, 0.57, and 0.53. Use the sign test at the 0.05 level of significance to test the null hypothesis $\tilde{\mu} = 0.55$ against the alternative hypothesis $\tilde{\mu} \neq 0.55$.

10.2 A random sample of nine women buying new eyeglasses tried on 12, 11, 14, 15, 10, 14, 11, 8, and 12 different frames. Use the sign test at the 0.05 level of significance to test the null hypothesis $\tilde{\mu} = 10$ (that the median number of different frames a woman buying new eyeglasses tries on, is 10) against the alternative hypothesis $\tilde{\mu} > 10$.

10.3 With reference to Exercise 2.3, which pertained to the boiling point of a certain silicon compound, use the sign test at the 0.05 level of significance to test the null hypothesis $\tilde{\mu} = 158$ (degrees Celsius) against the alternative hypothesis $\tilde{\mu} \neq 158$.

10.4 The quality control department of a large manufacturer obtained the following sample data (in pounds) on the breaking strength of a certain kind of 2-inch cotton ribbon: 153, 159, 144, 160, 158, 153, 171, 162, 159, 137, 159, 159, 148, 162, 154, 159, 160, 157, 140, 168, 163, 148, 151, 153, 157, 155, 148, 168, 162, and 149. Use the sign test at the 0.01 level of significance to test the null hypothesis $\tilde{\mu} = 150$ against the alternative hypothesis $\tilde{\mu} > 150$.

10.5 With reference to Exercise 2.10 on page 19, which pertained to the ignition times of certain upholstery materials, use the sign test at the 0.01 level of significance to test the null hypothesis $\tilde{\mu} = 6.50$ seconds against the alternative hypothesis $\tilde{\mu} < 6.50$ seconds.

10.6 The following are the number of speeding tickets issued by two police officers on 17 days: 7 and 10, 11 and 13, 14 and 14, 11 and 15, 12 and 9, 6 and 10, 9 and 13, 8 and 11, 10 and 11, 11 and 15, 13 and 11, 7 and 10, 8 and 8, 11 and 12, 9 and 14, 10 and 9, 13 and 16. Use the sign test at the 0.05 level of significance to test the null hypothesis that on the average the two police officer issue equally many speeding tickets per day against the alternative hypothesis that the second police officer tends to issue more than the first.

10.7 With reference to Exercise 7.71 on page 251, use the sign test at the 0.10 level of significance to check whether there is a systematic difference between weights obtained with the two scales.

10.8 With reference to Exercise 7.72 on page 251, use the sign test at the 0.01 level of significance to check whether the prescribed program of exercises is effective.

10.9 The following are the number of minutes it took a sample of 15 men and 12 women to complete the application form for a position:

Men: 16.5, 20.0, 17.0, 19.8, 18.5, 19.2, 19.0, 18.2, 20.8, 18.7, 16.7, 18.1, 17.9, 16.4, 18.9
Women: 18.6, 17.8, 18.3, 16.6, 20.5, 16.3, 19.3, 18.4, 19.7, 18.8, 19.9, 17.6

Use the U test at the 0.05 level of significance to test the null hypothesis that the two samples come from identical populations against the alternative that the two populations are not identical.

10.10 With reference to part (a) of Exercise 7.70 on page 250, use the U test at the 0.05 level of significance to test whether the populations of hardness of the two alloys are identical.

10.11 Comparing two kinds of emergency flares, a consumer testing service obtained the following burning times (rounded to the nearest tenth of a minute):

Brand C: 19.4, 21.5, 15.3, 17.4, 16.8, 16.6, 20.3, 22.5, 21.3, 23.4, 19.7, 21.0
Brand D: 16.5, 15.8, 24.7, 10.2, 13.5, 15.9, 15.7, 14.0, 12.1, 17.4, 15.6, 15.8

Use the U test at the 0.01 level of significance to check whether it is reasonable to say that the populations of burning times, of the two kinds of flares, are identical.

10.12 The following are the scores which random samples of students from two minority groups obtained on a current events test:

Minority Group 1: 73, 82, 39, 68, 91, 75, 89, 67, 50, 86, 57, 65, 70
Minority Group 2: 51, 42, 36, 53, 88, 59, 49, 66, 25, 64, 18, 76, 74

Use the U test at the 0.05 level of significance to test whether or not students from the two minority groups can be expected to score equally well on the test.

10.13 The following are data on the breaking strength (in pounds) of two kinds of material:

Material 1: 144, 181, 200, 187, 169, 171, 186, 194, 176, 182, 133, 183, 197, 165, 180, 198
Material 2: 175, 164, 172, 194, 176, 198, 154, 134, 169, 164, 185, 159, 161, 189, 170, 164

Use the U test at the 0.05 level of significance to test the claim that the strength of Material 1 is stochastically larger than the strength of Material 2.

10.14 The following are the numbers of misprints counted on pages selected at random from three Sunday editions of a newspaper:

April 11: 4, 10, 2, 6, 4, 12
April 18: 8, 5, 13, 8, 8, 10
April 25: 7, 9, 11, 2, 14, 7

Use the H test at the 0.05 level of significance to test the null hypothesis that the three samples come from identical populations.

10.15 So-called Franklin tests were performed to determine the insulation properties of grain-oriented silicon steel specimens that were annealed in five different atmospheres with the following results:

Atmosphere	*Test results (amperes)*							
1	0.58	0.61	0.69	0.79	0.61	0.59		
2	0.37	0.37	0.58	0.40	0.28	0.44	0.35	
3	0.29	0.19	0.34	0.17	0.29	0.16		
4	0.81	0.69	0.75	0.72	0.68	0.85	0.57	0.77
5	0.26	0.34	0.29	0.47	0.30	0.42		

Use the H test at the 0.05 level of significance to decide whether or not these five samples can be assumed to come from identical populations.

10.16 A panel of seven experts was asked to rate each of five industries on the likelihood that technological changes would produce improvement in environmental pollution over the next 10 years. Their ratings (in the form of judgmental probabilities) are as follows:

Expert	Industry A	B	C	D	E
1	0.15	0.75	0.10	0.00	0.30
2	0.30	0.60	0.20	0.05	0.25
3	0.20	0.80	0.30	0.00	0.50
4	0.00	0.50	0.25	0.10	0.60
5	0.10	0.55	0.15	0.15	0.40
6	0.25	0.70	0.35	0.25	0.45
7	0.40	0.95	0.45	0.20	0.35

(a) Test at the 0.05 level of significance whether or not the ratings given by the experts can be assumed to come from identical populations.
(b) Test at the 0.01 level of significance whether or not the ratings given to the industries can be assumed to come from identical populations. What argument(s) can you make against the validity of this test?

10.4 TESTS OF RANDOMNESS

When we discussed random sampling in Chapter 6, we gave several methods which provide some assurance in advance that a sample taken will be random. Since there are also situations in which we have no control over the way in which the data are selected, it is useful to have a technique for testing whether a sample may be looked upon as random after it has actually been obtained. One such technique is based on the order in which the sample values were obtained; more specifically, it is based on the number of **runs** exhibited in the sample results.

Given a sequence of two symbols, such as H and T (which might represent the occurrence of heads and tails in repeated tosses of a coin), a run is defined as a succession of identical symbols contained between different symbols or none at all. For example, the sequence

$$\underline{T\,T}\;\underline{H\,H}\;\underline{T\,T}\;\underline{H\,H\,H}\;\underline{T}\;\underline{H\,H\,H}\;\underline{T\,T\,T\,T}\;\underline{H\,H\,H}$$

contains 8 runs, as indicated by the underlines. The total number of runs in a sequence of n trials often serves as an indication that the arrangement is not random. For instance, if there had been only two runs consisting of 10 heads followed by 10 tails, we might have suspected that the probability of a success did not remain constant from trial to trial. On the other hand, had the sequence of 20 tosses consisted of alternating heads and tails, we might have suspected that the trials were not independent. In either case, there are grounds to suspect a lack of randomness. Note that our suspicion is not aroused by the numbers of H's and T's, but by the order in which they appeared.

If a sequence contains n_1 symbols of one kind and n_2 of another kind (and neither n_1 nor n_2 is less than 10), the sampling distribution of the **total number of runs**, u, can be approximated closely by a normal distribution with

Mean and standard deviation of u

$$\mu_u = \frac{2n_1n_2}{n_1 + n_2} + 1 \quad \textit{and} \quad \sigma_u = \sqrt{\frac{2n_1n_2(2n_1n_2 - n_1 - n_2)}{(n_1 + n_2)^2(n_1 + n_2 - 1)}}$$

Thus, the test of the null hypothesis that the arrangement of the symbols (and, hence, the sample) is random can be based on the statistic

Statistic for test of randomness

$$z = \frac{u - \mu_u}{\sigma_u}$$

which has approximately the standard normal distribution. Special tables are available for performing the test when n_1, n_2, or both are small.

EXAMPLE The following is the arrangement of defective, d, and nondefective, n, pieces produced in the given order by a certain machine:

$$\underline{n\,n\,n\,n\,n}\;\underline{d\,d\,d\,d}\;\underline{n\,n\,n\,n\,n\,n\,n\,n\,n\,n}\;\underline{d\,d}\;\underline{n\,n}\;\underline{d\,d\,d\,d}$$

Test for randomness at the 0.01 level of significance.

Solution

1. *Null hypothesis*: Arrangement is random.
   Alternative hypothesis: Arrangement is not random.
2. *Level of significance*: $\alpha = 0.01$
3. *Criterion*: Reject the null hypothesis if $z < -2.575$ or $z > 2.575$, where z is given by the above formula.
4. *Calculations*: Since $n_1 = 10$, $n_2 = 17$, and $u = 6$, we get

$$\mu_u = \frac{2 \cdot 10 \cdot 17}{10 + 17} + 1 = 13.59$$

$$\sigma_u = \sqrt{\frac{2 \cdot 10 \cdot 17(2 \cdot 10 \cdot 17 - 10 - 17)}{(10 + 17)^2(10 + 17 - 1)}} = 2.37$$

and

$$z = \frac{6 - 13.59}{2.37} = -3.20$$

5. *Decision*: Since $z = -3.20$ is less than -2.575, the null hypothesis must be rejected; we conclude that the arrangement is not random. Indeed, the total

number of runs is much smaller than expected and there is a strong indication that the defective pieces appear in clusters or groups; the reason for this will have to be uncovered by an engineer who is familiar with the process.

■

The run test can be used also to test the randomness of samples consisting of numerical data by counting **runs above and below the median**. Denoting an observation exceeding the median of the sample by the letter a and an observation less than the median by the letter b, we can use the resulting sequence of a's and b's to test for randomness by the method just indicated. A frequent application of this test is in quality control, where the means of successive small samples are exhibited on a graph in chronological order. The run test can then be used to check whether there might be a trend in the data, so that it is possible to adjust a machine setting or some other process variable before any serious damage occurs.

EXAMPLE An engineer is concerned about the possibility that too many changes are being made in the settings of an automatic lathe. Given the following mean diameters (in inches) of 40 successive shafts turned on the lathe

0.261	0.258	0.249	0.251	0.247	0.256	0.250	0.247	0.255	0.243
0.252	0.250	0.253	0.247	0.251	0.243	0.258	0.251	0.245	0.250
0.248	0.252	0.254	0.250	0.247	0.253	0.251	0.246	0.249	0.252
0.247	0.250	0.253	0.247	0.249	0.253	0.246	0.251	0.249	0.253

use the 0.01 level of significance to test the null hypothesis of randomness against the alternative that there is a frequently alternating pattern.

Solution

1. *Null hypothesis*: Arrangement of sample values is random.
   Alternative hypothesis: There is a frequently alternating pattern.
2. *Level of significance*: $\alpha = 0.01$
3. *Criterion*: Reject the null hypothesis if $z > 2.33$, where z is obtained by means of the formula on page 314 for the total number of runs above and below the median.
4. *Calculations*: The median of the 40 measurements is 0.250, so that we get the following arrangement of values above and below 0.250:

   a a b a b a b a b a a b a b a a b b a a b a a b b a b a b b a b a b a

   Thus, $n_1 = 19$, $n_2 = 16$, and $u = 27$, so that

$$\mu_u = \frac{2 \cdot 19 \cdot 16}{35} + 1 = 18.37$$

$$\sigma_u = \sqrt{\frac{2 \cdot 19 \cdot 16(2 \cdot 19 \cdot 16 - 19 - 16)}{(19 + 16)^2(19 + 16 - 1)}} = 2.89$$

and

$$z = \frac{27 - 18.37}{2.89} = 2.98$$

5. *Decision*: Since $z = 2.98$ exceeds 2.33, the null hypothesis of randomness must be rejected; since the number of runs is much larger than one might expect due to chance, it is reasonable to conclude that the lathe is being adjusted too often.

■

10.5 THE KOLMOGOROV–SMIRNOV TESTS

The **Kolmogorov–Smirnov tests** are nonparametric tests for differences between cumulative distributions. The one-sample test concerns the agreement between an observed cumulative distribution of sample values and a specified continuous distribution function; thus, it is a test of goodness of fit. The two-sample test concerns the agreement between two observed cumulative distributions; it tests the hypothesis whether two independent samples come from identical continuous distributions, and it is sensitive to population differences with respect to location, dispersion, or skewness.

The Kolmogorov–Smirnov one-sample test is generally more efficient than the chi-square test for goodness of fit for small samples, and it can be used for very small samples where the chi-square test does not apply. It must be remembered, however, that the chi-square test of Section 9.6 can be used in connection with discrete distributions whereas the Kolmogorov–Smirnov test cannot.

The one-sample test is based on the maximum absolute difference D between the values of the cumulative distribution of a random sample of size n and a specified theoretical distribution. To determine whether this difference is larger than can reasonably be expected for a given level of significance, we look up the critical value of D in Table 10.

EXAMPLE It is desired to check whether pinholes in electrolytic tin plate are uniformly distributed across a plated coil on the basis of the following distances in inches of 10 pinholes from one edge of a long strip of tin plate 30 inches wide:

4.8 14.8 28.2 23.1 4.4 28.7 19.5 2.4 25.0 6.2

Test this null hypothesis at the 0.05 level of significance.

Solution

1. *Null hypothesis*:

$$F(x) = \begin{cases} 0 & \text{for } x \leq 0 \\ \dfrac{x}{30} & \text{for } 0 < x < 30 \\ 1 & \text{for } x \geq 30 \end{cases}$$

where x is the distance of a pinhole from the edge.
Alternative hypothesis: The pinholes are not uniformly distributed across the tin plate.

2. *Level of significance*: $\alpha = 0.05$
3. *Criterion*: Reject the null hypothesis if $D > 0.410$, where D is the maximum difference between the observed cumulative distribution and the cumulative distribution assumed under the null hypothesis.
4. *Calculations*: Plotting the two cumulative distributions as in Figure 10.1, we find that the difference is greatest at $x = 6.2$, and that its value is $D = 0.40 - \dfrac{6.2}{30} = 0.193$.
5. *Decision*: Since $D = 0.193$ does not exceed 0.410, the null hypothesis (that the pinholes are uniformly distributed across the tin plate) cannot be rejected.

■

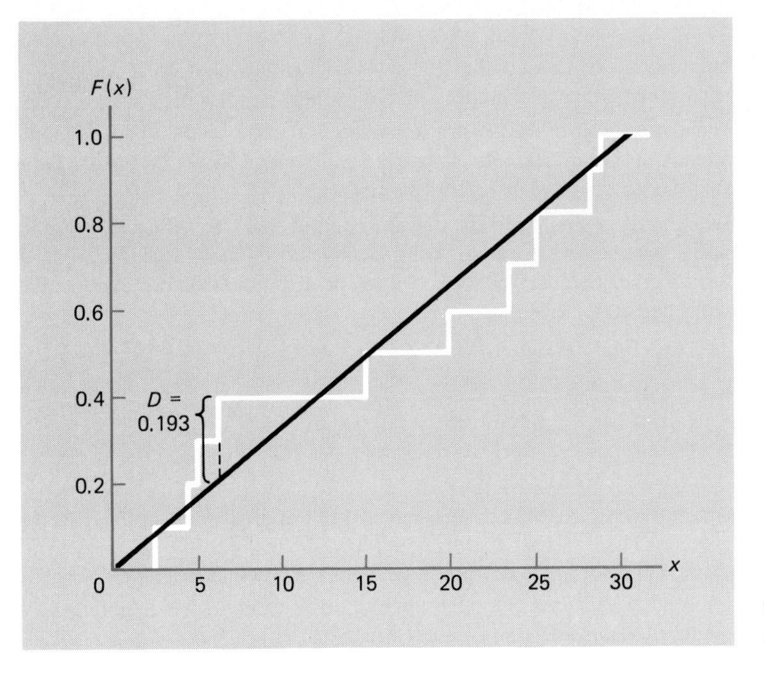

FIGURE 10.1
Diagram for Kolmogorov–Smirnov test.

EXERCISES

10.17 The following arrangement indicates whether 60 consecutive cars which went by the toll booth of a bridge had local plates, L, or out-of-state plates, O:

$$L\ L\ O\ L\ L\ L\ L\ O\ O\ L\ L\ L\ L\ O\ L\ O\ O\ L\ L\ L\ L\ O\ L\ O\ O\ L\ L\ L\ L\ L$$
$$O\ L\ L\ L\ O\ L\ O\ L\ L\ L\ L\ O\ O\ L\ O\ O\ O\ O\ L\ L\ L\ L\ O\ L\ O\ O\ L\ L\ L\ O$$

Test at the 0.05 level of significance whether this arrangement of L's and O's may be regarded as random.

10.18 The following arrangement indicates whether 50 consecutive persons interviewed by a pollster are for, F, or against, A, an increase in the state gasoline tax to build more roads:

$$A\ A\ A\ F\ A\ F\ A\ A\ A\ A\ F\ F\ A\ A\ F\ A\ A\ A\ A\ A\ F\ A\ A\ F\ F$$
$$A\ A\ F\ A\ A\ A\ A\ F\ A\ F\ F\ A\ A\ A\ A\ A\ F\ A\ A\ F\ A\ A\ A\ A\ F$$

Test at the 0.05 level of significance whether this arrangement of A's and F's may be regarded as random.

10.19 To check the randomness of the digits in Table 7, take those in any five rows, replace the odd digits by the letter O and the even digits by the letter E, and base your decision on the total number of runs in the resulting sequence of O's and E's. Use the 0.01 level of significance.

10.20 The following are the number of defective pieces turned out by a machine during 24 consecutive shifts: 15, 11, 17, 14, 16, 12, 19, 17, 21, 15, 17, 19, 21, 14, 22, 16, 19, 12, 16, 14, 18, 17, 24, and 13. Test for randomness at the 0.01 level of significance.

10.21 The following are 50 consecutive downtimes of a machine (in minutes) which were observed during a certain period of time: 22, 29, 32, 25, 33, 34, 38, 34, 29, 25, 27, 33, 34, 28, 39, 41, 24, 31, 34, 29, 34, 25, 30, 37, 40, 39, 35, 24, 32, 43, 44, 34, 40, 38, 39, 43, 46, 34, 39, 45, 42, 39, 54, 50, 38, 41, 43, 46, 52, and 55. Use the method of runs above and below the median and the 0.05 level of significance to test the null hypothesis of randomness against the alternative that there is a trend.

10.22 Using the result of the illustration given in the example on page 27, namely, that the median of the sulfur oxides emission data on page 8 is 19.05, test the original data for randomness at the 0.05 level of significance.

10.23 In a vibration study, certain airplane components were subjected to severe vibrations until they showed structural failures. Given the following failure times (in minutes), test whether they can be looked upon as a sample from an exponential population with the mean $\mu = 10$:

1.5	10.3	3.6	13.4	18.4	7.7	24.3	10.7	8.4
15.4	4.9	2.8	7.9	11.9	12.0	16.2	6.8	14.7

Use the 0.05 level of significance.

10.24 The following are 15 measurements of the boiling point of a silicon compound (in degrees Celsius): 166, 141, 136, 153, 170, 162, 155, 146, 183, 157, 148, 132, 160, 175, and 150. Use the Kolmogorov–Smirnov test at the 0.01 level of significance to test the null hypothesis that the boiling points come from a normal population with $\mu = 160$ degrees Celsius and $\sigma = 10$ degrees Celsius.

10.6

REVIEW EXERCISES

10.25 Use the sign test, with level of significance 0.063, to test that the observations in Figure 2.2 on page 7 have median 0 against the alternative $\tilde{\mu} > 0$.

10.26 According to Einstein's theory of relativity, light should bend when it passes through a gravitational field. This was first tested experimentally in 1919 when photographs were taken of stars near the sun during a total eclipse and again when the sun had moved to another part of the sky. These eclipse pictures should show the stars displaced outwards from the position of the sun. The direction, in the first of two coordinate axes, predicted by the theory was matched by the observed direction for 6 out of 7 stars. Record + for a match and − for a missmatch. Guessing would give probability $\frac{1}{2}$ of a match. Use the sign test with level 0.063 to support the claim that the theory holds with respect to matching the direction of displacement.

10.27 With reference to Exercise 7.68, use the U statistic at the 0.05 level of significance to test whether Method B is more effective.

10.28 To find the best arrangement of instruments on a control panel of an airplane, two different arrangements were compared by simulating an emergency condition and measuring the reaction time required to correct the condition. The reaction times (in tenths of a second) of 20 pilots (randomly assigned to the two different arrangements) were as follows:

Arrangement 1:	8,	15,	10,	13,	17,	10,	9,	11,	12,	15
Arrangement 2:	12,	7,	13,	8,	14,	6,	16,	7,	10,	9

Use the U test at the 0.05 level of significance to check the claim that the second arrangement is better than the first.

10.29 The following are the miles per gallon which a test driver got for 10 tankfuls each of three brands of gasoline:

Brand 1:	22,	25,	32,	18,	23,	15,	30,	27,	19,	23
Brand 2:	19,	22,	18,	29,	28,	32,	17,	33,	28,	20
Brand 3:	30,	29,	25,	24,	15,	27,	30,	27,	18,	32

Use the H test at the 0.05 level of significance to test whether there is a difference in the performance of the three brands of gasoline.

10.30 To test whether radio signals from deep space contain a message, an interval of time could be subdivided into a number of very short intervals and it could then be determined whether the signal strength exceeded a certain level (background noise) in each short interval. Suppose that the following is part of such a record, where H denotes a high signal strength and L denotes that the signal strength does not exceed a given noise level.

$$L\ L\ H\ L\ H\ L\ H\ L\ H\ H\ H\ L\ H\ H\ H\ L\ H\ H\ H\ L\ H\ L\ H\ L\ H\ L\ L\ L\ L$$

Test this sequence for randomness (using the 0.05 level of significance) and ascertain whether it is reasonable to assume that the signal contains a message.

10.31 The total number of retail stores opening for business and also quitting business within the calendar years 1948–1980 in a large city were 108, 103, 109, 107, 125, 142, 147, 122, 116, 153, 144, 162, 143, 126, 145, 129, 134, 137, 143, 150, 148, 152, 125, 106, 112, 139, 132, 122, 138, 148, 155, 146, and 158. Making use of the fact that the median is 138, test at the 0.05 level of significance whether there is a significant trend.

10.32 When two populations have the same probability density function, each outcome of n_1 ranks for the first sample, out of the possible values $1, 2, \ldots, n_1 + n_2$, is equally likely.

(a) Write out all of the possible outcomes when $n_1 = 3 = n_2$.

(b) Evaluate U_1 at each of the outcomes and construct its probability distribution.

10.33 With reference to the example on page 7, use the U statistic to test the null hypothesis of equality versus the alternative that the distribution of copper content from the first heat is stochastically larger than the distribution for the second heat. Following the approach in Exercise 10.32, it can be shown that the exact distribution gives $P(U_1 < 3) = 0.042$. Use this as the level of significance.

10.34 The difference between the observed flux and the theoretical value was observed at 20 points within a reactor. The values were 2, −2, −4, −6, −3, −6, 3, −5, 2, 6, 8, 5, 3, 9, 7, 3, 2, −1, −3, −1. Use a sign test at the 0.036 level to test the null hypothesis $\tilde{\mu} = 0$ versus the alternative hypothesis $\tilde{\mu} \neq 0$.

10.35 With reference to Exercise 10.34, test for randomness with level 0.05.

10.36 Survival times (days) of eight cancer-bearing mice that have been treated with a certain anticancer drug are as follows:

16, 11, 24, 18, 31, 15, 12, 21

Test at the 0.01 level of significance whether these data are consistent with the assumption of a log-normal distribution of survival times.

10.7 CHECK LIST OF KEY TERMS (with page references)

H test 309
Kolmogorov–Smirnov test 316
Kruskal–Wallis test 309
Mann–Whitney test 306
Nonparametric tests 304
Rank sums 306
Run 313
Runs above and below the median 315
Sign test 304
Stochastically larger 306
Total number of runs 314
U test 306
Wilcoxon test 306

CURVE FITTING

The main objective of many statistical investigations is to make predictions, preferably on the basis of mathematical equations. For instance, an engineer may wish to predict the amount of oxide that will form on the surface of a metal baked in an oven for a specified amount of time at 200 degrees Celsius, or the amount of deformation of a ring subjected to a compressive force of 1,000 pounds, or the time between recappings of a tire having a given tread thickness and composition. Usually, such predictions require that a formula be found which relates the dependent variable (whose value one wants to predict) to one or more independent variables.

Problems relating to predictions which are based on the known value of one variable are treated in Sections 11.1 through 11.3 and 11.6; the case where predictions are based on the known values of several variables is treated in Section 11.4. The importance of checking the assumptions concerning the prediction model is considered in Section 11.5.

11.1 THE METHOD OF LEAST SQUARES

In this section we begin our study of the case where a dependent variable is to be predicted in terms of a single independent variable. In many problems of this kind the independent variable is observed without error, or with an error which is negligible when compared with the error (chance variation) in the dependent variable. For example, in measuring the amount of oxide on the surface of a metal specimen, the baking temperature can usually be controlled with good precision, but the oxide-thickness measurement may be subject to considerable chance variation. Thus, even though the independent variable may be fixed at x, repeated measurements of the dependent variable may lead to y-values which differ considerably. Such differences among y-values can be attributed to several causes, chiefly to errors of measurement and to the existence of other, uncontrolled variables which may influence the value of y when x is fixed. Thus, measurements of the thickness of oxide layers may vary over several specimens baked for the same length of time at the same temperature because of the difficulty in measuring thickness as well as possible differences in the composition of the oven atmosphere, surface conditions of the specimens, and the like.

It should be apparent from this discussion that in this context y is the value of a random variable whose distribution depends on x. In most situations of this sort we are interested mainly in the relationship between x and the mean of the corresponding distribution of the y's, and we refer to this relationship as the **regression curve of y on x**. (For the time being we shall consider the case where x is fixed, that is, not random; in Section 11.6 we shall consider the case where x and y are both values of random variables.)

Let us first treat the case where the regression curve of y on x is **linear**, that is, where for any given x, the mean of the distribution of the y's is given by $\alpha + \beta x$. In general, an observed y will differ from this mean, and we shall denote this difference by ε, writing

$$y = \alpha + \beta x + \varepsilon$$

Thus, ε is a value of a random variable and we can always choose α so that the mean of the distribution of this random variable is equal to zero. The value of ε for any given observation will depend on a possible error of measurement and on the values of variables other than x which might have an influence on y.

To give an example where the regression curve of y on x can reasonably be assumed to be linear, suppose that a tensile ring is to be calibrated by measuring the deflection at various loads. In the following table, giving the results of 12 measurements, the x's are the load forces in thousands of pounds and the y's are the corresponding deflections in thousandths of an inch:

x	1	2	3	4	5	6	7	8	9	10	11	12
y	16	35	45	64	86	96	106	124	134	156	164	182

It is apparent from Figure 11.1, where these data have been plotted, that it is reasonable to assume that the relationship (regression curve) is linear; that is, a

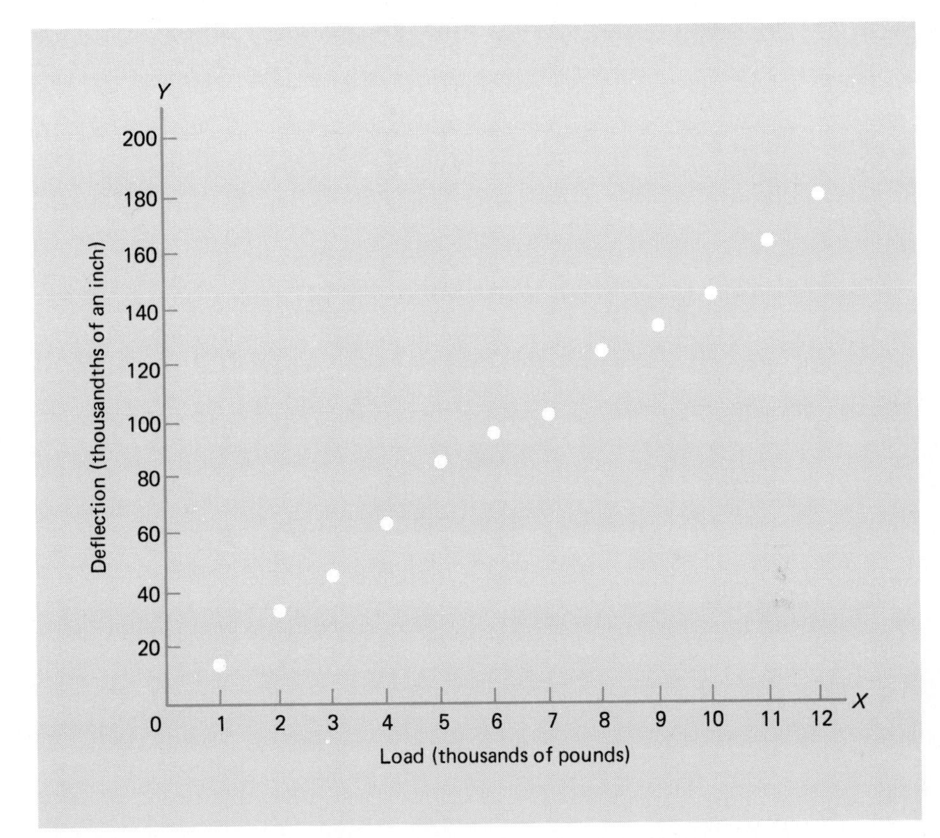

FIGURE 11.1
Scattergram indicating linear regression.

straight line gives a very good approximation over the range of the available data. This kind of diagram, which shows how the data points are scattered, is called a **scattergram**.

Now we face the problem of using the observed data to estimate the parameters α and β of the regression line, and this is equivalent to finding the equation of the straight line that somehow provides the best fit. In connection with Figure 11.1, it may well be satisfactory to do this by eye. If different experimenters were to fit a line in this way, they would probably all predict that for a load of 7,500 pounds the deflection should be close to 0.115 inch. However, if we have to deal with data such as those plotted in Figure 11.2, the problem of finding a best-fitting line is not so obvious. To handle problems of this kind, we must seek a nonsubjective method for fitting straight lines which reflects some desirable statistical properties.

To state the problem formally, we have n paired observations (x_i, y_i) for which it is reasonable to assume that the regression of y on x is linear, and we want to

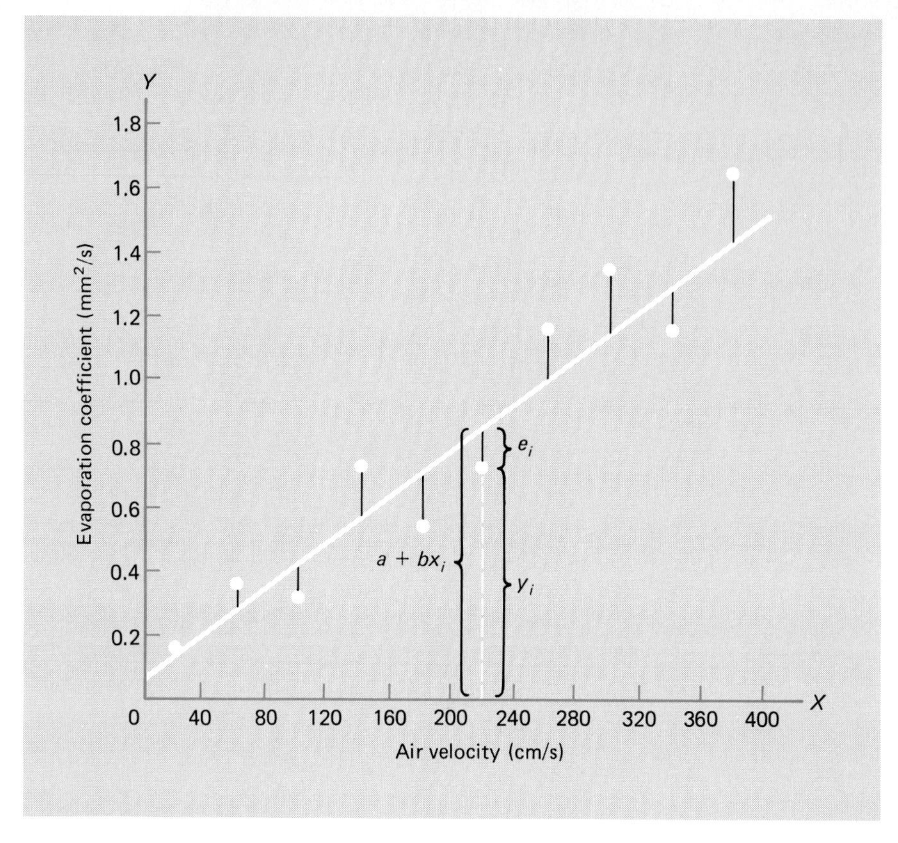

FIGURE 11.2
Diagram for least-squares criterion.

determine the line (that is, the equation of the line) which in some sense provides the best fit. There are several ways in which we can interpret the word "best," and the meaning we shall give it here may be explained as follows. If we predict y by means of the equation

$$\hat{y} = a + bx$$

where a and b are constants, then e_i, the error in predicting the value of y corresponding to the given x_i, is

$$e_i = y_i - \hat{y}_i$$

and we shall want to determine a and b so that these errors are in some sense as small as possible.

Since we cannot minimize each of the e_i individually, we might try to make their sum $\sum_{i=1}^{n} e_i$ as close as possible to zero. However, since this sum can be made equal to zero by many choices of totally unsuitable lines for which the positive and negative errors cancel, we shall minimize the sum of the squares of the e_i (for the same reason we worked with the squares of the deviations from the mean in the definition of the standard deviation). In other words, we shall choose a and b so that

$$\sum_{i=1}^{n} [y_i - (a + bx_i)]^2$$

is a minimum. Note from Figure 11.2 that this is equivalent to minimizing the sum of the squares of the vertical distances from the points to the line.

This method of finding the equation of the line which best fits a given set of paired data, called the **method of least squares**, yields values for a and b (estimates of α and β) that have many desirable properties; some of these are mentioned on page 327.

A necessary condition for a relative minimum is the vanishing of the partial derivatives with respect to a and b. We thus have

$$2 \sum_{i=1}^{n} [y_i - (a + bx_i)](-1) = 0$$

$$2 \sum_{i=1}^{n} [y_i - (a + bx_i)](-x_i) = 0$$

and we can rewrite these two equations as

Normal equations

$$\sum_{i=1}^{n} y_i = an + b \sum_{i=1}^{n} x_i$$

$$\sum_{i=1}^{n} x_i y_i = a \sum_{i=1}^{n} x_i + b \sum_{i=1}^{n} x_i^2$$

This set of two linear equations in the unknowns a and b, called the **normal equations**, gives the values of a and b for the line which provides the best fit to a given set of paired data in accordance with the criterion of least squares.

EXAMPLE The following are measurements of the air velocity and evaporation coefficient of burning fuel droplets in an impulse engine:

Air velocity (*cm/sec*) x	*Evaporation coefficient* (mm^2/sec) y
20	0.18
60	0.37
100	0.35
140	0.78
180	0.56
220	0.75
260	1.18
300	1.36
340	1.17
380	1.65

Fit a straight line to these data by the method of least squares, and use it to estimate the evaporation coefficient of a droplet when the air velocity is 190 cm/s.

Solution The quantities needed for substitution into the two normal equations are $n = 10$ and

$$\sum_{i=1}^{n} x_i = 2{,}000, \qquad \sum_{i=1}^{n} x_i^2 = 532{,}000$$

$$\sum_{i=1}^{n} y_i = 8.35, \qquad \sum_{i=1}^{n} x_i y_i = 2{,}175.40$$

and we get

$$8.35 = 10a + 2{,}000b$$

$$2{,}175.40 = 2{,}000a + 532{,}000b$$

Solving this system of equations by use of determinants or the method of elimination (or a statistical calculator which directly yields the values of a and b after we have punched in the data), we obtain $a = 0.069$ and $b = 0.0038$. Thus, the equation of the straight line that best fits the given data in the sense of least squares is

$$\hat{y} = 0.069 + 0.0038x$$

and for $x = 190$ we predict that the evaporation coefficient will be $\hat{y} = 0.069 + 0.0038(190) = 0.79$ mm^2/sec.

■

It is impossible to make any exact statements about the "goodness" of an estimate like this unless we make some assumptions about the underlying distributions of the random variables with which we are concerned and about the true nature of the regression. Looking upon the values of a and b obtained from the normal equations as estimates of the actual regression coefficients α and β, the reader will be asked to show in Exercises 11.22 and 11.23 on page 339 that these estimates are linear in the observations y_i and that they are values of unbiased estimators of α and β. With these properties, we can refer to the remarkable **Gauss-Markov theorem**, which states that among all unbiased estimators for α and β which are linear in the y_i, the least-squares estimators have the smallest variance. In other words, the least-squares estimators are the most reliable in the sense that they are subject to the smallest chance variations. A proof of the Gauss-Markov theorem may be found in the book by H. Scheffe referred to in the bibliography.

11.2 INFERENCES BASED ON THE LEAST-SQUARES ESTIMATORS

The method of the preceding section is used when the relationship between x and the mean of y is linear or close enough to a straight line so that the least-squares line yields reasonably good predictions. In what follows we shall assume that the regression *is* linear and furthermore that the n random variables having the values y_i $(i = 1, \ldots, n)$ are independently normally distributed with the means $\alpha + \beta x_i$ and the common variance σ^2. If we write

$$y_i = \alpha + \beta x_i + \varepsilon_i$$

it follows from these assumptions that the ε_i are values of independent normally distributed random variables having zero means and the common variance σ^2. The various assumptions we have made here are illustrated in Figure 11.3, showing the distributions of values of y_i for several values of the x_i. Note that these additional assumptions are required to discuss the goodness of predictions based on least-squares equations, the properties of a and b as estimates of α and β, and so on; they were not required to obtain the original estimates based on the method of least squares.

Before we state a theorem concerning the distribution of the least-squares estimators of α and β, it will be convenient to introduce some special notation. The

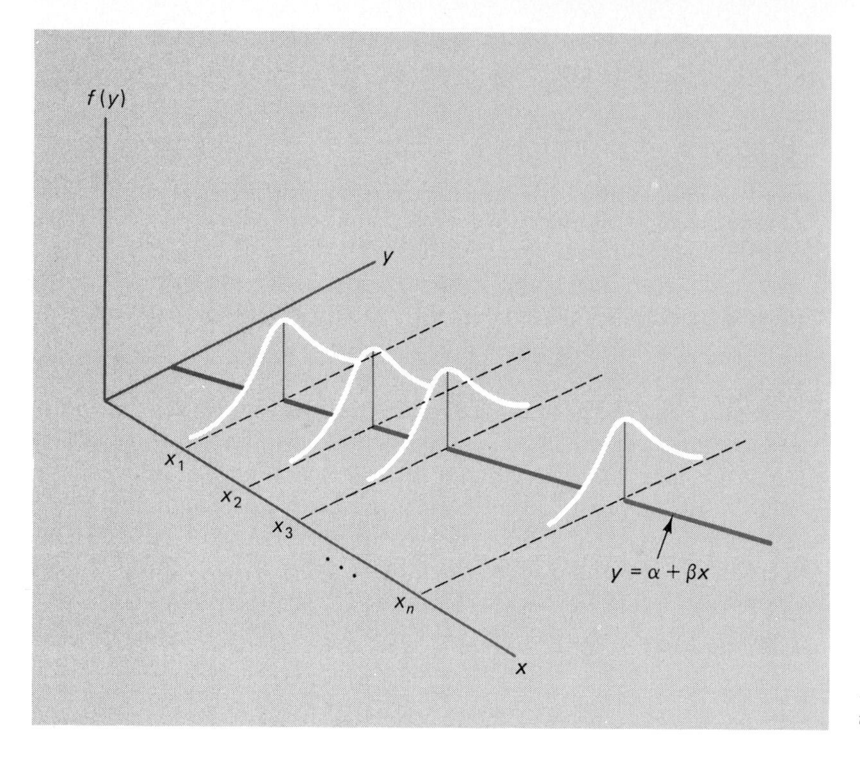

FIGURE 11.3
Diagram showing assumptions underlying Theorem 11.1.

following expressions pertaining to the sample values (x_i, y_i) occur so often that it is convenient to write them as

$$S_{xx} = \sum_{i=1}^{n} (x_i - \bar{x})^2 = \sum_{i=1}^{n} x_i^2 - \left(\sum_{i=1}^{n} x_i\right)^2 \Big/ n$$

$$S_{yy} = \sum_{i=1}^{n} (y_i - \bar{y})^2 = \sum_{i=1}^{n} y_i^2 - \left(\sum_{i=1}^{n} y_i\right)^2 \Big/ n$$

$$S_{xy} = \sum_{i=1}^{n} (x_i - \bar{x})(y_i - \bar{y}) = \sum_{i=1}^{n} x_i y_i - \left(\sum_{i=1}^{n} x_i\right)\left(\sum_{i=1}^{n} y_i\right) \Big/ n$$

The first expressions are preferred on conceptual grounds because they highlight deviations from the mean and on computing grounds because they are less susceptible to roundoff error. The second expressions are for hand-held calculators.

In Exercise 11.24 on page 339 the reader will be asked to show that, in this notation, the solutions of the two normal equations on page 325 can be written as

Solutions of normal equations

$$a = \bar{y} - b \cdot \bar{x} \quad \text{and} \quad b = \frac{S_{xy}}{S_{xx}}$$

where $\bar{x}$ and $\bar{y}$ are, respectively, the means of the x's and the y's. Note also the close relationship between S_{xx} and S_{yy} and the respective sample variances of the x's and the y's; in fact, $s_x^2 = S_{xx}/(n-1)$ and $s_y^2 = S_{yy}/(n-1)$, and we shall sometimes use this alternative notation.

The variance σ^2 defined on page 327 is usually estimated in terms of the vertical deviations of the sample points from the least-squares line. The ith such deviation is $y_i - \hat{y}_i = y_i - (a + bx_i)$ and the estimate of σ^2 is

$$s_e^2 = \frac{1}{n-2} \sum_{i=1}^{n} [y_i - (a + bx_i)]^2$$

where s_e is, traditionally, referred to as the **standard error of estimate**; also, the sum of squares given by $(n-2)s_e^2$ is referred to as the **residual sum of squares** or the **error sum of squares**. An equivalent formula for this estimate of σ^2, which is more convenient for hand-held calculators, is given by

Estimate of σ^2

$$s_e^2 = \frac{S_{yy} - (S_{xy})^2/S_{xx}}{n-2}$$

In these formulas the divisor $n-2$ is used to make the resulting estimator for σ^2 unbiased. The "loss" of two degrees of freedom is explained by the fact that the two regression coefficients α and β had to be replaced by their least-squares estimates. It can also be shown that under the given assumptions $(n-2)s_e^2/\sigma^2$ is a value of a random variable having the chi-square distribution with $n-2$ degrees of freedom.

Based on the assumptions made concerning the distribution of the y's, one can prove the following theorem concerning the distributions of the least-squares estimators of the regression coefficients α and β.

Statistics for inferences about α and β

Theorem 11.1 ***Under the assumptions given on page 327, the statistics***

$$t = \frac{(a-\alpha)}{s_e}\sqrt{\frac{nS_{xx}}{S_{xx} + n(\bar{x})^2}}$$

and

$$t = \frac{(b-\beta)}{s_e}\sqrt{S_{xx}}$$

are values of random variables having the t distribution with $n-2$ degrees of freedom.

To construct confidence intervals for the regression coefficients α and β, we substitute for the middle term of $-t_{\alpha/2} < t < t_{\alpha/2}$ the appropriate t statistic of Theorem 11.1. Then, simple algebra leads to

Confidence limits for regression coefficients

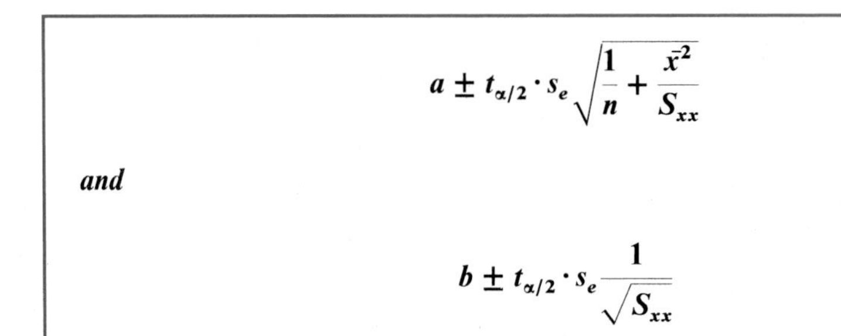

$$a \pm t_{\alpha/2} \cdot s_e \sqrt{\frac{1}{n} + \frac{\bar{x}^2}{S_{xx}}}$$

and

$$b \pm t_{\alpha/2} \cdot s_e \frac{1}{\sqrt{S_{xx}}}$$

EXAMPLE With reference to the example on page 326, construct a 95% confidence interval for the regression coefficient α.

Solution Using the numerical results on page 326 together with

$$\sum_{i=1}^{n} y_i^2 = 9.1097$$

we first obtain

$$S_{xx} = 532{,}000 - (2{,}000)^2/10 = 132{,}000$$
$$S_{yy} = 9.1097 - (8.35)^2/10 = 2.13745$$
$$S_{xy} = 2{,}175.40 - (2{,}000)(8.35)/10 = 505.40$$

and, hence,

$$s_e^2 = \frac{2.13745 - (505.40)^2/132{,}000}{8} = 0.0253$$

Since $t_{0.025} = 2.306$ for $10 - 2 = 8$ degrees of freedom, we get the 95% confidence limits

$$0.069 \pm (2.306)(0.159)\sqrt{\frac{1}{10} + \frac{(200)^2}{132{,}000}}$$

and, hence, the 95% confidence interval

$$-0.164 < \alpha < 0.302$$

■

In connection with tests of hypotheses concerning the regression coefficients α and β, those concerning β are of special importance because β is the **slope of the regression line**; that is, β is the change in the mean of the y's corresponding to a unit increase in x. If $\beta = 0$, the regression line is horizontal and the mean of the y's does not depend linearly on x. For tests of the null hypothesis $\beta = \beta_0$ we use the second statistic of Theorem 11.1 and the criteria are like those in the table on page 229 with t and β substituted for z and μ.

EXAMPLE

With reference to the example on page 326, test the null hypothesis $\beta = 0$ against the alternative hypothesis $\beta \neq 0$ at the 0.05 level of significance.

Solution

1. *Null hypothesis*: $\beta = 0$
   Alternative hypothesis: $\beta \neq 0$
2. *Level of significance*: $\alpha = 0.05$
3. *Criterion*: Reject the null hypothesis if $t < -2.306$ or $t > 2.306$, where 2.306 is the value of $t_{0.025}$ for $10 - 2 = 8$ degrees of freedom, and t is given by the second formula of Theorem 11.1.
4. *Calculations*: Using the quantities obtained on pages 326 and 330, we get

$$t = \frac{0.0038 - 0}{0.159} \sqrt{132{,}000} = 8.36$$

5. *Decision*: Since $t = 8.36$ exceeds 2.306, the null hypothesis must be rejected; we conclude that there is a relationship between air velocity and the average evaporation coefficient. (The relationship is linear by the assumptions that underly the test.)

■

Most statistical software packages include a least-squares fit of a straight line. When the number of (x_i, y_i) pairs is moderate to large, a computer should be used. We illustrate with the deflection-strength data on page 322. The first step is to plot the data (see Exercise 11.26), but this was already done on page 323. Typical output from a regression analysis program includes:

```
THE REGRESSION EQUATION IS
Y = 4.35 + 14.8 X

PREDICTOR        COEF          STDEV     T-RATIO
CONSTANT        4.348          2.244        1.94
X             14.8182         0.3049       48.60

S = 3.646
```

The least squares line is $\hat{y} = 4.35 + 14.8x$ and the estimate of σ^2 is $s_e^2 = (3.646)^2 = 13.29$.

Since $t = 48.60$, with $n - 2 = 10$ degrees of freedom, is highly significant, the slope is different from zero. We would expect the deflection to be 0 at 0 load but this is outside of the range of experimentation. Even so, in this case, our least squares analysis yields $t = 1.94$, which suggests that the constant term α is not needed in the equation.

In Exercise 11.26, more output is described and the least-squares analysis is modified to treat a possible outlier at $x = 5$.

We will return to this least squares fit in Section 11.5, where we investigate the assumptions of a straight-line model and the normal distribution for errors.

Figure 11.4 gives the *SAS* output for a least-squares fit with the deflection-strength data. Notice that more decimal places are given, for instance $s_e^2 = ROOT\ MSE = 3.646085$, and the P-value $0.0001 = PROB > |T|$ is given for testing $\beta = 0$ versus $\beta \neq 0$.

Another problem, closely related to the problem of estimating the regression coefficients α and β, is that of estimating $\alpha + \beta x$, namely, the mean of the distribution of the y's for a given value of x. If x is held fixed at x_0, the quantity we want to estimate is $\alpha + \beta x_0$ and it would seem reasonable to use $a + bx_0$, where a and b are again the values obtained by the method of least squares. In fact, it can be shown that this estimator is unbiased, has the variance

$$\sigma^2\left[\frac{1}{n} + \frac{(x_0 - \bar{x})^2}{S_{xx}}\right]$$

```
DEP VARIABLE: Y

                         SUM OF             MEAN
SOURCE     DF           SQUARES           SQUARE     F VALUE    PROB > F

MODEL       1       31399.72727      31399.72727    2361.958      0.0001
ERROR      10      132.93939394      13.29393939
C TOTAL    11       31532.66667

      ROOT MSE         3.646085         R-SQUARE      0.9958
      DEP MEAN         100.6667         ADJ-R-SQ      0.9954
      C.V.             3.621939

                         PARAMETER ESTIMATES

                    PARAMETER        STANDARD    T FOR H0:
VARIABLE   DF        ESTIMATE           ERROR    PARAMETER=0    PROB > |T|

INTERCEP    1      4.34848485      2.24401050          1.938        0.0814
X           1     14.81818182      0.30490099         48.600        0.0001

SUM OF RESIDUALS                  1.59872E-14
SUM OF SQUARED RESIDUALS             132.9394
```

FIGURE 11.4
Selected SAS output for a regression analysis using the data in the example on page 322.

and that $(1-\alpha)100\%$ confidence limits for $\alpha + \beta x_0$ are given by

Confidence limits for $\alpha + \beta x_0$

$$(a + bx_0) \pm t_{\alpha/2} \cdot s_e \sqrt{\frac{1}{n} + \frac{(x_0 - \bar{x})^2}{S_{xx}}}$$

where the number of degrees of freedom for $t_{\alpha/2}$ is $n-2$.

EXAMPLE With reference to the example on page 326, construct a 95% confidence interval for the mean evaporation coefficient when the air velocity is 190 cm/s.

Solution Substituting the various quantities already calculated on pages 326 and 330 into the above formula, we get

$$0.79 \pm (2.306)(0.159)\sqrt{\frac{1}{10} + \frac{(190-200)^2}{132{,}000}}$$

and, hence,

$$0.67 < \alpha + 190\beta < 0.91$$

■

Of even greater importance than the estimation of $\alpha + \beta x_0$ is the prediction of a future value of y when $x = x_0$, where x_0 is within the range of experimentation. (We added "within the range of experimentation" because, here as elsewhere extrapolation is risky and a linear relationship we have observed may not continue beyond that range.) Thus, for our example we already showed on page 327 that for an air velocity of 190 cm/s, a value well within the range of experimentation, the predicted evaporation coefficient is 0.79 mm^2/s.

To emphasize the danger inherent when extrapolating beyond the range of experimentation, consider the plot of conductivity versus temperature (*Source*: K. Onnes (1912) *Communications of the Physical Laboratory at the University of Leiden*, no. 124, unnumbered figure) in Figure 11.5. If the line is extended to predict conductivity at 4.10° Kelvin, we would predict 0.10 ohms. This is much greater than the value of 0, which Onnes observed when he discovered superconductivity at 4.19° Kelvin. That is, the physical model changes drastically outside of the experimental range shown.

Now let us indicate a method of constructing an interval in which a future observation, y, can be expected to lie with a given probability (or confidence) when $x = x_0$. If α and β were known, we could use the fact that y is a value of a random variable having a normal distribution with the mean $\alpha + \beta x_0$ and the variance σ^2 (or that $y - \alpha - \beta x_0$ is a value of a random variable having a normal distribution with zero mean and the variance σ^2). However, if α and β are not known, we must

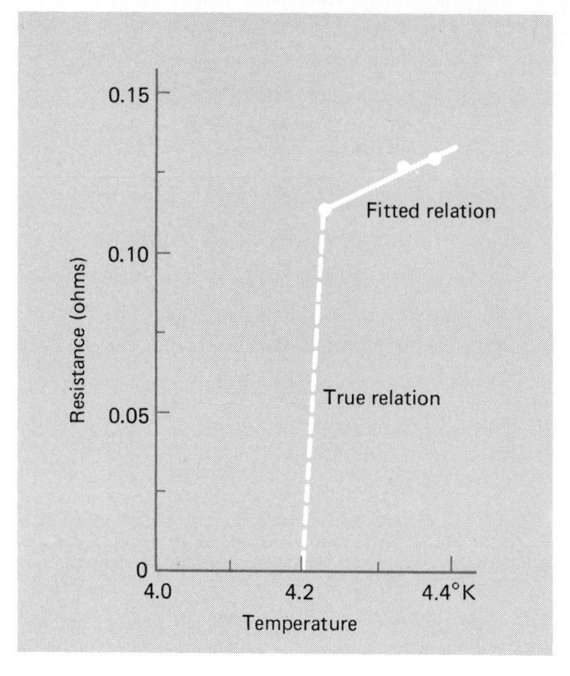

FIGURE 11.5
Resistance versus temperature for a specimen of mercury. A model change invalidates extrapolation.

consider the quantity $y - a - bx_0$, where y, a, and b are all values of random variables, and the resulting theory leads to the following **limits of prediction** for y when $x = x_0$:

Limits of prediction

$$(a + bx_0) \pm t_{\alpha/2} \cdot s_e \sqrt{1 + \frac{1}{n} + \frac{(x_0 - \bar{x})^2}{S_{xx}}}$$

The number of degrees of freedom for $t_{\alpha/2}$ is again $n - 2$.

EXAMPLE With reference to the example on page 326, find 95% limits of prediction for an observation of the evaporation coefficient when the air velocity is 190 cm/s.

Solution Adding a 1 under the radical sign in the preceding example, we get

$$0.79 \pm (2.306)(0.159)\sqrt{1 + \frac{1}{10} + \frac{(190 - 200)^2}{132{,}000}}$$

and, hence, the limits of prediction are $0.79 - 0.39 = 0.40$ and $0.79 + 0.39 = 1.18$. ∎

Note that although the mean of the distribution of y's when $x = 190$ can be estimated fairly closely, the value of a single future observation cannot be predicted with much precision. Indeed, even as $n \to \infty$ the difference between the limits of prediction does not approach zero; the limiting width of the interval of prediction depends on s_e, which measures the inherent variability of the data.

Note, further, that if we do wish to extrapolate, the interval of prediction (and also the confidence interval for $\alpha + \beta x_0$) becomes increasingly wide.

EXAMPLE With reference to the example on page 326, assume that the linear relationship continues beyond the range of experimentation and find 95% limits of prediction for an observation of the evaporation coefficient when the air velocity is 450 cm/s.

Solution Substituting the various quantities already calculated on pages 326 and 330 into the formula on page 334, we get

$$1.78 \pm (2.306)(0.159)\sqrt{1 + \frac{1}{10} + \frac{(450 - 200)^2}{132{,}000}}$$

and, hence, the limits of prediction are $1.78 - 0.46 = 1.32$ and $1.78 + 0.46 = 2.24$ mm^2/s. The width of this interval of prediction is $2(0.46) = 0.92$ compared to the width of $2(0.39) = 0.78$ obtained in the preceding example.

■

EXERCISES

11.1 A chemical company, wishing to study the effect of extraction time on the efficiency of an extraction operation, obtained the data shown in the following table:

Extraction time (minutes) x	*Extraction efficiency (%)* y
27	57
45	64
41	80
19	46
35	62
39	72
19	52
49	77
15	57
31	68

(a) Draw a scattergram to verify that a straight line will provide a good fit to the data, draw a straight line by eye, and use it to predict the extraction efficiency one can expect when the extraction time is 35 minutes.

(b) Fit a straight line to the given data by the method of least squares and use it to predict the extraction efficiency one can expect when the extraction time is 35 minutes.

11.2 Solve the normal equations on page 325 symbolically to show that

$$a = \frac{(\sum x^2)(\sum y) - (\sum x)(\sum xy)}{n(\sum x^2) - (\sum x)^2} \qquad b = \frac{n(\sum xy) - (\sum x)(\sum y)}{n(\sum x^2) - (\sum x)^2}$$

where the subscripts and limits of summation have been omitted for simplicity. Also use these formulas to verify the values obtained for a and b in part (b) of Exercise 11.1.

11.3 In the accompanying table, x is the tensile force applied to a steel specimen in thousands of pounds, and y is the resulting elongation in thousandths of an inch:

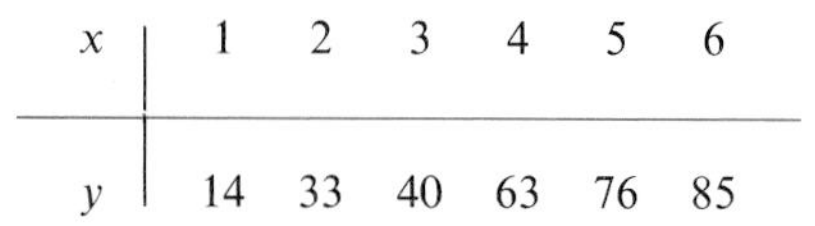

x	1	2	3	4	5	6
y	14	33	40	63	76	85

(a) Graph the data to verify that it is reasonable to assume that the regression of y on x is linear.

(b) Use the formulas of the preceding exercise to find the equation of the least-squares line, and use it to predict the elongation when the tensile force is 3.5 thousand pounds.

11.4 With reference to the preceding exercise, construct a 95% confidence interval for β, the elongation per thousand pounds of tensile stress.

11.5 With reference to Exercise 11.3, find 95% limits of prediction for the elongation of a specimen when $x = 3.5$ thousand pounds.

11.6 The following table shows how many weeks a sample of six persons have worked at an automobile inspection station and the number of cars each one inspected between noon and 2 P.M. on a given day:

Number of weeks employed x	*Number of cars inspected* y
2	13
7	21
9	23
1	14
5	15
12	21

(a) Use the formulas of Exercise 11.2 or directly solve the normal equations to find the equation of the least-squares line which will enable us to predict y in terms of x.

(b) Use the result of part (a) to estimate how many cars someone who has been working at the inspection station for 8 weeks can be expected to inspect during the given 2-hour period.

11.7 With reference to the preceding exercise, test the null hypothesis $\beta = 1.2$ against the alternative hypothesis $\beta < 1.2$ at the 0.05 level of significance.

11.8 With reference to Exercise 11.6 find

(a) a 95% confidence interval for the average number of cars inspected in the given period of time by a person who has been working at the inspection station for 8 weeks;

(b) 95% limits of prediction for the number of cars that will be inspected in the given period of time by a person who has worked at the inspection station for 8 weeks.

11.9 The following data pertain to the number of jobs per day and the central processing unit (CPU) time required.

Number of jobs x	*CPU time* y
1	2
2	5
3	4
4	9
5	10

(a) Use the first set of expressions on page 328, involving the deviations from the mean, to obtain a least squares fit of a line to the observations on CPU time.

(b) Use the equation of the least-squares line to estimate the mean CPU time at $x = 3.5$.

11.10 With reference to Exercise 11.9, construct a 95% confidence interval for α.

11.11 With reference to the Exercise 11.9, test the null hypothesis $\beta = 2$ against the alternative hypothesis $\beta > 2$ at the 0.05 level of significance.

11.12 Raw material used in the production of a synthetic fiber is stored in a place which has no humidity control. Measurements of the relative humidity in the storage place and the moisture content of a sample of the raw material (both in percentages) on 12 days yielded the following results:

Humidity x	*Moisture content* y
42	12
35	8
50	14
43	9
48	11
62	16
31	7
36	9
44	12
39	10
55	13
48	11

(a) Plot a scattergram to verify that it is reasonable to assume that the regression of y on x is linear.
(b) Fit a straight line by the method of least squares.
(c) Find a 99% confidence interval for the mean moisture content of the raw material when the humidity of the storage place is 40%.

11.13 With reference to the preceding exercise, find 95% limits of prediction for the moisture content of the raw material when the humidity of the storage place is 40%. Also indicate to what extent the width of the interval is affected by the size of the sample and to what extent it is affected by the inherent variability of the data.

11.14 The following show the improvement (gain in reading speed) of eight students in a speed-reading program, and the number of weeks they have been in the program:

Number of weeks	*Speed gain (words per minute)*
3	86
5	118
2	49
8	193
6	164
9	232
3	73
4	109

(a) Plot a scattergram to verify that it is reasonable to assume that the regression of speed gain on the number of weeks is linear.
(b) Fit a straight line by the method of least squares.

11.15 With reference to Exercise 11.14, find a 90% confidence interval for β.

11.16 With reference to Exercise 11.1, express 95% limits of prediction for the extraction efficiency in terms of the extraction time x_0. Choosing suitable values of x_0, sketch graphs of the loci of the upper and lower limits of prediction on the diagram of part (a) of Exercise 11.1. Note that since any two sets of limits of prediction obtained from these bands are dependent, they should be used only once for one extraction time x_0.

11.17 In Exercises 11.3 and 11.14 it would have been entirely reasonable to impose the condition $\alpha = 0$ before fitting a straight line by the method of least squares.
(a) Use the method of least squares to derive a formula for estimating β when the regression line has the form $y = \beta x$.
(b) With reference to Exercise 11.3, use the formula obtained in part (a) to estimate β and compare the result with the estimate previously obtained without the condition that the line must pass through the origin.
(c) With reference to Exercise 11.14, use the formula obtained in part (a) to estimate β and compare the result with the estimate previously obtained without the condition that the line must pass through the origin.

11.18 The cost of manufacturing a lot of a certain product depends on the lot size, as shown by the following sample data:

Cost (dollars)	30	70	140	270	530	1,010	2,500	5,020
Lot size	1	5	10	25	50	100	250	500

(a) Draw a scattergram to verify the assumption that the relationship is linear, letting lot size be x and cost y.

(b) Fit a straight line to these data by the method of least squares, using lot size as the independent variable, and draw its graph on the diagram obtained in part (a).

11.19 With reference to Exercise 11.18, find a 90% confidence interval for α, which can be interpreted here as the fixed overhead cost of manufacturing.

11.20 With reference to Exercise 11.18, fit a straight line to the data by the method of least squares, using cost as the independent variable, and draw its graph on the diagram obtained in part (a) of Exercise 11.18. Note that the two estimated regression lines do not coincide.

11.21 When the sum of the x values is equal to zero, the calculation of the coefficients of the regression line of y on x is greatly simplified; in fact, their estimates are given by

$$a = \frac{\sum y}{n} \quad \text{and} \quad b = \frac{\sum xy}{\sum x^2}$$

This simplification can also be attained when the x's are equally spaced, that is, when they are in arithmetic progression. We then code the data by substituting for the x's the values $\ldots, -2, -1, 0, 1, 2, \ldots$, when n is odd, or the values $\ldots, -3, -1, 1, 3, \ldots$, when n is even. The preceding formulas are then used in connection with the coded data.

(a) During its first 7 years of operation, a company's gross income from sales was 1.4, 2.1, 2.6, 3.5, 3.7, 4.9, and 5.5 million dollars. Fit a least-squares line and, assuming that the trend continues, predict the company's gross income from sales during the eighth year of operation.

(b) At the ends of the years 1977–1984 a manufacturing company had the following net investments in plants and equipments: 1.0, 1.7, 2.3, 3.1, 3.5, 3.4, 3.9, and 4.7 million dollars. Fit a least-squares line and, assuming that the trend continues, predict the company's net investment in plant and equipment at the end of 1986.

11.22 Using the formulas obtained in Exercise 11.2, show that

(a) the expression for a is linear in the y_i;

(b) a is an unbiased estimate of α.

11.23 Using the formulas obtained in Exercise 11.2, show that

(a) the expression for b is linear in the y_i;

(b) b is an unbiased estimate of β.

11.24 Show that the least-squares estimates of the coefficients of the regression line of y on x can be written in the form

$$a = \bar{y} - b \cdot \bar{x} \quad \text{and} \quad b = \frac{S_{xy}}{S_{xx}}$$

11.25 The decomposition of the sums of squares into a contribution due to error and a contribution due to regression underlies the least squares analysis. Consider the identity

$$y_i - \bar{y} - (\hat{y}_i - \bar{y}) = (y_i - \hat{y}_i)$$

Note that $\hat{y}_i = a + bx_i = \bar{y} - b\bar{x} + bx_i = \bar{y} + b(x_i - \bar{x})$ so $\hat{y}_i - \bar{y} = b(x_i - \bar{x})$. Using this last expression, then the definition of b and again the last expression,

$$\sum (y_i - \bar{y})(\hat{y}_i - \bar{y}) = b \sum (y_i - \bar{y})(x_i - \bar{x}) = b^2 \sum (x_i - \bar{x})^2 = \sum (\hat{y}_i - \bar{y})^2$$

and the sum of squares about the mean can be decomposed as

$$\underbrace{\sum_{i=1}^{n} (y_i - \bar{y})^2}_{\textit{total sum of squares}} = \underbrace{\sum_{i=1}^{n} (y_i - \hat{y}_i)^2}_{\textit{error sum of squares}} + \underbrace{\sum_{i=1}^{n} (\hat{y}_i - \bar{y})^2}_{\textit{regression sum of squares}}$$

Generally, we find the straight-line fit acceptable if the ratio

$$r^2 = \frac{\text{regression sum of squares}}{\text{total sum of squares}} = 1 - \frac{\sum_{i=1}^{n} (y_i - \hat{y}_i)^2}{\sum_{i=1}^{n} (y_i - \bar{y})^2}$$

is near 1.

Calculate the decomposition of the sum of squares and calculate r^2 using the observations in Exercise 11.9.

11.26 It is tedious to perform a least-squares analysis without using a computer. We illustrate a computer-based analysis using the *MINITAB* package. The observations on page 322 are entered and the scatter plot obtained using the commands

```
READ INTO C1 C2
 1    16
 2    35
 3    45
 4    64
 5    86
 6    96
 7    106
 8    124
 9    134
 10   156
 11   164
 12   182
 PLOT C2 VS C1
```

Then, the command

```
REGRESS C2 ON 1 PREDICTOR IN C1
```

produces the output

```
The regression equation is
y = 4.35 + 14.8 x

Predictor         Coef           Stdev     t-ratio
Constant         4.348           2.244        1.94
x              14.8182          0.3049       48.60

s = 3.646        R-square = 99.6%

Analysis of Variance

SOURCE        DF          SS            MS
Regression     1       31400         31400
Error         10         133            13
Total         11       31533

Unusual Observations
Obs.      x        y         Fit     Residual
   5    5.0     86.00      78.44       7.56
```

(a) The output flags a possible outlier at $x = 5$. Remove that pair of observations and repeat the analysis.

(b) Referring to previous exercise, identify the decomposition of sum of squares given as the analysis of variance.

11.3 CURVILINEAR REGRESSION

So far we have studied only the case where the regression curve of y on x is linear; that is, where for any given x, the mean of the distribution of the y's is given by $\alpha + \beta x$. In this section we first investigate cases where the regression curve is nonlinear but where the methods of Section 11.1 can nevertheless be applied; then we take up the problem of polynomial regression, that is, problems where for any given x the mean of the distribution of the y's is given by

$$\beta_0 + \beta_1 x + \beta_2 x^2 + \cdots + \beta_p x^p$$

Polynomial curve fitting is also used to obtain approximations when the exact functional form of the regression curve is unknown.

It is common practice for engineers to plot paired data on various kinds of graph paper, in order to determine whether for suitably transformed scales the points will fall close to a straight line. If that is the case, the nature of the transformation used leads to a functional form of the regression equation, and the necessary constants (parameters) can be determined by applying the method of Section 11.1 to the transformed data. For instance, if a set of paired data consisting of n points (x_i, y_i) "straightens out" when $\log y_i$ is plotted versus x_i, this indicates that the regression curve of y on x is **exponential**, namely, that for any given x, the mean of the distribution of the y's is given by $\alpha \cdot \beta^x$. If we take logarithms to the base 10 (or any other convenient base), the predicting equation $y = \alpha \cdot \beta^x$ becomes

$$\log y = \log \alpha + x \cdot \log \beta$$

and we can now get estimates of $\log \alpha$ and $\log \beta$, and hence of α and β, by applying the method of Section 11.1 to the n pairs of values $(x_i, \log y_i)$.

EXAMPLE The following are data on the percentage of the radial tires made by a certain manufacturer that are still usable after having been driven for the given numbers of miles:

Miles driven (thousands) x	*Percentage usable* y
1	98.2
2	91.7
5	81.3
10	64.0
20	36.4
30	32.6
40	17.1
50	11.3

(a) Plot $\log y_i$ versus x_i to verify that it is reasonable to assume that the relationship is exponential.
(b) Fit an exponential curve by applying the method of least squares to the data points $(x_i, \log y_i)$.
(c) Use the result of part (b) to estimate what percentage of the manufacturer's radial tires will last at least 25,000 miles.

Solution (a) As can be seen from Figure 11.6, the overall pattern is linear and this justifies fitting an exponential curve. (b) Determining first the logarithms of the eight y's, we get 1.9921, 1.9624, 1.9101, 1.8062, 1.5611, 1.5132, 1.2330, and 1.0531, and the summations required for substitution into the normal equations are $\sum x = 158$, $\sum x^2 = 5{,}530$, $\sum \log y = 13.0312$, and $\sum x \cdot \log y = 212.1224$, where the subscripts and limits of summation have been omitted for simplicity. Again using a and b for the least-squares estimates of α and β, we obtain the normal equations

$$13.0312 = 8(\log a) + 158(\log b)$$

$$212.1224 = 158(\log a) + 5{,}530(\log b)$$

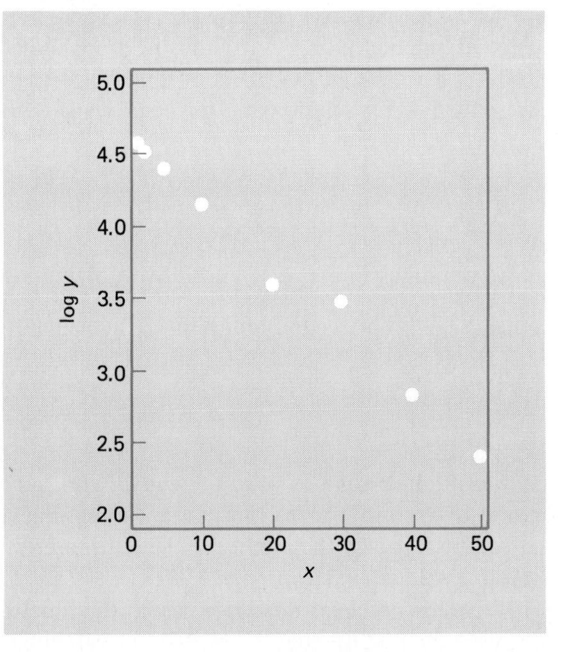

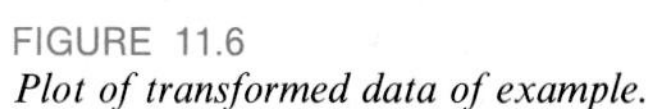
FIGURE 11.6
Plot of transformed data of example.

whose solution is $\log a = 2.0002$ and $\log b = -0.0188$, and hence $a = 100.0$ and $b = 0.96$. Thus, the equation of the estimated regression curve can be written as

$$\log \hat{y} = 2.0002 - 0.0188x$$

in **logarithmic form**, or as

$$\hat{y} = 100.0(0.96)^x$$

in **exponential form**. (c) Using the logarithmic form, which is more convenient, we get

$$\begin{aligned}\log \hat{y} &= 2.0002 - 0.0188(25)\\ &= 1.5302\end{aligned}$$

and, hence, $\hat{y} = 33.9\%$. The analysis of transformed relationships is easily implemented on a computer (see Exercise 11.43).

■

Two other relationships that frequently arise in engineering applications and can be fitted by the method of Section 11.1 after suitable transformations are the

reciprocal function $y = \frac{1}{\alpha + \beta x}$, and the **power function** $y = \alpha \cdot x^{\beta}$. The first of these represents a linear relationship between x and $\frac{1}{y}$, namely,

$$\frac{1}{y} = \alpha + \beta x$$

and we obtain estimates of α and β by applying the method of Section 11.1 to the points $\left(x_i, \frac{1}{y_i}\right)$. The second represents a linear relationship between $\log x$ and $\log y$, namely,

$$\log y = \log \alpha + \beta \cdot \log x$$

and we obtain estimates of $\log \alpha$ and β, and hence of α and β, by applying the method of Section 11.1 to the points $(\log x_i, \log y_i)$. Another example of a curve that can be fitted by the method of least squares after a suitable transformation is given in Exercise 11.31 on page 355.

If there is no clear indication about the functional form of the regression of y on x, we often assume that the underlying relationship is at least "well behaved" to the extent that it has a Taylor series expansion and that the first few terms of this expansion will yield a fairly good approximation. We thus fit to our data a **polynomial**, that is, a predicting equation of the form

$$y = \beta_0 + \beta_1 x + \beta_2 x^2 + \cdots + \beta_p x^p$$

where the degree is determined by inspection of the data or by a more rigorous method to be discussed below.

Given a set of data consisting of n points (x_i, y_i), we estimate the coefficients β_0, $\beta_1, \beta_2, \ldots, \beta_p$ of the pth-degree polynomial by minimizing

$$\sum_{i=1}^{n} [y_i - (\beta_0 + \beta_1 x_i + \beta_2 x_i^2 + \cdots + \beta_p x_i^p)]^2$$

In other words, we are now applying the least-squares criterion by minimizing the sum of the squares of the vertical distances from the points to the curve (see Figure 11.7). Differentiating partially with respect to $\beta_0, \beta_1, \beta_2, \ldots, \beta_p$, equating these partial derivatives to zero, rearranging some of the terms, and letting b_i be the estimate of β_i, we obtain the $p + 1$ normal equations

Normal equations for polynomial regression

$$\begin{aligned} \sum y &= nb_0 + b_1 \sum x + \cdots + b_p \sum x^p \\ \sum xy &= b_0 \sum x + b_1 \sum x^2 + \cdots + b_p \sum x^{p+1} \\ &\vdots \\ \sum x^p y &= b_0 \sum x^p + b_1 \sum x^{p+1} + \cdots + b_p \sum x^{2p} \end{aligned}$$

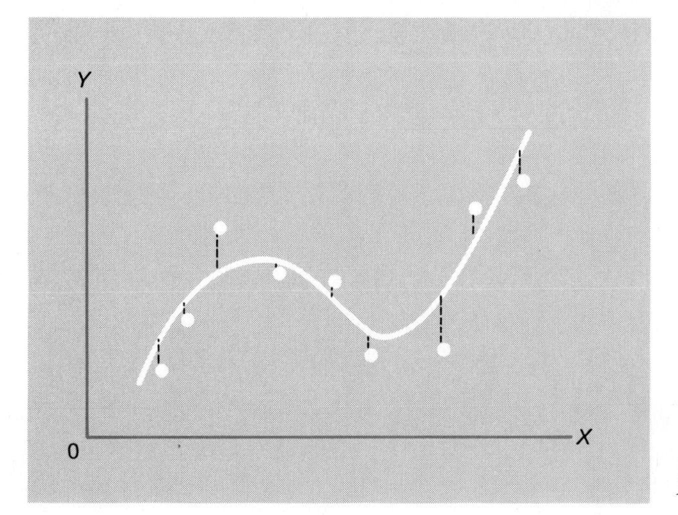

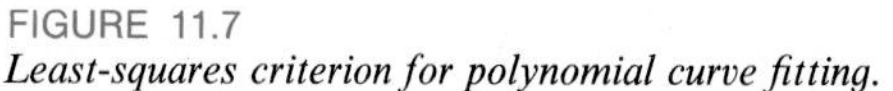
FIGURE 11.7
Least-squares criterion for polynomial curve fitting.

where the subscripts and limits of summation are omitted for simplicity. Note that this is a system of $p + 1$ linear equations in the $p + 1$ unknowns $b_0, b_1, b_2, \ldots,$ and b_p. Unless the various summations have very unusual values, this system of equations will have a unique solution.

EXAMPLE The following are data on the drying time of a certain varnish and the amount of an additive that is intended to reduce the drying time:

Amount of varnish additive (grams) x	*Drying time (hours)* y
0	12.0
1	10.5
2	10.0
3	8.0
4	7.0
5	8.0
6	7.5
7	8.5
8	9.0

(a) Draw a scattergram to verify that it is reasonable to assume that the relationship is parabolic.
(b) Fit a second-degree polynomial by the method of least squares.
(c) Use the result of part (b) to predict the drying time of the varnish when 6.5 grams of the additive is being used.

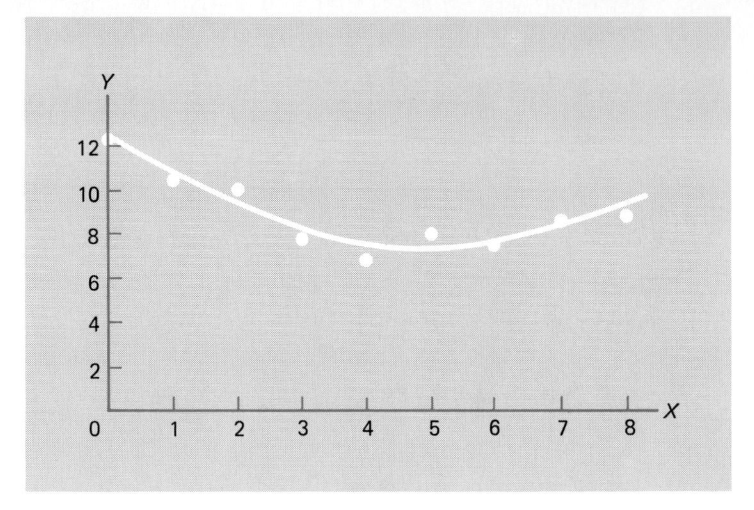

FIGURE 11.8
Parabola fitted to data of example.

Solution (a) As can be seen from Figure 11.8, the overall pattern suggests fitting a second-degree polynomial having one relative minimum. (b) The summations required for substitution into the normal equations are $\sum x = 36$, $\sum x^2 = 204$, $\sum x^3 = 1{,}296$, $\sum x^4 = 8{,}772$, $\sum y = 80.5$, $\sum xy = 299.0$, and $\sum x^2y = 1{,}697.0$, and we thus have to solve the following system of three linear equations in the unknowns b_0, b_1, and b_2:

$$80.5 = 9b_0 + 36b_1 + 204b_2$$
$$299.0 = 36b_0 + 204b_1 + 1296b_2$$
$$1697.0 = 204b_0 + 1296b_1 + 8772b_2$$

Getting $b_0 = 12.2$, $b_1 = -1.85$, and $b_2 = 0.183$, we find that the equation of the least-squares polynomial is

$$\hat{y} = 12.2 - 1.85x + 0.183x^2$$

(c) Substituting $x = 6.5$ into this equation, we get

$$\hat{y} = 12.2 - 1.85(6.5) + 0.183(6.5)^2$$
$$= 7.9$$

that is, a predicted drying time of 7.9 hours.

■

Note that it would have been rather dangerous in the preceding example to predict the drying time that corresponds to, say, 24.5 grams of the additive. The risks inherent in extrapolation, discussed on page 333 in connection with fitting straight lines, increase greatly when polynomials are used to approximate unknown regression functions.

In actual practice, it may be difficult to determine the degree of the polynomial to fit to a given set of paired data. As it is always possible to find a polynomial of degree at most $n - 1$ that will pass through each of n points corresponding to n

distinct values of x, it should be clear that what we actually seek is a polynomial of lowest possible degree that "adequately" describes the data. As we did in our example, it is often possible to determine the degree of the polynomial by inspection of the data.

There also exists a more rigorous method for determining the degree of the polynomial to be fitted to a given set of data. Essentially, it consists of first fitting a straight line as well as a second-degree polynomial and testing the null hypothesis $\beta_2 = 0$, namely, that nothing is gained by including the quadratic term. If this null hypothesis can be rejected, we then fit a third-degree polynomial and test the hypothesis $\beta_3 = 0$, namely, that nothing is gained by including the cubic term. This procedure is continued until the null hypothesis $\beta_i = 0$ cannot be rejected in two successive steps and there is, thus, no apparent advantage to carrying the extra terms. Note that in order to perform these tests it is necessary to impose the assumptions of normality, independence, and equal variances introduced in Section 11.2. Also, these tests should never be used "blindly," that is, without inspection of the overall pattern of the data.

The use of this technique is fairly tedious and we shall not illustrate it in the text. In Exercise 11.35 on page 356 the reader will be given detailed instructions to apply it to the varnish-additive, drying-time data in order to check whether it was really worthwhile to carry the quadratic term.

11.4 MULTIPLE REGRESSION

Before we extend the methods of the preceding sections to problems involving more than one independent variable, let us point out that the curves obtained (and the surfaces we will obtain) are not used only to make predictions. They are often used also for purposes of optimization namely, to determine for what values of the independent variable (or variables) the dependent variable is a maximum or minimum. For instance, in the example on page 345 we might use the polynomial fitted to the data to conclude that the drying time is a minimum when the amount of varnish additive used is 5.1 grams (see Exercise 11.36 on page 357).

Statistical methods of prediction and optimization are often referred to under the general heading of **response surface analysis**. Within the scope of this text, we shall be able to introduce two further methods of response surface analysis: **multiple regression** here and related problems of **factorial experimentation** in Chapter 13.

In multiple regression, we deal with data consisting of n $(r + 1)$-tuples $(x_{1i}, x_{2i}, \ldots, x_{ri}, y_i)$, where the x's are again assumed to be known without error while the y's are values of random variables. Data of this kind arise, for example, in studies designed to determine the effect of various climatic conditions on a metal's resistance to corrosion; the effect of kiln temperature, humidity, and iron content on the strength of a ceramic coating; or the effect of factory production, consumption level, and stocks in storage on the price of a product.

As in the case of one independent variable, we shall first treat the problem where the regression equation is linear, namely, where for any given set of values x_1, $x_2, \ldots,$ and x_r the mean of the distribution of the y's is given by

$$\beta_0 + \beta_1 x_1 + \beta_2 x_2 + \cdots + \beta_r x_r$$

For two independent variables, this is the problem of fitting a plane to a set of n points with coordinates (x_{1i}, x_{2i}, y_i) as is illustrated in Figure 11.9. Applying the method of least squares to obtain estimates of the coefficients β_0, β_1, and β_2, we minimize the sum of the squares of the vertical distances from the points to the plane (see Figure 11.9); symbolically, we minimize

$$\sum_{i=1}^{n} [y_i - (\beta_0 + \beta_1 x_{1i} + \beta_2 x_{2i})]^2$$

and it will be left to the reader to verify in Exercise 11.37 on page 357 that the resulting normal equations are

Normal equations for multiple regression with $r = 2$

$$\begin{aligned} \sum y &= nb_0 + b_1 \sum x_1 + b_2 \sum x_2 \\ \sum x_1 y &= b_0 \sum x_1 + b_1 \sum x_1^2 + b_2 \sum x_1 x_2 \\ \sum x_2 y &= b_0 \sum x_2 + b_1 \sum x_1 x_2 + b_2 \sum x_2^2 \end{aligned}$$

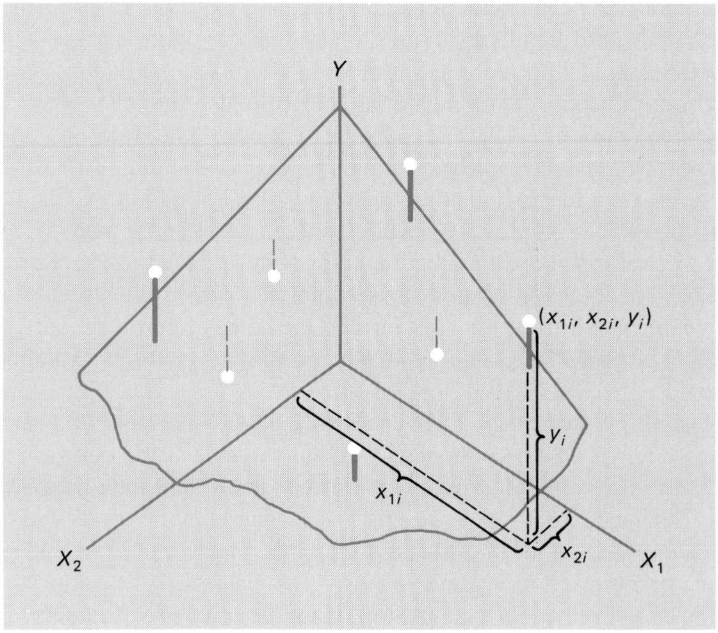

FIGURE 11.9
Regression plane.

As before, we write the least-squares estimates of β_0, β_1, and β_2 as b_0, b_1, and b_2. Note that in the abbreviated notation $\sum x_1$ stands for $\sum_{i=1}^{n} x_{1i}$, $\sum x_1 x_2$ stands for $\sum_{i=1}^{n} x_{1i}x_{2i}$, $\sum x_1 y$ stands for $\sum_{i=1}^{n} x_{1i}y_i$, and so forth.

EXAMPLE The following are data on the numbers of twists required to break a certain kind of forged alloy bar and the percentages of two alloying elements present in the metal:

Number of twists y	*Percent of element A* x_1	*Percent of element B* x_2
41	1	5
49	2	5
69	3	5
65	4	5
40	1	10
50	2	10
58	3	10
57	4	10
31	1	15
36	2	15
44	3	15
57	4	15
19	1	20
31	2	20
33	3	20
43	4	20

Fit a least-squares regression plane and use its equation to estimate the number of twists required to break one of the bars when $x_1 = 2.5$ and $x_2 = 12$.

Solution Substituting $\sum x_1 = 40$, $\sum x_2 = 200$, $\sum x_1^2 = 120$, $\sum x_1 x_2 = 500$, $\sum x_2^2 = 3{,}000$, $\sum y = 723$, $\sum x_1 y = 1{,}963$, and $\sum x_2 y = 8{,}210$ into the normal equations, we get

$$723 = 16b_0 + 40b_1 + 200b_2$$

$$1{,}963 = 40b_0 + 120b_1 + 500b_2$$

$$8{,}210 = 200b_0 + 500b_1 + 3{,}000b_2$$

The unique solution of this system of equations is $b_0 = 46.4$, $b_1 = 7.78$, $b_2 = -1.65$, and the equation of the estimated regression plane is

$$\hat{y} = 46.4 + 7.78x_1 - 1.65x_2$$

Finally, substituting $x_1 = 2.5$ and $x_2 = 12$ into this equation, we get

$$\hat{y} = 46.4 + 7.78(2.5) - 1.65(12)$$
$$= 46.0$$

Note that b_1 and b_2 are estimates of the average change in y resulting from a unit increase in the corresponding independent variable when the other independent variable is held fixed.

■

Computers remove the drudgery of calculation from a multiple regression analysis. (See Exercise 11.42.) Typical output includes

```
THE REGRESSION EQUATION IS
Y = 46.4 + 7.78 X1 - 1.65 X2  ①

PREDICTOR          COEF            STDEV      T-RATIO
CONSTANT          46.438           3.517        13.20 ③
X1                7.7750 ②        0.9485         8.20
X2               -1.6550          0.1897        -8.72

S = 4.242 ⑥        R-SQ = 91.7% ⑤

ANALYSIS OF VARIANCE

SOURCE        DF        ④    SS             MS
REGRESSION     2         2578.5         1289.3
ERROR         13          233.9           18.0
TOTAL         15         2812.4
```

We now identify some important parts of the output.

1. The least squares regression plane is

$$① \qquad \hat{y} = 46.4 + 7.78x_1 - 1.65x_2$$

   This means that the number of twists required to break a bar increases by 7.78 if the percent of element A is increased by 1% and x_2 remains fixed.

2. The least-squares estimates and their corresponding estimated standard errors are:

$$b_0 = 46.438 \quad \text{estimated standard error } 3.517$$
$$② \quad b_1 = 7.7750 \quad \text{estimated standard error } 0.9485$$
$$b_2 = -1.6550 \quad \text{estimated standard error } 0.1897$$

3. The t-ratios 13.20, 8.20, -8.72 are all highly significant, so all the terms are needed in the model. ③

4. In any regression analysis having a β_0 term, the decomposition

$$y_i - \bar{y} = (y_i - \hat{y}_i) + (\hat{y}_i - \bar{y})$$

produces the decomposition of the sum of squares

$$④ \quad \underset{\text{total sum of squares}}{\sum_{i=1}^{n} (y_i - \bar{y})^2} = \underset{\text{error sum of squares}}{\sum_{i=1}^{n} (y_i - \hat{y}_i)^2} + \underset{\text{regression sum of squares}}{\sum_{i=1}^{n} (\hat{y}_i - \bar{y})^2}$$

or

$$2812.4 = 233.9 + 2578.5$$

Thus, the proportion of variability explained by the regression is (see Exercise 11.63)

$$⑤ \quad r^2 = \frac{2578.5}{2812.4} = 1 - \frac{233.9}{2812.4} = 0.917$$

5. The estimate of σ^2 is $s_e^2 = 233.9/13 = 18.0 s_0 s_e = 4.242$. ⑥

For comparative purposes, the output from the *SAS* regression program is presented in Figure 11.10. The p-values given as $PROB > |T|$ confirm the significance of the t-ratios and thus the fact that all the terms are required in the model.

```
DEP VARIABLE: Y

                            SUM OF             MEAN
SOURCE   DF              ④ SQUARES            SQUARE   F VALUE   PROB > F

MODEL     2          2578.52500       1289.26250    71.652     0.0001
ERROR    13        233.91250000      17.99326923
C TOTAL  15          2812.43500

        ROOT MSE          ⑥ 4.241847      R-SQUARE       0.9168 ⑤
        DEP MEAN             45.1875      ADJ-R-SQ       0.9040
        C.V.                9.387214

                            PARAMETER ESTIMATES

                       PARAMETER         STANDARD      T FOR H0:
VARIABLE      DF        ESTIMATE            ERROR    PARAMETER=0    PROB > |T|

INTERCEP       1     46.43750000       3.51715405         13.203        0.0001
X1             1      7.77500000       0.94850591          8.197        0.0001
X2             1          -1.655       0.18970118         -8.724        0.0001
                             ②                               ③

SUM OF RESIDUALS                      -6.21725E-15
SUM OF SQUARED RESIDUALS                  233.9125
```

FIGURE 11.10
Selected SAS output for a multiple regression analysis using the data in the example on page 349.

11.5
CHECKING THE ADEQUACY OF THE MODEL

Assuming that the regression model is adequate, we can use the fitted equation to make inferences. Before doing so, it is imperative that we check the assumptions underlying the analysis. In the context of the regression model with two predictors, we question whether y_i is equal to $\beta_0 + \beta_1 x_{1i} + \beta_2 x_{2i} + \varepsilon_i$, where the errors ε_i are independent and have the same variance σ^2.

All of the information on lack of fit is contained in the residuals

$$
\begin{aligned}
e_1 &= y_1 - \hat{y}_1 = y_1 - b_0 - b_1 x_{11} - b_2 x_{21} \\
e_2 &= y_2 - \hat{y}_2 = y_2 - b_0 - b_1 x_{12} - b_2 x_{22} \\
&\vdots \\
e_n &= y_n - \hat{y}_n = y_n - b_0 - b_1 x_{1n} - b_2 x_{2n}
\end{aligned}
$$

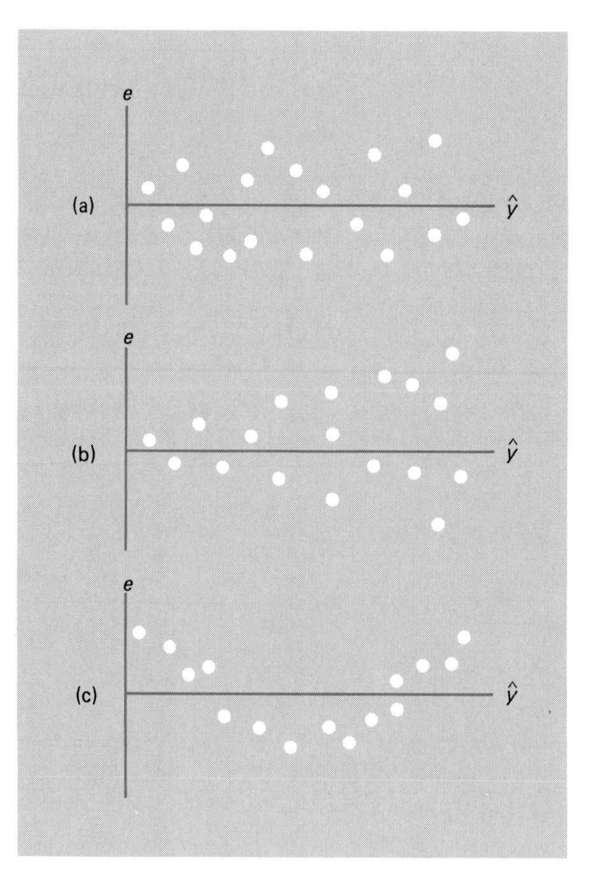

FIGURE 11.11
Residual plots.

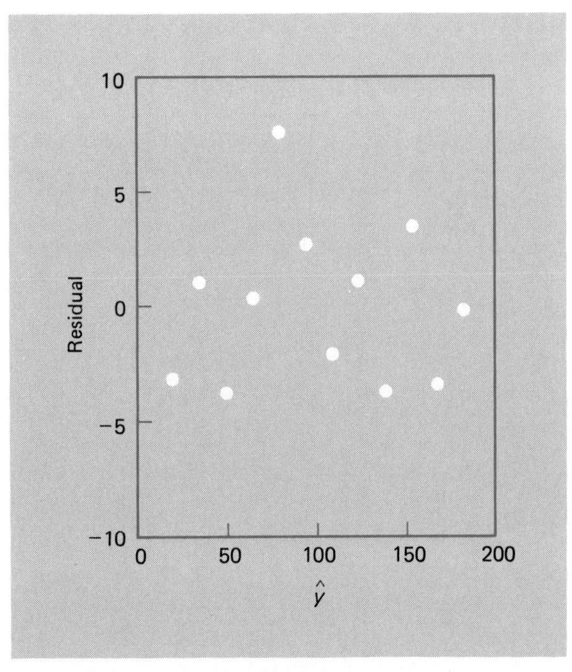

FIGURE 11.12
A plot of the residuals versus predicted values $\hat{y}$.

The residuals should be plotted in various ways to detect systematic departures from the assumptions.

A plot of the residuals versus the predicted values is a major diagnostic tool. Figure 11.11 shows (a) the ideal constant band and two typical violations; (b) variance increases with the response and a transformation is needed; and (c) the model $\beta_0 + \beta_1 x_1 + \beta_2 x_2$ is not adequate. In the latter case, terms with x_1^2 and x_2^2 may be needed.

We also recommend plotting the residuals versus time in order to detect possible trends over time.

The residual plot for the example on page 322, given in Figure 11.12, has the appearance of a horizontal band so no violations of the model are indicated. The one large residual was already flagged as a possible outlier in Exercise 11.26. It is a good idea to drop this observation and redo the analysis. If there is not much change, then we may wish to leave it in. Otherwise, we must look deeper for reasons before discarding it. This same residual sticks out in the normal-scores plot in Figure 11.13.

Although it is difficult to assess normality with only 12 residuals, the largest and smallest values give the plot in Figure 11.13 a curved appearance and make the normal assumption suspect. Fortunately, the normal assumption is generally not critical for inference so long as serious outliers are not present.

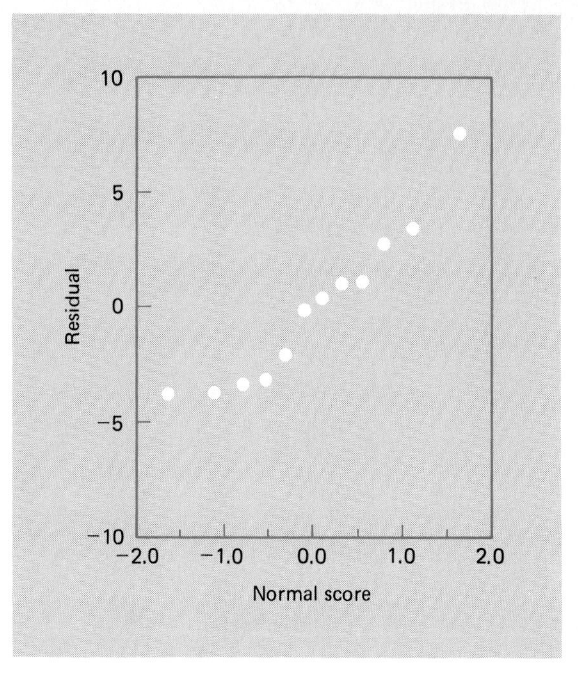

FIGURE 11.13
A normal-scores plot of the residuals.

EXERCISES

11.27 The following data pertain to the growth of a colony of bacteria in a culture medium:

Days since inoculation x	*Bacteria count* y
3	115,000
6	147,000
9	239,000
12	356,000
15	579,000
18	864,000

(a) Plot log y_i versus x_i to verify that it is reasonable to fit an exponential curve.
(b) Fit an exponential curve to the given data.
(c) Use the result obtained in part (b) to estimate the bacteria count at the end of 20 days.

11.28 The following data pertain to the cosmic ray doses measured at various altitudes:

Altitude (*feet*) x	*Dose rate* (*mrem/year*) y
50	28
450	30
780	32
1,200	36
4,400	51
4,800	58
5,300	69

(a) Fit an exponential curve.
(b) Use the result obtained in part (a) to estimate the mean dose at an altitude of 3,000 feet.

11.29 With reference to the preceding exercise, change the equation obtained in part (a) to the form $\hat{y} = a \cdot e^{-cx}$, and use the result to rework part (b).

11.30 The following data pertain to the demand for a product (in thousands of units) and its price (in cents) charged in five different market areas:

Price x	*Demand* y
20	22
16	41
10	120
11	89
14	56

Fit a power function and use it to estimate the demand when the price of the product is 12 cents.

11.31 Fit a **Gompertz curve** of the form

$$y = e^{e^{\alpha x + \beta}}$$

to the data of Exercise 11.28.

11.32 Plot the curve obtained in the preceding exercise and the one obtained in Exercise 11.28 on one diagram and compare the fit of these two curves.

11.33 The number of inches which a newly built structure is settling into the ground is given by

$$y = 3 - 3e^{-\alpha x}$$

where x is its age in months.

x	2	4	6	12	18	24
y	1.07	1.88	2.26	2.78	2.97	2.99

Use the method of least squares to estimate α. [*Hint*: Note that the relationship between ln $(3 - y)$ and x is linear.]

11.34 The following data pertain to the amount of hydrogen present (y, in parts per million) in core drillings made at 1-foot intervals along the length of a vacuum-cast ingot (x, core location in feet from base):

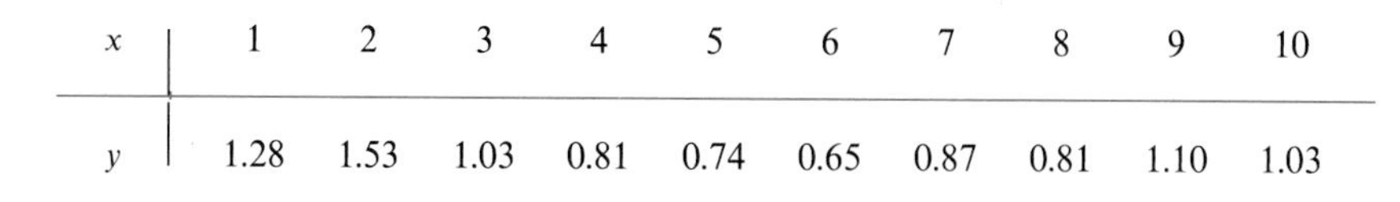

x	1	2	3	4	5	6	7	8	9	10
y	1.28	1.53	1.03	0.81	0.74	0.65	0.87	0.81	1.10	1.03

(a) Draw a scattergram to check whether it is reasonable to fit a parabola to the given data.
(b) Fit a parabola by the method of least squares.
(c) Use the equation obtained in part (b) to estimate the amount of hydrogen present at $x = 7.5$.

11.35 When fitting a polynomial to a set of paired data, we usually begin by fitting a straight line and using the method on page 331 to test the null hypothesis $\beta_1 = 0$. Then we fit a second-degree polynomial and test whether it is worthwhile to carry the quadratic term by comparing $\hat{\sigma}_1^2$, the **residual variance** after fitting the straight line, with $\hat{\sigma}_2^2$, the residual variance after fitting the second-degree polynomial. Each of these residual variances is given by the formula

$$\frac{\sum (y - \hat{y})^2}{\text{degrees of freedom}} = \frac{\text{SSE}}{v}$$

with $\hat{y}$ determined, respectively, from the equation of the line and the equation of the second-degree polynomial. The decision whether to carry the quadratic term is based on the statistic

$$F = \frac{\text{SSE}_1 - \text{SSE}_2}{\hat{\sigma}_2^2} = \frac{v_1\hat{\sigma}_1^2 - v_2\hat{\sigma}_2^2}{\hat{\sigma}_2^2}$$

which (under the assumptions of Section 11.2) is a value of a random variable having the F distribution with 1 and $n - 3$ degrees of freedom.

(a) Fit a straight line to the varnish-additive and drying-time data on page 345, test the null hypothesis $\beta_1 = 0$ at the 0.05 level of significance, and calculate $\hat{\sigma}_1^2$.
(b) Using the result on page 346 calculate $\hat{\sigma}_2^2$ for the given data and test at the 0.05 level whether we should carry the quadratic term. (Note that we could continue this procedure and test whether to carry a cubic term by means of a corresponding comparison of residual variances. Then we could test

whether to carry a fourth-degree term, and so on. It is customary to terminate this procedure after two successive steps have not produced significant results.)

11.36 With reference to the example on page 345, verify that the drying time is a minimum when the amount of additive used is 5.1 grams.

11.37 Verify that the system of normal equations on page 348 corresponds to the minimization of the sum of squares.

11.38 Twelve specimens of cold-reduced sheet steel, having different copper contents and annealing temperatures, are measured for hardness with the following results:

Hardness (Rockwell 30-T)	*Copper content (%)*	*Annealing temperature (degrees F)*
78.9	0.02	1000
65.1	0.02	1100
55.2	0.02	1200
56.4	0.02	1300
80.9	0.10	1000
69.7	0.10	1100
57.4	0.10	1200
55.4	0.10	1300
85.3	0.18	1000
71.8	0.18	1100
60.7	0.18	1200
58.9	0.18	1300

Fit an equation of the form $y = \beta_0 + \beta_1 x_1 + \beta_2 x_2$, where x_1 represents the copper content, x_2 represents the annealing temperature, and y represents the hardness.

11.39 With reference to Exercise 11.38, estimate the hardness of a sheet of steel with a copper content of 0.05% and an annealing temperature of 1150 degrees Fahrenheit.

11.40 The following are data on the ages and incomes of five executives working for the same company and the number of years they went to college:

Age x_1	*Years college* x_2	*Income (dollars)* y
37	4	51,200
45	0	46,800
38	5	55,000
42	2	50,300
31	4	45,400

Fit an equation of the form $y = \beta_0 + \beta_1 x_1 + \beta_2 x_2$ to the given data, and use it to estimate how much on the average an executive working for this company will make if he is 40 years old and has had 4 years of college.

11.41 The following sample data were collected to determine the relationship between two processing variables and the current gain of a certain kind of transistor:

Diffusion time (hours) x_1	*Sheet resistance* (Ω*-cm*) x_2	*Current gain* y
1.5	66	5.3
2.5	87	7.8
0.5	69	7.4
1.2	141	9.8
2.6	93	10.8
0.3	105	9.1
2.4	111	8.1
2.0	78	7.2
0.7	66	6.5
1.6	123	12.6

Fit a regression plane and use its equation to estimate the expected current gain when the diffusion time is 2.2 hours and the sheet resistance is 90 Ω-cm.

11.42 Multiple regression is best implemented on a computer. The following *MINITAB* command fits the y-values in C1 to 2 predictor values in C2 and C3.

```
REGRESS C1 ON 2 PREDICTORS IN C2 C3
```

It produces output like that on page 350. Use a computer to perform the multiple regression analysis in Exercise 11.38.

11.43 Using *MINITAB* we can transform the x values in C1 and/or the y values in C2. For instance, the commands

```
LOGTEN C2 SET IN C3
PLOT C3 VS C1
REGRESS C3 ON 1 PREDICTOR IN C1
```

give the plot of log y_i versus x_i and the corresponding least-squares analysis. Use the computer to do Exercise 11.31.

11.44 To fit the quadratic regression model using *MINITAB*, when the x values are in C1 and the y values are in C2, use the commands

```
MULTIPLY C1 BY C1 SET IN C3
REGRESS C2 ON 2 PREDICTORS IN C1 C3
```

Use the computer to repeat the analysis of the example on page 345.

11.45 With reference to Exercise 11.42, in order to plot residuals, you must use the modified command

```
REGRESS C1 ON 2 PREDICTORS IN C2 C3 STORE C4 FIT IN C5;
RESIDUALS C6.
```

The fitted values are stored in C5 and the residuals in C6. Then the commands

```
PLOT C6 VS C5
NSCORE C6 SET C7
PLOT C6 VS C7
```

will produce a plot of the residuals versus $\hat{y}$ and a normal-scores plot of the residuals.

Use a computer to analyze the residuals from the multiple regression analysis in

(a) the example on page 349; (b) Exercise 11.38.

11.46 With reference to Exercise 11.41, analyze the residuals from the regression plane.

11.47 The following residuals and predicted values were obtained from an experiment that related yield of a chemical process (y) to the initial concentration (x) of a component (the time order of the experiments is given in parentheses):

Predicted	*residual*	*Predicted*	*residual*
4.1 (5)	−2	3.5 (3)	0
3.2 (9)	−1	4.0 (12)	3
3.5 (13)	3	4.2 (4)	−2
4.3 (1)	−3	3.9 (11)	2
3.3 (7)	−1	4.3 (2)	−5
4.6 (14)	5	3.7 (10)	0
3.6 (8)	0	3.2 (6)	1

Examine the residuals for evidence of a violation of the assumptions.

11.6 CORRELATION

So far in this chapter, we have studied problems where the independent variable (or variables) was assumed to be known without error. Although this applies to many experimental situations, there are also problems where the x's as well as the y's are values assumed by random variables. This would be the case, for instance, if we studied the relationship between rainfall and the yield of a certain crop, the relationship between the tensile strength and the hardness of aluminum, or the relationship between impurities in the air and the incidence of a certain disease. Problems like these are referred to as problems of **correlation analysis**, where it is assumed that the data points (x_i, y_i) for $i = 1, 2, \ldots, n$ are values of a pair of random variables whose joint density is given by $f(x, y)$.

The bivariate density that is most commonly used in problems of correlation analysis is the **bivariate normal distribution**. We introduce it here in terms of the conditional density $g_2(y|x)$ and the marginal density $f_1(x)$, as defined in Section 5.10. So far as $g_2(y|x)$ is concerned, the conditions we shall impose are practically identical with the ones we used in connection with the sampling theory of Section 11.2. For any given x, it will be assumed that $g_2(y|x)$ is a normal distribution with the mean $\alpha + \beta x$ and the variance σ^2. Thus, the regression of y on x is linear and the variance of the conditional density does not depend on x. Furthermore, we shall assume that the marginal density $f_1(x)$ is normal with the mean μ_1 and the variance σ_1^2. Making use of the relationship $f(x, y) = f_1(x) \cdot g_2(y|x)$ given on page 168, we thus obtain

$$f(x, y) = \frac{1}{\sqrt{2\pi}\sigma_1} e^{-\frac{(x-\mu_1)^2}{2\sigma_1^2}} \cdot \frac{1}{\sqrt{2\pi}\sigma} e^{-\frac{[y-(\alpha+\beta x)]^2}{2\sigma^2}}$$

$$= \frac{1}{2\pi \cdot \sigma \cdot \sigma_1} e^{-\left\{\frac{[y-(\alpha+\beta x)]^2}{2\sigma^2} + \frac{(x-\mu_1)^2}{2\sigma_1^2}\right\}}$$

for $-\infty < x < \infty$ and $-\infty < y < \infty$. Note that this joint distribution involves the *five* parameters μ_1, σ_1, α, β, and σ.

For reasons of symmetry and other considerations to be explained later, it is customary to express the bivariate normal density in terms of the parameters μ_1, σ_1, μ_2, σ_2, and ρ. Here μ_2 and σ_2^2 are the mean and the variance of the marginal distribution $f_2(y)$, while ρ (rho), called the **population correlation coefficient**, is defined by

$$\rho^2 = 1 - \frac{\sigma^2}{\sigma_2^2}$$

with ρ taken to be positive when β is positive and negative when β is negative. Leaving it to the reader to show in Exercise 11.59 on page 367 that

$$\mu_2 = \alpha + \beta\mu_1 \quad \text{and} \quad \sigma_2^2 = \sigma^2 + \beta^2\sigma_1^2$$

we then substitute into the preceding expression for $f(x, y)$ and obtain the following form of the bivariate normal distribution:

$$f(x, y) = \frac{1}{2\pi \cdot \sigma_1 \sigma_2 \sqrt{1 - \rho^2}} e^{-\left[\left(\frac{x-\mu_1}{\sigma_1}\right)^2 - 2\rho\left(\frac{x-\mu_1}{\sigma_1}\right)\left(\frac{y-\mu_2}{\sigma_2}\right) + \left(\frac{y-\mu_2}{\sigma_2}\right)^2\right]/2(1-\rho^2)}$$

for $-\infty < x < \infty$ and $-\infty < y < \infty$ (see Exercise 11.60 on page 367).

Concerning the correlation coefficient ρ, note that $-1 \leq \rho \leq +1$ since $\sigma_2^2 = \sigma^2 + \beta^2\sigma_1^2$ and, hence, $\sigma_2^2 \geq \sigma^2$. Furthermore, ρ can equal -1 or $+1$ only when $\sigma^2 = 0$, which represents the degenerate case where all the probability is concentrated along the line $y = \alpha + \beta x$ and there is, thus, a perfect linear relationship between the two random variables. (That is, for a given value of x, y must equal $\alpha + \beta x$.) The correlation coefficient is equal to zero if and only if $\sigma^2 = \sigma_2^2$, and it follows from the identity $\sigma_2^2 = \sigma^2 + \beta^2\sigma_1^2$ that this can happen only when $\beta = 0$. Thus, $\rho = 0$ implies that the regression line of y on x is a horizontal line and, hence, that knowledge of x does not help in the prediction of y. Thus, when $\rho = \pm 1$ we say that there is a perfect linear correlation (relationship, or association) between the two random variables; when $\rho = 0$ we say that there is no correlation (relationship, or association) between the two random variables. In fact, $\rho = 0$ implies for the bivariate normal density that the two random variables are independent (see Exercise 11.61 on page 367).

For values between 0 and $+1$ or 0 and -1, we interpret ρ by referring back to the identity

$$\rho^2 = 1 - \frac{\sigma^2}{\sigma_2^2} = \frac{\sigma_2^2 - \sigma^2}{\sigma_2^2}$$

given above. Since σ^2 is a measure of the variation of the y's when x is known while σ_2^2 is a measure of the variation of the y's when x is unknown, $\sigma_2^2 - \sigma^2$ measures the variation of the y's that is accounted for by the linear relationship with x, and ρ^2 **tells us what proportion of the variation of the y's can be attributed to the linear relationship with x.**

The same argument applies also to r^2, the square of the **sample correlation coefficient**, given by

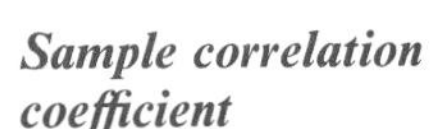
Sample correlation coefficient

$$\boldsymbol{r = \frac{S_{xy}}{\sqrt{S_{xx} \cdot S_{yy}}}}$$

for any n pairs of values (x_i, y_i); S_{xy}, S_{xx}, and S_{yy} are as defined on page 328. This estimator is not unbiased and, except for the factor $\sqrt{\frac{n-1}{n-2}}$, it may be obtained by substituting for σ_2^2 the sample variance of the y's and for σ^2 the square of the standard error of estimate, s_e^2 (see also Exercise 11.62 on page 368).

EXAMPLE The following are the numbers of minutes it took 10 mechanics to assemble a piece of machinery in the morning, x, and in the late afternoon, y:

x	y
11.1	10.9
10.3	14.2
12.0	13.8
15.1	21.5
13.7	13.2
18.5	21.1
17.3	16.4
14.2	19.3
14.8	17.4
15.3	19.0

Calculate r.

Solution First we determine the necessary summations, getting $\sum x = 142.3$, $\sum y = 166.8$, $\sum x^2 = 2{,}085.31$, $\sum xy = 2{,}434.69$, $\sum y^2 = 2{,}897.80$. Then, substituting into the formulas on page 328, we get

$$S_{xx} = 2{,}085.31 - (142.3)^2/10 = 60.381$$

$$S_{xy} = 2{,}434.69 - (142.3)(166.8)/10 = 61.126$$

$$S_{yy} = 2{,}897.80 - (166.8)^2/10 = 115.576$$

and, hence,

$$r = \frac{61.126}{\sqrt{(60.381)(115.576)}} = 0.73$$

The result we have obtained here implies that $100r^2 = 53\%$ of the variation among the afternoon times is explained by (is accounted for or may be attributed to) corresponding differences among the morning times. ■

Whenever a value of r is based on a random sample from a bivariate normal population, we can perform a test of significance (a test of the null hypothesis $\rho = \rho_0$) or construct a confidence interval for ρ on the basis of the following transformation:

Fisher Z transformation

$$Z = \frac{1}{2} \ln \frac{1 + r}{1 - r}$$

This statistic is a value of a random variable having approximately a normal distribution with the mean $\mu_Z = \frac{1}{2} \ln \frac{1+\rho}{1-\rho}$ and the variance $\frac{1}{n-3}$. Thus, we can base inferences about ρ on

Statistic for inferences about ρ

$$z = \frac{Z - \mu_Z}{1/\sqrt{n-3}} = \frac{\sqrt{n-3}}{2} \cdot \ln \frac{(1+r)(1-\rho)}{(1-r)(1+\rho)}$$

which is a value of a random variable having approximately the standard normal distribution.

In particular, we can test the null hypothesis of no correlation, namely, the null hypothesis $\rho = 0$, with the statistic

Statistic for test of null hypothesis ρ = 0

$$z = \sqrt{n-3} \cdot Z = \frac{\sqrt{n-3}}{2} \cdot \ln \frac{1+r}{1-r}$$

These tests are facilitated by the use of Table 11, which gives the values of Z corresponding to $r = 0.00, 0.01, 0.02, \ldots,$ and 0.99. When r is negative, we look up the value of Z corresponding to $-r$, and then take $-Z$.

EXAMPLE With reference to the preceding example, where we had $n = 10$ and $r = 0.73$, test the null hypothesis $\rho = 0$ against the alternative hypothesis $\rho \neq 0$ at the 0.05 level of significance.

Solution

1. *Null hypothesis*: $\rho = 0$
   Alternative hypothesis: $\rho \neq 0$
2. *Level of significance*: $\alpha = 0.05$
3. *Criterion*: Reject the null hypothesis if $z < -1.96$ or $z > 1.96$, where $z = \sqrt{n-3} \cdot Z$.
4. *Calculations*: The value of Z corresponding to $r = 0.73$ is 0.929 according to Table 11, so that

$$z = \sqrt{10-3} \cdot (0.929) = 2.46$$

5. *Decision*: Since $z = 2.46$ exceeds 1.96, the null hypothesis must be rejected; we conclude that there is a relationship between the morning and late afternoon time it takes a mechanic to assemble the given kind of machinery.

■

To construct a confidence interval for ρ, we first construct a confidence interval for μ_Z, the mean of the sampling distribution of Z, and then convert to r and ρ by means of Table 11. Making use of the theory on page 363, we can write the first of these confidence intervals as

Confidence interval for μ_Z

$$Z - \frac{z_{\alpha/2}}{\sqrt{n-3}} < \mu_Z < Z + \frac{z_{\alpha/2}}{\sqrt{n-3}}$$

EXAMPLE If $r = 0.70$ for the mathematics and physics grades of 30 students, construct a 95% confidence interval for the population correlation coefficient.

Solution Reading the value of Z that corresponds to $r = 0.70$ from Table 11 and substituting it together with $n = 30$ and $z_{0.025} = 1.96$ into the preceding confidence-interval formula for μ_Z, we get

$$0.867 - \frac{1.96}{\sqrt{27}} < \mu_Z < 0.867 + \frac{1.96}{\sqrt{27}}$$

or

$$0.490 < \mu_Z < 1.244$$

Then, looking up the values of r for which Z is closest to 0.490 and 1.244 in Table 11, we get the 95% confidence interval

$$0.45 < \rho < 0.85$$

for the true strength of the linear relationship between grades of students in the two given subjects.

■

EXAMPLE If $r = 0.20$ for a random sample of $n = 40$ paired data, construct a 95% confidence interval for ρ.

Solution Reading the value of Z that corresponds to $r = 0.20$ from Table 11 and substituting it together with $n = 40$ and $z_{0.025} = 1.96$ into the preceding confidence-interval formula for μ_Z, we get

$$0.203 - \frac{1.96}{\sqrt{37}} < \mu_Z < 0.203 + \frac{1.96}{\sqrt{37}}$$

or

$$-0.119 < \mu_Z < 0.525$$

Then, looking up the values of r for which Z is closest to 0.119 and 0.525 in Table 11, we get the 95% confidence interval

$$-0.12 < \rho < 0.48$$

for the population correlation coefficient.

■

Note that in both of these examples the confidence intervals for ρ are fairly wide; this illustrates the fact that correlation coefficients based on relatively small samples are generally not very reliable.

There are several serious pitfalls in the interpretation of the coefficient of correlation. First, it must be emphasized that r is an estimate of the strength of the linear relationship between the values of two random variables; thus, as is shown in Figure 11.14, r may be close to 0 when there is actually a strong (but nonlinear) relationship. Second, and perhaps of greatest importance, a significant correlation does not necessarily imply a causal relationship between the two random variables. Although it would not be surprising, for example, to obtain a high positive correlation between the annual sales of chewing gum and the incidence of crime in the United States, one cannot conclude that crime might be reduced by prohibiting the sale of chewing gum. Both variables depend upon the size of the population, and it is this mutual relationship with a third variable (population size) which produces the positive correlation.

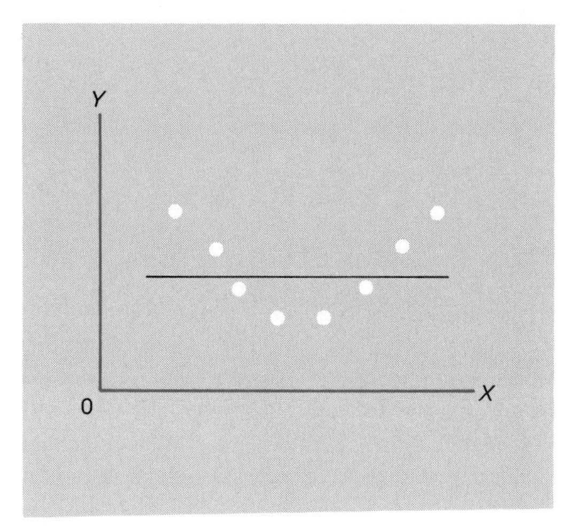

FIGURE 11.14
Nonlinear relationship where $r = 0$.

EXERCISES

11.48 Use the first set of expressions on page 328, involving the deviations from the mean, to calculate r for the following data.

x	y
8	3
1	4
5	0
4	2
7	1

11.49 Calculate r for the air velocities and evaporation coefficients of the example on page 326. Also, assuming that the necessary assumptions can be met, test the null hypothesis $\rho = 0$ against the alternative hypothesis $\rho \neq 0$ at the 0.05 level of significance.

11.50 The following data pertain to the resistance (ohms) and the failure time (minutes) of certain overloaded resistors:

Resistance	*Failure time*	*Resistance*	*Failure time*
43	32	36	36
29	20	39	33
44	45	36	21
33	35	47	44
33	22	28	26
47	46	40	45
34	28	42	39
31	26	33	25
48	37	46	36
34	33	28	25
46	47	48	45
37	30	45	36

Calculate r.

11.51 With reference to Exercise 11.50, test for significance at $\alpha = 0.01$.

11.52 Calculate r for the extraction times and extraction efficiencies of Exercise 11.1 on page 335. Assuming that the necessary assumptions can be met, test the null hypothesis $\rho = 0.75$ against the alternative hypothesis $\rho > 0.75$ at the 0.05 level of significance.

11.53 Calculate r for the humidities and moisture contents of Exercise 11.12 on page 337. Assuming that the necessary assumptions can be met, construct a 95% confidence interval for the population correlation coefficient ρ.

11.54 The following are measurements of the carbon content and the permeability index of 22 sinter mixtures:

Carbon content (%)	*Permeability index*	*Carbon content (%)*	*Permeability index*
4.4	12	4.1	13
5.5	14	4.9	19
4.2	18	4.7	22
3.0	35	5.0	20
4.5	23	4.6	16
4.9	29	3.6	27
4.6	16	4.9	21
5.0	12	5.1	13
4.7	18	4.8	18
5.1	21	5.2	17
4.4	27	5.2	11

(a) Calculate r.
(b) Find 99% confidence limits for ρ.

11.55 If $r = 0.83$ for one set of paired data and $r = 0.60$ for another, compare the strengths of the two relationships.

11.56 If data on the ages and prices of 25 pieces of equipment yielded $r = -0.58$, test the null hypothesis $\rho = -0.40$ against the alternative hypothesis $\rho < -0.40$ at the 0.05 level of significance.

11.57 Assuming that the necessary assumptions are met, construct a 95% confidence interval for ρ when

(a) $r = 0.72$ and $n = 19$; (b) $r = 0.35$ and $n = 25$;
(c) $r = 0.57$ and $n = 40$.

11.58 Assuming that the necessary assumptions are met, construct 99% confidence intervals for ρ when

(a) $r = -0.87$ and $n = 19$; (b) $r = 0.39$ and $n = 24$;
(c) $r = 0.16$ and $n = 40$.

11.59 Evaluating the necessary integrals, verify the identities

$$\mu_2 = \alpha + \beta\mu_1 \quad \text{and} \quad \sigma_2^2 = \sigma^2 + \beta^2\sigma_1^2$$

on page 360.

11.60 Substitute $\mu_2 = \alpha + \beta\mu_1$ and $\sigma_2^2 = \sigma^2 + \beta^2\sigma_1^2$ into the formula for the bivariate density given on page 360, and show that this gives the final form shown on page 361.

11.61 Show that for the bivariate normal distribution

(a) independence implies zero correlation;
(b) zero correlation implies independence.

11.62 By substituting the sample variance of the y's for σ_2^2 and s_e^2 for σ^2 in the formula

$$\rho^2 = 1 - \frac{\sigma^2}{\sigma_2^2}$$

verify the formula for r on page 361 (except for a multiplicative constant).

11.63 Instead of using the computing formula on page 361, we can obtain the correlation coefficient r with the formula

$$r = \pm\sqrt{1 - \frac{\sum (y - \hat{y})^2}{\sum (y - \bar{y})^2}}$$

which is analogous to the formula used to define ρ. Although the computations required by the use of this formula are tedious, the formula has the advantage that it can be used also to measure the strength of nonlinear relationships or relationships in several variables. For instance, in the multiple linear regression example on page 349, one could calculate the predicted values by means of the equation

$$\hat{y} = 46.4 + 7.78x_1 - 1.65x_2$$

and then determine r as a measure of how strongly y, the twist required to break one of the forged alloy bars, depends on both percentages of alloying elements present.

(a) Using the data on page 349, find $\sum (y - \bar{y})^2$ by means of the formula $\sum (y - \bar{y})^2 = \sum y^2 - n\bar{y}^2$.

(b) Using the regression equation obtained on page 349, calculate $\hat{y}$ for the sixteen points and then determine $\sum (y - \hat{y})^2$.

(c) Substitute the results obtained in (a) and (b) into the above formula for r. The result is called the **multiple correlation coefficient**.

11.64 With reference to Exercise 11.41 on page 358, use the theory of the preceding exercise to calculate the multiple correlation coefficient (which measures how strongly the current gain is related to the two independent variables).

11.65 When we substitute for the actual values of paired data the ranks which the values occupy in the respective samples, the correlation coefficient can be written in the form

$$r_S = 1 - \frac{6 \cdot \sum d_i^2}{n(n^2 - 1)}$$

in which it is called the **rank-correlation coefficient** or **Spearman's rank-correlation coefficient**. In this formula, d_i is the difference between the ranks of the paired observations (x_i, y_i) and n is the number of pairs in the sample. The formula for r_S follows from that for r when there are no ties in rank; if there are ties, we substitute for each of the tied observations the mean of the ranks that they jointly occupy, and there may be a small difference between the values of r and r_S obtained for the two sets of ranks.

(a) Calculate r_S for the example on page 362, which deals with the time it takes mechanics to assemble a certain piece of machinery in the morning and in the late afternoon. Compare the result with the value of r obtained for the original data.

(b) Calculate r_S for the data of Exercise 11.50 on page 366 and compare the result with the value of r obtained for the original data.

(c) Calculate r_S for the air velocities and evaporation coefficients of the example on page 326 and compare the result with the value of r obtained for the original data in Exercise 11.49 on page 366.

11.66 Since the assumptions underlying the significance test of Section 11.6 are rather stringent, it is often desirable to use a nonparametric alternative based on the rank-correlation coefficient defined in the preceding exercise. It is based on the theory that if there is no relationship between the x's and the y's (in fact, if they are randomly matched), the sampling distribution of r_S can be approximated closely with a normal distribution having zero mean and the standard deviation

$$\sigma_{r_S} = \frac{1}{\sqrt{n-1}}$$

Use the 0.01 level of significance to test the significance of the values of r_S obtained in the three parts of Exercise 11.65.

11.67 If a sample of $n = 18$ pairs of data yielded $r_S = 0.39$, is this rank-correlation coefficient significant at the 0.05 level of significance? [*Hint*: Refer to the two preceding exercises.]

11.68 If a sample of $n = 40$ pairs of data yielded $r_S = 0.48$, is this rank-correlation coefficient significant at the 0.01 level of significance? [*Hint*: Refer to Exercises 11.65 and 11.66.]

11.69 To calculate r using *MINITAB* when the x values are in C1 and the y values are in C2, use the command

```
CORR C1 C2
```

Use the computer to do Exercise 11.50.

11.7

MULTIPLE LINEAR REGRESSION (Matrix Notation).†

The model we are using in multiple linear regression lends itself uniquely to a unified treatment in matrix notation. This notation makes it possible to state general results in compact form and to use to great advantage many of the results of matrix theory.

† It is assumed for this section that the reader is familiar with the material ordinarily covered in a first course on matrix algebra. Since matrix notation is not used elsewhere in this book, this section may be omitted without loss of continuity.

It is customary to denote matrices by capital letters in boldface type and vectors by lower case boldface type. To express the normal equations on page 348 in matrix notation, let us define the following three matrices.

$$\mathbf{X} = \begin{bmatrix} 1 & x_{11} & x_{12} \\ 1 & x_{21} & x_{22} \\ \vdots & \vdots & \vdots \\ 1 & x_{n1} & x_{n2} \end{bmatrix}$$

$$\mathbf{y} = \begin{bmatrix} y_1 \\ y_2 \\ \vdots \\ y_n \end{bmatrix} \quad \text{and} \quad \mathbf{b} = \begin{bmatrix} b_0 \\ b_1 \\ b_2 \end{bmatrix}$$

The first one, **X** is an $n \times (2 + 1)$ matrix consisting essentially of the given values of the x's, with the column of 1's appended to accommodate the constant term. **y** is an $n \times 1$ matrix (or column vector) consisting of observed values of the response variable and **b** is the $(2 + 1) \times 1$ matrix (or column vector) consisting of the least squares estimates of the regression coefficients.

Using these matrices, we can now write the following symbolic solutions of the normal equations on page 348. The least squares estimates of the multiple regression coefficients are given by

$$\mathbf{b} = (\mathbf{X}'\mathbf{X})^{-1}\mathbf{X}'\mathbf{y}$$

where $\mathbf{X}'$ is the transpose of **X** and $(\mathbf{X}'\mathbf{X})^{-1}$ is the inverse of $\mathbf{X}'\mathbf{X}$.

To verify this relation, we first determine $\mathbf{X}'\mathbf{X}$, $\mathbf{X}'\mathbf{X}\mathbf{b}$, and $\mathbf{X}'\mathbf{y}$.

$$\mathbf{X}'\mathbf{X} = \begin{bmatrix} n & \sum x_1 & \sum x_2 \\ \sum x_1 & \sum x_1^2 & \sum x_1 x_2 \\ \sum x_2 & \sum x_2 x_1 & \sum x_2^2 \end{bmatrix}$$

$$\mathbf{X}'\mathbf{X}\mathbf{b} = \begin{bmatrix} b_0 n + b_1 \sum x_1 + b_2 \sum x_2 \\ b_0 \sum x_1 + b_1 \sum x_1^2 + b_2 \sum x_1 x_2 \\ b_0 \sum x_2 + b_1 \sum x_2 x_1 + b_2 \sum x_2^2 \end{bmatrix}$$

$$\mathbf{X}'\mathbf{y} = \begin{bmatrix} \sum y \\ \sum x_1 y \\ \sum x_2 y \end{bmatrix}$$

Identifying the elements of $\mathbf{X'Xb}$ as the expressions on the right-hand side of the normal equation on page 348 and those of $\mathbf{X'y}$ as the expressions on the left-hand side, we can write

$$\mathbf{X'Xb} = \mathbf{X'y}$$

Multiplying on the left by $(\mathbf{X'X})^{-1}$, we get

$$(\mathbf{X'X})^{-1}\mathbf{X'Xb} = (\mathbf{X'X})^{-1}\mathbf{X'y}$$

and, finally,

$$\mathbf{b} = (\mathbf{X'X})^{-1}\mathbf{X'y}$$

since $(\mathbf{X'X})^{-1}\mathbf{X'X}$ equals the $(2 + 1) \times (2 + 1)$ identity matrix $\mathbf{I}$, and by definition $\mathbf{Ib} = \mathbf{b}$. We have assumed here that $\mathbf{X'X}$ is nonsingular, so that its inverse exists.

EXAMPLE With reference to the example on page 349, use the matrix expressions to determine the least squares estimates of the multiple regression coefficients.

Solution Substituting $\sum x_1 = 40$, $\sum x_2 = 200$, $\sum x_1^2 = 120$, $\sum x_1 x_2 = 500$, $\sum x_2^2 = 3{,}000$, and $n = 16$ into the expression for $\mathbf{X'X}$ on page 370, we get

$$\mathbf{X'X} = \begin{bmatrix} 16 & 40 & 200 \\ 40 & 120 & 500 \\ 200 & 500 & 3{,}000 \end{bmatrix}$$

Then, the inverse of this matrix can be obtained by any one of a number of different techniques; using the one based on cofactors, we find that

$$(\mathbf{X'X})^{-1} = \frac{1}{160{,}000}\begin{bmatrix} 110{,}000 & -20{,}000 & -4{,}000 \\ -20{,}000 & 8{,}000 & 0 \\ -4{,}000 & 0 & 320 \end{bmatrix}$$

where 160,000 is the value of $|\mathbf{X'X}|$, the determinant of $\mathbf{X'X}$.

Substituting $\sum y = 723$, $\sum x_1 y = 1963$, and $\sum x_2 y = 8{,}210$ into the expression for $\mathbf{X'y}$ on page 370, we then get

$$\mathbf{X'y} = \begin{bmatrix} 723 \\ 1{,}963 \\ 8{,}210 \end{bmatrix}$$

and, finally,

$$\mathbf{b} = (\mathbf{X}'\mathbf{X})^{-1}\mathbf{X}'\mathbf{y} = \frac{1}{160{,}000}\begin{bmatrix} 110{,}000 & -20{,}000 & -4{,}000 \\ -20{,}000 & 8{,}000 & 0 \\ -4{,}000 & 0 & 320 \end{bmatrix}\begin{bmatrix} 723 \\ 1{,}963 \\ 8{,}210 \end{bmatrix}$$

$$= \frac{1}{160{,}000}\begin{bmatrix} 7{,}430{,}000 \\ 1{,}244{,}000 \\ -264{,}800 \end{bmatrix}$$

$$= \begin{bmatrix} 46.4375 \\ 7.7750 \\ -1.6550 \end{bmatrix}$$

Note that the results obtained here are identical with those shown in the computer printout on page 351.

■

The residual sum of squares also has a convenient matrix expression. The predicted values $\hat{y}_i = b_0 + b_1 x_{1i} + b_2 x_{2i}$ can be collected as a matrix (column vector).

$$\hat{\mathbf{y}} = \begin{bmatrix} \hat{y}_1 \\ \hat{y}_2 \\ \vdots \\ \hat{y}_n \end{bmatrix} = \mathbf{Xb}$$

Then, the residual sum of squares

$$\sum_{i=1}^{n} (y_i - \hat{y}_i)^2 = (\mathbf{y} - \hat{\mathbf{y}})'(\mathbf{y} - \hat{\mathbf{y}}) = (\mathbf{y} - \mathbf{Xb})'(\mathbf{y} - \mathbf{Xb})$$

Consequently, the estimate s_e^2 of σ^2 can be expressed as

$$s_e^2 = \frac{1}{n-3}(\mathbf{y} - \mathbf{Xb})'(\mathbf{y} - \mathbf{Xb})$$

The same matrix expressions for **b** and the residual sum of squares hold for any number of predictor variables. If the mean of y_i has the form $\beta_0 + \beta_1 x_1 + \beta_2 x_2 + \cdots + \beta_k x_k$, then we define the matrices

$$\mathbf{X} = \begin{bmatrix} 1 & x_{11} & x_{12} & \cdots & x_{1k} \\ 1 & x_{21} & x_{22} & \cdots & x_{2k} \\ \vdots & \vdots & \vdots & \ddots & \vdots \\ 1 & x_{n1} & x_{n2} & \cdots & x_{nk} \end{bmatrix}, \quad \mathbf{b} = \begin{bmatrix} b_0 \\ b_1 \\ b_2 \\ \vdots \\ b_k \end{bmatrix}, \quad \mathbf{y} = \begin{bmatrix} y_1 \\ y_2 \\ \vdots \\ y_n \end{bmatrix}$$

Then,

$$\mathbf{b} = (\mathbf{X}'\mathbf{X})^{-1}\mathbf{X}'\mathbf{y} \quad \text{and} \quad s_e^2 = \frac{1}{n-k-1}(\mathbf{y} - \mathbf{X}\mathbf{b})'(\mathbf{y} - \mathbf{X}\mathbf{b})$$

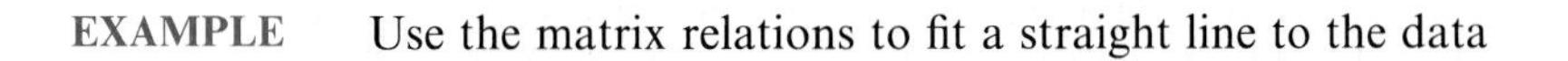

EXAMPLE Use the matrix relations to fit a straight line to the data

x	0	1	2	3	4
y	8	9	4	3	1

Solution Here $k = 1$ and dropping the subscript 1 we have

$\mathbf{X}'$	$\mathbf{y}$	$\mathbf{X}'\mathbf{X}$	$(\mathbf{X}'\mathbf{X})^{-1}$	$\mathbf{X}'\mathbf{y}$
$\begin{bmatrix} 1 & 1 & 1 & 1 & 1 \\ 0 & 1 & 2 & 3 & 4 \end{bmatrix}$	$\begin{bmatrix} 8 \\ 9 \\ 4 \\ 3 \\ 1 \end{bmatrix}$	$\begin{bmatrix} 5 & 10 \\ 10 & 30 \end{bmatrix}$	$\begin{bmatrix} 0.6 & -0.2 \\ -0.2 & 0.1 \end{bmatrix}$	$\begin{bmatrix} 25 \\ 30 \end{bmatrix}$

Consequently

$$\mathbf{b} = (\mathbf{X}'\mathbf{X})^{-1}\mathbf{X}'\mathbf{y} = \begin{bmatrix} 0.6 & -0.2 \\ -0.2 & 0.1 \end{bmatrix}\begin{bmatrix} 25 \\ 30 \end{bmatrix} = \begin{bmatrix} 9 \\ -2 \end{bmatrix}$$

and the fitted equation is

$$\hat{y} = 9 - 2x$$

The vector of fitted values is

$$\hat{\mathbf{y}} = \mathbf{X}\mathbf{b} = \begin{bmatrix} 10 \\ 11 \\ 12 \\ 13 \\ 14 \end{bmatrix}\begin{bmatrix} 9 \\ -2 \end{bmatrix} = \begin{bmatrix} 9 \\ 7 \\ 5 \\ 3 \\ 1 \end{bmatrix}$$

so the vector of residuals

$$\mathbf{y} - \hat{\mathbf{y}} = \begin{bmatrix} 8 \\ 9 \\ 4 \\ 3 \\ 1 \end{bmatrix} - \begin{bmatrix} 9 \\ 7 \\ 5 \\ 3 \\ 1 \end{bmatrix} = \begin{bmatrix} -1 \\ 2 \\ -1 \\ 0 \\ 0 \end{bmatrix}$$

and the residual sum of squares is

$$[-1 \quad 2 \quad -1 \quad 0 \quad 0]\begin{bmatrix} -1 \\ 2 \\ -1 \\ 0 \\ 0 \end{bmatrix} = 6$$

Finally,

$$s_e^2 = \frac{1}{n-k-1}(\mathbf{y} - \hat{\mathbf{y}})'(\mathbf{y} - \hat{\mathbf{y}}) = \frac{1}{5-2}(6) = 2.00.$$

■

The elegance of the expressions using matrices goes one step further. We can express the estimated variances and covariances of the least squares estimators as

$$\begin{bmatrix} \hat{var}(b_0) & \hat{cov}(b_0, b_1) & \cdots & \hat{cov}(b_0, b_k) \\ \hat{cov}(b_1, b_0) & \hat{var}(b_1) & \cdots & \hat{cov}(b_1, b_k) \\ \vdots & \vdots & \ddots & \vdots \\ \hat{cov}(b_k, b_0) & \hat{cov}(b_k, b_1) & \cdots & \hat{var}(b_k) \end{bmatrix} = s_e^2\,(\mathbf{X}'\mathbf{X})^{-1}$$

That is, to obtain the estimated variance, $\hat{var}(b_i)$, of b_i, we multiply the corresponding entry of $(\mathbf{X}'\mathbf{X})^{-1}$ by s_e^2 which is the estimate of σ^2.

EXAMPLE With reference to the preceding example, use the matrix relations to obtain the estimated variances $\hat{var}(b_0)$ and $\hat{var}(b_1)$.

Solution We have

$$\begin{bmatrix} \hat{var}(b_0) & \hat{cov}(b_0, b_1) \\ \hat{cov}(b_1, b_0) & \hat{var}(b_1) \end{bmatrix} = s_e^2(\mathbf{X}'\mathbf{X})^{-1}$$

$$= (2.00)\begin{bmatrix} 0.6 & -0.2 \\ -0.2 & 0.1 \end{bmatrix} = \begin{bmatrix} 1.2 & -0.4 \\ -0.4 & 0.2 \end{bmatrix}$$

where the values for $(\mathbf{X}'\mathbf{X})^{-1}$ and s_e^2 are those obtained in the preceding example.

Therefore the estimates are $\hat{var}(b_0) = 1.2$ and $\hat{var}(b_1) = 0.2$. Note also that the estimated covariance of b_0 and b_1 is $\hat{cov}(b_0, b_1) = -0.4$.

■

11.8 REVIEW EXERCISES

11.70 The following data pertain to the number of hours jet aircraft engines have been used and the number of hours required for repair.

Number of hours (hundreds)	*Repair time (hours)*
1	10
2	40
3	30
4	80
5	90

(a) Use the first set of expressions on page 328, involving the deviations from the mean, to fit a least-squares line to the observations.

(b) Use the equation of the least-squares line to estimate mean repair time at $x = 4.5$.

(c) What difficulty might you encounter if you use the least-squares line to predict the mean repair time for a jet aircraft engine with 700 hours?

11.71 With reference to Exercise 11.70, construct a 95% confidence interval for α.

11.72 With reference to Exercise 11.70, test the null hypothesis $\beta = 15$ against the alternative hypothesis $\beta > 15$ at the 0.05 level of significance.

11.73 With reference to Exercise 11.70,

(a) find a 95% confidence interval for the mean repair time at $x = 4.5$;

(b) find a 95% limits of prediction for the time to repair an engine that will be run for $x = 4.5$ hundred hours.

11.74 The following data pertain to the chlorine residual in a swimming pool at various times after it has been treated with chemicals:

Number of hours	*Chlorine residual (parts per million)*
2	1.8
4	1.5
6	1.4
8	1.1
10	1.1
12	0.9

(a) Fit a least-squares line from which we can predict the chlorine residual in terms of the number of hours since the pool has been treated with chemicals.

(b) Use the equation of the least-squares line to estimate the chlorine residual in the pool 5 hours after it has been treated with chemicals.

11.75 With reference to the preceding exercise, construct a 95% confidence interval for α.

11.76 With reference to Exercise 11.74, test the null hypothesis $\beta = -0.12$ against the alternative hypothesis $\beta > -0.12$ at the 0.01 level of significance.

11.77 In an experiment designed to determine the specific heat ratio γ for a certain gas, the gas was compressed adiabatically to several predetermined volumes V, and the corresponding pressure p was measured with the following results:

p (lb/in.^{2})	16.6	39.7	78.5	115.5	195.3	546.1
V (in.^{3})	50	30	20	15	10	5

Assuming the ideal gas law $p \cdot V^\gamma = C$, use these data to estimate γ for this gas.

11.78 With reference to Exercise 11.77, use the method of Section 11.2 to construct a 95% confidence interval for γ. State what assumptions will have to be made.

11.79 The rise of current in an inductive circuit having the time constant τ is given by

$$I = 1 - e^{-t/\tau}$$

where t is the time measured from the instant the switch is closed, and I is the ratio of the current at time t to the full value of the current given by Ohm's law. Given the measurements

I	0.073	0.220	0.301	0.370	0.418	0.467	0.517	0.578
t (sec)	0.1	0.2	0.3	0.4	0.5	0.6	0.7	0.8

estimate the time constant of this circuit from the experimental results given. [*Hint*: Note that the relationship between ln $(1 - I)$ and t is linear.]

11.80 The following are sample data provided by a moving company on the weights of six shipments, the distances they were moved, and the damage that was incurred:

Weight (1,000 *pounds*) x_1	*Distance* (1,000 *miles*) x_2	*Damage* (*dollars*) y
4.0	1.5	160
3.0	2.2	112
1.6	1.0	69
1.2	2.0	90
3.4	0.8	123
4.8	1.6	186

(a) Fit an equation of the form $y = \beta_0 + \beta_1 x_1 + \beta_2 x_2$.
(b) Use the equation obtained in part (a) to estimate the damage when a shipment weighing 2,400 pounds is moved 1,200 miles.

11.81 With reference to Exercise 11.9,
(a) find a 95% confidence interval for the mean CPU time required for $x = 3.0$ jobs;
(b) find a 95% limits of prediction for the CPU time required on a future day when $x = 3.0$ jobs must be run.

11.82 Use the first set of expressions on page 328, involving the deviations from the mean, to calculate r for the following data.

x	y
8	4
5	5
6	1
2	3
9	2

11.83 If $r = 0.41$ for one set of paired data and $r = 0.29$ for another, compare the strengths of the two relationships.

11.84 If for certain paired data $n = 18$ and $r = 0.44$, test the null hypothesis $\rho = 0.30$ against the alternative hypothesis $\rho > 0.30$ at the 0.01 level of significance.

11.85 Assuming that the necessary assumptions are met, construct 95% confidence intervals for ρ when
(a) $r = 0.78$ and $n = 15$; (b) $r = -0.62$ and $n = 32$;
(c) $r = 0.17$ and $n = 35$.

11.86 With reference to Exercise 11.80 on page 376, use the theory of Exercise 11.63 to calculate the multiple correlation coefficient (which measures how strongly the damage is related to both weight and distance).

11.87 Robert A. Millikan (1865–1953) produced the first accurate measurements on the charge e of an electron. He devised a method to observe a single drop of water or oil under the influence of both electric and gravitational fields. Usually, a droplet carried multiple electrons and direct calculations based on voltage, time of fall, etc. provided an estimate of the total charge. (Source: *Philosophical Magazine* 19 (1910): 209–228.)

x *(No. of e's)*	*Observations* ($10^9 \times$ *charge*)
3	1.392, 1.392, 1.398, 1.368, 1.368, 1.368, 1.345
4	1.768, 1.768, 1.910, 1.768, 1.746, 1.746, 1.886, 1.768, 1.768, 1.768
5	2.471, 2.471, 2.256, 2.256, 2.471
2	0.944, 0.992
6	2.981, 2.688

(a) Find the equation of the least-squares line.
(b) Find a 95% confidence interval for the slope β, the charge e on a single electron.
(c) Test the null hypothesis $\alpha = 0$ against the alternative hypothesis $\alpha \neq 0$.
(d) Examine the residuals.

11.88 Robert Boyle (1627–1691) established the law that (pressure $\times$ volume) = constant for a gas at a constant pressure. By pouring mercury into the open top of the long side of a J-shaped tube, he increased the pressure on the air trapped in the short leg. The volume of trapped air = $h \times$ cross section, where h is the height of the air in the short leg. If y = height of mercury, adjusted for the pressure of the atmosphere on the open end, then y and $x = 1/h$ should obey a straight-line relationship. (Source: *The Laws of Gases*, edited by Carl Barus (1899), New York: Harper and Brothers Publishers.)

h	48	46	44	42	40	38	36	34	32	30	28	26	24
y	$29\frac{2}{16}$	$30\frac{9}{16}$	$31\frac{15}{16}$	$33\frac{8}{16}$	$35\frac{5}{16}$	37	$39\frac{5}{16}$	$41\frac{10}{16}$	$44\frac{3}{16}$	$47\frac{1}{16}$	$50\frac{5}{16}$	$54\frac{5}{16}$	$58\frac{13}{16}$

h	23	22	21	20	19	18	17	16	15	14	13	12
y	$61\frac{5}{16}$	$64\frac{1}{16}$	$67\frac{1}{16}$	$70\frac{11}{16}$	$74\frac{2}{16}$	$77\frac{14}{16}$	$82\frac{12}{16}$	$87\frac{14}{16}$	$93\frac{1}{16}$	$100\frac{7}{16}$	$107\frac{13}{16}$	$117\frac{9}{16}$

(a) Fit a straight line by least squares.
(b) Check the residuals for a possible violation of the assumptions.

11.9

CHECK LIST OF KEY TERMS (with page references)

ANALYSIS OF VARIANCE

Some of the examples of Chapter 11 have already taught us that considerable economies in calculation can result by appropriately planning an experiment in advance. What is even more important, proper experimental planning can give a reasonable assurance that the results of experiment will provide clear-cut answers to questions under investigation. While it is impossible to give in this chapter a complete discussion of experimental design, including the many pitfalls to which the experimenter is exposed, we shall begin by presenting some of the general principles of experimental design. Several of the designs most frequently used in engineering and other applied research will be given in subsequent sections.

In Sections 12.2 and 12.3 we shall discuss the often used one- and two-way classification designs, in Section 12.5 we introduce Latin-square and Graeco-Latin-square designs, and in the remainder of the chapter we introduce tests for comparing several means as well as the analysis of a balanced experiment in the presence of a concomitant variable, or covariate.

12.1 SOME GENERAL PRINCIPLES

Many of the most important aspects of **experimental design** can be illustrated by means of an example drawn from the important field of engineering measurement.

Suppose that a steel mill supplies tin plate to three can manufacturers, the major specification being that the tin-coating weight should be at least 0.25 pound per base box. The mill and each can manufacturer has a laboratory where measurements are made of the tin-coating weights of samples taken from each shipment. Suppose, also, that some disagreement has arisen about the actual tin-coating weights of the tin plate being shipped, and it is decided to plan an experiment to determine whether the four laboratories are making consistent measurements. A complicating factor is that part of the measuring process consists of the chemical removal of the tin from the surface of the base metal; thus, it is impossible to have the identical sample measured by each laboratory to determine how closely the measurements correspond.

One possibility is to send several samples (in the form of circular discs having equal areas) to each of the laboratories. Although these discs may not actually have identical tin-coating weights, it is hoped that such differences will be small and that they will more or less "average out." In other words, it will be assumed that whatever differences there may be among the means of the four samples can be attributed to no other causes but systematic differences in measuring techniques and chance variability. This would make it possible to determine whether the results produced by the laboratories are consistent by comparing the variability of the four sample means with an appropriate measure of chance variation.

Now there remains the problem of deciding how many discs are to be sent to each laboratory and how the discs are actually to be selected. The question of sample size can be answered in many different ways, one of which is to use the formula on page 241 for the standard deviation of the sampling distribution of the difference between two means. Substituting known values of σ_1 and σ_2 and specifying what differences between the true means of any two of the laboratories should be detected with a probability of at least 0.95 (or 0.98, or 0.99), it is possible to determine $n_1 = n_2 = n$ (see Exercise 12.12 on page 395). Suppose that this method and, perhaps, also considerations of cost and availability of the necessary specimens lead to the decision to send a sample of 12 discs to each laboratory.

The problem of selecting the required 48 discs and allocating 12 to each laboratory is not as straightforward as it may seem at first. To begin with, suppose that a sheet of tin plate, sufficiently long and sufficiently wide, is selected and that the 48 discs are cut as shown in Figure 12.1. The 12 discs cut from strip 1 are sent to the first laboratory, the 12 discs from strip 2 are sent to the second laboratory, and so forth. If the four mean coating weights subsequently obtained were then found to differ significantly, would this allow us to conclude that these differences can be attributed to lack of consistency in the measuring techniques? Suppose, for instance, that additional investigation shows that the amount of tin deposited electrolytically on a long sheet of steel has a distinct and repeated pattern of variation perpendicular to the direction in which it is rolled. (Such a pattern might be caused by the placement of electrodes, "edge effects," and so forth.) Thus, even if all four laboratories measured the amount of tin consistently and without error, there could be cause for differences in the tin-coating weight determinations. The allocation of an entire strip of discs to each laboratory is such that inconsistencies among the laboratories' measuring techniques are inseparable from (or **confounded**

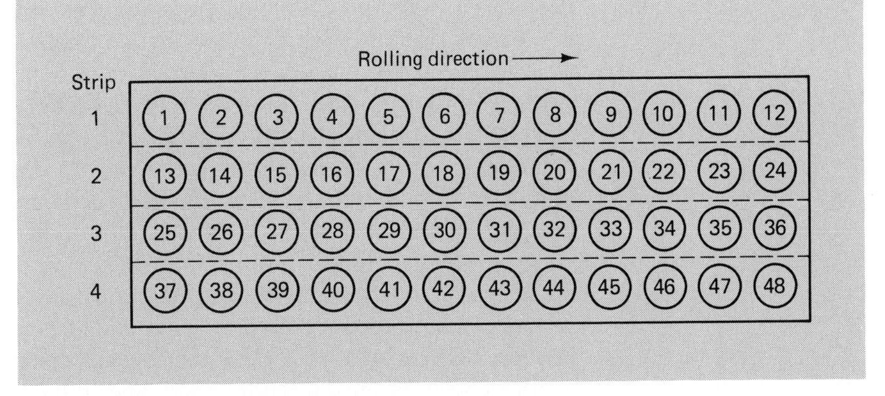

FIGURE 12.1
Numbering of tin-plate samples.

with) whatever differences there may be in the actual amount of tin deposited perpendicular to the direction in which the sheet of steel is rolled.

One way to avoid this kind of confounding is to number the discs and allocate them to the four laboratories at random, such as in the following arrangement, which was obtained with the aid of a table of random numbers:

Laboratory A:	3, 38, 17, 32, 24, 30, 48, 19, 11, 31, 22, 41
Laboratory B:	44, 20, 15, 25, 45, 4, 14, 5, 39, 7, 40, 34
Laboratory C:	12, 21, 42, 8, 27, 16, 47, 46, 18, 43, 35, 26
Laboratory D:	9, 2, 28, 23, 37, 1, 10, 6, 29, 36, 33, 13

If there were any actual pattern of tin-coating thickness on the sheet of tin plate, it would be "broken up" by the **randomization**.

Although we have identified and counteracted one possible systematic pattern of variation, there is no assurance that there can be no others. For instance, there may be systematic differences in the areas of the discs caused by progressive wear of the cutting instrument, or there may be scratches or other imperfections on one part of the sheet which could affect the measurements. Thus, there is always the possibility that differences in means attributed to inconsistencies among the laboratories are actually caused by some other uncontrolled variable, and it is the purpose of randomization to avoid confounding the variable under investigation with such other variables.

By distributing the 48 discs among the four laboratories entirely at random, we have no choice but to include whatever variation may be attributable to extraneous causes under the heading of "chance variation." This may give us an excessively large estimate of chance variation which, in turn, may make it difficult to detect differences between the true laboratory means. In order to avoid this, we could, perhaps, use only discs cut from the same strip (or from an otherwise homogeneous region). Unfortunately, this kind of **controlled experimentation** presents us with new complications. Of what use would it be, for example, to perform an experiment which allows us to conclude that the laboratories are consistent (or inconsistent), *if such a conclusion is limited to measurements made at a fixed distance from one edge of a sheet*? To consider a more poignant example, suppose that a manufacturer of plumbing materials wishes to compare the performance of several kinds of material to be used in underground water pipes. If such conditions as soil acidity, depth of pipe, and mineral content of water were all held fixed, any conclusions as to which material is best would be valid only for the given set of conditions. What the manufacturer really wants to know is which material is best over a fairly wide variety of conditions, and in designing a suitable experiment it would be advisable (indeed, necessary) to specify that pipe of each material be buried at each of several depths in each of several kinds of soil, and in locations where the water varies in hardness.

This example serves to illustrate that it is seldom desirable to hold all or most extraneous factors fixed throughout an experiment in order to obtain an estimate of chance variation that is not "inflated" by variations due to other causes. (In fact, it is rarely, if ever, possible to exercise such strict control, that is, to hold *all* extraneous variables fixed.) In actual practice, experiments should be planned so that known sources of variability are deliberately varied over as wide a range as necessary; furthermore, they should be varied in such a way that their variability can be eliminated from the estimate of chance variation. One way to accomplish this is to repeat the experiment in several **blocks**, where known sources of variability (that is, extraneous variables) are held fixed in each block, but vary from block to block.

In the tin-plating problem we might thus account for variations across the sheet of steel by randomly allocating three discs from each strip to each of the laboratories as in the following arrangement:

	Strip 1	*Strip 2*	*Strip 3*	*Strip 4*
Laboratory A	8, 4, 10	23, 24, 19	26, 29, 35	37, 44, 48
Laboratory B	2, 6, 12	21, 15, 22	34, 33, 32	45, 43, 46
Laboratory C	1, 5, 11	16, 20, 13	36, 27, 30	41, 38, 47
Laboratory D	7, 3, 9	17, 18, 14	28, 31, 25	39, 40, 42

In this experimental layout, the strips form the blocks, and if we base our estimate of chance variation on the variability *within* each of the 16 sets of three discs, this estimate will not be inflated by the extraneous variable, that is, differences among the strips. (Note also that, with this arrangement, differences among the means obtained for the four laboratories cannot be attributed to differences among the strips. The arrangement on page 381 does not have this property, since, for instance, 5 discs from strip 1 are allocated to Laboratory D.)

The analysis of experiments in which **blocking** is used to eliminate one source of variability is discussed in Section 12.3. The analysis of experiments in which two or three sources of variability are thus eliminated is treated in Section 12.5.

12.2 COMPLETELY RANDOMIZED DESIGNS

In this section we consider, in general, the statistical analysis of the **completely randomized design**, or **one-way classification**. We shall suppose that the experimenter has available the results of k independent random samples, each of size n, from k different populations (that is, data concerning k treatments, k groups, k methods of production, etc.); and he or she is concerned with testing the hypothesis that the means of these k populations are all equal. An example of such an experiment, with $k = 4$, is given by the layout on page 381. If we denote the jth observation in the ith sample by y_{ij}, the general schema for a one-way classification is as follows:

		Means
Sample 1:	$y_{11}, y_{12}, \ldots, y_{1j}, \ldots, y_{1n}$	$\bar{y}_1$
Sample 2:	$y_{21}, y_{22}, \ldots, y_{2j}, \ldots, y_{2n}$	$\bar{y}_2$
$\vdots$	$\vdots$	$\vdots$
Sample i:	$y_{i1}, y_{i2}, \ldots, y_{ij}, \ldots, y_{in}$	$\bar{y}_i$
$\vdots$	$\vdots$	$\vdots$
Sample k:	$y_{k1}, y_{k2}, \ldots, y_{kj}, \ldots, y_{kn}$	$\bar{y}_k$
		$\bar{y}.$

With reference to the experimental layout on page 381, y_{ij} ($i = 1, 2, 3, 4; j = 1, 2, \ldots, 12$) is the jth tin-coating weight measured by the ith laboratory, $\bar{y}_i$ is the mean of the measurements obtained by the ith laboratory, and $\bar{y}.$ is the overall mean (or **grand mean**) of all 48 observations.

To be able to test the hypothesis that the samples were obtained from k populations with equal means, we shall make several assumptions. Specifically, it will be assumed that we are dealing with *normal populations* having *equal variances.* There exist methods for testing the reasonableness of this last assumption (see the book by A. M. Mood, F. A. Graybill and D. C. Boes mentioned in the bibliography), but the methods we develop in this chapter are fairly **robust**; that is, they are relatively insensitive to violations of the assumption of normality as well as the assumption of equal variances.

If μ_i denotes the mean of the ith population and σ^2 denotes the common variance of the k populations, we can express each observation y_{ij} as μ_i plus the value of a random component; that is, we can write

$$y_{ij} = \mu_i + \varepsilon_{ij} \qquad \text{for } i = 1, 2, \ldots, k; j = 1, 2, \ldots, n$$

In accordance with the preceding assumptions, the ε_{ij} are values of independent, normally distributed random variables with zero means and the common variance σ^2.†

To attain uniformity with corresponding equations for more complicated kinds of designs, it is customary to replace μ_i by $\mu + \alpha_i$, where μ is the mean of the μ_i and α_i is the **effect** of the ith treatment; hence, $\sum_{i=1}^{k} \alpha_i = 0$ (see Exercise 12.13 on page 395). Using these new parameters, we can write the model equation for the one-way classification as

Model equation for one-way classification

$$y_{ij} = \mu + \alpha_i + \varepsilon_{ij} \qquad \text{for } i = 1, 2, \ldots, k; j = 1, 2, \ldots, n$$

and the null hypothesis that the k population means are all equal can be replaced by the null hypothesis that $\alpha_1 = \alpha_2 = \cdots = \alpha_k = 0$. The alternative hypothesis that at least two of the population means are unequal is equivalent to the alternative hypothesis that $\alpha_i \neq 0$ for some i.

To test the null hypothesis that the k population means are all equal, we shall compare two estimates of σ^2—one based on the variation among the sample means, and one based on the variation within the samples. Since, by assumption, each sample comes from a population having the variance σ^2, this variance can be estimated by any one of the sample variances

$$s_i^2 = \sum_{j=1}^{n} \frac{(y_{ij} - \bar{y}_i)^2}{n - 1}$$

and, hence, also by their mean

$$\hat{\sigma}_W^2 = \sum_{i=1}^{k} \frac{s_i^2}{k} = \sum_{i=1}^{k} \sum_{j=1}^{n} \frac{(y_{ij} - \bar{y}_i)^2}{k(n - 1)}$$

Note that each of the sample variances s_i^2 is based on $n - 1$ degrees of freedom ($n - 1$ independent deviations from $\bar{y}_i$) and, hence, $\hat{\sigma}_W^2$ is based on $k(n - 1)$ degrees of freedom. Now, the variance of the k sample means is given by

$$s_{\bar{x}}^2 = \sum_{i=1}^{k} \frac{(\bar{y}_i - \bar{y}_.)^2}{k - 1}$$

† Note that this equation, or model, can be regarded as a multiple regression equation; introducing the variables x_{il} which equal 0 or 1 depending on whether the two subscripts are unequal or equal, we can write

$$y_{ij} = \mu_1 x_{i1} + \mu_2 x_{i2} + \cdots + \mu_k x_{ik} + \varepsilon_{ij}$$

The parameters μ_i can now be interpreted as regression coefficients, and they can be estimated by the least-squares methods of Chapter 11.

and *if the null hypothesis is true* it estimates σ^2/n. Thus, an estimate of σ^2 based on the differences among the sample means is given by

$$\hat{\sigma}_B^2 = n \cdot s_{\bar{x}}^2 = n \cdot \sum_{i=1}^{k} \frac{(\bar{y}_i - \bar{y}_.)^2}{k - 1}$$

and it is based on $k - 1$ degrees of freedom.

If the null hypothesis is true, it can be shown that $\hat{\sigma}_W^2$ and $\hat{\sigma}_B^2$ are independent estimates of σ^2, and it follows that

$$F = \frac{\hat{\sigma}_B^2}{\hat{\sigma}_W^2}$$

is a value of a random variable having the F distribution with $k - 1$ and $k(n - 1)$ degrees of freedom. Since the **between-sample variance,** $\hat{\sigma}_B^2$, can be expected to exceed the **within-sample variance,** $\hat{\sigma}_W^2$, when the null hypothesis is *false*, the null hypothesis will be rejected if F exceeds F_α, where F_α is obtained from Table 6 with $k - 1$ and $k(n - 1)$ degrees of freedom.

The preceding argument has shown how the test of the equality of k means can be based on the comparison of two variance estimates. More remarkable, perhaps, is the fact that the two estimates in question [except for the divisors $k - 1$ and $k(n - 1)$] can be obtained by "breaking up" or analyzing the total variance of all nk observations into two parts. The sample variance of all nk observations is given by

$$s^2 = \sum_{i=1}^{k} \sum_{j=1}^{n} \frac{(y_{ij} - \bar{y}_.)^2}{nk - 1}$$

and with reference to its numerator, called the **total sum of squares,** we shall now prove the following theorem.

Identity for one-way analysis of variance

Theorem 12.1

$$\sum_{i=1}^{k} \sum_{j=1}^{n} (y_{ij} - \bar{y}_.)^2 = \sum_{i=1}^{k} \sum_{j=1}^{n} (y_{ij} - \bar{y}_i)^2 + n \cdot \sum_{i=1}^{k} (\bar{y}_i - \bar{y}_.)^2$$

The proof of this theorem is based on the identity

$$y_{ij} - \bar{y}_. = (y_{ij} - \bar{y}_i) + (\bar{y}_i - \bar{y}_.)$$

Squaring both sides and summing on i and j, we obtain

$$\sum_{i=1}^{k}\sum_{j=1}^{n}(y_{ij}-\bar{y}_{.})^2 = \sum_{i=1}^{k}\sum_{j=1}^{n}(y_{ij}-\bar{y}_i)^2 + \sum_{i=1}^{k}\sum_{j=1}^{n}(\bar{y}_i-\bar{y}_{.})^2 + 2\sum_{i=1}^{k}\sum_{j=1}^{n}(y_{ij}-\bar{y}_i)(\bar{y}_i-\bar{y}_{.})$$

Next, we observe that

$$\sum_{i=1}^{k}\sum_{j=1}^{n}(y_{ij}-\bar{y}_i)(\bar{y}_i-\bar{y}_{.}) = \sum_{i=1}^{k}(\bar{y}_i-\bar{y}_{.})\sum_{j=1}^{n}(y_{ij}-\bar{y}_i) = 0$$

since $\bar{y}_i$ is the mean of the ith sample and, hence, $\sum_{j=1}^{n}(y_{ij}-\bar{y}_i)=0$ for all i. To complete the proof of Theorem 12.1, we have only to observe that the summand of the second sum on the right-hand side of the above identity does not involve the subscript j and that, consequently,

$$\sum_{i=1}^{k}\sum_{j=1}^{n}(\bar{y}_i-\bar{y}_{.})^2 = n\cdot\sum_{i=1}^{k}(\bar{y}_i-\bar{y}_{.})^2$$

It is customary to denote the total sum of squares, the left-hand member of the identity of Theorem 12.1, by SST. The first term on the right-hand side is $\hat{\sigma}_W^2$ times its degrees of freedom, and we refer to this sum as the **error sum of squares**, SSE. The term "error sum of squares" expresses the idea that the quantity estimates random (or chance) error. The second term on the right-hand side of the identity of Theorem 12.1 is $\hat{\sigma}_B^2$ times its degrees of freedom, and we refer to it as the **between-samples sum of squares** or the **treatment sum of squares**, $SS(Tr)$. (Most of the early applications of this kind of analysis were in the field of agriculture, where the k populations represented different **treatments**, such as fertilizers, applied to agricultural plots.) Note that in this notation the F ratio on page 385 can be written

F ratio for treatments

$$F = \frac{SS(Tr)/(k-1)}{SSE/k(n-1)}$$

The sums of squares required for substitution into this last formula are usually obtained by means of the following shortcut formulas, which the reader will be asked to verify in Exercise 12.14 on page 395. We first calculate SST and $SS(Tr)$ by means of the formulas

Sums of squares—equal sample sizes

$$SST = \sum_{i=1}^{k} \sum_{j=1}^{n} y_{ij}^2 - C$$

$$SS(Tr) = \frac{\sum_{i=1}^{k} T_{i\cdot}^2}{n} - C$$

where C, called the correction term, is given by

$$C = \frac{T_{\cdot\cdot}^2}{kn}$$

In these formulas, $T_{i\cdot}$ is the total of the n observations in the ith sample, whereas $T_{\cdot\cdot}$ is the grand total of all kn observations. The error sum of squares, SSE, is then obtained by subtraction; according to Theorem 12.1 we can write

Error sum of squares

$$SSE = SST - SS(Tr)$$

The results obtained in analyzing the total sum of squares into its components are conveniently summarized by means of the following kind of **analysis-of-variance table**:

Source of variation	*Degrees of freedom*	*Sum of squares*	*Mean square*	*F*
Treatments	$k - 1$	$SS(Tr)$	$MS(Tr) = SS(Tr)/(k - 1)$	$\frac{MS(Tr)}{MSE}$
Error	$k(n - 1)$	SSE	$MSE = SSE/k(n - 1)$	
Total	$nk - 1$	SST		

Note that each **mean square** is obtained by dividing the corresponding sum of squares by its degrees of freedom.

EXAMPLE To illustrate the **analysis of variance** (as this technique is appropriately called) for a one-way classification, suppose that in accordance with the layout on page 381 each laboratory measures the tin-coating weights of 12 discs and that the results are as follows:

Laboratory A	*Laboratory B*	*Laboratory C*	*Laboratory D*
0.25	0.18	0.19	0.23
0.27	0.28	0.25	0.30
0.22	0.21	0.27	0.28
0.30	0.23	0.24	0.28
0.27	0.25	0.18	0.24
0.28	0.20	0.26	0.34
0.32	0.27	0.28	0.20
0.24	0.19	0.24	0.18
0.31	0.24	0.25	0.24
0.26	0.22	0.20	0.28
0.21	0.29	0.21	0.22
0.28	0.16	0.19	0.21

Construct an analysis-of-variance table.

Solution The totals for the four samples are, respectively, 3.21, 2.72, 2.76, and 3.00, the grand total is 11.69, and the calculations required to obtain the necessary sums of squares are as follows:

$$C = \frac{(11.69)^2}{48} = 2.8470$$

$$SST = (0.25)^2 + (0.27)^2 + \cdots + (0.21)^2 - 2.8470 = 0.0809$$

$$SS(Tr) = \frac{(3.21)^2 + (2.72)^2 + (2.76)^2 + (3.00)^2}{12} - 2.8470 = 0.0130$$

$$SSE = 0.0809 - 0.0130 = 0.0679$$

Thus, we get the following analysis-of-variance table:

Source of variation	*Degrees of freedom*	*Sum of squares*	*Mean square*	*F*
Laboratories	3	0.0130	0.0043	2.87
Error	44	0.0679	0.0015	
Total	47	0.0809		

Since the value obtained for F exceeds 2.82, the value of $F_{0.05}$ with 3 and 44 degrees of freedom, the null hypothesis can be rejected at the 0.05 level of significance; we conclude that the laboratories are *not* obtaining consistent results.

■

To estimate the parameters μ, α_1, α_2, α_3, and α_4 (or μ_1, μ_2, μ_3, and μ_4), we can use the method of least squares, minimizing

$$\sum_{i=1}^{k} \sum_{j=1}^{n} (y_{ij} - \mu - \alpha_i)^2$$

with respect to μ and the α_i, subject to the restriction that $\sum_{i=1}^{k} \alpha_i = 0$. This may be done by eliminating one of the α's or, better, by using the method of Lagrange multipliers, which is treated in most texts on advanced calculus. In either case we obtain the "intuitively obvious" estimates $\hat{\mu} = \bar{y}_{.}$ and $\hat{\alpha}_i = \bar{y}_i - \bar{y}_{.}$ for $i = 1, 2, \ldots, k$, and the corresponding estimates for the μ_i are given by $\hat{\mu}_i = \bar{y}_i$.

EXAMPLE Estimate the parameters of the one-way classification model for the tin-coating weights given in the preceding example.

Solution For the data from the four laboratories we get

$$\hat{\mu} = \frac{11.69}{48} = 0.244, \qquad \hat{\alpha}_1 = \frac{3.21}{12} - 0.244 = 0.024$$

$$\hat{\alpha}_2 = \frac{2.72}{12} - 0.244 = -0.017, \qquad \hat{\alpha}_3 = \frac{2.76}{12} - 0.244 = -0.014$$

and

$$\hat{\alpha}_4 = \frac{3.00}{12} - 0.244 = 0.006$$

■

The analysis of variance described in this section applies to one-way classifications where each sample has the same number of observations. If this is not the case and the sample sizes are $n_1, n_2, \ldots,$ and n_k we have only to substitute $N = \sum_{i=1}^{k} n_i$ for nk throughout and write the computing formulas for SST and $SS(Tr)$ as

Sums of squares—unequal sample sizes

$$SST = \sum_{i=1}^{k} \sum_{j=1}^{n_i} y_{ij}^2 - C$$

$$SS(Tr) = \sum_{i=1}^{k} \frac{T_i^2}{n_i} - C$$

Otherwise, the procedure is the same as before. (See also Exercise 12.15 on page 395.)

EXAMPLE As part of the investigation of the collapse of the roof of a building, a testing laboratory is given all the available bolts that connected the steel structure at three different positions on the roof. The forces required to shear each of these bolts (coded values) are as follows:

Position 1: 90, 82, 79, 98, 83, 91
Position 2: 105, 89, 93, 104, 89, 95, 86
Position 3: 83, 89, 80, 94

Perform an analysis of variance to test at the 0.05 level of significance whether the differences among the sample means at the three positions are significant.

Solution Using the same steps as in previous chapters for tests of hypotheses, we get

1. *Null hypothesis:* $\mu_1 = \mu_2 = \mu_3$
   Alternative hypothesis: The μ's are not all equal.
2. *Level of significance*: $\alpha = 0.05$
3. *Criterion*: Reject the null hypothesis if $F > 3.74$, the value of $F_{0.05}$ for $k - 1 = 3 - 1 = 2$ and $N - k = 17 - 3 = 14$ degrees of freedom, where F is to be determined by an analysis of variance; otherwise, accept it.
4. *Calculations*: Substituting $n_1 = 6$, $n_2 = 7$, $n_3 = 4$, $N = 17$, $T_1 = 523$, $T_2 = 661$, $T_3 = 346$, $T_. = 1530$, and $\sum \sum y_{ij}^2 = 138{,}638$ into the computing formulas for the sums of squares, we get

$$SST = 138{,}638 - \frac{1530^2}{17} = 938$$

$$SS(Tr) = \frac{523^2}{6} + \frac{661^2}{7} + \frac{346^2}{4} - \frac{1530^2}{17}$$
$$= 234$$

and

$$SSE = 938 - 234 = 704$$

The remainder of the work is shown in the following analysis-of-variance table:

Source of variation	*Degrees of freedom*	*Sum of squares*	*Mean square*	*F*
Positions	2	234	117	2.33
Error	14	704	50.3	
Total	16	938		

5. *Decision*: Since $F = 2.33$ does not exceed 3.74, the value of $F_{0.05}$ for 2 and 14 degrees of freedom, the null hypothesis cannot be rejected; in other words, we cannot conclude that there is a difference in the mean shear strengths of the bolts at the three different positions on the roof.

■

It is instructive to review the manner in which the one-way analysis of variance follows from the decomposition of each observation. Suppose three drying formulas for curing a glue are studied and the following times observed.

Formula A: 13, 10, 8, 11, 8
Formula B: 13, 11, 14, 14
Formula C: 4, 1, 3, 4, 2, 4

Let the grand mean $\bar{y} = T_{.}/N = \sum_{i=1}\sum_{j=1}^{n_i} y_{ij}/N$. Each observation y_{ij} will be decomposed as

$$\underset{\text{observation}}{y_{ij}} = \underset{\text{grand mean}}{\bar{y}} + \underset{\text{deviation due to treatment}}{(\bar{y}_i - \bar{y})} + \underset{\text{error}}{(y_{ij} - \bar{y}_i)}$$

For instance $13 = 8 + (10 - 8) + (13 - 10) = 8 + 2 + 3$. Repeating this decomposition for each observation, we obtain the arrays

$$\underset{\substack{\text{observation}\\ y_{ij}}}{\begin{bmatrix} 13 & 10 & 8 & 11 & 8 & \\ 13 & 11 & 14 & 14 & & \\ 4 & 1 & 3 & 4 & 2 & 4 \end{bmatrix}} = \underset{\substack{\text{grand mean}\\ \bar{y}}}{\begin{bmatrix} 8 & 8 & 8 & 8 & 8 & \\ 8 & 8 & 8 & 8 & & \\ 8 & 8 & 8 & 8 & 8 & 8 \end{bmatrix}}$$

$$+ \underset{\substack{\text{treatment effects}\\ \bar{y}_i - \bar{y}}}{\begin{bmatrix} 2 & 2 & 2 & 2 & 2 & \\ 5 & 5 & 5 & 5 & & \\ -5 & -5 & -5 & -5 & -5 & -5 \end{bmatrix}} + \underset{\substack{\text{error}\\ y_{ij} - \bar{y}_i}}{\begin{bmatrix} 3 & 0 & -2 & 1 & -2 & \\ 0 & -2 & 1 & 1 & & \\ 1 & -2 & 0 & 1 & -1 & 1 \end{bmatrix}}$$

Taking the sum of squares as a measure of variation for the whole array.

$$\textit{treatment sum of squares} = \sum_{i=1}^{k} n_i(\bar{y}_i - \bar{y})^2 = 5(2)^2 + 4(5)^2 + 6(-5)^2 = 270$$

$$\textit{error sum of squares} = \sum_{i=1}^{k}\sum_{j=1}^{n_i}(y_{ij} - \bar{y}_i)^2$$

$$= 3^2 + 0^2 + (-2)^2 + \cdots + (-1)^2 + 1^2 = 32$$

These are the quantities that are entered in the body of the analysis-of-variance table (see Exercise 12.8). Their sum, $302 = 270 + 32$, the total sum of squares, also equals the sum of squared entries in the observation array minus the sum of squares of the entries in the grand mean array.

The decomposition also provides us with another interpretation of the degrees of freedom associated with each sum of squares. In this example, the treatment effects array has only three possibly distinct entries: $\bar{y}_1 - \bar{y}$, $\bar{y}_2 - \bar{y}$, and $\bar{y}_3 - \bar{y}$. Further, the sum of entries is always zero so, for instance, the third value is determined by the first two. Consequently, there are $3 - 1 = 2$ degrees of freedom associated with treatments. In the general case, there are $k - 1$ degrees of freedom.

Among the entries of the error array, each row sums to zero, so 1 degree of freedom is lost for each row. The array has $n_1 + n_2 + n_3 - 3 = 5 + 4 + 6 - 3 = 12$ degrees of freedom. In the general case, there are $n_1 + n_2 + \cdots + n_k - k$ degrees of freedom.

The grand mean array has a single value $\bar{y}$ and hence has 1 degree of freedom, whereas the observation array has $n_1 + n_2 + \cdots + n_k$ possibly distinct entries and hence that number of degrees of freedom. The total sum of squares, based on the difference of these last two arrays, has

$$n_1 + n_2 + \cdots + n_k - 1 = 5 + 4 + 6 - 1 = 14$$

degrees of freedom.

EXERCISES

12.1 An experiment is performed to compare the cleansing action of two detergents, Detergent A and Detergent B. Twenty swatches of cloth are soiled with dirt and grease, each is washed with one of the detergents in an agitator-type machine, and then measured for "whiteness." Criticize the following aspects of the experiment:

(a) The entire experiment is performed with soft water.

(b) Fifteen of the swatches are washed with Detergent A and five with Detergent B.

(c) To accelerate the testing procedure, very hot water and 30-second washing times are used in the experiment.

(d) The "whiteness" readings of all swatches washed with Detergent A are taken first.

12.2 A certain *bon vivant*, wishing to ascertain the cause of his frequent hangovers, conducted the following experiment. On the first night, he drank nothing but whiskey and water; on the second night he drank vodka and water; on the third night he drank gin and water; and on the fourth night he drank rum and water. On each of the following mornings he had a hangover, and he concluded that it was the common factor, the water, that made him ill.

(a) This conclusion is obviously unwarranted, but can you state what principles of sound experimental design are violated?

(b) Give a less obvious example of an experiment having the same shortcoming.

(c) Suppose that our friend had modified his experiment so each of the four alcoholic beverages was used both with and without water, so that the experiment lasted eight nights. Could the results of this enlarged experiment serve to support or refute the hypothesis that water was the cause of the hangovers? Explain.

12.3 To compare the effectiveness of three methods of teaching the programming of a certain computer—Method A, which is straight teaching-machine instruction, Method B, which involves the personal attention of an instructor and some direct experience working with the computer, and Method C, which involves the personal attention of an instructor but no work with the computer—random samples of size four are taken from large groups of persons taught by the three methods and the following are the scores which they obtained in an appropriate achievement test:

Method A	*Method B*	*Method C*
73	91	72
77	81	77
67	87	76
71	85	79

(a) Without using the shortcut formulas, calculate

$$\sum_{i=1}^{k}\sum_{j=1}^{n}(y_{ij}-\bar{y}_{\cdot})^2, \quad \sum_{i=1}^{k}\sum_{j=1}^{n}(y_{ij}-\bar{y}_{i})^2, \quad \text{and} \quad n\cdot\sum_{i=1}^{k}(\bar{y}_i-\bar{y}_{\cdot})^2$$

and verify the identity of Theorem 12.1.

(b) Verify the results obtained for the three sums of squares by using the shortcut formulas on page 387.

12.4 Using the sums of squares obtained in Exercise 12.3, test at the level of significance $\alpha = 0.05$ whether the differences among the means obtained for the three samples are significant.

12.5 The following are the numbers of mistakes made in five successive days by four technicians working for a photographic laboratory:

Technician I	*Technician II*	*Technician III*	*Technician IV*
6	14	10	9
14	9	12	12
10	12	7	8
8	10	15	10
11	14	11	11

Test at the level of significance $\alpha = 0.01$ whether the differences among the four sample means can be attributed to chance.

12.6 The following are the weight losses of certain machine parts (in milligrams) due to friction, when three different lubricants were used under controlled conditions:

Lubricant A:	12.2, 11.8, 13.1, 11.0, 3.9, 4.1, 10.3, 8.4
Lubricant B:	10.9, 5.7, 13.5, 9.4, 11.4, 15.7, 10.8, 14.0
Lubricant C:	12.7, 19.9, 13.6, 11.7, 18.3, 14.3, 22.8, 20.4

(a) Test at the 0.01 level of significance whether the differences among the sample means can be attributed to chance.

(b) Estimate the parameters of the model used in the analysis of this experiment.

12.7 Given the following observations collected according to the one-way analysis of variance design:

Treatment 1:	6, 4, 5
Treatment 2:	13, 10, 13, 12
Treatment 3:	7, 9, 11
Treatment 4:	3, 6, 1, 4, 1

(a) Decompose each observation y_{ij} as

$$y_{ij} = \bar{y} + (\bar{y}_i - \bar{y}) + (y_{ij} - \bar{y}_i)$$

and obtain the sum of squares and degrees of freedom for each component.

(b) Construct the analysis-of-variance table and test the equality of treatments using $\alpha = 0.05$.

12.8 With reference to the example on page 391, construct the analysis-of-variance table and test the equality of the mean curing times using $\alpha = 0.05$.

12.9 To find the best arrangement of instruments on a control panel of an airplane, three different arrangements were tested by simulating an emergency condition and observing the reaction time required to correct the condition. The reaction times (in *tenths* of a second) of 28 pilots (randomly assigned to the different arrangements) were as follows:

Arrangement 1:	14, 13, 9, 15, 11, 13, 14, 11
Arrangement 2:	10, 12, 9, 7, 11, 8, 12, 9, 10, 13, 9, 10
Arrangement 3:	11, 5, 9, 10, 6, 8, 8, 7

Test at the level of significance $\alpha = 0.01$ whether we can reject the null hypothesis that the differences among the arrangements have no effect.

12.10 Several different aluminum alloys are under consideration for use in heavy-duty circuit wiring applications. Among the desired properties is low electrical resistance, and a number of specimens of each wire are tested by applying a fixed voltage to a given length of wire and measuring the current passing through the wire. Given the following results, would you conclude that these alloys differ in resistance? (Use the 0.01 level of significance.)

Alloy	Current (amperes)				
1	1.085	1.016	1.009	1.034	
2	1.051	0.993	1.022		
3	0.985	1.001	0.990	0.988	1.011
4	1.101	1.015			

12.11 Two tests are made of the compressive strength of each of six samples of poured concrete. The force required to crumble each cylindrical specimen, measured in kilograms, is as follows:

	Sample					
	A	*B*	*C*	*D*	*E*	*F*
Test 1	110	125	98	95	104	115
Test 2	105	130	107	92	96	121

Test at the 0.05 level of significance whether these samples differ in compressive strength.

12.12 Referring to the discussion on page 380, assume that the standard deviations of the tin-coating weights determined by any one of the three laboratories have the common value $\sigma = 0.012$, and that it is desired to be 95% confident of detecting a difference in means between any two of the laboratories in excess of 0.01 pound per base box. Show that these assumptions lead to the decision to send a sample of 12 discs to each laboratory.

12.13 Show that if $\mu_i = \mu + \alpha_i$, and μ is the mean of the μ_i, it follows that $\sum_{i=1}^{k} \alpha_i = 0$.

12.14 Verify the shortcut formulas for computing SST and $SS(Tr)$ given on page 387.

12.15 State and prove a theorem analogous to Theorem 12.1 for the case where the size of the ith sample is n_i, that is, where the sample sizes are not necessarily equal.

12.16 Samples of peanut butter produced by two different manufacturers are tested for aflatoxin content, with the following results:

Aflatoxin content (ppb)	
Brand A	Brand B
0.5	4.7
0.0	6.2
3.2	0.0
1.4	10.5
0.0	2.1
1.0	0.8
8.6	
2.9	

(a) Use analysis of variance to test whether the two brands differ in aflatoxin content.

(b) Test the same hypothesis using a two-sample t test.

(c) It can be shown that the t statistic with v degrees of freedom and the F statistic with 1 and v degrees of freedom are related by the formula

$$F(1, v) = t^2(v)$$

where v = degrees of freedom. Using this result, prove that the analysis of variance and two-sample t-test methods are equivalent in this case.

12.17 The one-way analysis of variance is conveniently implemented using *MINITAB*. With reference to the example on page 391, we first set the observations in columns.

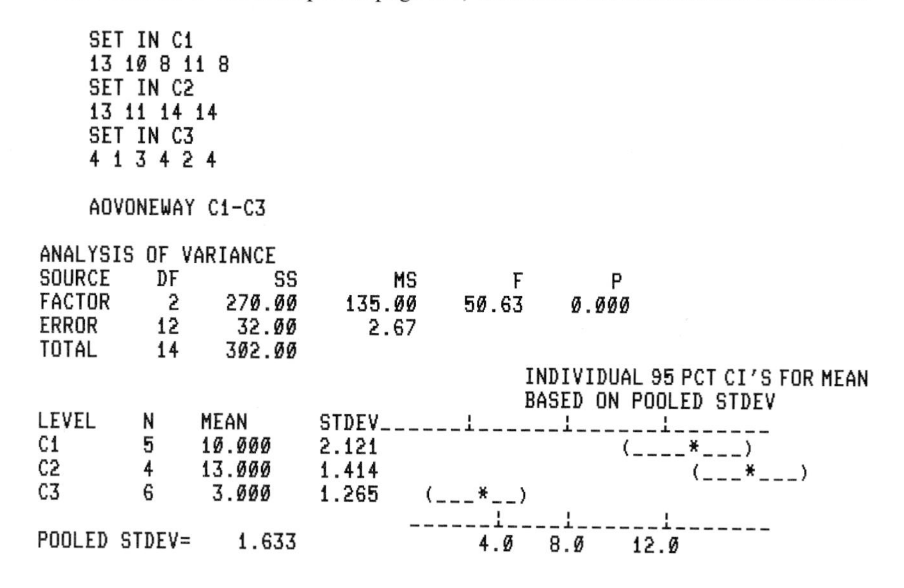

```
SET IN C1
13 10 8 11 8
SET IN C2
13 11 14 14
SET IN C3
4 1 3 4 2 4

AOVONEWAY C1-C3

ANALYSIS OF VARIANCE
SOURCE     DF        SS         MS         F        P
FACTOR      2    270.00     135.00     50.63    0.000
ERROR      12     32.00       2.67
TOTAL      14    302.00
                                       INDIVIDUAL 95 PCT CI'S FOR MEAN
                                       BASED ON POOLED STDEV
LEVEL      N     MEAN      STDEV------I-------I-------I-------
C1         5    10.000     2.121                 (----*---)
C2         4    13.000     1.414                      (---*---)
C3         6     3.000     1.265    (---*--)
                                 -------I-----I--------I-------
POOLED STDEV=    1.633                 4.0   8.0     12.0
```

Use the computer to perform the analysis of variance suggested in Exercise 12.9.

12.3
RANDOMIZED-BLOCK DESIGNS

As we observed in Section 12.1, the estimate of chance variation (the experimental error) can often be reduced, that is, freed of variability due to extraneous causes, by dividing the observations in each classification into blocks. This is accomplished when known sources of variability (that is, extraneous variables) are fixed in each block, but vary from block to block.

In this section we shall suppose that the experimenter has available measurements pertaining to a treatments distributed over b blocks. First, we shall consider the case where there is exactly one observation from each treatment in each block;

with reference to the illustration on page 382, this case would arise if each laboratory tested one disc from each strip. Letting y_{ij} denote the observation pertaining to the ith treatment and the jth block, $\bar{y}_{i.}$ the mean of the b observations for the ith treatment, $\bar{y}_{.j}$ the mean of the a observations in the jth block, and $\bar{y}_{..}$ the grand mean of all the ab observations, we shall use the following layout for this kind of **two-way classification**:

	Blocks	
	$B_1 \quad B_2 \quad \cdots \quad B_j \quad \cdots \quad B_b$	*Means*
Treatment 1:	$y_{11}, y_{12}, \ldots, y_{1j}, \ldots, y_{1b}$	$\bar{y}_{1.}$
Treatment 2:	$y_{21}, y_{22}, \ldots, y_{2j}, \ldots, y_{2b}$	$\bar{y}_{2.}$
$\vdots$	$\vdots$	$\vdots$
Treatment i:	$y_{i1}, y_{i2}, \ldots, y_{ij}, \ldots, y_{ib}$	$\bar{y}_{i.}$
$\vdots$	$\vdots$	$\vdots$
Treatment a:	$y_{a1}\; y_{a2}, \ldots, y_{aj}, \ldots, y_{ab}$	$\bar{y}_{a.}$
Means	$\bar{y}_{.1} \quad \bar{y}_{.2} \quad \cdots \quad \bar{y}_{.j} \quad \cdots \quad \bar{y}_{.b}$	$\bar{y}_{..}$

This kind of arrangement is also called a **randomized-block design**, provided the treatments are allocated at random *within* each block. Note that when a dot is used in place of a subscript, this means that the mean is obtained by summing over that subscript.

The underlying model which we shall assume for the analysis of this kind of experiment with one observation per "cell" (that is, there is one observation corresponding to each treatment within each block) is given by

Model equation for randomized-block design

$$y_{ij} = \mu + \alpha_i + \beta_j + \varepsilon_{ij} \qquad \text{for } i = 1, 2, \ldots, a;\ j = 1, 2, \ldots, b$$

Here μ is the grand mean, α_i is the effect of the ith treatment, β_j is the effect of the jth block, and the ε_{ij} are values of *independent, normally distributed* random variables having *zero means* and the *common variance* σ^2. Analogous to the model for the one-way classification, we restrict the parameters by imposing the conditions that $\sum_{i=1}^{a} \alpha_i = 0$ and $\sum_{j=1}^{b} \beta_j = 0$ (see Exercise 12.27 on page 409).

In the analysis of a two-way classification where each treatment is represented once in each block, the major objective is to test for the significance of the difference among the $\bar{y}_{i.}$, that is, to test the null hypothesis

$$\alpha_1 = \alpha_2 = \cdots = \alpha_a = 0$$

In addition, it may also be desirable to test whether the blocking has been effective, that is, whether the null hypothesis

$$\beta_1 = \beta_2 = \cdots = \beta_b = 0$$

can be rejected. In either case, the alternative hypothesis is that at least one of the effects is different from zero.

As in the one-way analysis of variance, we shall base these significance tests on comparisons of estimates of σ^2—one based on the variation among treatments, one based on the variation among blocks, and one measuring the experimental error. Note that only the latter is an estimate of σ^2 when either (or both) of the null hypotheses do not hold. The required sums of squares are given by the three components into which the total sum of squares is partitioned by means of the following theorem:

Identity for analysis of two-way classification

Theorem 12.2

$$\sum_{i=1}^{a}\sum_{j=1}^{b}(y_{ij}-\bar{y}_{..})^2 = \sum_{i=1}^{a}\sum_{j=1}^{b}(y_{ij}-\bar{y}_{i.}-\bar{y}_{.j}+\bar{y}_{..})^2 + b\sum_{i=1}^{a}(\bar{y}_{i.}-\bar{y}_{..})^2 + a\sum_{j=1}^{b}(\bar{y}_{.j}-\bar{y}_{..})^2$$

The left-hand side of this identity represents the total sum of squares, SST, and the terms on the right-hand side are, respectively, the error sum of squares, SSE, the treatment sum of squares, $SS(Tr)$, and the **block sum of squares** $SS(Bl)$. To prove this theorem, we make use of the identity

$$y_{ij}-\bar{y}_{..} = (y_{ij}-\bar{y}_{i.}-\bar{y}_{.j}+\bar{y}_{..}) + (\bar{y}_{i.}-\bar{y}_{..}) + (\bar{y}_{.j}-\bar{y}_{..})$$

and follow essentially the same argument as in the proof of Theorem 12.1.

Convenient formulas are available to calculate SST, $SS(Tr)$, and $SS(Bl)$ using hand held calculators.

Sums of squares for two-way analysis of variance

$$SST = \sum_{i=1}^{a}\sum_{j=1}^{b} y_{ij}^2 - C$$

$$SS(Tr) = \frac{\sum_{i=1}^{a} T_{i.}^2}{b} - C$$

$$SS(Bl) = \frac{\sum_{j=1}^{b} T_{.j}^2}{a} - C$$

where C, the correction term, is given by

$$C = \frac{T_{..}^2}{ab}$$

In these formulas $T_{i.}$ is the sum of the b observations for the ith treatment, $T_{.j}$ is the sum of the a observations in the jth block, and $T_{..}$ is the grand total of all the observations. Note that the divisors for $SS(Tr)$ and $SS(Bl)$ are the number of observations in the respective totals, $T_{i.}$ and $T_{.j}$. The error sum of squares is then obtained by subtraction; according to Theorem 12.2 we can write

Error sum of squares

$$\boldsymbol{SSE = SST - SS(Tr) - SS(Bl)}$$

In Exercise 12.28 on page 409, the reader will be asked to verify that all these computing formulas are, indeed, equivalent to the corresponding terms of the identity of Theorem 12.2.

Using these sums of squares, we can reject the null hypothesis that the α_i are all equal to zero at the level of significance α if

F-ratio for treatments

$$F_{Tr} = \frac{MS(Tr)}{MSE} = \frac{SS(Tr)/(a-1)}{SSE/(a-1)(b-1)}$$

exceeds F_α with $a - 1$ and $(a-1)(b-1)$ degrees of freedom. The null hypothesis that the β_j are all equal to zero can be rejected at the level of significance α if

F-ratio for blocks

$$F_{Bl} = \frac{MS(Bl)}{MSE} = \frac{SS(Bl)/(b-1)}{SSE/(a-1)(b-1)}$$

exceeds F_α with $b - 1$ and $(a-1)(b-1)$ degrees of freedom. Note that the mean squares, $MS(Tr)$, $MS(Bl)$, and MSE, are again defined as the corresponding sums of squares divided by their degrees of freedom.

The results obtained in this analysis are summarized in the following analysis-of-variance table:

Source of variation	*Degrees of freedom*	*Sum of squares*	*Mean square*	*F*
Treatments	$a-1$	$SS(Tr)$	$MS(Tr) = \frac{SS(Tr)}{(a-1)}$	$F_{Tr} = \frac{MS(Tr)}{MSE}$
Blocks	$b-1$	$SS(Bl)$	$MS(Bl) = \frac{SS(Bl)}{(b-1)}$	$F_{Bl} = \frac{MS(Bl)}{MSE}$
Error	$(a-1)(b-1)$	SSE	$MSE = \frac{SSE}{(a-1)(b-1)}$	
Total	$ab-1$	SST		

It is instructive to see how the analysis of variance for a randomized block experiment follows from the decomposition of each observation. Given the observations

	Blocks			
Treatment 1:	13	7	9	3
Treatment 2:	6	6	3	1
Treatment 3:	11	5	15	5

Each observation y_{ij} will be decomposed as

$$\underset{\text{observation}}{y_{ij}} = \underset{\substack{\text{grand}\\ \text{mean}}}{\bar{y}_{..}} + \underset{\substack{\text{deviation}\\ \text{due to}\\ \text{treatment}}}{(\bar{y}_{i.} - \bar{y}_{..})} + \underset{\substack{\text{deviation}\\ \text{due to}\\ \text{blocks}}}{(\bar{y}_{.j} - \bar{y}_{..})} + \underset{\text{error}}{(y_{ij} - \bar{y}_{i.} - \bar{y}_{.j} + \bar{y}_{..})}$$

For instance $13 = 7 + (8 - 7) + (10 - 7) + (13 - 8 - 10 + 7) = 7 + 1 + 3 + 2$. Repeating this decomposition for each observation, we obtain the arrays

$$\underset{\substack{\text{observation}\\ y_{ij}}}{\begin{bmatrix} 13 & 7 & 9 & 3 \\ 6 & 6 & 3 & 1 \\ 11 & 5 & 15 & 5 \end{bmatrix}} = \underset{\substack{\text{mean}\\ \bar{y}_{..}}}{\begin{bmatrix} 7 & 7 & 7 & 7 \\ 7 & 7 & 7 & 7 \\ 7 & 7 & 7 & 7 \end{bmatrix}} + \underset{\substack{\text{treatment}\\ \bar{y}_{i.} - \bar{y}_{..}}}{\begin{bmatrix} 1 & 1 & 1 & 1 \\ -3 & -3 & -3 & -3 \\ 2 & 2 & 2 & 2 \end{bmatrix}}$$

$$+ \underset{\substack{\text{block}\\ \bar{y}_{.j} - \bar{y}_{..}}}{\begin{bmatrix} 3 & -1 & 2 & -4 \\ 3 & -1 & 2 & -4 \\ 3 & -1 & 2 & -4 \end{bmatrix}} + \underset{\substack{\text{error}\\ y_{ij} - \bar{y}_{i.} - \bar{y}_{.j} + \bar{y}_{..}}}{\begin{bmatrix} 2 & 0 & -1 & -1 \\ -1 & 3 & -3 & 1 \\ -1 & -3 & 4 & 0 \end{bmatrix}}$$

Taking the sum of squares for each array,

$$\textit{treatment sum of squares} = b \sum_{i=1}^{a} (\bar{y}_{i.} - \bar{y}_{..})^2 = 4(1)^2 + 4(-3)^2 + 4(2)^2 = 56$$

$$\textit{block sum of squares} = a \sum_{j=1}^{b} (\bar{y}_{.j} - \bar{y}_{..})^2$$

$$= 3(3)^2 + 3(-1)^2 + 3(2)^2 + 3(-4)^2 = 90$$

$$\textit{error sum of squares} = \sum_{i=1}^{a} \sum_{j=1}^{b} (y_{ij} - \bar{y}_{i.} - \bar{y}_{.j} + \bar{y}_{..})^2$$

$$= 2^2 + 0^2 + \cdots + 4^2 + 0^2 = 52$$

we obtain the entries in the body of the analysis of variance table. Their sum, $90 + 56 + 52 = 198$, the total sum of squares, is also equal to the sum of squares of the observations, 786, minus the sum of squares $12 \times 7^2 = 588$ for the grand mean array.

The mean array, which has a single entry $\bar{y}_{..}$, has 1 degree of freedom. The 3 distinct values in the treatment array always sum to zero, so it has $3 - 1 = 2$ degrees of freedom. In general, it has $a - 1$ degrees of freedom when there are a treatments. Similarly, the block array has $4 - 1 = 3$ degrees of freedom in this example and $b - 1$ when there are b blocks.

The number of degrees of freedom associated with the error array is $(a - 1)(b - 1) = (2)(3) = 6$. Because every row sum is zero, the last column is always determined from the first $b - 1$. Similarly, the last row is always determined by the first $a - 1$. Thus there are $(a - 1)(b - 1)$ unconstrained entries. In summary, the degrees of freedom can be decomposed as

$$\underset{\textit{total}}{ab - 1} = \underset{\textit{treatment}}{(a - 1)} + \underset{\textit{blocks}}{(b - 1)} + \underset{\textit{error}}{(a - 1)(b - 1)}$$

In practice, for hand calculations, we calculate the necessary sums of squares using shortcut formulas on page 398.

EXAMPLE Construct the analysis of variance from the decomposition of the observations just given.

Solution Using the sums of squares and their associated degrees of freedom, we have

Source of variation	*Degrees of freedom*	*Sum of squares*	*Mean square*	*F*
Treatments	2	56.0	28.00	3.23
Blocks	3	90.0	30.00	3.46
Error	6	52.0	8.67	
Total	11	198.0		

■

EXAMPLE An experiment was designed to study the performance of four different detergents. The following "whiteness" readings were obtained with specially designed equipment for 12 loads of washing distributed over three different models of washing machines:

	Machine 1	*Machine 2*	*Machine 3*	*Totals*
Detergent A	45	43	51	139
Detergent B	47	46	52	145
Detergent C	48	50	55	153
Detergent D	42	37	49	128
Totals	182	176	207	565

Looking on the detergents as treatments and the machines as blocks, obtain the appropriate analysis-of-variance table and test at the 0.01 level of significance whether there are differences in the detergents or in the washing machines.

Solution

1. *Null hypotheses*: $\alpha_1 = \alpha_2 = \alpha_3 = \alpha_4 = 0$; $\beta_1 = \beta_2 = \beta_3 = 0$
   Alternative hypotheses: The α's are not all equal to zero; the β's are not all equal to zero.
2. *Level of significance*: $\alpha = 0.01$
3. *Criteria*: For treatments, reject the null hypothesis if $F > 9.78$, the value of $F_{0.01}$ with $a - 1 = 4 - 1 = 3$ and $(a - 1)(b - 1) = (4 - 1)(3 - 1) = 6$ degrees of freedom; for blocks, reject the null hypothesis if $F > 10.9$, the value of $F_{0.01}$ for $b - 1 = 3 - 1 = 2$ and $(a - 1)(b - 1) = (4 - 1)(3 - 1) = 6$ degrees of freedom.
4. *Calculations*: Substituting $a = 4$, $b = 3$, $T_{1.} = 139$, $T_{2.} = 145$, $T_{3.} = 153$, $T_{4.} = 128$, $T_{.1} = 182$, $T_{.2} = 176$, $T_{.3} = 207$, $T_{..} = 565$, and $\sum \sum y_{ij}^2 = 26{,}867$ into the formulas for the sums of squares, we get

$$C = \frac{(565)^2}{12} = 26{,}602$$

$$SST = 45^2 + 43^2 + \cdots + 49^2 = 26{,}867 - 26{,}602 = 265$$

$$SS(Tr) = \frac{139^2 + 145^2 + 153^2 + 128^2}{3} - 26{,}602 = 111$$

$$SS(Bl) = \frac{182^2 + 176^2 + 207^2}{4} - 26{,}602 = 135$$

$$SSE = 265 - 111 - 135 = 19$$

Then dividing the sums of squares by their respective degrees of freedom to obtain the appropriate mean squares, we get the results shown in the following analysis-of-variance table:

Source of variation	*Degrees of freedom*	*Sum of squares*	*Mean square*	*F*
Detergents	3	111	37.0	11.6
Machines	2	135	67.5	21.1
Error	6	19	3.2	
Total	11	265		

5. *Decisions*: Since $F_{Tr} = 11.6$ exceeds 9.78, the value of $F_{0.01}$ with 3 and 6 degrees of freedom, we conclude that there are differences in the effectiveness of the four detergents. Also, since $F_{Bl} = 21.1$ exceeds 10.9, the value of $F_{0.01}$ with 2 and 6 degrees of freedom, we conclude that the differences among the results obtained for the three washing machines are significant, namely, that the blocking has been effective. To make the effect of this blocking even more evident, the reader will be asked to verify in Exercise 12.23 on page 408 that the test for differences among the detergents would *not* yield significant results if we looked upon the data as a one-way classification.

■

The effect of the *i*th detergent can be estimated by means of the formula $\hat{\alpha}_i = \bar{y}_{i.} - \bar{y}_{..}$, which may be obtained by the method of least squares. The resulting estimates are

$$\hat{\alpha}_1 = 46.3 - 47.1 = -0.8, \qquad \hat{\alpha}_2 = 48.3 - 47.1 = 1.2,$$

$$\hat{\alpha}_3 = 51.0 - 47.1 = 3.9, \qquad \hat{\alpha}_4 = 42.7 - 47.1 = -4.4$$

Similar calculations lead to $\hat{\beta}_1 = -1.6$, $\hat{\beta}_2 = -3.1$, and $\hat{\beta}_3 = 4.7$ for the estimated effects of the washing machines.

It should be observed that a two-way classification automatically allows for repetitions of the experimental conditions; for example, in the preceding experiment each detergent was tested three times. Further repetitions may be handled in several ways, and care must be taken that the model used appropriately describes the situation. One way to provide further repetition in a two-way classification is to include additional blocks—for example, to test each detergent using several additional washing machines, randomizing the order of testing for each machine.

Note that the model remains essentially the same as before, the only change being an increase in b, and a corresponding increase in the degrees of freedom for blocks and for error. The latter is important, because an increase in the degrees of freedom for error makes the test of the null hypothesis $\alpha_i = 0$ for all i *more sensitive* to small differences among the treatment means. In fact, the real purpose of this kind of repetition is to increase the degrees of freedom for error, thereby increasing the sensitivity of the F tests (see Exercise 12.26 on page 409).

A second method is to repeat the entire experiment, using a new pattern of randomization to obtain $a \cdot b$ additional observations. This is possible only if the blocks are *identifiable*, that is, if the conditions defining each block can be repeated. For example, in the tin-coating weight experiment described in Section 12.1, the blocks are strips across the rolling direction of a sheet of tin plate, and, given a new sheet, it is possible to identify which is strip 1, which is strip 2, and so forth. In the example of this section, this kind of repetition (usually called **replication**) would require that the settings of the machines be exactly duplicated. This kind of repetition will be used in connection with Latin-square designs in Section 12.5; see also Exercises 12.24 and 12.25 on page 409.

A third method of repetition is to include n observations for each treatment in each block. When an experiment is designed in this way, the n observations in each "cell" are regarded as duplicates, and it is to be expected that their variability will be somewhat less than experimental error. To illustrate this point, suppose that the tin-coating weights of three discs from adjacent positions in a strip are measured in sequence by one of the laboratories, using the same chemical solutions. The variability of these measurements will probably be considerably less than that of three discs from the same strip measured in that laboratory at different times, using different chemical solutions, and perhaps different technicians. The analysis-of-variance appropriate for this kind of repetition reduces essentially to a two-way analysis of variance applied to the *means* of the n duplicates in the $a \cdot b$ cells; thus, *there would be no gain in degrees of freedom for error*, and, consequently, *no gain in sensitivity of the F tests.* It can be expected, however, that there will be some reduction in the error mean square, since it now measures the residual variance of the *means* of several observations.

12.4 MULTIPLE COMPARISONS

The F tests used so far in this chapter showed whether differences among several means are significant, but they did not tell us whether a given mean (or group of means) differs significantly from another given mean (or group of means). In actual practice, the latter is the kind of information an investigator really wants; for instance, having determined on page 388 that the means of the tin-coating weights obtained by the four laboratories differ significantly, it may be important to find out which laboratory (or laboratories) differs from which others.

If an experimenter is confronted with k means, it may seem reasonable at first to test for significant differences between all possible pairs, that is, to perform

$$\binom{k}{2} = \frac{k(k-1)}{2}$$

two-sample t tests as described on page 245. Aside from the fact that this would require a large number of tests even if k is relatively small, these tests would not be independent, and it would be virtually impossible to assign an overall level of significance to this procedure.

Several **multiple-comparisons tests** have been proposed to overcome these difficulties, among them the **Duncan multiple-range test**, which we shall study in this section. (References to other multiple-comparisons tests are mentioned in the book by H. Scheffe listed in the bibliography.) The assumptions underlying the Duncan multiple-range test are, essentially, those of the one-way analysis of variance for which the sample sizes are equal. The test compares the range of any set of p means with an appropriate **least significant range**, R_p, given by

Least significant range

$$R_p = s_{\bar{x}} \cdot r_p$$

Here $s_{\bar{x}}$ is an estimate of $\sigma_{\bar{x}} = \sigma/\sqrt{n}$, and it is computed by means of the formula

Standard error of the mean

$$s_{\bar{x}} = \sqrt{\frac{MSE}{n}}$$

where MSE is the error mean square in the analysis of variance. The value of r_p depends on the desired level of significance and the number of degrees of freedom corresponding to MSE, and it may be obtained from Tables 12(a) and (b) for $\alpha = 0.05$ and 0.01, for $p = 2, 3, \ldots, 10$, and for various degrees of freedom from 1 to 120.

EXAMPLE Referring to the tin-coating weight data on page 388, perform a Duncan multiple-range test to determine which laboratory means differ from which others, using the 0.05 level of significance.

Solution First we arrange the four sample means as follows in an increasing order of magnitude:

Laboratory	*B*	*C*	*D*	*A*
Mean	0.227	0.230	0.250	0.268

Next, we compute $s_{\bar{x}}$, using the error mean square of 0.0015 in the analysis of variance on page 388, and we obtain

$$s_{\bar{x}} = \sqrt{\frac{0.0015}{12}} = 0.011$$

Then, we get (by linear interpolation) from Table 12(a) the following values of r_p for $\alpha = 0.05$ and 44 degrees of freedom:

p	2	3	4
r_p	2.85	3.00	3.09

Multiplying each value of r_p by $s_{\bar{x}} = 0.011$, we finally obtain

p	2	3	4
R_p	0.031	0.033	0.034

The range of *all four means* is $0.268 - 0.227 = 0.041$, which exceeds $R_4 = 0.034$, the least significant range. This result should have been expected, since the F test on page 388 showed that the differences among all four means are significant at $\alpha = 0.05$. To test for significant differences among *three adjacent means*, we obtain ranges of 0.038 and 0.023, respectively, for 0.230, 0.250, 0.268 and 0.227, 0.230, 0.250. Since the first of these values exceeds $R_3 = 0.033$, the differences observed in the first set are significant; since the second value does not exceed 0.033, the corresponding differences are not significant. Finally, for *adjacent pairs* of means we find that no adjacent pair has a range greater than the least significant range $R_2 = 0.031$. All these results can be summarized by writing

0.227 0.230 0.250 0.268

where *a line is drawn under any set of adjacent means for which the range is less than the appropriate value of* R_p, *that is, under any set of adjacent means for which differences are not significant.* We thus conclude in our example that Laboratory A averages higher tin-coating weights than Laboratories B and C.

■

If we apply this same method to the example of Section 12.3, where we compared the four detergents, we obtain (see also Exercise 12.30 on page 409)

Detergent

D	*A*	*B*	*C*
42.7	46.3	48.3	51.0

In other words, among triplets of adjacent means both sets of differences are significant. So far as pairs of adjacent means are concerned, we find that only the difference between 42.7 and 46.3 is significant. Interpreting these results, we conclude that detergent D is significantly inferior to any of the others, and detergent A is significantly inferior to detergent C.

EXERCISES

12.18 A laboratory technician measures the breaking strength of each of five kinds of linen threads by means of four different instruments, and obtains the following results (in ounces):

	Measuring instrument I_1	I_2	I_3	I_4
Thread 1	20.6	20.7	20.0	21.4
Thread 2	24.7	26.5	27.1	24.3
Thread 3	25.2	23.4	21.6	23.9
Thread 4	24.5	21.5	23.6	25.2
Thread 5	19.3	21.5	22.2	20.6

Looking upon the threads as treatments and the instruments as blocks, perform an analysis of variance at the level of significance $\alpha = 0.01$.

12.19 Looking at the days (rows) as blocks, rework Exercise 12.5 on page 393 by the method of Section 12.3.

12.20 Four different, though supposedly equivalent, forms of a standardized reading achievement test were given to each of five students, and the following are the scores which they obtained:

	Student 1	*Student 2*	*Student 3*	*Student 4*	*Student 5*
Form A	75	73	59	69	84
Form B	83	72	56	70	92
Form C	86	61	53	72	88
Form D	73	67	62	79	95

Perform a two-way analysis of variance to test at the level of significance $\alpha = 0.01$ whether it is reasonable to treat the four forms as equivalent.

12.21 Given the observations

	Blocks				
Treatment 1:	14	6	11	0	9
Treatment 2:	14	10	16	9	16
Treatment 3:	12	7	10	9	12
Treatment 4:	12	9	11	6	7

(a) Decompose each observation y_{ij} as

$$y_{ij} = \bar{y}_{..} + (\bar{y}_{i.} - \bar{y}_{..}) + (\bar{y}_{.j} - \bar{y}_{..}) + (y_{ij} - \bar{y}_{i.} - \bar{y}_{.j} + \bar{y}_{..})$$

(b) Obtain the sums of squares and degrees of freedom for each component.

(c) Construct the analysis-of-variance table and test for differences among the treatments using $\alpha = 0.05$.

12.22 An industrial engineer tests four different shop-floor layouts by having each of six work crews construct a subassembly and measuring the construction times (minutes) as follows:

	Layout 1	*Layout 2*	*Layout 3*	*Layout 4*
Crew A	48.2	53.1	51.2	58.6
Crew B	49.5	52.9	50.0	60.1
Crew C	50.7	56.8	49.9	62.4
Crew D	48.6	50.6	47.5	57.5
Crew E	47.1	51.8	49.1	55.3
Crew F	52.4	57.2	53.5	61.7

Test at the 0.01 level of significance whether the four floor layouts produce different assembly times and whether some of the work crews are consistently faster in constructing this subassembly than the others.

12.23 To emphasize the importance of "blocking," reanalyze the "whiteness" data on page 402 as a one-way classification with the four detergents being the different treatments.

12.24 If, in a two-way classification, the entire experiment is repeated r times, the model becomes

$$y_{ijk} = \mu + \alpha_i + \beta_j + \rho_k + \varepsilon_{ijk}$$

for $i = 1, 2, \ldots, a, j = 1, 2, \ldots, b$, and $k = 1, 2, \ldots, r$, where the sum of the α's, the sum of the β's, and the sum of the ρ's are equal to zero, and where the ρ's represent the effects of the replications. The ε_{ijk} are again values of independent normally distributed random variables with zero means and the common variance σ^2.

(a) Write down (but do not prove) an identity analogous to the one of Theorem 12.2, subdividing the total sum of squares into components attributable to treatments, blocks, replications, and error.

(b) Generalize the computing formulas on page 398 so that they apply to a replicated randomized-block design. Note that the divisor in each case equals the number of observations in the respective totals.

(c) If the number of degrees of freedom for the replicate sum of squares is $r - 1$, how many degrees of freedom are there for the error sum of squares?

12.25 The following are the number of defectives produced by four workers operating, in turn, three different machines; in each case, the first figure represents the number of defectives produced on a Friday and the second figure represents the number of defectives produced on the following Monday:

	Worker B_1	B_2	B_3	B_4
Machine A_1	37, 43	38, 44	38, 40	32, 36
Machine A_2	31, 36	40, 44	43, 41	31, 38
Machine A_3	36, 40	33, 37	41, 39	38, 45

Use the theory developed in Exercise 12.24 to analyze the combined figures for the two days as a two-way classification with replication. Use the level of significance $\alpha = 0.05$.

12.26 As was pointed out on page 403, two ways of increasing the size of a two-way classification experiment are (a) to double the number of blocks, and (b) to replicate the entire experiment. Discuss and compare the gain in degrees of freedom for the error sum of squares by the two methods.

12.27 Show that if $\mu_{ij} = \mu + \alpha_i + \beta_j$, the mean of the μ_{ij} (summed on j) is equal to $\mu + \alpha_i$, and the mean of the μ_{ij} (summed on i and j) is equal to μ, it follows that $\sum_{i=1}^{a} \alpha_i = \sum_{j=1}^{b} \beta_j = 0$.

12.28 Verify that the computing formulas for SST, $SS(Tr)$, $SS(Bl)$, and SSE, given on page 398, are equivalent to the corresponding terms of the identity of Theorem 12.2.

12.29 Use the Duncan test with $\alpha = 0.05$ to test for differences among the treatments in exercise 12.21 on page 408.

12.30 Verify the results of Duncan's test for the comparison of the four detergents, given on page 406.

12.31 Use the Duncan test with $\alpha = 0.05$ to compare the effectiveness of the three methods of teaching the programming of the computer in Exercise 12.3 on page 393.

12.32 Use the Duncan test with $\alpha = 0.05$ to compare the effectiveness of the three lubricants in Exercise 12.6 on page 394.

12.33 Use the Duncan test with $\alpha = 0.01$ to compare the strength of the five linen threads in Exercise 12.18 on page 407.

12.34 Determine which of the shop-floor layouts of Exercise 12.22 on page 408 (if any) differ from the others in the time it takes the average crew to construct the given subassembly. (Use the 0.01 level of significance.)

12.5
SOME FURTHER EXPERIMENTAL DESIGNS

The randomized-block design of Section 12.4 is appropriate when one extraneous source of variability is to be eliminated in comparing a set of sample means. An important feature of this kind of design is its **balance**, achieved by assigning the same number of observations for each treatment to each block. (In this connection, see also the comment on page 382, where we pointed out that differences due to blocks will not affect the means obtained for the different treatments.) The same kind of balance can be attained in more complicated kinds of designs, where it is desired to eliminate the effect of several extraneous sources of variability. In this section we shall introduce two further balanced designs, the Latin-square design and the Graeco-Latin-square design, which are used to eliminate the effects of two and three extraneous sources of variability, respectively.

To introduce the **Latin-square design**, suppose that it is desired to compare three treatments, A, B, and C, in the presence of two other sources of variability. For example, the three treatments may be three methods for soldering copper electrical leads, and the two extraneous sources of variability may be (1) different operators doing the soldering, and (2) the use of different solder fluxes. If three operators and three fluxes are to be considered, the experiment might be arranged in the following pattern:

	Flux 1	*Flux 2*	*Flux 3*
Operator 1	A	B	C
Operator 2	C	A	B
Operator 3	B	C	A

Here each soldering method is applied once by each operator in conjunction with each flux, and if there are systematic effects due to differences among operators or

4 X 4

A	B	C	D
B	C	D	A
C	D	A	B
D	A	B	C

5 X 5

A	B	C	D	E
B	A	E	C	D
C	D	A	E	B
D	E	B	A	C
E	C	D	B	A

FIGURE 12.2
Latin squares.

differences among fluxes, these effects are present equally for each treatment, that is, for each method of soldering.

An experimental arrangement such as the one just described is called a **Latin square**. An $n \times n$ Latin square is a square array of n distinct letters, with each letter appearing once and only once in each row and in each column. Examples of Latin squares with $n = 4$ and $n = 5$ are shown in Figure 12.2, and larger ones may be obtained from the book by W. G. Cochran and G. M. Cox mentioned in the bibliography. Note that in a Latin-square experiment involving n treatments, it is necessary to include n^2 observations, n for each treatment.

As we shall see on page 413, a Latin-square experiment without replication provides only $(n-1)(n-2)$ degrees of freedom for estimating the experimental error. Thus, such experiments are rarely run without replication if n is small, that is, without repeating the entire Latin-square pattern several times. If there is a total of r replicates, the analysis of the data presumes the following model, where $y_{ij(k)l}$ is the observation in the ith row and the jth column of the lth replicate, and the subscript k, in parentheses, indicates that it pertains to the kth treatment:

Model equation for Latin square

$$y_{ij(k)l} = \mu + \alpha_i + \beta_j + \gamma_k + \rho_l + \varepsilon_{ij(k)l}$$

for $i, j, k = 1, 2, \ldots, n$, and $l = 1, 2, \ldots, r$, subject to the restrictions that

$$\sum_{i=1}^{n} \alpha_i = 0, \quad \sum_{j=1}^{n} \beta_j = 0, \quad \sum_{k=1}^{n} \gamma_k = 0 \quad \text{and} \quad \sum_{l=1}^{r} \rho_l = 0$$

Here μ is the grand mean, α_i is the effect of the ith row, β_j is the effect of the jth column, γ_k is the effect of the kth treatment, ρ_l is the effect of the lth replicate, and the $\varepsilon_{ij(k)l}$ are values of independent, normally distributed random variables with zero means and the common variance σ^2. Note that by "effects of the rows" and "effects of the columns" we mean the effects of the two extraneous variables and that we are including replicate effects, since, as we shall see, replication can introduce a third extraneous variable. Note also that the subscript k is in parentheses in $y_{ij(k)l}$ because, for a given Latin-square design, k is automatically determined when i and j are known.

The main hypothesis we shall want to test is the null hypothesis $\gamma_k = 0$ for all k, namely, the null hypothesis that there is no difference in the effectiveness of the n treatments. However, we can also test whether the "cross blocking" of the Latin-square design has been effective; that is, we can test the two null hypotheses $\alpha_i = 0$ for all i and $\beta_j = 0$ for all j (against suitable alternatives) to see whether the two extraneous variables actually have an effect on the phenomenon under consideration. Furthermore, we can test the null hypothesis $\rho_l = 0$ for all l against the alternative that the ρ_l are not all equal to zero, and this test for effects of the replicates may be important if the parts of the experiment representing the individual Latin squares were performed on different days, by different technicians, at different temperatures, and so on.

The sums of squares required to perform these tests are usually obtained by means of the following shortcut formulas, where $T_{i..}$ is the total of the $r \cdot n$ observations in all of the ith rows, $T_{.j.}$ is the total of the $r \cdot n$ observations in all of the jth columns, $T_{..l}$ is the total of the n^2 observations in the lth replicate, $T_{(k)}$ is the total of all the $r \cdot n$ observations pertaining to the kth treatment, and $T_{...}$ is the grand total of all the $r \cdot n^2$ observations:

Sums of squares—Latin square

$$C = \frac{(T_{...})^2}{r \cdot n^2}$$

$$SS(Tr) = \frac{1}{r \cdot n} \sum_{k=1}^{n} T_{(k)}^2 - C$$

$$SSR = \frac{1}{r \cdot n} \sum_{i=1}^{n} T_{i..}^2 - C \qquad (\textit{for rows})$$

$$SSC = \frac{1}{r \cdot n} \sum_{j=1}^{n} T_{.j.}^2 - C \qquad (\textit{for columns})$$

$$SS(Rep) = \frac{1}{n^2} \sum_{l=1}^{r} T_{..l}^2 - C \qquad (\textit{for replicates})$$

$$SST = \sum_{i=1}^{n} \sum_{j=1}^{n} \sum_{l=1}^{r} y_{ij(k)l}^2 - C$$

$$SSE = SST - SS(Tr) - SSR - SSC - SS(Rep)$$

Note that again each divisor equals the number of observations in the corresponding squared totals. Finally, the results of the analysis are as shown in the following analysis-of-variance table:

Source of variation	Degrees of freedom	Sum of squares	Mean square	F
Treatments	$n-1$	$SS(Tr)$	$MS(Tr) = \frac{SS(Tr)}{n-1}$	$\frac{MS(Tr)}{MSE}$
Rows	$n-1$	SSR	$MSR = \frac{SSR}{n-1}$	$\frac{MSR}{MSE}$
Columns	$n-1$	SSC	$MSC = \frac{SSC}{n-1}$	$\frac{MSC}{MSE}$
Replicates	$r-1$	$SS(Rep)$	$MS(Rep) = \frac{SS(Rep)}{r-1}$	$\frac{MS(Rep)}{MSE}$
Error	$(n-1)(rn+r-3)$	SSE	$MSE = \frac{SSE}{(n-1)(rn+r-3)}$	
Total	rn^2-1	SST		

As before, the degrees of freedom for the total sum of squares equals the *sum* of the degrees of freedom for the individual components; thus, the degrees of freedom for error are usually found last, by subtraction.

EXAMPLE Suppose that two replicates of the aforementioned soldering experiment were run, using the following arrangement:

Replicate I — *flux*

	1	*2*	*3*
Operator 1	*A*	*B*	*C*
Operator 2	*C*	*A*	*B*
Operator 3	*B*	*C*	*A*

Replicate II — *flux*

	1	*2*	*3*
Operator 1	*C*	*B*	*A*
Operator 2	*A*	*C*	*B*
Operator 3	*B*	*A*	*C*

The results, showing the number of pounds tensile force required to separate the soldered leads, were as follows:

Replicate I

14.0	16.5	11.0
9.5	17.0	15.0
11.0	12.0	13.5

Replicate II

10.0	16.5	13.0
12.0	12.0	14.0
13.5	18.0	11.5

Analyze this experiment as a Latin square and test at the 0.01 level of significance whether there are differences in the methods, the operators, the fluxes, or the replicates.

Solution

1. *Null hypotheses*: $\alpha_1 = \alpha_2 = \alpha_3 = 0$; $\beta_1 = \beta_2 = \beta_3 = 0$; $\gamma_1 = \gamma_2 = \gamma_3 = 0$; $\rho_1 = \rho_2 = 0$.
   Alternate hypotheses: The α's are not all equal to zero; the β's are not all equal to zero; the γ's are not all equal to zero; the ρ's are not all equal to zero.
2. *Levels of significance*: $\alpha = 0.01$ for each test.
3. *Criteria*: For treatments, rows, or columns, reject the null hypothesis if $F > 7.56$, the value of $F_{0.01}$ for $n - 1 = 3 - 1 = 2$ and $(n-1)(rn + r - 3) = (3-1)(2 \times 3 + 2 - 3) = 10$ degrees of freedom. For replicates, reject the null hypothesis if $F > 10.00$, the value of $F_{0.01}$ for $r - 1 = 2 - 1 = 1$ and 10 degrees of freedom.
4. *Calculations*: Substituting $n = 3$, $r = 2$, $T_{1..} = 81.0$, $T_{2..} = 79.5$, $T_{3..} = 79.5$, $T_{.1.} = 70.0$, $T_{.2.} = 92.0$, $T_{.3.} = 78.0$, $T_{..1} = 119.5$, $T_{..2} = 120.5$, $T_{(A)} = 87.5$, $T_{(B)} = 86.5$, $T_{(C)} = 66.0$, $T_{...} = 240.0$, and $\sum\sum\sum y_{ij(k)l}^2 = 3{,}304.5$ into the formulas for the sums of squares, we get

$$C = \frac{(240)^2}{18} = 3200.0$$

$$SS(Tr) = \tfrac{1}{6}[(87.5)^2 + (86.5)^2 + (66.0)^2] - 3200.0 = 49.1$$

$$SSR = \tfrac{1}{6}[(81.0)^2 + (79.5)^2 + (79.5)^2] - 3200.0 = 0.2$$

$$SSC = \tfrac{1}{6}[(70.0)^2 + (92.0)^2 + (78.0)^2] - 3200.0 = 41.3$$

$$SS(Rep) = \tfrac{1}{9}[(119.5)^2 + (120.5)^2] - 3200.0 = 0.1$$

$$SST = (14.0)^2 + (16.5)^2 + \cdots + (11.5)^2 - 3200.0 = 104.5$$

$$SSE = 104.5 - 49.1 - 41.3 - 0.2 - 0.1 = 13.8$$

and the results are as shown in the following analysis-of-variance table:

Source of variation	*Degrees of freedom*	*Sum of squares*	*Mean square*	*F*
Treatments (methods)	2	49.1	24.6	17.6
Rows (operators)	2	0.2	0.1	0.1
Columns (fluxes)	2	41.3	20.6	14.7
Replicates	1	0.1	0.1	0.1
Error	10	13.8	1.4	
Total	17	104.5		

5. *Decisions*: For treatments (methods) and columns (fluxes), since $F = 17.6$ and 14.7 exceed 7.56, the corresponding null hypotheses must be rejected; for rows (operators), since $F = 0.1$ does not exceed 7.56, and for replicates, since $F = 0.1$ does not exceed 10.00, the corresponding null hypotheses cannot be rejected. In other words, we conclude that differences in methods and fluxes, but not differences in operators or replications, affect the solder strengths of the electrical leads. To go one step further, the Duncan multiple-range test of Section 12.4 gives the following *decision pattern* at the 0.01 level of significance:

	Method C	*Method B*	*Method A*
Mean	11.0	14.4	14.6

(Methods B and A are joined by a common underline.)

Thus, we conclude that Method *C* definitely yields weaker solder bonds than Methods *A* or *B*.

■

The elimination of *three* extraneous sources of variability can be accomplished by means of a design called a **Graeco-Latin square**. This design is a square array of n Latin letters and n Greek letters, with the Latin and Greek letters each forming a Latin square; furthermore, each Latin letter appears once and only once in conjunction with each Greek letter. The following is an example of a 4×4 Graeco-Latin square:

$A\alpha$	$B\beta$	$C\gamma$	$D\delta$
$B\delta$	$A\gamma$	$D\beta$	$C\alpha$
$C\beta$	$D\alpha$	$A\delta$	$B\gamma$
$D\gamma$	$C\delta$	$B\alpha$	$A\beta$

Graeco-Latin squares are also called **orthogonal Latin squares**.

To give an example where the use of a Graeco-Latin square might be appropriate, suppose that in the soldering example an additional source of variability is the temperature of the solder. If three solder temperatures, denoted by α, β, and γ, are to be used together with the three methods, three operators (rows), and three solder fluxes (columns), a replicate of a suitable Graeco-Latin-square experiment could be laid out as follows:

	Flux 1	*Flux 2*	*Flux 3*
Operator 1	$A\alpha$	$B\gamma$	$C\beta$
Operator 2	$C\gamma$	$A\beta$	$B\alpha$
Operator 3	$B\beta$	$C\alpha$	$A\gamma$

Thus, Method A would be used by operator 1 using solder flux 1 and temperature α, by operator 2 using solder flux 2 and temperature β, and by operator 3 using solder flux 3 and temperature γ. Similarly, Method B would be used by operator 1 using solder flux 2 and temperature γ, and so forth.

In a Graeco-Latin square each variable (represented by rows, columns, Latin letters, or Greek letters) is "distributed evenly" over the other variables. Thus, in comparing the means obtained for one variable, the effects of the other variables are all averaged out. The analysis of a Graeco-Latin square is similar to that of a Latin square, with the addition of an extra source of variability corresponding to the Greek letters.

There exists a wide variety of experimental designs, other than the ones discussed in this chapter, that are suitable for many diverse purposes. Among the more widely used designs are the **incomplete-block designs**, which are characterized by the feature that each treatment is not represented in each block. If the number of treatments under investigation in an experiment is large, it often happens that it is impossible to find homogeneous blocks such that all of the treatments can be accommodated in each block.

For instance, if n paints are to be compared by applying each paint to a sheet of metal and then baking the sheets in an oven, it may be impossible to put all of the sheets in the oven at one time. Consequently, it would be necessary to use an experimental design in which $k < n$ treatments (paints) are included in each block (oven run). One way of doing this is to assign the treatments to each oven run in such a way that each treatment occurs together with each other treatment in the same number of blocks. For example, for $n = 4$ and $k = 2$ we might use the following scheme:

Oven run	*Paints*
1	1 and 2
2	3 and 4
3	1 and 3
4	2 and 4
5	1 and 4
6	2 and 3

This kind of design is called a **balanced incomplete-block design**, and it has the important feature that *comparisons between any two treatments can be made with equal precision.*

Since balanced incomplete-block designs may require too many blocks, many other schemes have been developed. Most of these experimental designs arose to meet the specific needs of experimenters, notably in the field of agriculture. As we have pointed out earlier, much of the language of experimental design, including such terms as "treatments," "blocks," "plots," etc., has been borrowed from agriculture. Only in recent years have the more sophisticated designs been applied to industrial and engineering experimentation, and, with more widespread application, it is to be expected that many new designs will be developed to meet the requirements in these fields.

EXERCISES

12.35 Referring to the problem in which samples of tin plate are to be distributed among four laboratories (Section 12.1) suppose that we are concerned with systematic differences in tin-coating weight along the direction of rolling as well as across the rolling direction. To eliminate these two sources of variability, each of two sheets of tin plate is divided into 16 parts, representing four positions across and four positions along the rolling direction. Then, four samples from each sheet are sent to each of the Laboratories A, B, C, and D, as shown, and the resulting tin-coating weights are determined.

Replicate I

C 0.20	*A* 0.24	*D* 0.20	*B* 0.27
B 0.28	*C* 0.19	*A* 0.22	*D* 0.28
D 0.34	*B* 0.23	*C* 0.21	*A* 0.28
A 0.32	*D* 0.22	*B* 0.16	*C* 0.27

Rolling direction ↓

Replicate II

B 0.29	*A* 0.25	*C* 0.18	*D* 0.28
D 0.28	*B* 0.18	*A* 0.21	*C* 0.25
C 0.28	*D* 0.23	*B* 0.20	*A* 0.28
A 0.30	*C* 0.19	*D* 0.24	*B* 0.25

Determine from these data whether the laboratories were obtaining consistent results. Also determine whether there are actual tin-coating weight differences across and along the rolling direction. (Use the 0.05 level of significance.)

12.36 A Latin-square design was used to compare the bond strengths of gold semiconductor lead wires bonded to the lead terminal by five different methods, *A*, *B*, *C*, *D*, and *E*. The bonds were made by five different operators, and the devices were encapsulated using five different plastics, with the following results, expressed as pounds force required to break the bond.

Plastic	Operator O_1	O_2	O_3	O_4	O_5
P_1	*A* 3.0	*B* 2.4	*C* 1.9	*D* 2.2	*E* 1.7
P_2	*B* 2.1	*C* 2.7	*D* 2.3	*E* 2.5	*A* 3.1
P_3	*C* 2.1	*D* 2.6	*E* 2.5	*A* 2.9	*B* 2.1
P_4	*D* 2.0	*E* 2.5	*A* 3.2	*B* 2.5	*C* 2.2
P_5	*E* 2.1	*A* 3.6	*B* 2.4	*C* 2.4	*D* 2.1

Analyze these results and apply the Duncan multiple-range test with $\alpha = 0.01$ to the mean breaking strengths of the five bonding methods.

12.37 To study the effectiveness of five different kinds of front-seat passenger restraint systems in automobiles, A, B, C, D, and E, the following Graeco-Latin-square experiment was performed. The rows represent different automotive size classes (from subcompact to full size), the columns represent different barrier impact speeds, and the Greek letters α, β, γ, δ, and ε represent different impact angles. The experimental results are given in terms of an index of forces at critical points on the test dummy that relates to the probability of fatal injury.

$A\alpha$ 0.50	$B\beta$ 0.21	$C\gamma$ 0.43	$D\delta$ 0.35	$E\varepsilon$ 0.46
$B\gamma$ 0.51	$C\delta$ 0.20	$D\varepsilon$ 0.40	$E\alpha$ 0.25	$A\beta$ 0.39
$C\varepsilon$ 0.45	$D\alpha$ 0.07	$E\beta$ 0.29	$A\gamma$ 0.20	$B\delta$ 0.31
$D\beta$ 0.39	$E\gamma$ 0.10	$A\delta$ 0.31	$B\varepsilon$ 0.24	$C\alpha$ 0.27
$E\delta$ 0.43	$A\varepsilon$ 0.17	$B\alpha$ 0.31	$C\beta$ 0.22	$D\gamma$ 0.32

Analyze this experiment.

12.38 A clothing manufacturer wishes to determine which of four different needle designs is best for the sewing machines. The sources of variability that must be eliminated to make this comparison are the actual sewing machine used, the operator, and the type of thread. Using the design shown (the rows represent the operators, the columns represent the machines, the Latin letters represent the needles, and the Greek letters stand for the types of thread), the manufacturer recorded the number of rejected garments at the end of each of two weeks, with the following results:

First week

$D\gamma$ 47	$B\alpha$ 40	$A\beta$ 23	$C\delta$ 72
$C\beta$ 74	$A\delta$ 37	$B\gamma$ 28	$D\alpha$ 75
$A\alpha$ 52	$C\gamma$ 95	$D\delta$ 57	$B\beta$ 15
$B\delta$ 10	$D\beta$ 45	$C\alpha$ 93	$A\gamma$ 52

Second week

$C\alpha$ 105	$A\gamma$ 38	$B\delta$ 15	$D\beta$ 60
$D\delta$ 70	$B\beta$ 20	$A\alpha$ 60	$C\gamma$ 85
$B\gamma$ 13	$D\alpha$ 82	$C\beta$ 90	$A\delta$ 28
$A\beta$ 33	$C\delta$ 75	$D\gamma$ 53	$B\alpha$ 31

Using the 0.05 level of significance, determine whether there is a difference in the effectiveness of the needles. Also, determine whether there are significant differences attributable to the operators, the machines, and the types of thread.

12.39 Give an example of a balanced incomplete-block design for which

(a) $n = 3$ and $k = 2$; (b) $n = 4$ and $k = 3$;
(c) $n = 6$ and $k = 4$.

What is the minimum number of times each treatment must be replicated in each of these designs?

12.6 ANALYSIS OF COVARIANCE

The purpose of the methods of Sections 12.3 and 12.5 was to free the experimental error from variability due to identifiable and controllable extraneous causes. In this section, we shall introduce a method, called the **analysis of covariance**, which applies when such extraneous, or **concomitant**, variables cannot be held fixed, but can be measured, nevertheless. This would be the case, for example, if we wanted to compare the effectiveness of several industrial training programs and the results depended on the trainees' IQs; if we wanted to compare the durability of several kinds of leather soles and the results depended on the weight of the persons wearing the shoes; or if we wanted to compare the merits of several cleaning agents and the results depended on the original condition of the surfaces cleaned.

The method by which we analyze data of this kind is a combination of the linear regression method of Section 11.1 and the analysis of variance of Section 12.2. The underlying model is given by

Model equation for analysis of covariance

$$y_{ij} = \mu + \alpha_i + \delta x_{ij} + \varepsilon_{ij}$$

for $i = 1, 2, \ldots, k; j = 1, 2, \ldots, n$. As in the model on page 384, μ is the grand mean, α_i is the effect of the ith treatment, and the ε_{ij} are values of independent, normally distributed variables with zero means and the common variance σ^2; as in the model on page 327 where we called it β, δ is the slope of the linear regression equation.

In the analysis of such data, the values of the concomitant variable, the x_{ij}, are eliminated by regression methods, namely, by estimating δ by the method of least squares, and then an analysis of variance is performed on the adjusted y's, namely, on the quantities $y'_{ij} = y_{ij} - \hat{\delta} x_{ij}$. This procedure is referred to as an analysis of covariance, as it involves a partitioning of the total **sum of products**

$$SPT = \sum_{i=1}^{k} \sum_{j=1}^{n} (y_{ij} - \bar{y}_{..})(x_{ij} - \bar{x}_{..})$$

in the same way as an ordinary analysis of variance involves the partitioning of the total sum of squares. In actual practice, the calculations are performed as follows:

1. The total, treatment, and error sums of squares are calculated for the x's by means of the formulas for a one-way classification on page 387; they will be denoted by SST_x, $SS(Tr)_x$, and SSE_x.
2. The total, treatment, and error sums of squares are calculated for the y's by means of the formulas for a one-way classification on page 387; they will be denoted by SST_y, $SS(Tr)_y$, and SSE_y.
3. The total, treatment, and error sums of products are calculated by means of the formulas

Sums of products—analysis of covariance

$$SPT = \sum_{i=1}^{k} \sum_{j=1}^{n} x_{ij} \cdot y_{ij} - C$$

$$SP(Tr) = \frac{\sum_{i=1}^{k} T_{x_i} \cdot T_{y_i}}{n} - C$$

$$SPE = SPT - SP(Tr)$$

where the correction term, C, is given by

$$C = \frac{T_x \cdot T_y}{k \cdot n}$$

and where T_{x_i} is the total of the x's for the ith treatment, T_{y_i} is the total of the y's for the ith treatment, T_x is the total of all the x's, and T_y is the total of all the y's.

4. The total, error, and treatment sums of squares are calculated for the adjusted y's by means of the formulas

Adjusted sums of squares—analysis of covariance

$$SST_{y'} = SST_y - \frac{(SPT)^2}{SST_x}$$

$$SSE_{y'} = SSE_y - \frac{(SPE)^2}{SSE_x}$$

$$SS(Tr)_{y'} = SST_{y'} - SSE_{y'}$$

The results obtained in these calculations are conveniently summarized by means of the following kind of **analysis-of-covariance table**:

Source of variation	*Sum of squares for x*	*Sum of squares for y*	*Sum of products*	*Sum of squares for y'*	*Degrees of freedom*	*Mean square*
Treatments	$SS(Tr)_x$	$SS(Tr)_y$	$SP(Tr)$	$SS(Tr)_{y'}$	$k-1$	$MS(Tr)_{y'} = \frac{SS(Tr)_{y'}}{k-1}$
Error	SSE_x	SSE_y	SPE	$SSE_{y'}$	$nk-k-1$	$MSE_{y'} = \frac{SSE_{y'}}{nk-k-1}$
Total	SST_x	SST_y	SPT	$SST_{y'}$	$nk-2$	

Note that each mean square is obtained by dividing the corresponding sum of squares by its degrees of freedom.

Finally, the null hypothesis $\alpha_1 = \alpha_2 = \cdots = \alpha_k = 0$ is tested against the alternative hypothesis that the α_i are not all equal to zero on the basis of the statistic

F-ratio for adjusted treatments

$$F = \frac{MS(Tr)_{y'}}{MSE_{y'}}$$

It is rejected at the level of significance α if the value obtained for F exceeds F_α with $k-1$ and $nk-k-1$ degrees of freedom.

EXAMPLE

Suppose that a research worker has three different cleaning agents, A_1, A_2, and A_3, and he wishes to select the most efficient agent for cleaning a metallic surface. The cleanliness of a surface is measured by its reflectivity, expressed in arbitrary units as the ratio of the reflectivity observed to that of a standard mirror surface. Analysis of covariance must be used because the effect of a cleaning agent on reflectivity will depend on the original cleanliness, namely, the original reflectivity of the surface. The research worker obtained the following results:

A_1	Original reflectivity, x	0.50	0.55	0.60	0.35
	Final reflectivity, y	1.00	1.20	0.80	1.40
A_2	Original reflectivity, x	0.75	1.65	1.00	1.10
	Final reflectivity, y	0.75	0.60	0.55	0.50
A_3	Original reflectivity, x	0.60	0.90	0.80	0.70
	Final reflectivity, y	1.00	0.70	0.80	0.90

Perform an analysis of covariance to determine (at the 0.05 level of significance) whether there are differences in the reflectivity improvements produced by the three cleaning agents.

Solution

1. *Null hypothesis*: $\alpha_1 = \alpha_2 = \alpha_3 = 0$.
   Alternate hypothesis: The α's are not all equal to zero.
2. *Level of significance*: $\alpha = 0.05$
3. *Criterion*: Reject the null hypothesis if $F > 4.46$, the value of $F_{0.05}$ for $k - 1 = 3 - 1 = 2$ and $nk - k - 1 = 4 \times 3 - 3 - 1 = 8$ degrees of freedom.
4. *Calculations*: The totals are $T_{x_1} = 2.00$, $T_{x_2} = 4.50$, $T_{x_3} = 3.00$, $T_x = 9.50$, $T_{y_1} = 4.40$, $T_{y_2} = 2.40$, $T_{y_3} = 3.40$, and $T_y = 10.20$.

   For the x's, the correction term is $\frac{(9.50)^2}{3 \cdot 4} = 7.52$, and the sums of squares are

$$SST_x = 0.50^2 + 0.55^2 + \cdots + 0.70^2 - 7.52 = 1.31$$

$$SS(Tr)_x = \frac{2.00^2 + 4.50^2 + 3.00^2}{4} - 7.52 = 0.79$$

$$SSE_x = 1.31 - 0.79 = 0.52$$

   For the y's, the correction term is $\frac{(10.20)^2}{3 \cdot 4} = 8.67$, and the sums of squares are

$$SST_y = 1.00^2 + 1.20^2 + \cdots + 0.90^2 - 8.67 = 0.79$$

$$SS(Tr)_y = \frac{4.40^2 + 2.40^2 + 3.40^2}{4} - 8.67 = 0.50$$

$$SSE_y = 0.79 - 0.50 = 0.29$$

For the sums of products, the correction term is $\frac{(9.50)(10.20)}{3 \cdot 4} = 8.08$, and we get

$$SPT = (0.50)(1.00) + (0.55)(1.20) + \cdots + (0.70)(0.90) - 8.08$$
$$= -0.80$$

$$SP(Tr) = \frac{(2.00)(4.40) + (4.50)(2.40) + (3.00)(3.40)}{4} - 8.08$$
$$= -0.63$$

$$SPE = -0.80 - (-0.63) = -0.17$$

Finally, for the adjusted y's we get

$$SST_{y'} = 0.79 - \frac{(-0.80)^2}{1.31} = 0.30$$

$$SSE_{y'} = 0.29 - \frac{(-0.17)^2}{0.52} = 0.23$$

$$SS(Tr)_{y'} = 0.30 - 0.23 = 0.07$$

All these results are summarized in the following analysis-of-covariance table:

Source of variation	*Sum of squares for x*	*Sum of squares for y*	*Sum of products*	*Sum of squares for y'*	*Degrees of freedom*	*Mean squares*
Treatments	0.79	0.50	−0.63	0.07	2	0.035
Error	0.52	0.29	−0.17	0.23	8	0.026
Total	1.31	0.79	−0.80	0.30	10	

5. *Decision*: Since $F = \frac{0.035}{0.026} = 1.34$ does not exceed 4.46, the null hypothesis cannot be rejected. In other words, one cannot conclude that any one of the cleaning agents is more effective than the others.

■

Although the calculations may at first appear to be formidable, they are routine with many computer statistical packages. The output from the *SAS* program for this example is presented in Figure 12.3. Because only two decimal

```
DEPENDENT VARIABLE: Y

SOURCE     DF   SUM OF SQUARES   MEAN SQUARE   F VALUE   PR > F   R-SQUARE

MODEL       3       0.55421498    0.18473833      6.40   0.0161   0.706006

ERROR       8       0.23078502    0.02884813               ROOT MSE

C TOTAL    11       0.78500000                           0.16984737


SOURCE         DF      TYPE III SS      F VALUE      PR > F

TREATMENTS      2       0.07447755         1.29      0.3267
X               1       0.05421498         1.88      0.2076
```

FIGURE 12.3
Selected SAS output for the analysis of covariance using the data from the example on page 422.

points were retained in the preceding example, the computer calculations for the F *statistic* are more accurate. However, the conclusions are the same.

Analysis-of-covariance methods have not been widely used until recent years, due mainly to the rather extensive calculations that are required. Of course, with the widespread availability of computers and appropriate programs, this is no longer a problem. There are several ways in which the analysis-of-covariance method presented here can be generalized. First, there can be more than one concomitant variable; then, the method can be applied to more complicated kinds of designs, say, to a randomized block design, where the regression coefficient could even assume a different value for each block.

EXERCISES

12.40 To compare the life expectancy of a transistor under three storage conditions and account at the same time for a leakage current (collector to base), a laboratory technician obtained the following results, where the leakage current, x, is in microamperes, and the life times, y, are in hours:

Storage condition 1		*Storage condition 2*		*Storage condition 3*	
x	y	x	y	x	y
4.8	9,912	6.4	9,952	8.8	9,596
7.2	9,383	8.7	9,482	6.2	9,697
5.5	9,734	7.1	9,435	7.5	9,700
6.0	9,551	5.3	9,915	4.9	9,610
8.3	8,959	4.6	9,492	5.4	10,145
7.6	9,474	6.0	9,565	5.8	10,191
5.9	9,179	7.2	9,704	7.3	9,855
8.0	9,359	8.8	9,636	8.6	9,682
4.3	9,580	5.4	9,608	8.8	10,160
5.1	9,245	7.8	9,548	6.0	9,982

Perform an analysis of covariance, using the level of significance $\alpha = 0.05$. Also, estimate the value of the regression coefficient.

12.41 Four different railroad-track cross-section configurations were tested to determine which is most resistant to breakage under use conditions. Ten miles of each kind of track were laid in each of five locations, and the number of cracks and other fracture-related conditions (y) was measured over a two-year usage period. To compare these track designs adequately, however, it is necessary to correct for extent of usage (x), measured in terms of the average number of trains per day that ran over each section of track. Use the following experimental results to test (0.01 level of significance) whether the track designs were equally resistant to breakage and to estimate the effect of usage on breakage resistance.

Track design A		*Track design B*		*Track design C*		*Track design D*	
x	y	x	y	x	y	x	y
10.4	3	16.9	8	17.8	5	19.6	9
19.3	7	23.6	11	24.4	9	25.4	8
13.7	4	14.4	7	13.5	5	35.5	16
7.2	0	17.2	10	20.1	6	16.8	7
16.3	5	9.1	4	11.0	4	31.2	11

12.7 REVIEW EXERCISES

12.42 Assume the following data obey the one-way analysis-of-variance model.

Treatment 1: 13, 8, 10, 11, 8
Treatment 2: 1, 0, 3, 0
Treatment 3: 8, 5, 10, 3, 7, 3

(a) Decompose each observation y_{ij} as

$$y_{ij} = \bar{y} + (\bar{y}_i - \bar{y}) + (y_{ij} - \bar{y}_i)$$

(b) Obtain the sums of squares and degrees of freedom for each array.
(c) Construct the analysis-of-variance table and test for differences among the treatments with $\alpha = 0.05$.

12.43 To determine the effect on exit dust loading in a precipitator, the following measurements were made:

Total flow (ft^3/hr)	*Exit dust loading (grains per cubic yard in flue gas)*				
200	1.5	1.7	1.6	1.9	1.9
300	1.5	1.8	2.2	1.9	2.2
400	1.4	1.6	1.7	1.5	1.8
500	1.1	1.5	1.4	1.4	2.0

Use the level of significance $\alpha = 0.05$ to test whether the flow through the precipitator has an effect on the exit dust loading.

12.44 To study the performance of a newly designed motorboat it was timed over a marked course under various wind and water conditions. Use the following data (in minutes) to test the null hypothesis that the boat's performance is not affected by the differences in wind and water conditions:

Calm conditions: 20, 17, 14, 24
Moderate conditions: 21, 23, 16, 25, 18, 23
Choppy conditions: 26, 24, 23, 29, 21

Use the level of significance $\alpha = 0.05$.

12.45 Given the following data from a randomized block design

	Blocks			
Treatment 1	9	10	2	7
Treatment 2	6	13	1	12
Treatment 3	9	16	9	14

(a) Decompose each observation y_{ij} as

$$y_{ij} = \bar{y}_{..} + (\bar{y}_{i.} - \bar{y}_{..}) + (\bar{y}_{.j} - \bar{y}_{..}) + (y_{ij} - \bar{y}_{i.} - \bar{y}_{.j} + \bar{y}_{..})$$

(b) Obtain the sums of squares and degrees of freedom for each array.
(c) Construct the analysis of variance table and test for differences among the treatments with $\alpha = 0.05$.

12.46 Use the Duncan test with $\alpha = 0.05$ to compare the treatments in Exercise 12.45.

12.47 An experiment was performed to judge the effect of four different fuels and two different types of launchers on the range of a certain rocket. Test, on the basis of the following data (in nautical miles) whether there are significant differences (a) among the means obtained for the fuels, and (b) between the means obtained for the launchers:

	Fuel I	*Fuel II*	*Fuel III*	*Fuel IV*
Launcher X	62.5	49.3	33.8	43.6
Launcher Y	40.4	39.7	47.4	59.8

Use the level of significance $\alpha = 0.05$.

12.48 Samples of groundwater were taken from five different toxic-waste dump sites by each of three different agencies, the EPA, the company that owned each site, and an independent consulting engineer. Each sample was analyzed for the presence of a certain contaminant by whatever laboratory method was customarily used by the agency collecting the sample, with the following results:

	Concentration (parts per million)				
	Site A	*Site B*	*Site C*	*Site D*	*Site E*
Agency 1	23.8	7.6	15.4	30.6	4.2
Agency 2	19.2	6.8	13.2	22.5	3.9
Agency 3	20.9	5.9	14.0	27.1	3.0

Is there reason to believe that the agencies are not consistent with one another in their measurements? Do the dump sites differ from one another in their level of contamination? Use the 0.05 level of significance.

12.49 Referring to Exercise 12.43 on page 427, suppose that the data in each of the five columns were obtained from a different precipitator. Repeat the analysis of variance, treating the experiment as a two-way classification, and observe what change results in the error mean square.

12.50 Does it make sense to use the Duncan test with $\alpha = 0.05$ to compare the results obtained with the four fuels in Exercise 12.47 on page 427? Why?

12.51 Use the Duncan test with $\alpha = 0.05$ to compare the pollution levels of the five sites in Exercise 12.48.

12.52 Is one of the agencies in Exercise 12.48 on this page producing results that are consistently higher (or lower) than those of the other two? (Use $\alpha = 0.05$.)

12.53 A test was made to determine which of three different golf-ball designs, *A*, *B*, and *C* will give the greatest distance. Golf balls were driven by three different golf pros, P_1, P_2, and P_3, using three different drivers, D_1, D_2, and D_3. The experiment was performed on three different fairways, and the distances from the tee to the point where the ball came to rest were measured in yards, as follows:

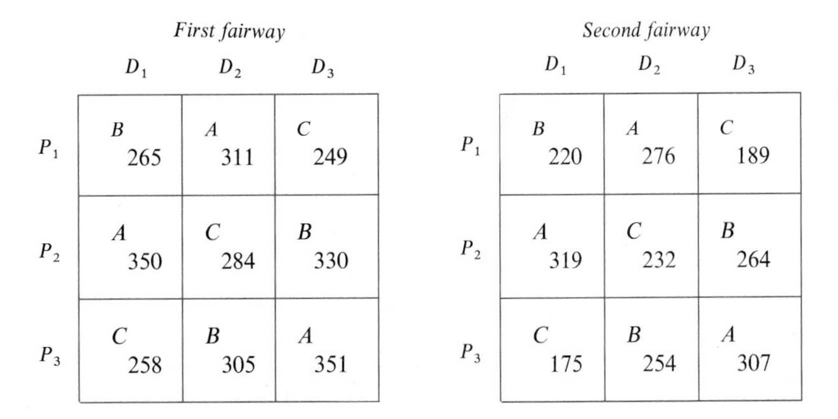

First fairway

	D_1	D_2	D_3
P_1	*B* 265	*A* 311	*C* 249
P_2	*A* 350	*C* 284	*B* 330
P_3	*C* 258	*B* 305	*A* 351

Second fairway

	D_1	D_2	D_3
P_1	*B* 220	*A* 276	*C* 189
P_2	*A* 319	*C* 232	*B* 264
P_3	*C* 175	*B* 254	*A* 307

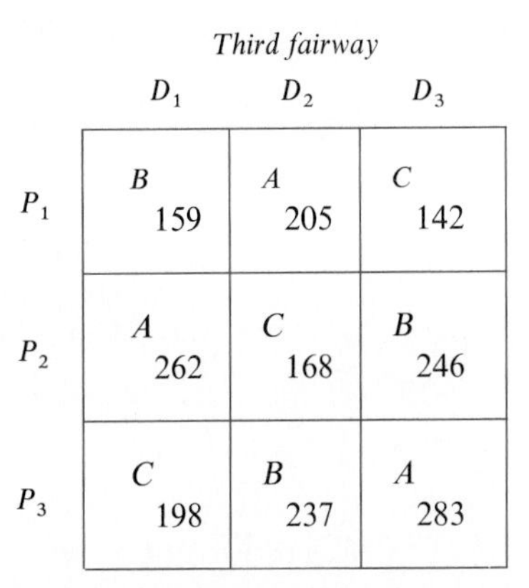

Third fairway

	D_1	D_2	D_3
P_1	*B* 159	*A* 205	*C* 142
P_2	*A* 262	*C* 168	*B* 246
P_3	*C* 198	*B* 237	*A* 283

Is any of the golf-ball designs superior to the others with respect to distance?

12.54 Three different instrument-panel configurations were tested by placing airline pilots in flight simulators and testing their reaction time to simulated flight emergencies. Each pilot was faced with ten emergency conditions in a randomized sequence, and the total time required to take corrective action for all eight conditions was measured, with the following results:

Instrument panel 1		*Instrument panel 2*		*Instrument panel 3*	
x	y	x	y	x	y
8.1	6.55	12.1	5.74	15.2	6.37
19.4	6.40	2.1	5.93	8.7	6.97
11.6	5.93	3.9	6.16	7.2	7.38
24.9	6.79	5.2	5.68	6.1	6.43
6.2	7.16	4.6	5.41	11.8	7.59
3.8	5.64	14.4	6.29	12.1	7.16
18.4	5.87	16.1	5.55	9.5	7.02
9.4	6.32	8.5	4.82	2.6	6.85

In this table, x is the number of years of experience of the pilot, and y is the total reaction time in seconds. Perform an analysis of covariance to test whether the instrument-panel configurations yield significantly different results ($\alpha = 0.05$). Also, perform a one-way analysis of variance (ignoring the covariate, x) and determine in that way what effect experience has on the results.

12.55 With reference to Exercise 10.15 on page 312, perform an analysis of variance using $\alpha = 0.05$.

12.56 Benjamin Franklin (1706–1790) conducted an experiment to study the effect of water depth on the amount of drag on a boat being pulled up a canal. He made a 14-foot trough and a model boat 6 inches long. A thread was attached to the bow, put through a pulley, and then a weight was attached. Not having a second hand on his watch, he counted as fast as he could to 10 repeatedly. These times, for the model boat to traverse the trough at the different water depths, are

Water 1.5 inches:	100	104	104	106	100	99	100	100
Water 2.0 inches:	94	93	91	87	88	86	90	88
Water 4.5 inches:	79	78	77	79	79	80	79	81

(Source: Letter to John Pringle, May 10, 1768)

(a) Perform an analysis of variance and test for differences due to water depth using $\alpha = 0.05$.
(b) Use the Duncan test with $\alpha = 0.05$ to investigate differences.

12.8 CHECK LIST OF KEY TERMS (with page references)

13 FACTORIAL EXPERIMENTATION

In Chapter 12 we were interested mainly in the effects of one variable, whose values we referred to as "treatments." In this chapter we shall be concerned with the individual and joint effects of several variables, and combinations of the values, or levels, of these variables will now play the roles of the different treatments. Extraneous variables, if any, will be handled as before by means of blocks.

In Sections 13.1 and 13.2 we deal with the analysis of experiments whose treatments can be regarded as combinations of the levels of two or more factors. In Sections 13.3 and 13.4 we study the special case of factors having two levels, and in the remainder of the chapter we take up the analysis of experiments where there are too many combinations of all the factors to be included in the same block or experimental program.

13.1

TWO-FACTOR EXPERIMENTS

To introduce the idea of a simple **two-factor (two-variable) experiment**, suppose that it is desired to determine the effects of flue temperature and oven width on the time required to make coke. The experimental conditions used are

Oven width (inches)	*Flue temperature (degrees F)*
4	1,600
4	1,900
8	1,600
8	1,900
12	1,600
12	1,900

and if several blocks (or replicates) were run, each consisting of these six "treatments," it would be possible to analyze the data as a two-way classification and test for significant differences among the six treatment means. In this instance, however, the experimenter is interested in knowing far more than that—he wishes to know whether variations in oven width or in flue temperature affect the coking time, and perhaps also whether any changes in coking time attributable to variations in oven width are the same at different temperatures.

It is possible to answer questions of this kind if the experimental conditions, the treatments, consist of appropriate combinations of the **levels** (or values) of the various **factors**. The factors in the preceding example are oven width and flue temperature; oven width has the *three levels* 4, 8, and 12 inches, while flue temperature has the *two levels* 1,600 and 1,900 degrees Fahrenheit. Note that the six treatments were chosen in such a way that each level of oven width is used once in conjunction with each level of flue temperature. In general, if two factors A and B are to be investigated at a levels and b levels, respectively, and if there are $a \cdot b$ experimental conditions (treatments) corresponding to all possible combinations of the levels of the two factors, the resulting experiment is referred to as a **complete $a \times b$ factorial experiment**. Note that if one or more of the $a \cdot b$ experimental conditions is omitted, the experiment can still be analyzed as a two-way classification, but it cannot readily be analyzed as a factorial experiment. It is customary to omit the word "complete," so that an **$a \times b$ factorial experiment** is understood to contain experimental conditions corresponding to all possible combinations of the levels of the two factors.

In order to obtain an estimate of the experimental error in a two-factor experiment it is necessary to replicate, that is, to repeat the entire set of $a \cdot b$

experimental conditions, say, a total of r times, randomizing the order of applying the conditions in each replicate. If y_{ijk} is the observation in the kth replicate, taken at the ith level of factor A and the jth level of factor B, the model assumed for the analysis of this kind of experiment is usually written as

Model equation for two-factor experiment

$$y_{ijk} = \mu + \alpha_i + \beta_j + (\alpha\beta)_{ij} + \rho_k + \varepsilon_{ijk}$$

for $i = 1, 2, \ldots, a, j = 1, 2, \ldots, b$, and $k = 1, 2, \ldots, r$. Here μ is the grand mean, α_i is the effect of the ith level of factor A, β_j is the effect of the jth level of factor B, $(\alpha\beta)_{ij}$ is the **interaction**, or joint effect, of the ith level of factor A and the jth level of factor B, and ρ_k is the effect of the kth replicate. As in the models used in Chapter 12 we shall assume that the ε_{ijk} are values of independent random variables having normal distributions with zero means and the common variance σ^2. Also, analogous to the restrictions imposed on the models on pages 384 and 397, we shall assume that

$$\sum_{i=1}^{a} \alpha_i = \sum_{j=1}^{b} \beta_j = \sum_{i=1}^{a} (\alpha\beta)_{ij} = \sum_{j=1}^{b} (\alpha\beta)_{ij} = \sum_{k=1}^{r} \rho_k = 0$$

It can be shown that these restrictions will assure unique definitions of the parameters μ, α_i, β_j, $(\alpha\beta)_{ij}$, and ρ_k.

To illustrate the model underlying a two-factor experiment, let us consider an experiment with two replicates in which factor A occurs at two levels, factor B occurs at two levels, and the replication effects are zero, that is, $\rho_1 = \rho_2 = 0$. In view of the restrictions on the parameters we also have

$$\alpha_2 = -\alpha_1, \beta_2 = -\beta_1 \quad \text{and} \quad (\alpha\beta)_{21} = (\alpha\beta)_{12} = -(\alpha\beta)_{22} = -(\alpha\beta)_{11}$$

and the population means corresponding to the four experimental conditions defined by the two levels of factor A and the two levels of factor B can be written as

$$\mu_{111} = \mu_{112} = \mu + \alpha_1 + \beta_1 + (\alpha\beta)_{11}$$
$$\mu_{121} = \mu_{122} = \mu + \alpha_1 - \beta_1 - (\alpha\beta)_{11}$$
$$\mu_{211} = \mu_{212} = \mu - \alpha_1 + \beta_1 - (\alpha\beta)_{11}$$
$$\mu_{221} = \mu_{222} = \mu - \alpha_1 - \beta_1 + (\alpha\beta)_{11}$$

Substituting for $\mu_{ij1} = \mu_{ij2}$ the mean of all observations obtained for the ith level of factor A and the jth level of factor B, we get four simultaneous linear equations which can be solved to yield estimates for the parameters μ, α_1, β_1, and $(\alpha\beta)_{11}$ (see Exercise 13.9 on page 453).

To continue our illustration, let us now suppose that $\mu = 10$. If all of the other effects equalled zero, each of the μ_{ijk} would equal 10, and the response surface would be the horizontal plane shown in Figure 13.1(a). If we now add an effect of

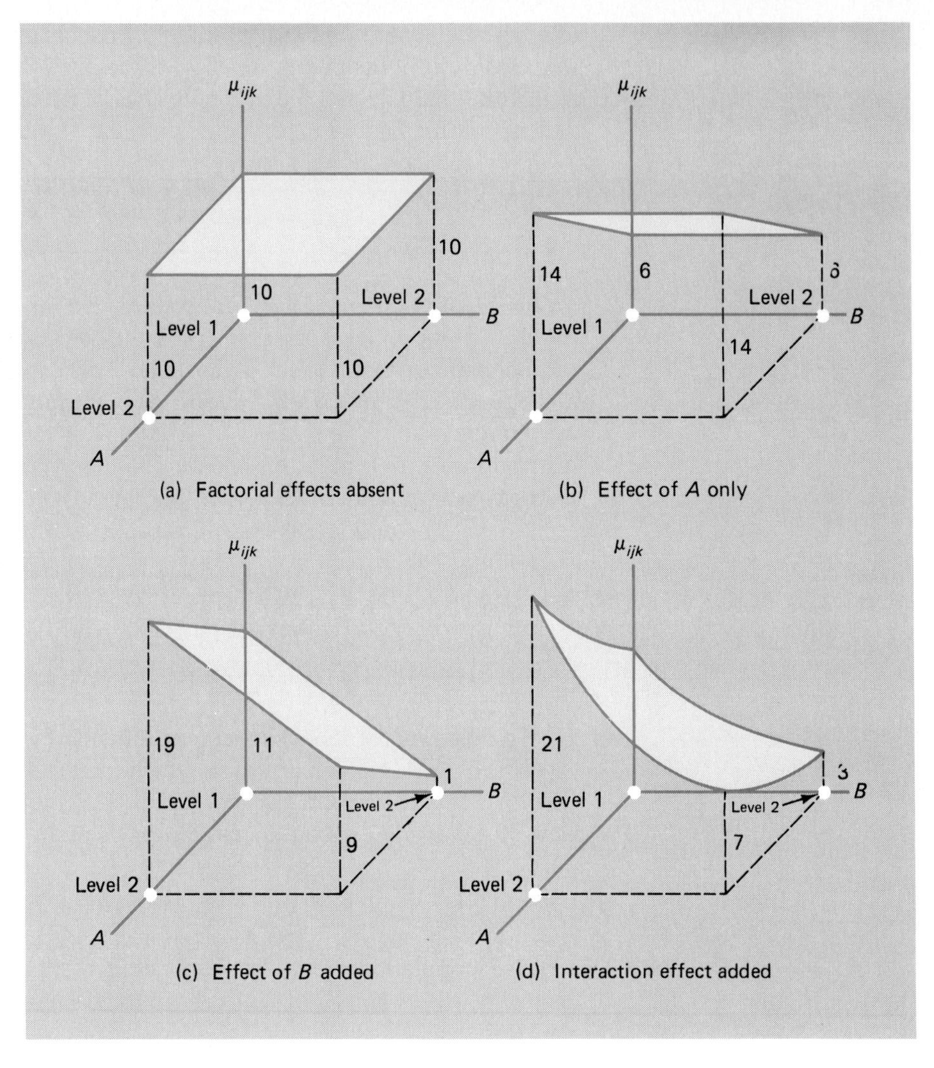

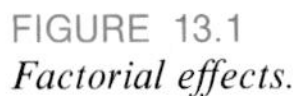

FIGURE 13.1
Factorial effects.

factor A, with $\alpha_1 = -4$, the response surface becomes the tilted plane shown in Figure 13.1(b), and if we add to this an effect of factor B, with $\beta_1 = 5$, we get the plane shown in Figure 13.1(c). Note that, so far, the effects of factors A and B are *additive*, that is, the change in the mean for either factor in going from level 1 to level 2 does not depend on the level of the other factor, and the response surface is a plane. If we now include an interaction, with $(\alpha\beta)_{11} = -2$, the plane becomes twisted as shown in Figure 13.1(d), the effects are no longer additive, and the response surface is no longer a plane. Note, also, that if the replication effects had not equalled zero, we would have obtained a different surface for each replicate; for each replicate the surface of Figure 13.1(d) would have been shifted an appropriate number of units up or down.

Generalizing these ideas from the 2×2 factorial experiment, the analysis of an $a \times b$ factorial experiment is based on the following breakdown of the total sum of squares. First, we partition SST into components attributed to treatments, replicates (or blocks), and error, by means of the identity

$$\sum_{i=1}^{a}\sum_{j=1}^{b}\sum_{k=1}^{r}(y_{ijk} - \bar{y}_{\cdots})^2 = r\sum_{i=1}^{a}\sum_{j=1}^{b}(\bar{y}_{ij\cdot} - \bar{y}_{\cdots})^2 + ab\sum_{k=1}^{r}(\bar{y}_{\cdot\cdot k} - \bar{y}_{\cdots})^2$$
$$+ \sum_{i=1}^{a}\sum_{j=1}^{b}\sum_{k=1}^{r}(y_{ijk} - \bar{y}_{ij\cdot} - \bar{y}_{\cdot\cdot k} + \bar{y}_{\cdots})^2$$

Except for notation, this identity is equivalent to that of Theorem 12.2. The total sum of squares, on the left-hand side of the identity, has $abr - 1$ degrees of freedom. The terms on the right are, respectively, the treatment sum of squares having $ab - 1$ degrees of freedom, the replicate (block) sum of squares having $r - 1$ degrees of freedom, and the error sum of squares having $(ab - 1)(r - 1)$ degrees of freedom. (Note that the various degrees of freedom are the same as those of the analysis-of-variance table on page 399 if we substitute ab for a and r for b.)

So far there is nothing new about the analysis of the data; it is the analysis of a two-way classification, *but the distinguishing feature of a factorial experiment is that the treatment sum of squares can be further subdivided into components corresponding to the various factorial effects.* Thus, for a two-factor experiment we have the following partition, or breakdown, of the treatment sum of squares:

$$r\sum_{i=1}^{a}\sum_{j=1}^{b}(\bar{y}_{ij\cdot} - \bar{y}_{\cdots})^2 = rb\sum_{i=1}^{a}(\bar{y}_{i\cdot\cdot} - \bar{y}_{\cdots})^2 + ra\sum_{j=1}^{b}(\bar{y}_{\cdot j\cdot} - \bar{y}_{\cdots})^2$$
$$+ r\sum_{i=1}^{a}\sum_{j=1}^{b}(\bar{y}_{ij\cdot} - \bar{y}_{i\cdot\cdot} - \bar{y}_{\cdot j\cdot} + \bar{y}_{\cdots})^2$$

The first term on the right measures the variability of the means corresponding to the different levels of factor A, and we refer to it as the **factor A sum of squares**, SSA. Similarly, the second term is the sum of squares for factor B, SSB, and the third term is the **interaction sum of squares** $SS(AB)$, which measures the variability of the means $\bar{y}_{ij}$ that is not attributable to the individual (or separate) effects of factors A and B. The $ab - 1$ degrees of freedom for treatments are, accordingly, subdivided into $a - 1$ degrees of freedom for the effect of factor A, $b - 1$ for the effect of factor B, and

$$ab - 1 - (a - 1) - (b - 1) = (a - 1)(b - 1)$$

degrees of freedom for interaction.

The perceptive reader will notice that the breakdown of total sum of squares corresponds to the breakdown

$$\underset{\text{observation}}{y_{ijk}} = \underset{\substack{\text{grand}\\ \text{mean}}}{\bar{y}_{...}} + \underset{\substack{\text{factor A}\\ \text{effect}}}{(\bar{y}_{i..} - \bar{y}_{...})} + \underset{\substack{\text{factor B}\\ \text{effect}}}{(\bar{y}_{.j.} - \bar{y}_{...})}$$

$$+ \underset{\text{AB interaction}}{(\bar{y}_{ij.} - \bar{y}_{i..} - \bar{y}_{.j.} + \bar{y}_{...})} + \underset{\text{replication}}{(\bar{y}_{..k} - \bar{y}_{...})} + \underset{\text{error}}{(y_{ijk} - \bar{y}_{ij.} - \bar{y}_{..k} + \bar{y}_{...})}$$

of the individual observations, which, in turn, corresponds to the terms in the population model on page 433. Transposing $\bar{y}_{...}$ to the left-hand side, the breakdown corresponds to the breakdown in variation.

The next example illustrates the successive breakdown of the sums of squares. From the amount of calculation involved, you can appreciate the widespread availability of computer programs for creating an analysis of variance table (see Exercise 13.10).

EXAMPLE Referring to the coking experiment described on page 432, suppose that three replicates of this experiment yield the following coking times (in hours):

Factor A *Oven width*	*Factor B* *Flue temp.*	*Rep. 1*	*Rep. 2*	*Rep. 3*	*Total*
4	1600	3.5	3.0	2.7	9.2
4	1900	2.2	2.3	2.4	6.9
8	1600	7.1	6.9	7.5	21.5
8	1900	5.2	4.6	6.8	16.6
12	1600	10.8	10.6	11.0	32.4
12	1900	7.6	7.1	7.3	22.0
	Total	36.4	34.5	37.7	108.6

Perform an analysis of variance based on this two-factor experiment and test for the significance of the factorial effects, using the 0.01 level of significance.

Solution Following steps analogous to those used in the analysis of a two-way classification, we get

1. *Null hypotheses*:

$$\alpha_1 = \alpha_2 = \alpha_3 = 0, \qquad \beta_1 = \beta_2 = 0$$

$$(\alpha\beta)_{11} = (\alpha\beta)_{12} = (\alpha\beta)_{21} = (\alpha\beta)_{22} = (\alpha\beta)_{31} = (\alpha\beta)_{32} = 0$$

$$\rho_1 = \rho_2 = \rho_3 = 0$$

   Alternative hypotheses: The α's are not all equal to zero; the β's are not all equal to zero; the $(\alpha\beta)$ terms are not all equal to zero; the ρ's are not all equal to zero.
2. *Levels of significance*: $\alpha = 0.01$ for all tests.
3. *Criteria*: For replications, reject the null hypothesis if $F > 7.56$, the value of $F_{0.01}$ for $r - 1 = 3 - 1 = 2$ and $(ab - 1)(r - 1) = (3 \times 2 - 1)(3 - 1) = 10$ degrees of freedom; for the main effect of factor A, reject the null hypothesis if $F > 7.56$, the value of $F_{0.01}$ for $a - 1 = 3 - 1 = 2$ and $(ab - 1)(r - 1) = (3 \times 2 - 1)(3 - 1) = 10$ degrees of freedom; for the main effect of factor B, reject if $F > 10.00$, the value of $F_{0.01}$ for $b - 1 = 2 - 1 = 1$ and $(ab - 1)(r - 1) = (3 \times 2 - 1)(3 - 1) = 10$ degrees of freedom; for the interaction effect, reject if $F > 7.56$, the value of $F_{0.01}$ for $(a - 1)(b - 1) = (3 - 1)(2 - 1) = 2$ and $(ab - 1)(r - 1) = (3 \times 2 - 1)(3 - 1) = 10$ degrees of freedom.
4. *Calculations*: Following the procedure used in analyzing a two-way classification, we first compute the correction term

$$C = \frac{(108.6)^2}{18} = 655.22$$

   Then, the total sum of squares is given by

$$SST = (3.5)^2 + (2.2)^2 + \cdots + (7.3)^2 - 655.22 = 149.38$$

   and the treatment and replicate (instead of blocks) sums of squares are given by

$$SS(Tr) = \tfrac{1}{3}[(9.2)^2 + (6.9)^2 + \cdots + (22.0)^2] - 655.22 = 146.05$$

$$SSR = \tfrac{1}{6}[(36.4)^2 + (34.5)^2 + (37.7)^2] - 655.22 = 0.86$$

   Finally, by subtraction, we obtain

$$SSE = 149.38 - 146.05 - 0.86 = 2.47$$

Subdivision of the treatment sum of squares into components for factors A and B, and for interaction, can be facilitated by constructing the following kind of two-way table, where the entries are the totals in the right-hand column of the table giving the original data:

		Factor B *Flue temperature* *1600*	*1900*	
Factor A *Oven width*	*4*	9.2	6.9	16.1
	8	21.5	16.6	38.1
	12	32.4	22.0	54.4
		63.1	45.5	108.6

Using formulas analogous to the ones with which we computed the sums of squares for various effects in Chapter 12, we now have for the two main effects

$$SSA = \frac{1}{b \cdot r} \sum_{i=1}^{a} T_{i..}^2 - C$$

$$= \tfrac{1}{6}[(16.1)^2 + (38.1)^2 + (54.4)^2] - 655.22 = 123.14$$

$$SSB = \frac{1}{a \cdot r} \sum_{j=1}^{b} T_{.j.}^2 - C$$

$$= \tfrac{1}{9}[(63.1)^2 + (45.5)^2] - 655.22 = 17.21$$

and for the interaction

$$SS(AB) = SS(Tr) - SSA - SSB$$

$$= 146.05 - 123.14 - 17.21 = 5.70$$

Finally, dividing the various sums of squares by their degrees of freedom, and dividing the appropriate mean squares by the error mean square, we obtain the results shown in the following analysis-of-variance table:

Source of variation	*Degrees of freedom*	*Sum of squares*	*Mean square*	*F*
Replication	2	0.86	0.43	1.72
Main effects:				
A	2	123.14	61.57	246
B	1	17.21	17.21	68.8
Interaction	2	5.70	2.85	11.4
Error	10	2.47	0.25	
Total	17	149.38		

5. *Decisions*: For replications, since $F = 1.72$ does not exceed 7.56, the value of $F_{0.01}$ for 2 and 10 degrees of freedom, we cannot reject the null hypothesis. For the main effect of factor A, since $F = 246$ exceeds 7.56, and for the main effect of factor B, since $F = 68.8$ exceeds 10.00, the null hypotheses must be rejected. For the interaction effect, since $F = 11.4$ exceeds 7.56 we must reject the null hypothesis. As illustrated in Figure 13.2, the trend of mean coking times for changing oven width is different each of the flue temperatures. It is apparent from this figure that the increase in

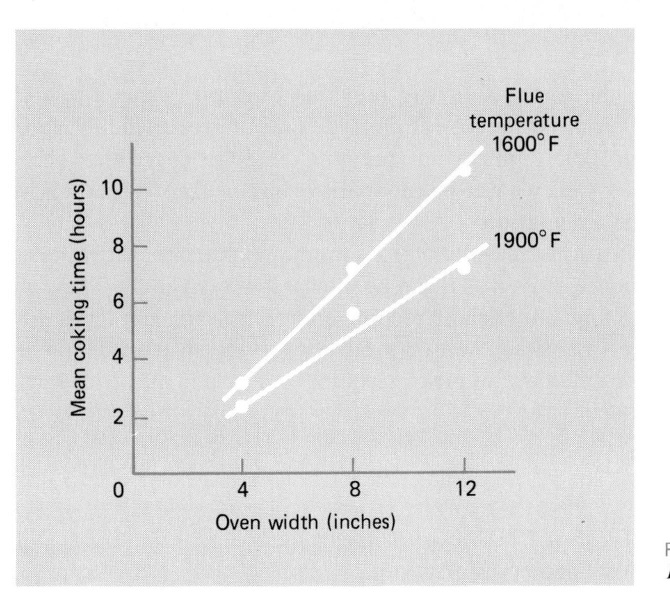

FIGURE 13.2
Results of coking experiment.

coking time for changing oven width is *greater* at the lower flue temperature. In view of this interaction, great care must be exercised in stating the results of this experiment. For instance, it would be very misleading to state merely that the effect of increasing flue temperature from 1,600 to 1,900 degrees Fahrenheit is to lower the coking time by $\frac{63.1}{9} - \frac{45.5}{9} = 1.96$ hours. In fact, the coking time is lowered on the average by as little as 0.77 hour when the oven width is 4 inches, and by as much as 3.47 hours when the oven width is 12 inches.

Alternatively, the summary can take the form of the two-way table of cell means.

Summary Table of Cell Means

		Factor B *Flue temperature* *1600*	*1900*	
Factor A *Oven width*	*4*	3.07	2.30	2.69
	8	7.17	5.53	6.35
	12	10.80	7.33	9.07
		7.01	5.05	6.03

■

Repeating the analysis in the previous example, using the *SAS* statistical package ANOVA program, we obtained the output presented in Figure 13.3. The P values less than 0.05 confirm our previous analysis. Note that the F values in Figure 13.3 differ somewhat from those above because more decimals were retained in the computer calculation.

When replications and interactions are not significant, the influence of Factor A and the influence of Factor B can be interpreted separately. Then, when a factor is significant, many statisticians recommend comparing the levels by calculating confidence intervals using the two sample intervals on page 246. The confidence interval, for the difference in mean response at levels i_1 and i_2 of Factor A is

$$\bar{y}_{i_1..} - \bar{y}_{i_2..} \pm t_{\alpha/2}\sqrt{s^2 \frac{2}{b \cdot r}}$$

where $s^2 = SSE/(ab - 1)(r - 1)$ is the mean square error and the $t_{\alpha/2}$ value is based on $(ab - 1)(r - 1)$ degrees of freedom.

```
DEPENDENT VARIABLE: Y

SOURCE      DF    SUM OF SQUARES    MEAN SQUARE  F VALUE   PR > F    R-SQUARE

MODEL        7      146.91666667    20.98809524    85.20   0.0001    0.983510

ERROR       10        2.46333333     0.24633333             ROOT MSE

C TOTAL     17      149.38000000                          0.49631979

SOURCE      DF        TYPE I SS    F VALUE    PR > F

REP          2       0.86333333       1.75    0.2226
A            2     123.14333333     249.95    0.0001
B            1      17.20888889      69.86    0.0001
A*B          2       5.70111111      11.57    0.0025
```

FIGURE 13.3
Selected SAS output for ANOVA using the data in the example on page 436.

Similarly, for levels j_1 and j_2 of Factor B, the confidence interval for the difference in mean response is

$$\bar{y}_{.j_1.} - \bar{y}_{.j_2.} \pm t_{\alpha/2}\sqrt{s^2\frac{2}{a\cdot r}}$$

EXAMPLE Illustrate the calculation of the confidence intervals for the difference in mean response using the means and s^2 from the previous example.

Solution From the analysis of variance table, $s^2 = 0.25$ is the mean square error based on 10 degrees of freedom. For these degrees of freedom, we find $t_{0.025} = 2.228$. Therefore, the confidence intervals for differences in mean due to the $a = 3$ levels of oven width, Factor A, are

$$\bar{y}_{1..} - \bar{y}_{2..} \pm t_{0.025}\sqrt{s^2\frac{2}{b\cdot r}} \qquad \bar{y}_{1..} - \bar{y}_{3..} \pm t_{0.025}\sqrt{s^2\frac{2}{b\cdot r}} \qquad \bar{y}_{2..} - \bar{y}_{3..} \pm t_{0.025}\sqrt{s^2\frac{2}{b\cdot r}}$$

$$2.69 - 6.35 \pm 2.228\sqrt{0.25\frac{2}{2\cdot 3}} \qquad 2.69 - 9.07 \pm 2.228\sqrt{0.25\frac{2}{2\cdot 3}} \qquad 6.35 - 9.07 \pm 2.228\sqrt{0.25\frac{2}{2\cdot 3}}$$

or -4.30 to -3.02 $\qquad$ -7.02 to -5.74 $\qquad$ -3.36 to -2.08

Because the interaction was significant, we cannot interpret these intervals on differences of mean coking times as due to changing oven width alone.

Similarly the single difference in mean due to the $b = 2$ flue temperatures is

$$\bar{y}_{\cdot 1 \cdot} - \bar{y}_{\cdot 2 \cdot} \pm t_{0.025}\sqrt{s^2 \frac{2}{a \cdot r}}$$

$$7.01 - 5.05 \pm 2.228\sqrt{0.25 \frac{2}{3 \cdot 3}}$$

$$\text{or} \quad 1.43 \text{ to } 2.49$$

■

13.2 MULTIFACTOR EXPERIMENTS

Much industrial research and experimentation is conducted to discover the individual and joint effects of several factors on variables which are most relevant to phenomena under investigation. The experimental designs most often used are of the simple randomized-block or two-way classification type, but the distinguishing feature of most of them is the factorial arrangement of the treatments, or experimental conditions. As we observed in the preceding section, r sets of data pertaining to $a \cdot b$ experimental conditions can be analyzed as a factorial experiment in r replicates if the experimental conditions represent all possible combinations of the levels of two factors A and B. In this section, we shall extend the discussion to factorial experiments involving more than two factors, that is, to experiments where the experimental conditions represent all possible combinations of the levels of three or more factors.

To illustrate the analysis of a **multifactor experiment**, let us consider the following situation. A warm sulfuric pickling bath is used to remove oxides from the surface of a metal prior to plating, and it is desired to determine what factors in addition to the concentration of the sulfuric acid might affect the electrical conductivity of the bath. As it is felt that the salt concentration as well as the bath temperature might also affect the electrical conductivity, an experiment is planned to determine the individual and joint effects of these three variables on the electrical conductivity of the bath. In order to cover the ranges of concentrations and temperatures normally encountered, it is decided to use the following levels of the three factors:

Factor	*Level 1*	*Level 2*	*Level 3*	*Level 4*
A. Acid concentration (percent)	0	6	12	18
B. Salt concentration (percent)	0	10	20	
C. Bath temperature (degrees F)	80	100		

The resulting factorial experiment requires $4 \cdot 3 \cdot 2 = 24$ experimental conditions in each replicate, where each experimental condition is a pickling bath made up according to specifications. The order in which these pickling baths are made up should be random. Let us suppose that two replicates of the experiment have actually been completed, that is, the electrical conductivities of the various pickling baths have been measured, and that the results are as shown in the following table:

Results of Acid-Bath Experiment

Level of factor			*Conductivity (mhos/cm³)*		
A	*B*	*C*	*Rep. 1*	*Rep. 2*	*Total*
1	1	1	0.99	0.93	1.92
1	1	2	1.15	0.99	2.14
1	2	1	0.97	0.91	1.88
1	2	2	0.87	0.86	1.73
1	3	1	0.95	0.86	1.81
1	3	2	0.91	0.85	1.76
2	1	1	1.00	1.17	2.17
2	1	2	1.12	1.13	2.25
2	2	1	0.99	1.04	2.03
2	2	2	0.96	0.98	1.94
2	3	1	0.97	0.95	1.92
2	3	2	0.94	0.99	1.93
3	1	1	1.24	1.22	2.46
3	1	2	1.12	1.15	2.27
3	2	1	1.15	0.95	2.10
3	2	2	1.11	0.95	2.06
3	3	1	1.03	1.01	2.04
3	3	2	1.12	0.96	2.08
4	1	1	1.24	1.20	2.44
4	1	2	1.32	1.24	2.56
4	2	1	1.14	1.10	2.24
4	2	2	1.20	1.19	2.39
4	3	1	1.02	1.01	2.03
4	3	2	1.02	1.00	2.02
		Total	25.53	24.64	50.17

The model we shall assume for the analysis of this experiment (or any similar three-factor experiment) is an immediate extension of the one used in Section 13.1. If y_{ijkl} is the conductivity measurement obtained at the ith level of acid concentration, the jth level of salt concentration, the kth level of bath temperature, in the lth replicate, we write

Model equation for three-factor experiment

$$y_{ijkl} = \mu + \alpha_i + \beta_j + \gamma_k + (\alpha\beta)_{ij} + (\alpha\gamma)_{ik} + (\beta\gamma)_{jk} + (\alpha\beta\gamma)_{ijk} + \rho_l + \varepsilon_{ijkl}$$

for $i = 1, 2, \ldots, a, j = 1, 2, \ldots, b, k = 1, 2, \ldots, c$, and $l = 1, 2, \ldots, r$. We also assume that the sums of the **main effects** (α's, β's, and γ's) as well as the sum of the replication effects are equal to zero, that the sums of the **two-way interaction effects** summed on either subscript equal zero for any value of the other subscript, and that the sum of the **three-way interaction effects** summed on any one of the subscripts is zero for any values of the other two subscripts. As before, the ε_{ijkl} are assumed to be values of independent random variables having zero means and the common variance σ^2.

We begin the analysis of the data by treating the experiment as a two-way classification with $a \cdot b \cdot c$ treatments and r replicates (blocks), and using the shortcut formulas on page 398, we obtain

$$C = \frac{(50.17)^2}{48} = 52.4381$$

$$SST = (0.99)^2 + (1.15)^2 + \cdots + (1.00)^2 - 52.4381 = 0.6624$$

$$SS(Tr) = \tfrac{1}{2}[(1.92)^2 + (2.14)^2 + \cdots + (2.02)^2] - 52.4381 = 0.5712$$

$$SSR = \tfrac{1}{24}[(25.53)^2 + (24.64)^2] - 52.4381 = 0.0165$$

$$SSE = 0.6624 - 0.5712 - 0.0165 = 0.0747$$

The degrees of freedom for these sums of squares are, respectively, 47, 23, 1, and 23.

Next, we shall want to subdivide the treatment sum of squares into the three **main effect sums of squares** SSA, SSB, SSC, the three **two-way interaction sums of squares**, $SS(AB)$, $SS(AC)$, and $SS(BC)$, and the **three-way interaction sum of squares** $SS(ABC)$. To facilitate the calculation of these sums of squares we first construct the following three tables analogous to the one on page 438:

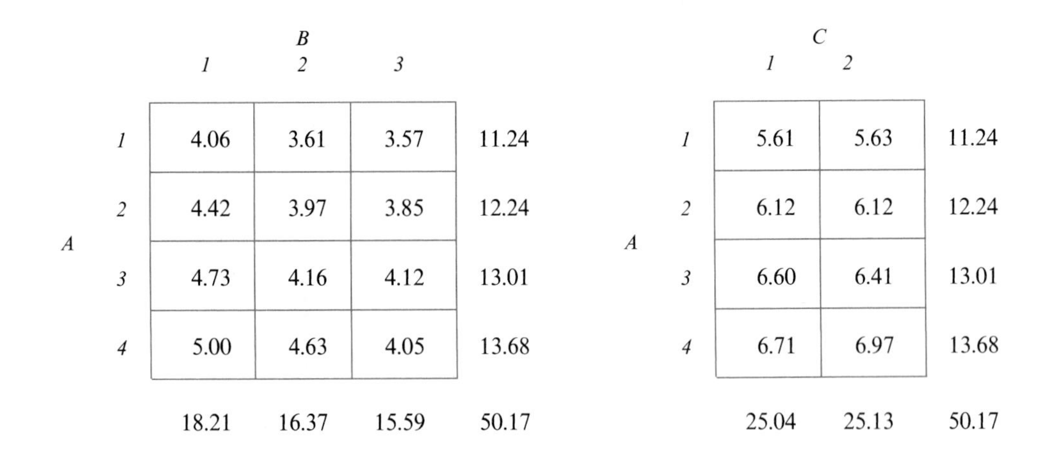

A \ B	1	2	3	
1	4.06	3.61	3.57	11.24
2	4.42	3.97	3.85	12.24
3	4.73	4.16	4.12	13.01
4	5.00	4.63	4.05	13.68
	18.21	16.37	15.59	50.17

A \ C	1	2	
1	5.61	5.63	11.24
2	6.12	6.12	12.24
3	6.60	6.41	13.01
4	6.71	6.97	13.68
	25.04	25.13	50.17

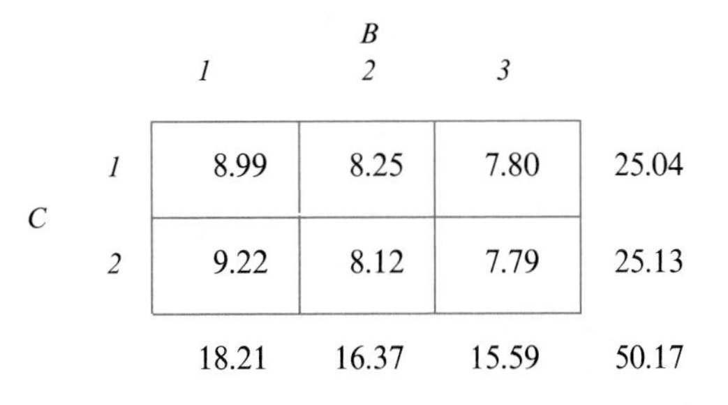

		B 1	B 2	B 3	
C	1	8.99	8.25	7.80	25.04
C	2	9.22	8.12	7.79	25.13
		18.21	16.37	15.59	50.17

The entries of these tables are the totals of all measurements obtained at the respective levels of the two variables. Note the self-checking feature of these tables; the same marginal totals appear several times, thus providing a rapid and effective check on the calculations.

To calculate SSA, SSB, and $SS(AB)$ we refer to the first of the preceding tables and an identity analogous to the one on page 435. As a matter of fact, the calculations parallel those for calculating SSA, SSB, and $SS(AB)$ in the two-factor experiment. To take the place of the treatment sum of squares, we first calculate

$$\begin{aligned} rc\sum_{i=1}^{a}\sum_{j=1}^{b}(\bar{y}_{ij\cdot\cdot}-\bar{y}_{\cdot\cdot\cdot\cdot})^2 &= \frac{1}{r\cdot c}\sum_{i=1}^{a}\sum_{j=1}^{b}T_{ij\cdot\cdot}^2 - C \\ &= \tfrac{1}{4}[(4.06)^2+(4.42)^2+\cdots+(4.05)^2] - 52.4381 \\ &= 0.5301 \end{aligned}$$

and we then obtain

$$\begin{aligned} SSA &= \frac{1}{bcr}\sum_{i=1}^{a}T_{i\cdot\cdot\cdot}^2 - C \\ &= \tfrac{1}{12}[(11.24)^2+\cdots+(13.68)^2] - 52.4381 \\ &= 0.2750 \end{aligned}$$

$$\begin{aligned} SSB &= \frac{1}{acr}\sum_{j=1}^{b}T_{\cdot j\cdot\cdot}^2 - C \\ &= \tfrac{1}{16}[(18.21)^2+(16.37)^2+(15.59)^2] - 52.4381 \\ &= 0.2262 \end{aligned}$$

and

$$\begin{aligned} SS(AB) &= 0.5301 - 0.2750 - 0.2262 \\ &= 0.0289 \end{aligned}$$

Performing the same calculations for the second of the tables on page 444, we obtain, similarly,

$$SSC = 0.0002 \quad \text{and} \quad SS(AC) = 0.0085$$

and the analysis of the third table yields

$$SS(BC) = 0.0042$$

For the three-way interaction sum of squares, we finally obtain by subtraction

$$\begin{aligned} SS(ABC) &= SS(Tr) - SSA - SSB - SSC \\ &\quad - SS(AB) - SS(AC) - SS(BC) \\ &= 0.5712 - 0.2750 - 0.2262 - 0.0002 \\ &\quad - 0.0289 - 0.0085 - 0.0042 \\ &= 0.0282 \end{aligned}$$

Note that the degrees of freedom for each main effect is one less than the number of levels of the corresponding factor. The degrees of freedom for each interaction is the *product* of the degrees of freedom for those factors appearing in the interaction. Thus, the degrees of freedom for the three main effects are 3, 2, and 1 in this example, while the degrees of freedom for the two-way interactions are 6, 3, and 2, and the degrees of freedom for the three-way interaction is 6.

The following table shows the complete analysis of variance for the acid-bath experiment:

Source of variation	*Degrees of freedom*	*Sum of squares*	*Mean square*	*F*
Replicates	1	0.0165	0.0165	5.16
Main effects:				
A	3	0.2750	0.0917	28.66
B	2	0.2262	0.1131	35.34
C	1	0.0002	0.0002	<1
Two-factor interactions:				
AB	6	0.0289	0.0048	1.50
AC	3	0.0085	0.0028	<1
BC	2	0.0042	0.0021	<1
Three-factor interaction:				
ABC	6	0.0282	0.0047	1.47
Error	23	0.0747	0.0032	
Total	47	0.6624		

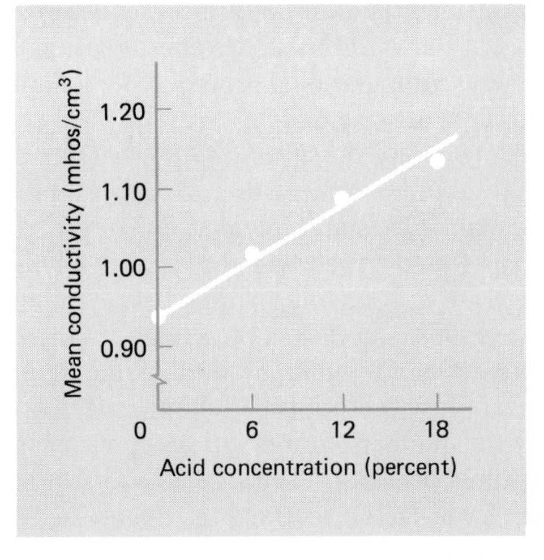

FIGURE 13.4
Effect of acid concentration.

Obtaining the appropriate values of $F_{0.05}$ and $F_{0.01}$ from Table 6, we find that the test for replicates is significant at the 0.05 level (perhaps the two replicates were performed under different atmospheric conditions or the thermometer used to measure bath temperatures went out of calibration, etc.), the tests for the factor A and factor B main effects are significant at the 0.01 level, while none of the other F's is significant at either level. We conclude from this analysis that variations in acid concentration and salt concentration affect the electrical conductivity, variations in bath temperature do not, and that there are no interactions. To go one step further, we might investigate the *magnitudes* of the effects by studying graphs of means like those shown in Figures 13.4 and 13.5. Here we find that the conductivity increases as acid is added and decreases as salt is added; using the methods of Chapter 11 we

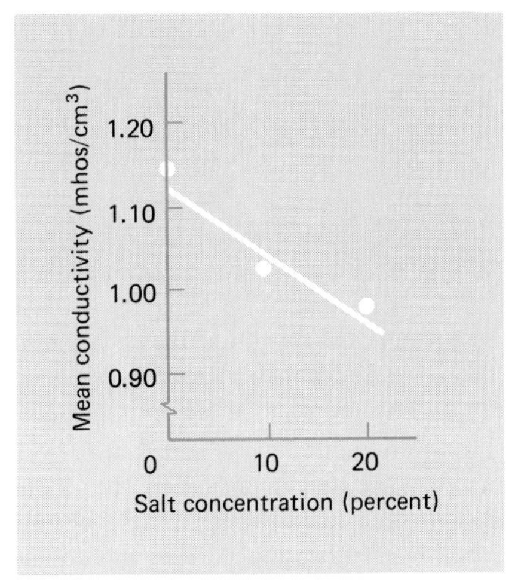

FIGURE 13.5
Effect of salt concentration.

might even fit lines, curves, or surfaces to describe the response surface relating conductivity to the variables under consideration.

The general computational procedure for a multifactor experiment is similar to the method illustrated here for a $4 \times 3 \times 2$ factorial experiment. We first analyze the data as a two-way classification (or whatever other design is being used) and then we analyze the treatment sum of squares into the components attributed to the various main effects and interactions. In general, the sum of squares for each main effect is found by adding the squares of the totals corresponding to the different levels of that factor, dividing by the number of observations comprising each of these totals, and then subtracting the correction term. The sum of squares for any interaction is found by adding the squares of all totals obtained by summing over those subscripts pertaining to factors not involved in the interaction, dividing by the number of observations comprising each of these totals, and then subtracting the correction term *and* all sums of squares corresponding to main effects and fewer-factor interactions involving the factors contained in that interaction.

EXERCISES

13.1 To determine optimum conditions for a plating bath, the effects of sulfone concentration and bath temperature on the reflectivity of the plated metal are studied in a 2×5 factorial experiment. The results of three replicates are as follows:

Concentration (grams/liter)	*Temperature (degrees F)*	*Reflectivity* Rep. 1	Rep. 2	Rep. 3
5	75	35	39	36
5	100	31	37	36
5	125	30	31	33
5	150	28	20	23
5	175	19	18	22
10	75	38	46	41
10	100	36	44	39
10	125	39	32	38
10	150	35	47	40
10	175	30	38	31

Analyze these results and determine the bath condition or conditions that produce the highest reflectivity. Also construct a 0.95 confidence interval for the reflectivity of the plating bath corresponding to these optimum conditions.

13.2 A spoilage retarding ingredient is added in brewing beer. To determine the extent to which the taste of the beer is affected by the amount of this ingredient added to each batch, and how such taste changes might depend on the age of the beer, a 3×4 factorial experiment in two replications was designed. The taste of the beer was rated on a scale of 0, 1, 2, or 3 (3 being the most desirable) by a panel of trained experts, who reported the following mean ratings:

Amount of ingredient (grams per batch)	*Aging period (weeks)*	*Mean ratings*	
		Rep. 1	*Rep. 2*
2	2	2.1	1.6
2	4	2.6	1.9
2	6	2.9	2.4
3	2	1.4	1.7
3	4	1.9	2.2
3	6	2.3	2.7
4	2	0.5	0.9
4	4	1.2	0.8
4	6	1.7	1.4
5	2	1.0	1.6
5	4	2.2	1.3
5	6	2.3	2.1

(a) Analyze the data first as a two-way classification with 12 treatments and two blocks (replicates).

(b) Calculate the sums of squares corresponding to the main effects and interaction, and present the results in an analysis-of-variance table.

(c) Interpret the results of the experiment.

13.3 Suppose that in the experiment described on page 401 it is desired to determine also whether there is an interaction between the detergents and the washing machines, that is, whether one detergent might perform better in Machine 1, another might perform better in Machine 2, and so forth. Combining the data on page 402 with the following replicate of the experiment, test for a significant interaction and discuss the results.

	Machine 1	*Machine 2*	*Machine 3*
Detergent A	39	42	58
Detergent B	44	46	48
Detergent C	34	47	45
Detergent D	47	45	57

13.4 Given the two replications of a 2 × 3 factorial experiment, calculate the analysis-of-variance table using the formulas on page 435 rather than the shortcut formulas in the example.

Factor A	*Factor B*	*Rep. 1*	*Rep. 2*
1	1	15	21
1	2	1	3
1	3	10	8
2	1	1	1
2	2	16	14
2	3	5	13

13.5 To test the design of a tennis racket, a subject plays tennis with a given racket for 1 hour and then rates the racket on a scale from 1 to 10. Two different sizes of racket face (standard and large) are tested, using three string tensions. Eighteen players, provided with rackets at random, yield the following results:

Racket face	*String tension*	*Rep. 1*	*Rep. 2*	*Rep. 3*
Standard	Low	5	6	5
Standard	Medium	6	6	7
Standard	High	5	4	6
Large	Low	6	7	5
Large	Medium	8	9	7
Large	High	3	5	2

Perform an appropriate analysis of variance and determine the conditions, if any, under which the large racket face is preferred.

13.6 The following tables give the weights (in grams) of food ingested by two different strains of rats after having been deprived of food for the stated number of hours and then having been given the stated dosage of a certain drug:

Dosage 0.2 mg/kg	Replicate 1		Replicate 2	
	Strain A	*Strain B*	*Strain A*	*Strain B*
1 hour	8.77	7.59	9.07	6.02
4 hours	11.82	9.21	9.16	7.05
8 hours	14.65	15.35	16.08	12.01

Dosage 0.4 mg/kg	Replicate 1		Replicate 2	
	Strain A	*Strain B*	*Strain A*	*Strain B*
1 hour	8.76	6.13	5.63	5.87
4 hours	11.53	8.30	11.57	9.56
8 hours	14.46	9.26	10.30	10.13

Dosage 0.8 mg/kg	Replicate 1		Replicate 2	
	Strain A	*Strain B*	*Strain A*	*Strain B*
1 hour	3.01	3.81	4.42	4.35
4 hours	9.21	10.10	5.22	8.01
8 hours	6.10	11.16	7.27	8.17

Perform an appropriate analysis of variance and interpret the results.

13.7 To study the effects of ingot location (A), slab position (B), specimen preparation (C), and twisting temperature (D) on the number of turns required to break a steel specimen by twisting, the following observations were recorded:

A	*B*	*C*	*D*	*No. of turns* Rep. 1	Rep. 2
Top	1	Turn	2,100°F	24	22
Top	1	Turn	2,200	25	28
Top	1	Turn	2,300	41	39
Top	1	Grind	2,100	18	18
Top	1	Grind	2,200	33	27
Top	1	Grind	2,300	35	41
Top	2	Turn	2,100	22	19
Top	2	Turn	2,200	26	31
Top	2	Turn	2,300	37	43
Top	2	Grind	2,100	23	7
Top	2	Grind	2,200	30	26
Top	2	Grind	2,300	34	30
Mid	1	Turn	2,100	26	19
Mid	1	Turn	2,200	30	31
Mid	1	Turn	2,300	39	42
Mid	1	Grind	2,100	19	19
Mid	1	Grind	2,200	31	31
Mid	1	Grind	2,300	26	35
Mid	2	Turn	2,100	30	26
Mid	2	Turn	2,200	31	34
Mid	2	Turn	2,300	39	42
Mid	2	Grind	2,100	22	20
Mid	2	Grind	2,200	32	26
Mid	2	Grind	2,300	38	22
Bot	1	Turn	2,100	18	21
Bot	1	Turn	2,200	35	32
Bot	1	Turn	2,300	34	37
Bot	1	Grind	2,100	21	19
Bot	1	Grind	2,200	20	29
Bot	1	Grind	2,300	44	31
Bot	2	Turn	2,100	23	22
Bot	2	Turn	2,200	31	26
Bot	2	Turn	2,300	38	41
Bot	2	Grind	2,100	18	19
Bot	2	Grind	2,200	31	24
Bot	2	Grind	2,300	35	41

Analyze this experiment.

13.8 A market test was performed to evaluate shelf position, label color, and package size for a canned food product, with the following results:

Shelf position	*Label color*	*Package size (oz)*	*Sales (dollars)* *Day 1*	*Day 2*	*Day 3*
Low	Red	10	70.10	68.00	69.50
Low	Red	12	72.25	71.90	74.40
Low	Red	16	78.05	74.85	82.60
Low	Red	24	61.50	62.10	59.15
Low	Green	10	65.75	62.35	68.60
Low	Green	12	69.45	71.05	75.45
Low	Green	16	75.15	70.70	71.25
Low	Green	24	64.80	60.85	59.90
Medium	Red	10	94.10	90.20	88.05
Medium	Red	12	104.85	99.55	96.80
Medium	Red	16	109.10	105.80	112.60
Medium	Red	24	59.90	62.50	54.75
Medium	Green	10	88.95	91.10	90.15
Medium	Green	12	100.60	94.05	101.35
Medium	Green	16	98.70	99.90	96.75
Medium	Green	24	62.50	53.85	59.40
High	Red	10	92.60	88.80	85.50
High	Red	12	100.55	102.15	99.10
High	Red	16	111.95	108.25	109.45
High	Red	24	61.40	65.20	59.70
High	Green	10	97.35	98.70	92.60
High	Green	12	120.65	115.45	108.65
High	Green	16	118.10	116.35	121.90
High	Green	24	70.30	65.05	71.40

Analyze this experiment.

13.9 Solve the four equations on page 433 for μ, α_1, β_1, and $(\alpha\beta)_{11}$ in terms of the population means μ_{ij1} corresponding to the four experimental conditions in the first replicate. Note that these equations serve as a guide for estimating the parameters in terms of the *sample means* corresponding to the various experimental conditions.

13.10 We illustrate computer calculation of the analysis-of-variance table given on page 436 using *MINITAB*. The three levels of A are coded as 1, 2, and 3, and the two levels of B are coded as 1 and 2.

```
READ  C1 C2 C3

  3.5    1    1
  3.0    1    1
  2.7    1    1
  2.2    1    2
  .  .  .

 TWOWAY obs in C1 factor levels in C2 C3

ANALYSIS OF VARIANCE

SOURCE         DF         SS         MS
FACTOR A        2    123.143     61.572
FACTOR B        1     17.209     17.209
INTERACTION     2      5.701      2.851
ERROR          12      3.327      0.277
TOTAL          17    149.380
```

The *TWOWAY* command combines the replicate sum of squares and the error sum of squares given in the example to obtain its ERROR sum of squares. (To test for differences among replicates, you must calculate $3 \times 2 \sum_{k=1}^{2} (\bar{y}_{..k} - \bar{y}_{...})^2$ separately and then subtract this value from 3.327, as in the example.) Repeat Exercise 13.5 using the computer.

13.3
2^n FACTORIAL EXPERIMENTS

Factorial experiments provide a powerful tool for understanding complex physical phenomena. Their key ingredient is the systematic variation of all input variables simultaneously. To see how important it is to change more than one variable at a time, suppose two input variables x_1 and x_2 are varied in an attempt to locate the maximum response. In the situation illustrated in Figure 13.6, moving (x_1, x_2) toward the upper right-hand corner will increase the response from 6 to 10 and even higher. However, if the experimenter fixes x_1 at 1.6 and varies x_2, it will look like a maximum occurs at $x_2 = 1.7$. If x_2 is then fixed at 1.7 and x_1 varied, it will then appear as if a maximum is confirmed. That is, the classical method of varying one variable at a time can lead to a false location for maximum response. Factorial

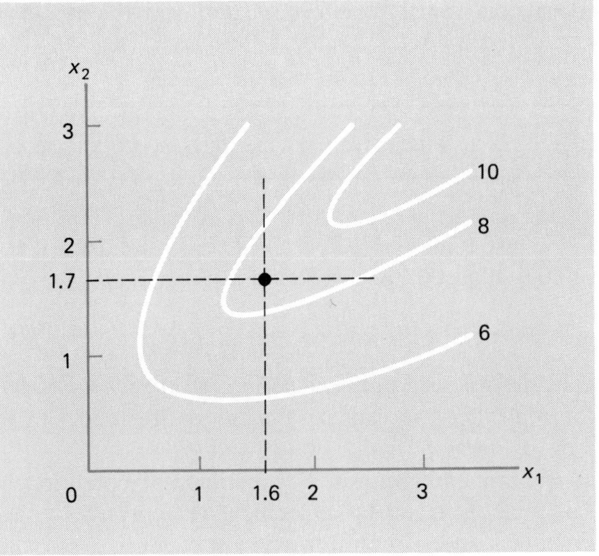

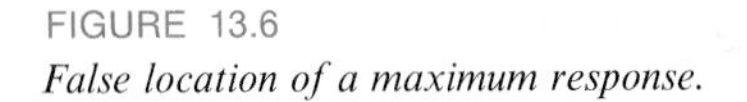
FIGURE 13.6
False location of a maximum response.

designs are well suited for studying the influence of several factors on a response. There are several reasons why factorial experiments are often performed with each factor taken at only two levels. Primarily, the number of experimental conditions in a factorial experiment increases multiplicatively with the number of levels of each factor; thus, if many factors are to be investigated simultaneously, it may be economically impossible to include more than two levels of each factor. Another important reason for treating **2^n factorial experiments** separately is that the analysis simplifies. Moreover, for $n = 2$ *or* 3, easily understood graphics can help communicate the main features of the data. Other advantages, such as the ease of confounding higher-order interactions and the adaptability of 2^n factorials to experiments involving fractional replication, will be discussed in succeeding sections.

Before we introduce some of the special notation used in connection with 2^n factorial experiments, let us point out that such experiments do have some drawbacks. Since each factor is measured only at two levels, it is impossible to judge whether the effects produced by variations in a factor are linear or, perhaps, parabolic or exponential. For this reason 2^n factorial experiments are often used in "screening experiments," which are followed up by experiments involving fewer factors (ordinarily those found to be "significant" individually or jointly in the screening experiment) taken at more than two levels.

In the analysis of a 2^n factorial experiment it is convenient to denote the two levels of each factor by 0 and 1 (instead of 1 and 2). Thus, the models used for the analysis of this kind of experiment differ from those of Section 13.2 only inasmuch as we now have $i = 0, 1$ instead of $i = 1, 2, \ldots, a$, $j = 0, 1$ instead of $j = 1, 2, \ldots, b$, and so forth. For instance, for a 2^3 factorial experiment the model on page 443 becomes

Model equation for 2^3 factorial experiment

$$y_{ijkl} = \mu + \alpha_i + \beta_j + \gamma_k + (\alpha\beta)_{ij} + (\alpha\gamma)_{ik} + (\beta\gamma)_{jk} + (\alpha\beta\gamma)_{ijk} + \rho_l + \varepsilon_{ijkl}$$

for $i = 0, 1, j = 0, 1, k = 0, 1$, and $l = 1, 2, \ldots, r$. The ε_{ijkl} are defined as before, and the parameters are now subject to the restrictions $\alpha_1 = -\alpha_0, \beta_1 = -\beta_0, \gamma_1 = -\gamma_0$, $(\alpha\beta)_{10} = (\alpha\beta)_{01} = -(\alpha\beta)_{11} = -(\alpha\beta)_{00}, \ldots$, and $\sum_{i=1}^{r} \rho_l = 0$. Note that besides the parameters for replicates we need only *one parameter of each kind*: that is, besides the parameters for replicates, we can express the entire model in terms of the parameters μ, α_0, β_0, γ_0, $(\alpha\beta)_{00}$, $(\alpha\gamma)_{00}$, $(\beta\gamma)_{00}$, and $(\alpha\beta\gamma)_{000}$.

A 2^n factorial experiment requires 2^n experimental conditions; since their number can be fairly large, it will be convenient to represent the experimental conditions by means of a special notation and list them in a so-called standard order. The notation consists of representing each experimental condition by the product of lowercase letters corresponding to the factors which are taken at level 1, called the "higher level." If a lowercase letter corresponding to a factor is missing, this means that the factor is taken at level 0, called the "lower level." Thus, in a three-factor experiment, *ac* represents the experimental condition where factors *A* and *C* are taken at the higher level and factor *B* is taken at the lower level, *c* represents the experimental condition where factor *C* is taken at the higher level and factors *A* and *B* are taken at the lower level, and so forth. The symbol "1" is used to denote the experimental condition in which all factors are taken at the lower level.

Although the experimental conditions are applied in a random order during the experiment itself, for the purpose of analyzing the results it is convenient to arrange them in a so-called **standard order**. For $n = 2$, this order is 1, *a*, *b*, *ab*, and for $n = 3$ it is the order shown in the following table:

Experimental condition	*Level of factor* *A*	*B*	*C*
1	0	0	0
a	1	0	0
b	0	1	0
ab	1	1	0
c	0	0	1
ac	1	0	1
bc	0	1	1
abc	1	1	1

Note that the symbols for the first four experimental conditions are like those for a two-factor experiment, and that the second four are obtained by multiplying each of the first four symbols by c. Similarly, the arrangement for $n = 4$ on page 459 is obtained by first listing the eight symbols for a three-factor experiment and then repeating the set with each symbol multiplied by d.

Throughout this and the preceding chapter we referred to the total of all observations corresponding to a given experimental condition as a treatment total, and we represented these totals by means of symbols such as $T_{i\cdot}$, $T_{ij\cdot\cdot}$, and so forth. Having introduced a special notation for the experimental conditions in a 2^n factorial experiment, we extend this notation by letting (1), (a), (b), (ab), (c), ... , be the treatment totals corresponding to experimental conditions 1, a, b, ab, c, ... Thus, in a three-factor experiment

$$(1) = \sum_{l=1}^{r} y_{000l} \qquad (a) = \sum_{l=1}^{r} y_{100l}$$

$$\vdots \qquad\qquad\qquad \vdots$$

$$(bc) = \sum_{l=1}^{r} y_{011l} \qquad (abc) = \sum_{l=1}^{r} y_{111l}$$

The simplification in the analysis referred to on page 455 consists of the fact that estimates of the various main effects and interactions, as well as the corresponding sums of squares can be expressed in terms of *linear combinations* of the treatment totals. To illustrate what we mean, let us consider the quantity

$$-(1) + (a) - (b) + (ab) - (c) + (ac) - (bc) + (abc)$$

which is a linear combination, with coefficients $+1$ and -1, of the treatment totals corresponding to the eight experimental conditions. Referring to the model equation on page 456 and making use of the relationships among the parameters (but leaving all details to the reader in Exercises 13.15, 13.16, and 13.17 on page 472) it can be shown that

$$-(1) + (a) - (b) + (ab) - (c) + (ac) - (bc) + (abc) = -8r\alpha_0 + \varepsilon_A$$

where ε_A is a corresponding linear combination of sums of the ε_{ijkl}. From Theorem 7.1 on page 241 it follows that ε_A is a value of a random variable whose distribution has zero mean; as a matter of fact, it can be shown that ε_A is a value of a random variable having a normal distribution with zero mean and the variance $8r\sigma^2$. Referring to the above linear combination as the **effect total** $[A]$ for factor A, we find that $-[A]/8r$ provides an estimate of α_0, the main effect for factor A, and it can be shown that $[A]^2/8r$ actually equals SSA, the sum of squares for the main effect of factor A.

Similarly analyzing the linear combination

$$(1) + (a) - (b) - (ab) - (c) - (ac) + (bc) + (abc)$$

the reader will be asked to show in Exercise 13.17 on page 473 that it equals $8r(\beta\gamma)_{00} + \varepsilon_{BC}$, where ε_{BC} is a corresponding linear combination of sums of the ε_{ijkl}. Referring to this linear combination of treatment totals as the effect total $[BC]$ for the two-way interaction of factors B and C, we find that $[BC]/8r$ provides an estimate of $(\beta\gamma)_{00}$, the effect of the BC interaction, and it can also be shown that $[BC]^2/8r$ equals $SS(BC)$, the sum of squares for the BC interaction. Proceeding in this fashion, we can present linear combinations of the treatment totals that yield estimates of the various other main effects and interactions, and whose squares, divided by $8r$, yield the corresponding sums of squares. These linear combinations, or effect totals, can easily be obtained with the use of the following **table of signs**:

(1)	(a)	(b)	(ab)	(c)	(ac)	(bc)	(abc)	*Effect totals*
1	1	1	1	1	1	1	1	$[I]$
−1	1	−1	1	−1	1	−1	1	$[A]$
−1	−1	1	1	−1	−1	1	1	$[B]$
1	−1	−1	1	1	−1	−1	1	$[AB]$
−1	−1	−1	−1	1	1	1	1	$[C]$
1	−1	1	−1	−1	1	−1	1	$[AC]$
1	1	−1	−1	−1	−1	1	1	$[BC]$
−1	1	1	−1	1	−1	−1	1	$[ABC]$

The entries of this table are the coefficients of the linear combinations of the treatment totals for the various main effects and interactions. As an aid for constructing similar tables for $n = 4$, $n = 5$, etc., note that for each main effect there is a "+1" when the factor is at the higher level and a "−1" when the factor is at the lower level. The signs for an interaction effect are obtained by multiplying the corresponding coefficients of *all* factors contained in the interaction. Thus, for $[AB]$ we multiply each sign for $[A]$ by the corresponding sign for $[B]$, getting

$$(-1)(-1) \quad (1)(-1) \quad (-1)(1) \quad (1)(1) \quad (-1)(-1) \quad (1)(-1) \quad (-1)(1) \quad (1)(1)$$

or

$$1 \quad -1 \quad -1 \quad 1 \quad 1 \quad -1 \quad -1 \quad 1$$

Note also that in the preceding table $[I]$ stands for the grand total of all the observations, so that $[I]^2/8r$ gives the correction term for calculating SST, SSE, SSR, and $SS(Tr)$.

Although we have illustrated the above shortcut method for obtaining the various main-effect and interaction sums of squares with reference to a 2^3 factorial experiment, the only difference in a 2^n factorial experiment with $n > 3$ is that we require a more extensive table of signs and that the respective sums of squares are obtained by dividing the squares of the effect totals by $r \cdot 2^n$.

To illustrate this technique and introduce a further simplification, let us consider the following 2^4 factorial experiment, designed to determine the effects of certain variables on the reliability of a rotary stepping switch. The factors studied were as follows:

Factor	*Low level*	*High level*
A. Lubrication	Dry	Lubricated
B. Dust protection	Unprotected	Enclosed in dust cover
C. Spark suppression	No	Yes
D. Current	0	0.5 Amps

Each switch was operated continuously until a malfunction occurred, and the number of hours of operation was recorded. The whole experiment was performed twice, with the following results:

Experimental condition	*Hours of operation* Rep. 1	*Rep. 2*	*Total*
1	828	797	1,625
a	997	948	1,945
b	735	776	1,511
ab	807	1,003	1,810
c	994	949	1,943
ac	1,069	1,094	2,163
bc	989	1,215	2,204
abc	889	1,010	1,899
d	593	813	1,406
ad	773	1,026	1,799
bd	740	922	1,662
abd	936	1,138	2,074
cd	748	970	1,718
acd	1,202	1,182	2,384
bcd	1,103	966	2,069
abcd	985	1,154	2,139
Total	14,388	15,963	30,351

Analyzing these data first as a two-way classification with 16 treatments and two replications (blocks), we obtain

$$C = \frac{(30{,}351)^2}{32} = 28{,}786{,}975$$

$$\begin{aligned} SST &= (828)^2 + (997)^2 + \cdots + (1{,}154)^2 - 28{,}786{,}975 \\ &= 744{,}876 \\ SS(Tr) &= \tfrac{1}{2}[(1{,}625)^2 + (1{,}945)^2 + \cdots + (2{,}139)^2] - 28{,}786{,}975 \\ &= 547{,}288 \\ SSR &= \tfrac{1}{16}[(14{,}388)^2 + (15{,}963)^2] - 28{,}786{,}975 \\ &= 77{,}520 \\ SSE &= 744{,}876 - 547{,}288 - 77{,}520 = 120{,}068 \end{aligned}$$

In order to subdivide the treatment sum of squares into SSA, $SSB, \ldots,$ and $SS(ABCD)$, we could construct a table of signs like the one on page 458, calculate the effect totals, and then divide the squares of the effect totals by $r \cdot 2^n = 2 \cdot 2^4 = 32$. For the A factor main effect we would thus obtain

$$\begin{aligned} [A] = &-1{,}625 + 1{,}945 - 1{,}511 + 1{,}810 - 1{,}943 + 2{,}163 - 2{,}204 \\ &+ 1{,}899 - 1{,}406 + 1{,}799 - 1{,}662 + 2{,}074 - 1{,}718 + 2{,}384 \\ &- 2{,}069 + 2{,}139 \\ = &\ 2{,}075 \end{aligned}$$

and

$$SSA = \frac{(2{,}075)^2}{32} = 134{,}551$$

These calculations are quite tedious, but they can be simplified considerably by using a further shortcut, called the **Yates' method**. This method of calculating the effect totals is illustrated in the following table. The experimental conditions and the corresponding totals are listed in standard order. In the column marked (1), the upper half is obtained by adding successive pairs of treatment totals, and the lower half is obtained by subtracting successive pairs. Thus, in column (1) we obtained

$$1{,}625 + 1{,}945 = 3{,}570$$
$$1{,}511 + 1{,}810 = 3{,}321$$
$$\vdots$$
$$2{,}069 + 2{,}139 = 4{,}208$$
$$1{,}945 - 1{,}625 = 320$$
$$1{,}810 - 1{,}511 = 299$$
$$\vdots$$
$$2{,}139 - 2{,}069 = 70$$

Experimental condition	*Treatment total*	(*1*)	(*2*)	(*3*)	(*4*)	*Identification*	*Sum of squares*
1	1,625	3,570	6,891	15,100	30,351	[*I*]	28,786,975
a	1,945	3,321	8,209	15,251	2,075	[*A*]	134,551
b	1,511	4,106	6,941	534	385	[*B*]	4,632
ab	1,810	4,103	8,310	1,541	−1,123	[*AB*]	39,410
c	1,943	3,205	619	−252	2,687	[*C*]	225,624
ac	2,163	3,736	−85	637	−773	[*AC*]	18,673
bc	2,204	4,102	805	−546	−179	[*BC*]	1,001
abc	1,899	4,208	736	−577	−1,119	[*ABC*]	39,130
d	1,406	320	−249	1,318	151	[*D*]	713
ad	1,799	299	−3	1,369	1,007	[*AD*]	31,689
bd	1,662	220	531	−704	889	[*BD*]	24,698
abd	2,074	−305	106	−69	−31	[*ABD*]	30
cd	1,718	393	−21	246	51	[*CD*]	81
acd	2,384	412	−525	−425	635	[*ACD*]	12,601
bcd	2,069	666	19	−504	−671	[*BCD*]	14,070
abcd	2,139	70	−596	−615	−111	[*ABCD*]	385

Note that, in the lower half of column (1), the first total in each pair is subtracted from the second. Column (2) is then obtained by performing the identical operations on the entries of column (1), and columns (3) and (4) are obtained in the same manner from the entries in columns (2) and (3), respectively. Column (4), and in general column (n), gives the effect totals in standard order, as shown. Each sum of squares is then obtained as before, by squaring the corresponding effect total and then dividing the result by $r \cdot 2^n = 2 \cdot 2^4 = 32$.

Dividing the sums of squares by their degrees of freedom to obtain the mean squares, and dividing the various mean squares by the error mean square, we get the following analysis-of-variance table for the 2^4 factorial experiment:

Source of variation	*Degrees of freedom*	*Sum of squares*	*Mean square*	*F*
Replicates	1	77,520	77,520	9.68
Main effects:				
A	1	134,551	134,551	16.81
B	1	4,632	4,632	<1
C	1	225,624	225,624	28.19
D	1	713	713	<1
Two-factor interactions:				
AB	1	39,410	39,410	4.92
AC	1	18,673	18,673	2.33
AD	1	31,689	31,689	3.96
BC	1	1,001	1,001	<1
BD	1	24,698	24,698	3.09
CD	1	81	81	<1
Three-factor interactions:				
ABC	1	39,130	39,130	4.89
ABD	1	30	30	<1
ACD	1	12,601	12,601	1.57
BCD	1	14,070	14,070	1.76
Four-factor interactions:				
ABCD	1	385	385	<1
Error	15	120,068	8,005	
Total	31	744,876		

Since $F_{0.05} = 4.54$ and $F_{0.01} = 8.68$ for 1 and 15 degrees of freedom, we find that the replication effects as well as the effects of lubrication and spark suppression are significant at the 0.01 level, and that there are significant interactions at the 0.05

level between lubrication, dust protection, and spark suppression. The reader will be asked to interpret these results and estimate the magnitude of some of the effects in Exercise 13.18 on page 473.

13.4

THE GRAPHIC PRESENTATION OF 2^2 AND 2^3 EXPERIMENTS

Both the 2^2 and 2^3 factorial designs have the added advantage that dramatic graphical displays are available to convey the results of the experiment. In spite of its simplicity, even the 2^2 design is a powerful tool to improve products and processes. The key is the decision to perform any designed experiment at all.

Suppose the yield of a new chemical process for growing crystals needs improvement. Because temperature and ph are thought to influence yield, a 2^2 design is attempted and the following results obtained for yield.

	Temperature	*ph*	*Rep. 1*	*Rep. 2*	*Total*
(1)	300	2	10	14	24
(*a*)	350	2	21	19	40
(*b*)	300	3	17	15	32
(*ab*)	350	3	20	24	44

The means for each experimental condition are written at the corners of the square, representing the design, in Figure 13.7. For instance, $\bar{y}_{00.} = 24/2 = 12$ is attached to the lower left-hand corner.

It is clear from Figure 13.7 that changing temperature from low (300°) to high (350°) increases the yield substantially at both levels of ph. Changing ph has little or no effect. This figure is a very effective way to present the information contained in the experiment.

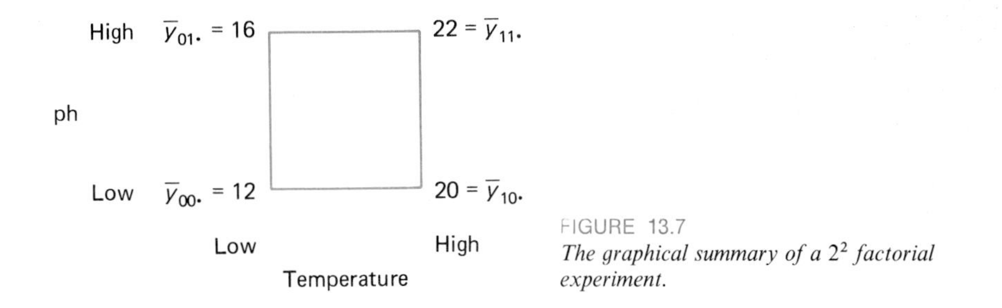

FIGURE 13.7
The graphical summary of a 2^2 factorial experiment.

If the interaction is significant, the square in Figure 13.7, with means at the corners, is an alternate display to the table of cell means (see page 440). If the interaction is not significant, we continue the analysis by finding point estimates and confidence intervals for the magnitude of the difference in the mean response that results when a factor is changed from the low to the high level. The population model, analogous to the 2^3 model on page 456, is

$$y_{ijk} = \mu + \alpha_i + \beta_j + (\alpha\beta)_{ij} + \rho_k + \varepsilon_{ijk}$$

with $\alpha_1 = -\alpha_0$. Consequently, the difference in mean response $\alpha_1 - \alpha_0 = 2\alpha_1 = -2\alpha_0$ is twice each factor A effect term. We estimate the difference in mean response $\alpha_1 - \alpha_0$ by

$$\bar{y}_{1..} - \bar{y}_{0..} = \tfrac{1}{2}(\bar{y}_{10.} + \bar{y}_{11.}) - \tfrac{1}{2}(\bar{y}_{01.} + \bar{y}_{00.}) = \tfrac{1}{2}(20 + 22) - \tfrac{1}{2}(16 + 12) = 7$$

That is, we take the average on the right side and subtract the average on the left side of the square, as indicated in Figure 13.8.

For the second factor, ph, we estimate the difference in mean response between the low and high level by

$$\bar{y}_{.1.} - \bar{y}_{.0.} = \tfrac{1}{2}(\bar{y}_{01.} + \bar{y}_{11.}) - \tfrac{1}{2}(\bar{y}_{10.} + \bar{y}_{00.}) = \tfrac{1}{2}(16 + 22) - \tfrac{1}{2}(20 + 12) = 3$$

To estimate the interaction, we note that $\bar{y}_{10.} - \bar{y}_{00.}$ gives the increase in yield at $ph = 2$, whereas $\bar{y}_{11.} - \bar{y}_{01.}$ gives the increase in yield at $ph = 3$. The average of their difference estimates the interaction

$$\begin{aligned}\tfrac{1}{2}(\bar{y}_{11.} - \bar{y}_{01.}) - \tfrac{1}{2}(\bar{y}_{10.} - \bar{y}_{00.}) &= \tfrac{1}{2}(\bar{y}_{11.} - \bar{y}_{10.} - \bar{y}_{01.} + \bar{y}_{00.}) \\ &= \tfrac{1}{2}(22 - 20 - 16 + 12) = -1\end{aligned}$$

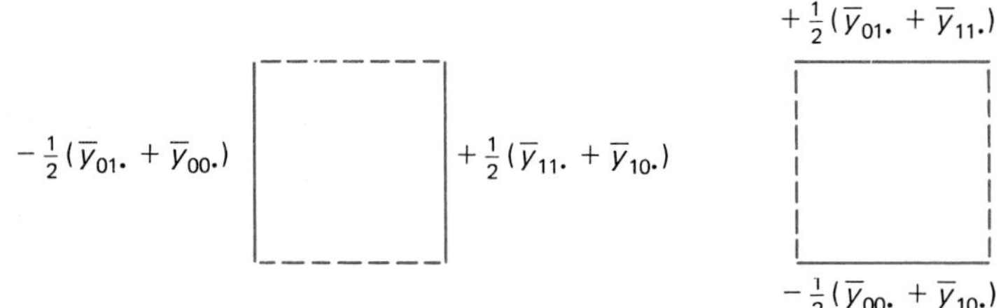

(a) Effect of A

(b) Effect of B

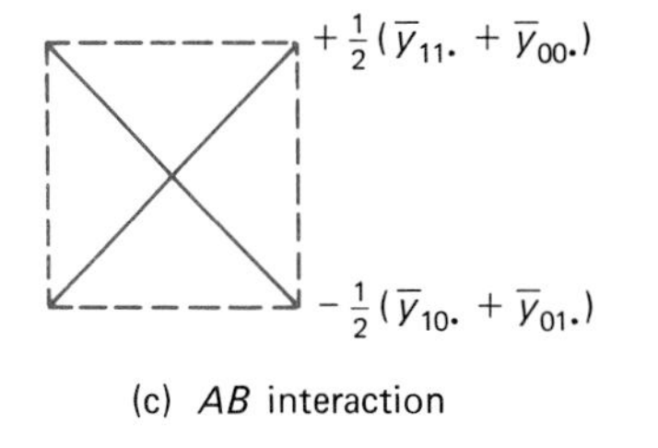

(c) AB interaction

FIGURE 13.8
The signs for the $\bar{y}_{ij.}$ to estimate effects in a 2^2 design.

In Exercise 13.21, the reader is asked to verify that the preceding estimates are equal to twice the effects estimated in the analysis of variance model, or $[A]/2r$, $[B]/2r$, and $[AB]/2r$, respectively.

Each estimator of an effect is a linear combination of four independent $\bar{y}_{ij.}$'s and so has the same variance. This common variance is estimated by s^2/r where s^2 is the error mean square which has $(2^2 - 1)(r - 1)$ degrees of freedom. Here $s^2 = 18/3 = 6$ and $t_{0.025} = 3.182$ with $(2^2 - 1)(2 - 1) = 3$ degrees of freedom (see Exercise 13.21), so the resulting confidence intervals are

$$\text{temperature effect:} \quad \bar{y}_{1..} - \bar{y}_{0..} \pm t_{0.025}\sqrt{\frac{s^2}{r}} = 7 \pm 5.51, \quad \text{or} \quad 1.49 \text{ to } 12.51.$$

$$\text{ph effect:} \quad \bar{y}_{.1.} - \bar{y}_{.0.} \pm t_{0.025}\sqrt{\frac{s^2}{r}} = 3 \pm 5.51, \quad \text{or} \quad -2.51 \text{ to } 8.51.$$

temperature × ph interaction:

$$\tfrac{1}{2}(\bar{y}_{11.} - \bar{y}_{10.} - \bar{y}_{01.} + \bar{y}_{00.}) \pm t_{0.025}\sqrt{\frac{s^2}{r}}$$

$$= -1 \pm 5.51, \quad \text{or} \quad -6.51 \text{ to } 4.51$$

Here the confidence interval for interaction covers 0, so we neglect interaction. The same is true for the ph effect. We are 95% confident that changing temperature from 300 degrees to 350 degrees will increase the mean yield by anywhere from 1.49 to 12.51 units.

We now turn to graphic displays for the 2^3 factorial design. In order to help expand the companies market for a plastic wrapping material, engineers were asked to improve the opacity. They felt that three factors, rate of extrusion, amount of an additive, and nozzle setting, might have an effect. Two levels were selected for each factor and a 2^3 factorial experiment produced the observations:

Factor A Rate	*Factor B Amount additive*	*Factor C Nozzle setting*	*Rep. 1*	*Rep. 2*
0	0	0	4.5	4.1
1	0	0	3.8	3.4
0	1	0	3.1	4.3
1	1	0	7.2	6.8
0	0	1	5.4	5.0
1	0	1	4.5	4.9
0	1	1	4.2	5.4
1	1	1	7.3	6.9

A cube representing the factors is shown in Figure 13.9. At the corners of the cube, we have attached the mean response for that set of experimental conditions. For instance, $(7.3 + 6.9)/2 = 7.1$ appears at the upper right corner of the front face. It is clear from Figure 13.9 that the response increases as both Factor A and Factor B are simultaneously changed from their low to high levels. This is an interaction effect since just changing one of the factors does not always increase the response. Factor C, nozzle setting, also seems to have an effect.

We now verify these conclusions. From the model on page 456, the difference in mean response between the low and high level of factor A is $\alpha_1 - \alpha_0 = 2\alpha_1 = -2\alpha_0$, or twice the effect term. We estimate this mean difference by

$$\begin{aligned}\bar{y}_{1\ldots} - \bar{y}_{0\ldots} &= \tfrac{1}{4}(\bar{y}_{100.} + \bar{y}_{101.} + \bar{y}_{110.} + \bar{y}_{111.}) - \tfrac{1}{4}(\bar{y}_{000.} + \bar{y}_{001.} + \bar{y}_{010.} + \bar{y}_{011.}) \\ &= \tfrac{1}{4}(3.6 + 4.7 + 7.0 + 7.1) - \tfrac{1}{4}(4.3 + 5.2 + 3.7 + 4.8) = 1.1\end{aligned}$$

which is the average of the four $\bar{y}_{ijk.}$ on the front face minus the average of the four observations on the back face (see Figure 13.10). Alternatively, it is the average of the four increases along each edge. Note that each $\bar{y}_{ijk.}$ in the first set of parentheses is associated with an experimental condition that has a term a in the special notation. Thus the signs are the same as those used with Yate's algorithm.

The estimates of the difference in mean response, when changing from the low to the high level, have a similar interpretation for the other two factors:

$$\bar{y}_{.1..} - \bar{y}_{.0..} = 1.2$$

$$\bar{y}_{..1.} - \bar{y}_{..0.} = 0.8$$

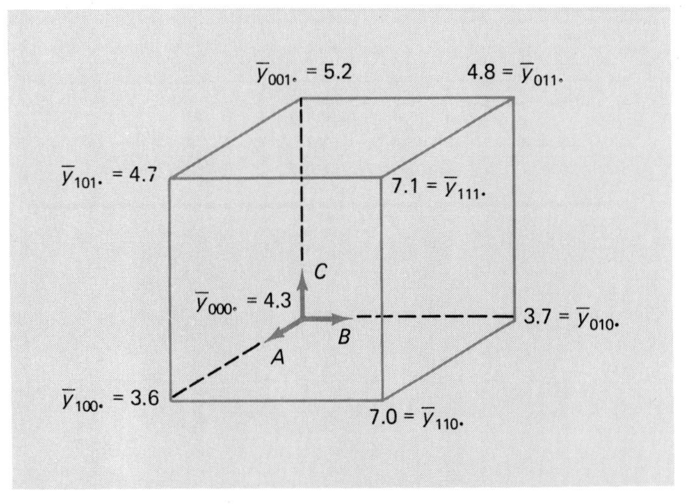

FIGURE 13.9
Graphic summary of a 2^3 design experiment.

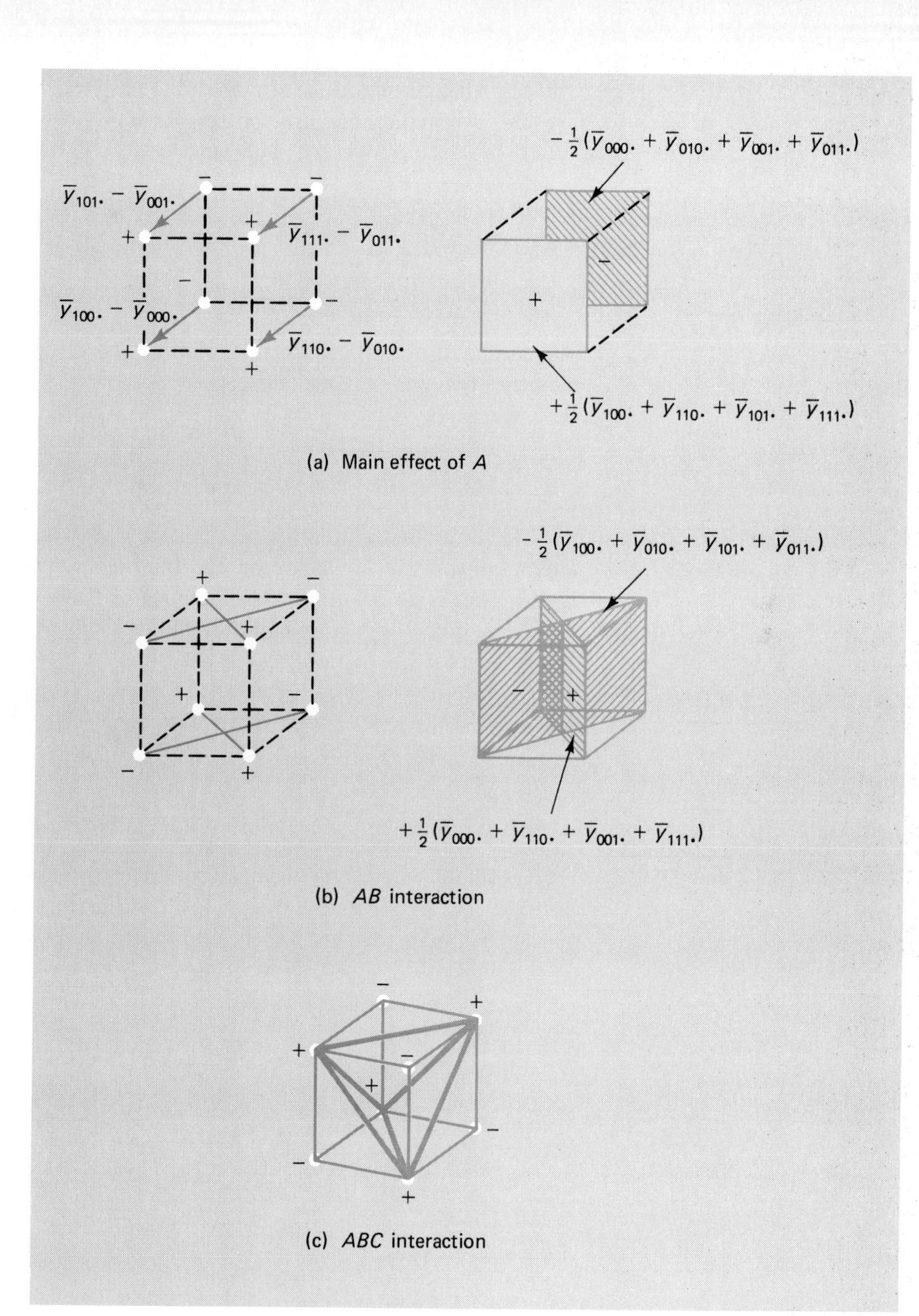

FIGURE 13.10
The signs for estimating effects in a 2^3 design.

The AB interaction is estimated as the average of the interaction on the top face and the interaction on the bottom face. Figure 13.10 gives the signs for the $\bar{y}_{ijk.}$. For the AB interaction, we estimate

$$\begin{aligned}
&\tfrac{1}{2}(\bar{y}_{11..} - \bar{y}_{10..} - \bar{y}_{01..} + \bar{y}_{00..}) \\
&\quad = \tfrac{1}{4}(\bar{y}_{111.} - \bar{y}_{101.} - \bar{y}_{011.} + \bar{y}_{001.}) + \tfrac{1}{4}(\bar{y}_{110.} - \bar{y}_{100.} - \bar{y}_{010.} + \bar{y}_{000.}) \\
&\quad = \tfrac{1}{4}(7.1 - 4.7 - 4.8 + 5.2) + \tfrac{1}{4}(7.0 - 3.6 - 3.7 + 4.3) = 1.7
\end{aligned}$$

The combination of signs for the eight means are obtained in a similar manner for the other two factor interactions.

$$\tfrac{1}{2}(\bar{y}_{1.1.} - \bar{y}_{1.0.} - \bar{y}_{0.1.} + \bar{y}_{0.0.}) = -0.2$$

$$\tfrac{1}{2}(\bar{y}_{.11.} - \bar{y}_{.10.} - \bar{y}_{.01.} + \bar{y}_{.00.}) = -0.2$$

The three factor interaction ABC is a measure of the difference of the AB interaction on the top face and the interaction on the bottom face (see Figure 13.10). That is, it quantifies the influence of C on the AB interaction. (There is a symmetry here and, up to a minus sign, the same result is obtained starting with any two-factor interaction.)

$$\begin{aligned}
&\tfrac{1}{4}(\bar{y}_{111.} - \bar{y}_{101.} - \bar{y}_{011.} - \bar{y}_{001.}) - \tfrac{1}{4}(\bar{y}_{110.} - \bar{y}_{100.} - \bar{y}_{010.} + \bar{y}_{000.}) \\
&\quad = \tfrac{1}{4}(7.1 - 4.7 - 4.8 + 5.2) - \tfrac{1}{4}(7.0 - 3.6 - 3.7 + 4.3) = -0.3
\end{aligned}$$

The confidence intervals are obtained by adding and subtracting $t_{\alpha/2}\sqrt{\frac{s^2}{2r}}$ to each estimate, where s^2 is the mean square error based on $(2^3 - 1)(r - 1)$ degrees of freedom (see Exercise 13.24), and $t_{\alpha/2}$ also has this number of degrees of freedom.

We summarize this experiment with the 95% confidence intervals. Here $s^2 = 1.88/7$.

rate effect: $\quad \bar{y}_{1...} - \bar{y}_{0...} \pm t_{0.025}\sqrt{\frac{s^2}{2r}} = 1.1 \pm 0.61, \quad$ or $\quad 0.49$ to 1.71

additive effect: $\quad \bar{y}_{.1..} - \bar{y}_{.0..} \pm t_{0.025}\sqrt{\frac{s^2}{2r}} = 1.2 \pm 0.61, \quad$ or $\quad 0.59$ to 1.81

nozzle effect: $\quad \bar{y}_{..1.} - \bar{y}_{..0.} \pm t_{0.025}\sqrt{\frac{s^2}{2r}} = 0.8 \pm 0.61, \quad$ or $\quad 0.19$ to 1.41

rate × additive interaction: $\quad \tfrac{1}{2}(\bar{y}_{11..} - \bar{y}_{01..} - \bar{y}_{10..} + \bar{y}_{00..}) \pm t_{0.025}\sqrt{\frac{s^2}{2r}}$

$= 1.7 \pm 0.61, \quad$ or $\quad 1.09$ to 2.31

rate × nozzle interaction: $\quad \tfrac{1}{2}(\bar{y}_{1.1.} - \bar{y}_{0.1.} - \bar{y}_{1.0.} + \bar{y}_{0.0.}) \pm t_{0.025}\sqrt{\frac{s^2}{2r}}$

$= -0.2 \pm 0.61, \quad$ or $\quad -0.81$ to 0.41

additive × nozzle interaction: $\frac{1}{2}(\bar{y}_{.11.} - \bar{y}_{.01.} - \bar{y}_{.10.} + \bar{y}_{.00.}) \pm t_{0.025}\sqrt{\frac{s^2}{2r}}$

$$= -0.2 \pm 0.61, \quad \text{or} \quad -0.81 \text{ to } 0.41$$

rate × additive × nozzle interaction: $\frac{1}{4}(\bar{y}_{111.} - \bar{y}_{101.} - \bar{y}_{011.} + \bar{y}_{001.})$

$$- \tfrac{1}{4}(\bar{y}_{110.} - \bar{y}_{100.} - \bar{y}_{010.} + \bar{y}_{000.})$$

$$\pm t_{0.025}\sqrt{\frac{s^2}{2r}}$$

$$= -0.3 \pm 0.61, \quad \text{or} \quad -0.91 \text{ to } 0.31.$$

Here the confidence interval for the three factor interaction covers 0, so we neglect this interaction. However, we are 95% confident that the interval from 1.09 to 2.31 contains the rate × additive interaction, so we cannot interpret the *rate* and *additive* factors individually. Factor C, nozzle setting, is not involved in any significant interactions. With 95% confidence, we conclude that using the high nozzle setting will increase the opacity of the plastic wrap from between 0.19 to 1.41 units.

EXERCISES

13.11 To determine the effect on taste of three different factors in manufacturing soft-drink cans, an experiment was performed where the taste of one soft drink was rated by a judge on a scale from 1 to 10. The results are as follows:

A *Lubricant*	*B* *Heat*	*C* *Resin*	*Ratings* *Rep. 1*	*Rep. 2*
Fresh	Unheated	*A*	6	8
Fresh	Unheated	*B*	8	7
Fresh	Heated	*A*	9	9
Fresh	Heated	*B*	1	2
Aged	Unheated	*A*	6	7
Aged	Unheated	*B*	6	8
Aged	Heated	*A*	9	8
Aged	Heated	*B*	2	3

(a) Analyze the results first as a two-way classification with 7 degrees of freedom for treatments and 1 degree of freedom for blocks (replicates).

(b) Use an appropriate table of signs to calculate the effect totals [*A*], [*B*], [*C*], [*AB*], [*AC*], [*BC*], [*ABC*].

(c) Using the results obtained in part (b) find the sums of squares corresponding to the main effects and interactions, and check their total against the treatment sum of squares obtained in part (a).

(d) Arrange the data with treatment combinations in standard order, and use the Yates method to find the effect totals. Compare with the results obtained in part (b).

(e) Construct an analysis of variance table and analyze the experiment.

13.12 A screening experiment was conducted to determine what factors are influential in controlling the final phosphorus content of steel produced in a converter. The levels of the factors studied and the experimental results are contained in the following table:

E Pouring temp. (°F)	*D* Lime ratio	*C* Oxygen	*B* Original phosphorus	*A* Original manganese	*Final phosphorus* Rep. 1	Rep. 2
2,400	3	5%	0.15%	1%	0.003%	0.001%
2,400	3	5	0.15	3	0.004	0.009
2,400	3	5	0.30	1	0.002	0.008
2,400	3	5	0.30	3	0.015	0.007
2,400	3	15	0.15	1	0.002	0.005
2,400	3	15	0.15	3	0.011	0.006
2,400	3	15	0.30	1	0.004	0.001
2,400	3	15	0.30	3	0.002	0.004
2,400	4	5	0.15	1	0.000	0.003
2,400	4	5	0.15	3	0.008	0.002
2,400	4	5	0.30	1	0.003	0.007
2,400	4	5	0.30	3	0.005	0.012
2,400	4	15	0.15	1	0.010	0.006
2,400	4	15	0.15	3	0.006	0.001
2,400	4	15	0.30	1	0.006	0.014
2,400	4	15	0.30	3	0.011	0.015
2,600	3	5	0.15	1	0.003	0.007
2,600	3	5	0.15	3	0.007	0.004
2,600	3	5	0.30	1	0.011	0.005
2,600	3	5	0.30	3	0.010	0.017
2,600	3	15	0.15	1	0.004	0.008
2,600	3	15	0.15	3	0.019	0.013
2,600	3	15	0.30	1	0.004	0.008
2,600	3	15	0.30	3	0.017	0.023
2,600	4	5	0.15	1	0.007	0.004
2,600	4	5	0.15	3	0.015	0.009
2,600	4	5	0.30	1	0.004	0.011
2,600	4	5	0.30	3	0.010	0.006
2,600	4	15	0.15	1	0.017	0.011
2,600	4	15	0.15	3	0.005	0.010
2,600	4	15	0.30	1	0.014	0.009
2,600	4	15	0.30	3	0.016	0.011

Analyze the results of this experiment.

13.13 An experiment was conducted to determine the effects of the following factors on the gain of a semiconductor device:

Factor	*Level 0*	*Level 1*
A. Location of assembly	Laboratory	Production line
B. Partial pressure of controlling material	10^{-15}	10^{-4}
C. Relative humidity	1%	30%
D. Aging time	72 hours	144 hours

The results were as follows:

Experimental condition	*Gain*	
	Rep. 1	*Rep. 2*
1	39.0	43.2
a	31.8	43.7
b	47.0	51.4
ab	40.9	40.3
c	43.8	40.5
ac	29.3	52.9
bc	34.8	48.2
abc	45.6	58.2
d	40.1	41.9
ad	42.0	40.5
bd	54.9	53.0
abd	39.9	40.2
cd	43.1	40.2
acd	30.1	39.9
bcd	35.6	53.7
abcd	41.4	49.5

Perform an appropriate analysis of variance and interpret the results.

13.14 A study was performed to examine the effect of five different factors on the time required for odorants in natural gas to reach the surface once a leak has occurred. The results of the experiment were reported as follows:

A Odorant	*B* Odorant concentration mg/ml	*C* Soil moisture	*D* Flow rate ml/min	*E* Tempera- ture	Time (hours) Rep. 1	Rep. 2
A	4	8%	20	50°F	20	20
A	4	8	20	70	17	18
A	4	8	80	50	20	19
A	4	8	80	70	18	19
A	4	16	20	50	12	13
A	4	16	20	70	14	16
A	4	16	80	50	15	14
A	4	16	80	70	17	16
A	16	8	20	50	10	9
A	16	8	20	70	9	8
A	16	8	80	50	8	9
A	16	8	80	70	7	8
A	16	16	20	50	7	6
A	16	16	20	70	6	4
A	16	16	80	50	5	6
A	16	16	80	70	4	5
B	4	8	20	50	28	29
B	4	8	20	70	27	26
B	4	8	80	50	25	27
B	4	8	80	70	28	28
B	4	16	20	50	24	25
B	4	16	20	70	26	25
B	4	16	80	50	23	24
B	4	16	80	70	24	25
B	16	8	20	50	26	24
B	16	8	20	70	23	25
B	16	8	80	50	27	26
B	16	8	80	70	25	26
B	16	16	20	50	19	18
B	16	16	20	70	16	17
B	16	16	80	50	20	19
B	16	16	80	70	17	15

Analyze the results of this experiment.

13.15 Writing the treatment total (a) as the sum of the corresponding observations y_{100l} and substituting for these observations the expressions given by the model equation on page 456, it can be shown that

$$(a) = r[\mu + \alpha_1 + \beta_0 + \gamma_0 + (\alpha\beta)_{10} + (\alpha\gamma)_{10} + (\beta\gamma)_{00} + (\alpha\beta\gamma)_{100}] + \sum_{l=1}^{r} \varepsilon_{100l}$$

Making use of the restrictions imposed on the parameters, rewrite this expression for (a) in terms of the parameters μ, α_0, β_0, γ_0, $(\alpha\beta)_{00}$, $(\alpha\gamma)_{00}$, $(\beta\gamma)_{00}$, and $(\alpha\beta\gamma)_{000}$.

13.16 Duplicating the work of Exercise 13.15, express (1), (b), (ab), (c), (ac), (bc), and (abc) in terms of the parameters μ, α_0, β_0, γ_0, $(\alpha\beta)_{00}$, $(\alpha\gamma)_{00}$, $(\beta\gamma)_{00}$, and $(\alpha\beta\gamma)_{000}$.

13.17 Using the results of Exercises 13.15 and 13.16, verify the expressions for $[A]$ and $[BC]$ obtained on page 457. Also express ε_A in terms of the quantities ε_{ijkl}.

13.18 Interpret the results of the analysis of variance given by the table on page 462, and estimate the magnitude of the significant effects.

13.19 A computational check on the sums of squares obtained for the various main effects and interactions is that their *sum* must equal the treatment sum of squares obtained by analyzing the data first as a two-way classification. Perform this check on the sums of squares given in the table on page 462.

13.20 If it is desired to find an expression for an effect total without constructing a complete table of signs, we can use the following method, illustrated by finding $[ABC]$ for a 2^4 factorial experiment. We take the expression $(a \pm 1)(b \pm 1)(c \pm 1)(d \pm 1)$ with a "+" if the corresponding letter does *not* appear in the symbol for the main effect or interaction for which we want to calculate an effect total, and a "−" if the corresponding letter does appear. Thus, for finding $[ABC]$ we write

$$\begin{aligned}(a-1)(b-1)(c-1)(d+1) &= abcd + abc - abd - acd - bcd \\ &\quad - ab - ac + ad - bc + bd + cd \\ &\quad + a + b + c - d - 1\end{aligned}$$

and after arranging the terms in standard order and adding parentheses, we finally obtain

$$\begin{aligned}[ABC] &= -(1) + (a) + (b) - (ab) + (c) - (ac) - (bc) + (abc) - (d) \\ &\quad + (ad) + (bd) - (abd) + (cd) - (acd) - (bcd) + (abcd)\end{aligned}$$

(a) Use this method to express $[B]$, $[AC]$, and $[ABC]$ in terms of the treatment totals in a 2^3 factorial experiment.

(b) Use this method to express $[AC]$ and $[BCD]$ in terms of the treatment totals in a 2^4 factorial experiment.

13.21 Referring to the 2^2 factorial design example on page 463, use the table of signs

(I)	*(a)*	*(b)*	*(ab)*	*Total*
1	1	1	1	$[I]$
−1	1	−1	1	$[A]$
−1	−1	1	1	$[B]$
1	−1	−1	1	$[AB]$

(a) to verify the calculation of the mean square error;
(b) to verify the expression $\bar{y}_{.1.} - \bar{y}_{.0.} = [B]/2r$;
(c) to verify the expression

$$\frac{1}{2}(\bar{y}_{11.} + \bar{y}_{00.}) - \frac{1}{2}(\bar{y}_{10.} + \bar{y}_{01.}) = \frac{[AB]}{2r}$$

13.22 Referring to the example on page 463, construct an analysis-of-variance table.

13.23 Given the following observations,

Factor A	*Factor B*	*Rep. 1*	*Rep. 2*
0	0	14	6
1	0	15	21
0	1	11	17
1	1	24	16

summarize the experiment according to the visual procedure given in Section 13.4. Interpret the effects based on the confidence intervals.

13.24 Referring to the example on page 465, the error sum of squares is

$$SSE = \sum_{i=0}^{1} \sum_{j=0}^{1} \sum_{k=0}^{1} \sum_{l=1}^{r} (y_{ijkl} - \bar{y}_{ijk.})^2 - 2^n \sum_{l=1}^{r} (\bar{y}_{...l} - \bar{y}_{....})^2$$

and $s^2 = SSE/(2^3 - 1)(r - 1)$. Verify

(a) the numerical value for s^2;
(b) the calculations for the confidence intervals.

13.25 The effect on engine wear of oil viscosity, temperature, and a special additive were tested using a 2^3 factorial design. Given the following results from the experiment,

Factor A Viscosity	*Factor B Temperature*	*Factor C Additive*	*Rep. 1*	*Rep. 2*
0	0	0	3.7	4.1
1	0	0	4.6	5.0
0	1	0	3.1	2.7
1	1	0	3.4	3.8
0	0	1	3.4	3.6
1	0	1	5.3	4.9
0	1	1	2.4	3.2
1	1	1	4.7	4.1

summarize the experiment according to the visual procedure given in Section 13.4. Interpret the effects based on the confidence intervals.

13.5

CONFOUNDING IN A 2^n FACTORIAL EXPERIMENT

In some experiments it is impossible to run all the required experimental conditions in one block. For example, if a 2^3 factorial experiment involves eight combinations of paint pigments that are to be applied to a surface and baked in an oven that can accommodate only four specimens, it becomes necessary to divide the eight treatments into two blocks (oven runs) in each replicate. As we have pointed out earlier, if the block size is too small to accommodate all treatments, this requires special **incomplete-block designs**.

When the experimental conditions are distributed over several blocks, one or more of the effects may become confounded (inseparable) with possible block effects, that is, between-block differences. For example, if in the 2^3 factorial experiment referred to in the preceding paragraph, experimental conditions *a*, *ab*, *ac*, and *abc* are included in one oven run (Block 1) and experimental conditions 1, *b*, *c*, and *bc* are included in a second oven run (Block 2), then the block effect, the difference between the two block totals, is given by

$$[(a) + (ab) + (ac) + (abc)] - [(1) + (b) + (c) + (bc)]$$

Referring to the table of signs on page 458, we observe that this quantity is, in fact, the effect total $[A]$, so that the estimate of the main effect of factor A is **confounded** with blocks. Note that all other factorial effects remain unconfounded; for each other effect total there are two $+1$ coefficients and two -1 coefficients in each block, so that the block effects cancel out. This kind of argument can also be used to decide what experimental conditions to put into each block to confound a given main effect or interaction. For instance, had we wanted to confound the ABC interaction with blocks in the above example, we could have put experimental conditions *a*, *b*, *c*, and *abc*, whose totals have $+1$ coefficients in $[ABC]$, into one block, and experimental conditions 1, *ab*, *ac*, and *bc*, whose totals have -1 coefficients, into another.

In general, confounding in a 2^n factorial experiment can be much more complicated than in the example just given. To avoid serious difficulties, we shall require that the number of blocks used is a power of 2, say 2^p. It turns out that the price paid for running a 2^n factorial experiment in 2^p blocks is that a total of $2^p - 1$ effects are confounded with blocks. To make it clear just which effects are confounded, and to indicate a method that can be used to confound only certain effects and no others, it is helpful to define the term "generalized interaction" as follows: the **generalized interaction** of two effects is the "product" of these effects, with like letters cancelled. Thus, the generalized interaction of AB and CD is $ABCD$, and the generalized interaction of ABC and BCD is $A\not{B}\not{C}\not{B}\not{C}D$, or AD. To confound a 2^n factorial experiment in 2^p blocks, the following method can be used: one selects any p effects for confounding, making sure that none is the generalized interaction of any of the others selected. Then, it can be shown that a further

$2^p - (p + 1)$ effects are automatically confounded with blocks; together with the p effects originally chosen, this gives a total of $2^p - 1$ confounded effects in the experiment. The additional confounded effects are, in fact, the generalized interactions of the p effects originally chosen.

To illustrate the construction of a confounded design, let us divide a 2^4 factorial experiment into four blocks so that desired effects are confounded with blocks. In actual practice one ordinarily confounds only the higher-order interactions (in the hope that they are nonexistent anyhow). Since we have decided upon four blocks, we have $2^p = 4$ and $p = 2$, and we shall arbitrarily select two higher-order interactions for confounding. If we were to select $ABCD$ and BCD, then their generalized interaction, A, would also be confounded. Thus, to avoid confounding any main effects, and to confound as few two-factor interactions as possible, we shall select ABD and ACD, noting that the BC interaction is also confounded. (Observe that it is impossible to avoid confounding at least one main effect or two-factor interaction in this experiment.)

In order to assign the 16 experimental conditions to four blocks, we first distribute them into two blocks, so that the ABD interaction is confounded with blocks. Referring to an appropriate table of signs, we put all treatments whose totals have a "+1" in the row for [ABD] into one block, all those whose totals have a "−1" into a second block, and we get the following blocks:

First block:	*a*	*b*	*ac*	*bc*	*d*	*abd*	*cd*	*abcd*
Second block:	1	*ab*	*c*	*abc*	*ad*	*bd*	*acd*	*bcd*

Note that each experimental condition in the first block has an *odd number* of letters in common with ABD, whereas each experimental condition in the second block has an *even number* of letters in common with ABD. This **odd-even rule** provides an alternative way of distributing experimental conditions among two blocks to confound a given effect, and it has the advantage that it does not require the construction of a complete table of signs.

So far, we have confounded the ABD interaction by dividing the 16 experimental conditions into two blocks; now we shall confound the ACD interaction by dividing each of these blocks into two blocks of four conditions each. Using the odd-even rule just described (or a table of signs), we obtain the following four blocks:

Block 1:	*a*	*bc*	*d*	*abcd*
Block 2:	*b*	*ac*	*abd*	*cd*
Block 3:	*ab*	*c*	*bd*	*acd*
Block 4:	*1*	*abc*	*ad*	*bcd*

By comparing these blocks with a table of signs, or, equivalently, by applying the odd-even rule, the reader will be asked to verify in Exercise 13.30 on page 484 that the BC interaction is also confounded with blocks, whereas all other effects are left unconfounded.

The analysis of a confounded 2^n factorial experiment is similar to that of an unconfounded experiment, with the exception that the sums of squares for the confounded effects are not computed, and we compute a block sum of squares as if the experiment consisted of br blocks rather than b blocks in each of r replicates.

Referring to our example of a 2^4 factorial experiment with the ABD, ACD, and BC interactions confounded, and using two replicates, we have the following *dummy* analysis-of-variance table:

Source of variation	*Degrees of freedom*
Blocks	7
Main effects	4
Unconfounded two-factor interactions	5
Unconfounded three-factor interactions	2
Four-factor interaction	1
Intrablock error	12
Total	31

The sum of squares for blocks is obtained, as usual, by adding the squares of the eight block totals, dividing the result by 4 (the number of observations in each block), and subtracting the correction term. The total sum of squares and the sums of squares for the unconfounded factorial effects are obtained in the usual way, and the sum of squares for the **intrablock error**, a measure of the variability *within blocks*, is obtained by subtraction.

To illustrate the analysis of a confounded 2^n factorial experiment, let us suppose that each replicate of the stepping-switch experiment described in the preceding section was actually run in four blocks, because only four mountings were available for the 16 switches. (The order of running the blocks is assumed to have been randomized within each replicate, and the assignment of switches is assumed to have been randomized within each block.) Assuming also that the

ABD, *ACD*, and *BC* interactions were confounded with the blocks as shown on page 476, we obtain the following block totals from the data on page 459:

	Block 1	*Block 2*	*Block 3*	*Block 4*
Replicate 1	3,564	3,488	3,743	3,593
Replicate 2	4,130	3,978	4,056	3,799

Thus, the sum of squares for blocks is given by

$$SS(Bl) = \frac{(3{,}564)^2 + (3{,}488)^2 + \cdots + (3{,}799)^2}{4} - 28{,}786{,}975$$

$$= 101{,}240$$

where the correction factor is the same as in the analysis on page 460.

Copying the total sum of squares and the sums of squares for the various unconfounded effects from the table on page 462, we obtain the analysis-of-variance table for the following confounded factorial experiment. In this analysis, the *A* and *C* main effects are again significant at the 0.01 level, but none of the other main effects or interactions is significant.

Source of variation	*Degrees of freedom*	*Sum of squares*	*Mean square*	*F*
Blocks	7	101,240	14,463	1.58
Main effects:				
A	1	134,551	134,551	14.68
B	1	4,632	4,632	<1
C	1	225,624	225,624	24.62
D	1	713	713	<1
Unconfounded two-factor interactions:				
AB	1	39,410	39,410	4.30
AC	1	18,673	18,673	2.04
AD	1	31,689	31,689	3.46
BD	1	24,698	24,698	2.69
CD	1	81	81	<1

(*continued*)

Source of variation	*Degrees of freedom*	*Sum of squares*	*Mean square*	*F*
Unconfounded three-factor interactions:				
ABC	1	39,130	39,130	4.27
BCD	1	14,070	14,070	1.54
Four-factor interaction:				
$ABCD$	1	385	385	<1
Intrablock error	12	109,980	9,165	
Total	31	744,876		

If there is replication, some of the lost information about the confounded effects can be recovered by a further breakdown of the above blocks sum of squares. This analysis consists of dividing the sum of squares for blocks into a component for each of the confounded effects, a component for replications, and a residual component called the *interblock error*, which is a measure of the variability *between blocks.* Copying the sum of squares for replicates as well as those for the BC, ABD, and ACD interactions from the analysis-of-variance table on page 462 and copying the blocks sum of squares from the preceding **intrablock analysis-of-variance** table, we obtain the following **interblock analysis-of-variance** table:

Source of variation	*Degrees of freedom*	*Sum of squares*	*Mean square*	*F*
Replicates	1	77,520	77,520	23.05
Confounded effects:				
BC	1	1,001	1,001	<1
ABD	1	30	30	<1
ACD	1	12,601	12,601	3.75
Interblock error	3	10,088	3,363	
Total (blocks)	7	101,240		

Note that the **interblock error** is obtained by subtraction and that the F-ratios are obtained by dividing the mean squares for the confounded effects and the mean square for replicates by the mean square for the interblock error. Only the F test for replicates is significant (at the 0.05 level). The small number of degrees of freedom for the interblock error implies that the sensitivity of these significance tests is *very poor*; in fact, it rarely pays to make this kind of interblock analysis unless the number of replications is relatively large.

13.6 FRACTIONAL REPLICATION

In studies involving complex production lines, chemical processes such as may be encountered in the petroleum, plastics, or metals industries, physical-chemical processes such as may be encountered in the electronics or space-technology industries, and in many other engineering studies, experimenters are often faced with a large and bewildering array of interrelated variables. The principles of factorial experimentation treated so far in this chapter help them to "sort out" these variables, to discover which of them have the greatest influence on the process under consideration, and to find what important interrelationships may exist.

There are, however, some serious limitations to the simultaneous study of a large number of factors. Even if each factor is assigned only two levels, one replicate of a six-factor experiment requires 64 observations; there are 128 observations in one replicate of a seven-factor experiment, and 1,024 observations in one replicate of a ten-factor experiment. The economic and practical limitations of these large numbers make it necessary to seek out ways in which the size of factorial experiments can be kept within manageable bounds. Of course, it must be emphasized that there is no substitute for careful preliminary planning which, coupled with engineering insight, can result in the elimination of many needless factors.

In spite of the most careful preliminary planning, however, it is often difficult to avoid having to include as many as six or ten (or more) factors in a single experiment. One way to reduce the size of such an experiment would be to break it up into several parts, each part involving the deliberate variation of one factor while all others are held fixed. This would have the undesirable consequence that we could not study any of the interactions. Even if we were to include half the factors in one part and the other half in another, such as replacing a ten-factor experiment (requiring 1,024 observations) by two five-factor experiments (each requiring 32 observations), any interaction between factors in the first part and factors in the second part would be irretrievably lost. It is possible to overcome some of these difficulties by observing that most of the time we are not interested in *all* the interactions. For instance, it is possible to perform only a fraction of a 2^n factorial and yet obtain most of the desired information, say, about the main effects and two-factor interactions (but not the higher interactions).

The principles involved in **fractional replication**, that is, in performing only a fraction of a complete 2^n factorial experiment, are similar to those used in confounding. To obtain a **half-replicate** one selects only one of the two blocks into which the experimental conditions have been divided by confounding one effect; to obtain a **quarter-replicate** one selects only one of the four blocks into which the experimental conditions have been divided by confounding two effects, and so forth. In contrast to a confounded experiment as discussed in Section 13.5, we find that in a fractional replicate *the effects are confounded, not with blocks, but with each other.*

To illustrate, suppose that only the experimental conditions a, b, c, and abc are included in a half-replicate of a 2^3 factorial experiment. (This is one block of a 2^3 factorial experiment with the ABC interaction confounded.) Considering the table of signs on page 458 with all columns except those corresponding to a, b, c, and abc crossed out, we find that the effect total for factor A is now given by

$$[A] = (a) - (b) - (c) + (abc)$$

If we write $(a) = \sum_{l=1}^{r} y_{100l}$, $(b) = \sum_{l=1}^{r} y_{010l}, \ldots$, and substitute for the y_{ijkl} the expressions given by the model equation on page 456, we obtain

$$[A] = -4r[\alpha_0 - (\beta\gamma)_{00}] + \varepsilon$$

where ε is the value of a random variable having zero mean (see Exercise 13.37 on page 486). We thus find that $[A]$ measures the main effect of factor A as well as the BC interaction, so that these two effects have become inseparable, or confounded. Note also that in the reduced table of signs (having columns only corresponding to a, b, c, and abc) the signs for $[A]$ and $[BC]$ are identical, and hence $[A] = [BC]$. In Exercise 13.38 on page 486, the reader will be asked to show that for the given fractional replicate the main effect for factor B is, similarly, confounded, or **aliased**, with the AC interaction, while the main effect for factor C is confounded, or aliased, with the AB interaction. The ABC interaction cannot be estimated.

With careful design, it is generally possible to confound all main effects and two-factor interactions *only* with interactions of higher order. This is illustrated in the following example, where we shall construct a half-replicate of a 2^5 factorial. First, we select an effect (usually a higher-order interaction) to split the experiment into two blocks, as in confounding. The effect chosen is called the **defining contrast**, and it cannot be estimated at all by the fractional replicate. Every other effect is aliased with another effect, namely, its generalized interaction (see page 475) with the defining contrast. Thus, if the defining contrast is $ABCDE$, the main effect for factor A has the four-factor interaction $BCDE$ as its alias, BC and ADE are an alias pair, and so forth. As we have seen, only the combined effect (the sum or difference of the aliased effects) can be estimated in the experiment. However, if it can be assumed that there are no higher-order interactions, one can attribute the effect of the alias pair BC and ADE entirely to the two-factor interaction BC, one can attribute the effect of the alias pair *A* and *BCDE* entirely to the main effect of factor *A*, and so forth. A complete listing of the aliases in a half-replicate of a 2^5 factorial experiment having the defining contrast *ABCDE* is as follows:

A and *BCDE*,	*B* and *ACDE*,	*C* and *ABDE*,	*D* and *ABCE*
E and *ABCD*,	*AB* and *CDE*,	*AC* and *BDE*,	*AD* and *BCE*
AE and *BCD*,	*BC* and *ADE*,	*BD* and *ACE*,	*BE* and *ACD*
CD and *ABE*,	*CE* and *ABD*,	*DE* and *ABC*	

Note that no main effect or two-factor interaction is aliased with another main effect or two-factor interaction.

The 16 experimental conditions to be included in the half-replicate are given by those in either of the two blocks obtained by confounding the defining contrast. Choosing "evens" in the odd-even rule, namely, those conditions which have an even number of letters in common with the defining contrast *ABCDE*, we obtain the following half-replicate:

1	*ad*	*ae*	*de*
ab	*bd*	*be*	*abde*
ac	*cd*	*ce*	*acde*
bc	*abcd*	*abce*	*bcde*

To go one step further, let us illustrate how to construct a quarter-replicate of the given 2^5 factorial. We shall do this by dividing the above half-replicate in half, confounding the three-factor interaction *ABC*. Again using "evens," we get the following eight experimental conditions:

1	*de*	*ab*	*abde*
ac	*acde*	*bc*	*bcde*

Since *DE*, the generalized interaction of the two confounded effects, is also confounded, we now have the three defining contrasts *ABCDE*, *ABC*, and *DE*. None of these effects can be estimated in the quarter-replicate, and each other effect is aliased with its three generalized interactions with the three defining contrasts. The complete aliasing is as follows:

Alias sets			
A	*BCDE*	*BC*	*ADE*
B	*ACDE*	*AC*	*BDE*
C	*ABDE*	*AB*	*CDE*
D	*ABCE*	*ABCD*	*E*
AD	*BCE*	*BCD*	*AE*
BD	*ACE*	*ACD*	*BE*
CD	*ABE*	*ABD*	*CE*

We have given this quarter-replicate merely as an illustration; it would hardly seem useful in actual practice because of the hopeless aliasing of main effects and two-factor interactions. Nevertheless, quarter-replicates of six- and seven-factor experiments (and even eighth-replicates of seven- and eight-factor experiments) can often provide much useful information.

The analysis of a fractional factorial is practically the same as that of a fully replicated factorial experiment. Given the fraction $1/2^p$ of a 2^n factorial, there are 2^{n-p} experimental conditions, and the Yates' method can be used as if the experiment were a 2^{n-p} factorial. Some care must be exercised to arrange the experimental conditions in a modified standard order as indicated in Exercise 13.35 on page 485.

When dealing with fractional factorials, there is the problem of obtaining an estimate of the experimental error. For example, in the half-replicate of a 2^5 factorial described on page 482, the breakdown of the total sum of squares having 15 degrees of freedom yields sums of squares for main effects (5 degrees of freedom), sums of squares for two-factor interactions (10 degrees of freedom), but no component (0 degrees of freedom) for the experimental error. In a situation like this, and in all other cases where the number of degrees of freedom for error is small, it is best to include a limited amount of replication. This may be accomplished by randomly selecting several experimental conditions, and making additional observations corresponding to these conditions. Furthermore, if it can be assumed that there are no higher-order interactions, the total of the sums of squares corresponding to higher-order interactions *which are not aliased with main effects or lower-order interactions* can be attributed to "error," and used in the denominator of the F test (see Exercise 13.36). In this way, use can be made of the "hidden replication" inherent in most large factorial experiments.

To summarize, fractional replication is useful whenever the number of factors to be included in an experiment is large and it is not economically feasible to include all possible experimental conditions. The reduction in size (and, therefore, in cost) of a fractionally replicated experiment is partially offset by the loss of information caused by aliasing, and by the difficulties inherent in estimating the experimental error. A more detailed discussion of fractional replication, including fractional replicates of 3^n experiments and a variety of other designs, can be found in the book by W. G. Cochran and G. M. Cox listed in the bibliography.

EXERCISES

13.26 A 2^5 factorial experiment, having the factors A, B, C, D, and E, is to be run in several blocks. Show which treatments are assigned to each block if

(a) there are to be two blocks, with the $ACDE$ interaction confounded;

(b) there are to be four blocks, with the BDE and $ABCE$ interactions confounded. What other factorial effect or effects are also confounded?

13.27 List the experimental conditions included in each block if a 2^4 factorial experiment is to be confounded

(a) in two blocks on $ABCD$; (b) in four blocks on ABC and BCD.

13.28 List the experimental conditions included in each block if a 2^6 factorial experiment is to be confounded in 8 blocks on $ABDE$, $BCDF$, and ABC. What other factorial effects are confounded?

13.29 Four new drugs are to be investigated to determine their effectiveness as tranquilizers, individually and in combination with each other. Each patient was given regular doses of one of the sixteen tranquilizers formed from these drugs (including a placebo corresponding to the 0 level for each drug) and, after a two-week period, the effect of these tranquilizers on the emotional stability of each patient was judged (on a scale from 1 to 5) by five psychiatrists. To keep the staff work load within reasonable bounds, two hospitals were used in this experiment, and eight patients were selected by the staff of each hospital for each trial (replicate); thus, the experiment involved two replicates of two blocks each, with the *ABCD* interaction confounded. The results of the two 2-weeks trials were as follows:

FIRST HOSPITAL			SECOND HOSPITAL		
Treatment combination	*Mean rating* *Trial 1*	*Trial 2*	*Treatment combination*	*Mean rating* *Trial 1*	*Trial 2*
1	2.0	2.6	*a*	2.8	2.6
ab	3.8	3.4	*b*	3.6	2.0
ac	4.2	4.8	*c*	2.4	1.8
bc	4.8	4.0	*abc*	4.0	3.8
ad	1.8	2.4	*d*	1.8	2.2
bd	3.4	3.8	*abd*	1.6	2.0
cd	4.6	2.8	*acd*	3.6	2.4
abcd	4.2	4.6	*bcd*	3.4	3.8

If a high rating indicates satisfactory progress, which drug combination (or combinations) seems to be the most promising? Perform an intrablock analysis of variance to test for significant effects.

13.30 Verify that in the example on page 476 all treatment totals having $+1$ as a coefficient in $[BC]$ are in Blocks 1 and 4, that all treatment totals having -1 as a coefficient in $[BC]$ are in Blocks 2 and 3, and, hence, that the *BC* interaction is confounded with blocks.

13.31 Show that the "odd-even" rule for assigning treatments to blocks, given on page 476, is equivalent to the method described in Exercise 13.20 on page 473.

13.32 Referring to Exercise 13.27, list the alias sets if an experiment consists of

(a) a half-replicate of a 2^4 factorial experiment with *ABCD* confounded;

(b) a quarter-replicate of a 2^4 factorial experiment with *ABC* and *BCD* confounded.

13.33 Construct a half-replicate of a 2^6 factorial having the defining contrast *ABCDEF*. List the 32 treatments in the block containing experimental condition 1, and show the alias pairs.

13.34 Design an experiment consisting of a quarter-replicate of a 2^7 factorial with *ABCDE*, *ABCFG*, and *DEFG* confounded, if the experimental condition with each factor at the 0 level is to be included. Also exhibit all alias sets.

13.35 Referring to Exercise 13.33, we can define a **modified standard order** for the 32 treatment combinations in the half-replicate as follows. First, we list the 32 treatment combinations corresponding to the five factors A, B, C, D, and E in standard order. Then we append the letter f to 16 of these treatment combinations, so that the list contains the same ones as the block chosen for the half-replicate.

(a) Use this method to list the treatment combinations obtained in Exercise 13.33 in modified standard order.

(b) Generalize the above rule for arranging the treatment combinations in modified standard order so that it applies to a half-replicate of a 2^n factorial experiment. (For a $1/2^p$ fractional replicate of a 2^n factorial, a modified standard order can be obtained by noting that any block chosen contains a subset of $n - p$ letters which form a complete replicate of a 2^{n-p} factorial. The modified standard order is then obtained by using these letters only and later appending the remaining p letters to get the required treatment combinations.)

13.36 The following factors are to be studied in a half-replicate of a 2^6 factorial experiment (defining contrast $ABCDEF$), designed to evaluate several chemicals as insecticides.

Factor	*Level 0*	*Level 1*
A. BMC	0%	5%
B. Malathion	3%	6%
C. Tedion	1%	2%
D. Chlordane	2%	5%
E. Lindane	1%	4%
F. Pyrethrum	2%	4%

Each experimental unit consists of 10 insects, and the average lifetimes (in seconds) after application of the respective insecticides are as follows, in the random order in which they are obtained:

ce	181	*acdf*	162	*bd*	135	*abdf*	131
ae	172	1	182	*df*	171	*ab*	136
abef	140	*bf*	171	*acef*	159	*bcde*	105
bcdf	165	*cf*	176	*bc*	179	*abcdef*	109
acde	139	*be*	187	*ac*	165	*af*	176
ef	186	*abce*	131	*bcef*	181	*ad*	150
de	164	*abcf*	125	*cdef*	163	*abde*	115
abcd	112	*adef*	158	*bdef*	128	*cd*	166

(a) Write down the alias pairs.

(b) Arrange the results in modified standard order (see Exercise 13.35) and use the Yates' method to find the effect totals.

(c) Identify the effect totals as follows. In the last column of the Yates table, write down the 32 combinations [I], [A], [B], [AB], ..., [$ABCDE$] in standard order. Each of these is aliased with another one; identify the alias pairs by using the main effect or lower-order interaction. (For example, $ABCDE$ is aliased with F—identify the effect total [$ABCDE$] as that of the main effect F.) Note that 10 of the alias pairs are three-factor interactions aliased with three-factor interactions—label these "error."

(d) Obtain the mean squares and complete the analysis of variance. Note that the error mean square, having 10 degrees of freedom, is the average of the mean squares of the 10 effects labeled "error."

13.37 Verify the expression obtained for [A] on page 481.

13.38 Duplicating the steps indicated on page 481, show that in the given design the main effect for factor B is confounded with the AC interaction. [*Hint*: Express [B] and [AC] in terms of the parameters of the model.]

13.39 In Exercise 13.13 on page 471 suppose that only one-half of one of the replicates can be performed in the time available. Construct the half-replicate having the defining contrast $ABCD$. Using the appropriate data from Rep. 1, together with replicates of experimental conditions 1, *ab*, *bc*, and *abcd* taken from Rep. 2, perform an analysis of this half-replicate identifying those effects attributable to "error." [*Hint*: Each of the four pairs of replicated treatments provides one additional degree of freedom for error, and contributes $(y_1 - y_2)^2/2$ to the error sum of squares, where y_1 is the observation resulting from applying the treatment in the first replicate, and y_2 is the corresponding observation in the second replicate.]

13.7 REVIEW EXERCISES

13.40 Referring to Exercise 12.47 on page 427, it can be argued that the rocket-launcher experiment was poorly designed, as it was not replicated, and, thus, it is impossible to test whether there is an interaction between launchers and rocket fuels. Suppose that a second replicate were performed, with the following results:

	Fuel I	*Fuel II*	*Fuel III*	*Fuel IV*
Launcher X	66.8	51.1	40.1	49.2
Launcher Y	47.7	38.2	50.6	64.3

Combining the results of both replicates, perform an appropriate analysis of variance, and test for the presence of an interaction.

13.41 A study was conducted to measure the effect of three different meat tenderizers on the weight loss of steaks having the same initial (precooked) weights. The effects of cooking temperatures and cooking times also were measured by performing a $3 \times 2 \times 2$ factorial experiment in 3 replicates. The results are as follows:

Tenderizer	*Cooking time (minutes)*	*Cooking temperature (°F)*	*Weight loss (ounces)*		
			Rep. 1	*Rep. 2*	*Rep. 3*
A	20	350	1.5	1.3	1.4
A	20	400	1.6	1.4	1.5
A	30	350	1.7	1.8	1.7
A	30	400	1.8	1.9	2.0
B	20	350	1.9	2.1	2.0
B	20	400	2.2	2.4	2.5
B	30	350	2.6	2.3	2.4
B	30	400	2.6	2.7	2.5
C	20	350	0.9	0.8	0.8
C	20	400	1.1	1.0	0.9
C	30	350	0.8	0.9	1.0
C	30	400	1.2	1.0	1.1

(a) Analyze this experiment first as a two-way classification with 12 treatments and three replicates (blocks).

(b) Complete the analysis by computing the sums of squares corresponding to the various main effects and interactions.

(c) Present the results in an analysis-of-variance table and interpret the experiment.

13.42 An experiment was conducted to determine the effects of certain alloying elements on the ductility of a metal, and the following results were obtained:

Nickel	*Carbon*	*Manganese*	*Breaking strength (ft-lb)*		
			Rep. 1	*Rep. 2*	*Rep. 3*
0.0%	0.3%	0.5%	36.7	39.6	38.2
0.0	0.3	1.0	47.5	43.5	45.9
0.0	0.6	0.5	40.6	36.8	36.0
0.0	0.6	1.0	41.1	45.8	46.4
4.0	0.3	0.5	37.8	32.7	31.6
4.0	0.3	1.0	34.2	37.2	36.5
4.0	0.6	0.5	39.5	41.7	39.1
4.0	0.6	1.0	46.4	43.7	49.4

Perform an appropriate analysis of variance and interpret the results.

13.43 Given the two replications of a 2×3 factorial experiment, calculate the analysis of variance table using the formulas on page 435 rather than the shortcut formulas in the example.

Factor A	*Factor B*	*Rep. 1*	*Rep. 2*
1	1	29	35
1	2	15	17
1	3	14	22
2	1	15	13
2	2	27	25
2	3	16	24

13.44 Given the following observations

Factor A	*Factor B*	*Rep. 1*	*Rep. 2*
0	0	14	6
1	0	15	21
0	1	11	17
1	1	24	16

summarize the experiment according to the visual procedure given in Section 13.4. Interpret the effects based on the confidence intervals.

13.45 With reference to the example on page 463, suppose a third replicate

	Temperature	*ph*	*Rep. 3*
(1)	300	2	9
(*a*)	350	2	16
(*b*)	300	3	23
(*ab*)	350	3	25

is run. Analyze the experiment, using all three replicates, according to the visual procedure given in Section 13.4. Interpret the effects based on the confidence intervals.

13.46 Given the following results from a 2^3 factorial experiment,

Factor A	*Factor B*	*Factor C*	*Rep. 1*	*Rep. 2*
0	0	0	13.8	14.6
1	0	0	10.8	8.4
0	1	0	9.0	9.8
1	1	0	10.1	10.9
0	0	1	14.4	13.6
1	0	1	6.2	8.6
0	1	1	7.7	7.9
1	1	1	9.0	8.2

summarize the experiment according to the visual procedure given in Section 13.4. Interpret the effects based on the confidence intervals.

13.47 With reference to Exercise 13.45, create the analysis-of-variance table.

13.48 With reference to Exercise 13.46, create the analysis-of-variance table.

13.49 What is the largest number of blocks in which one can perform a 2^6 factorial experiment without confounding any main effects?

13.50 Suppose that in Exercise 13.11 on page 469 the judge rates the soft drinks in sets of four, with a rest period in between, and that the experiment was actually performed so that each replicate consisted of two blocks with ABC confounded. Perform an intrablock analysis of the data.

13.51 Suppose that in Exercise 13.12 on page 470 only eight specimens could be tested in any one shift, and that the experiment was actually performed so that each replicate consists of four blocks with ABC, ADE, and $BCDE$ confounded.

(a) If for each factor levels 0 and 1 are, respectively, the lower and higher values, construct a table showing which experimental conditions go into each of the four blocks.

(b) Perform an intrablock analysis of the experiment.

13.52 Referring to Exercise 13.26, suppose that in each case the block containing treatment combination 1 was chosen as a fractional replicate of a 2^5 factorial.

(a) Show the alias pairs in the resulting half-replicate.

(b) Show the alias sets in the resulting quarter-replicate. (In practice, random selection should be used to choose the block to be included in a fractional replicate.)

13.53 In Exercise 13.14 on page 471 construct and analyze that half-replicate having the defining contrast $ABCDE$, using the appropriate experimental results from Rep. 1.

13.8

CHECK LIST OF KEY TERMS (with page references)

Alias 481
Complete factorial experiment 432
Confounding 475
Defining contrast 481
Effect total 457
Factorial experiment 432
Factors 432
Fractional replicate 481
Generalized interaction 475
Half-replicate 481
Incomplete block design 475
Interaction 433
Interaction sum of squares 435
Interblock analysis of variance 479
Interblock error 480
Intrablock analysis of variance 479
Interblock error 477
Levels of factors 432
Main effects 444
Model equation for three-factor experiment 443
Model equation for two-factor experiment 433
Model equation for 2^3 *factorial experiment* 456
Multifactor experiment 442
Odd-even rule 476
Quarter-replicate 481
Standard order 456
Table of signs 458
Two-factor experiment 432
2^n *factorial experiment* 455
Yates' method 460

14 THE STATISTICAL CONTENT OF QUALITY-IMPROVEMENT PROGRAMS

Although there is a tendency to think of the subject of quality as a recent development, there is nothing new about the basic idea of making a quality product characterized by a high degree of uniformity. For centuries skilled artisans have striven to make products distinctive through superior quality, and once a standard of quality was achieved, to eliminate insofar as possible all variability between products that were nominally alike.

What is new in **quality improvement** is the idea that a product is never good enough and should be continually improved. This concept, honed to a fine edge in Japan, has created a crisis in the international market place for firms that do not follow suit. In quality-improvement programs, the emphasis is on employing designed experiments to improve the product in the design, production, and assembly stages rather than in futile attempts to inspect quality into a product after it is produced. We introduce these ideas in Sections 14.1 through 14.3.

Three special techniques of (statistical) **quality assurance** are also treated in this chapter; quality control is discussed in Sections 14.4 through 14.6, the establishment of tolerance limits in Section 14.7, and acceptance sampling in Section 14.8. Note that the word quality, when used technically as in this discussion, refers to some measurable or countable property of a product, such as the outside diameter of a ball bearing, the breaking strength of a yarn, the number of imperfections in a piece of cloth, the potency of a drug, and so forth.

14.1
QUALITY-IMPROVEMENT PROGRAMS

What is a quality-improvement program? To answer this question, we present a scenario of what happens when action is taken to improve quality. In the context of a machine tooling operation, the fraction of defective pieces per day were plotted [see Figure 14.1(a)] for each day over a 5-week period. This plot reveals stable variation about a value of nearly 15% defective. That is, the process is predictable. We can estimate the mean by the average over days and we can also estimate the amount of variation (see Section 14.4). The fact that the process is stable does not make it good! It is turning out too many defective pieces.

Once it is realized the process needs improvement, action can be taken. Data collected on several possible sources of variation are displayed in the Pareto chart (see Section 2.1) in Figure 14.2. Based on these data, it was decided to give the operators more training on the use of the machine. The record for the next 5 weeks of the daily fraction defectives, after the training, is plotted in Figure 14.1(b). The new process also appears stable but this time about a lower mean. Should we be satisfied? No. The central precept of quality improvement calls for the process of improvement to be continual. Maybe the gains will be smaller at each progressive stage, but efforts must be continued to reduce the amount of variation and the proportion of defectives. Further substantial improvements will come only by taking action on the system. However, since the process is stable, the effects of changes can be observed. Engineers can make innovations to improve the process. The two-level factorial designs discussed in the previous chapter are particularly relevant.

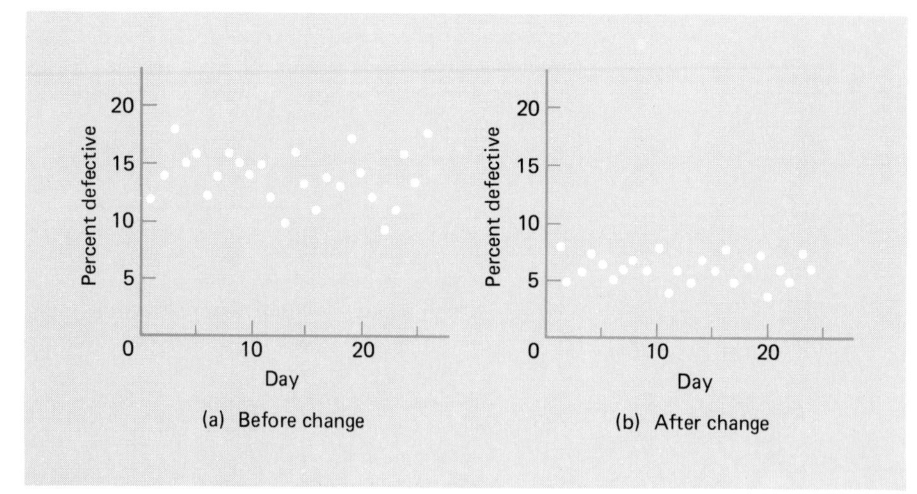

FIGURE 14.1
Fraction defective per day.

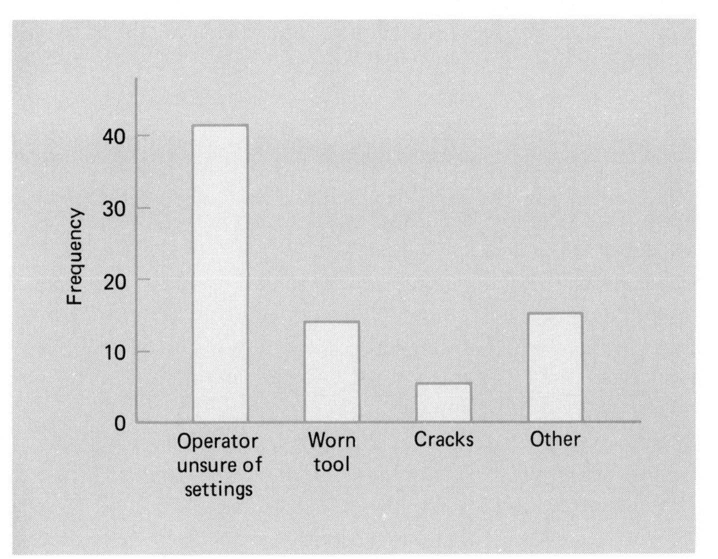

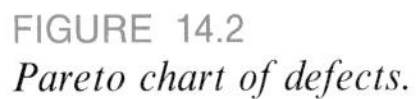
FIGURE 14.2
Pareto chart of defects.

There is some folklore that high quality and high production rates are incompatible. But time and again it is the cost of reworking bad product that is a major component of production cost. It is really low quality that results in high costs. Besides the high costs of reworking pieces to make them usable, there are also high costs associated with lost customers who were sold inferior products.

The transformation to quality production in Japan, starting in 1950, has created a new economic age and a crisis for American businesses. Briefly, Japanese merchandise of the time was known to be shoddy. Several highly placed engineers in Japan studied the literature on quality control produced at Bell Laboratories by Walter Shewhart and others. W. Edwards Deming was brought in as a foreign expert. Unlike in America, where the applications of statistical methods to problems of quality fell far below from their wartime successes because managers did not fully appreciate them, in Japan the top managers came with their engineers to learn about the techniques. What followed was company- and industrywide commitments to improve quality through education which included statistical methods. This transformation has taken many Japanese companies to world leadership. In order to compete in the international marketplace, other companies and countries must also stress quality improvement.

A theory of management for product or service improvement has been pioneered by W. Edwards Deming. It contains concise statements of the elements of the transformation that must take place. Deming sums up his ideas on the transformation of American industry in 14 points for management. They apply not only to manufacturing but also service industries and pertain to organizations of all sizes, large and small. We present **Deming's 14 points** as an overview of his philosophy and heartily recommend his book, listed in the bibliography, to all interested readers.

1. Create constancy of purpose toward improvement of the product and service, with the aim to become competitive and to stay in business and to provide jobs.
2. Adopt the new philosophy. We are in a new economic age. Western management must awaken to the challenge, must learn their responsibilities, and take on leadership for change.
3. Cease dependence on inspection to achieve quality. Eliminate the need for inspection on a mass basis by building quality into the product in the first place.
4. End the practice of awarding business on the basis of price tag. Instead, minimize total cost. Move toward a single supplier for any one item, on a long-term relationship of loyalty and trust.
5. Improve constantly and forever the system of production and service, to improve quality and productivity, and thus constantly decrease costs.
6. Institute training on the job.
7. Institute leadership. The aim of supervision should be to help people and machines and gadgets to do a better job.
8. Drive out fear, so that every one may work effectively for the company.
9. Break down barriers between departments. People in research, design, sales, and production must work as a team, to foresee problems of production and in use that may be encountered with the product or service.
10. Eliminate slogans, exhortations, and targets for the work force asking for zero defects and new levels of productivity. Such exhortations only create adversarial relationships, as the bulk of the causes of low quality and low productivity belong to the system and thus lie beyond the power of the work force.

11a. Eliminate work standards (quotas) on the factory floor. Substitute leadership.

11b. Eliminate management by objective. Eliminate management by numbers, numerical goals. Substitute leadership.

12a. Remove barriers that rob the hourly worker of his right to pride of workmanship. The responsibility of supervisors must be changed from sheer number to quality.

12b. Remove barriers that rob people in management and in engineering of their right to pride of workmanship. This means, *inter alia*, abolishment of the annual or merit rating and of management by objective.

13. Institute a vigorous program of education and self-improvement.
14. Put everybody in the company to work to accomplish the transformation. The transformation is everybody's job.†

It requires a method to achieve quality improvement, not just hopes. A central problem of management is to understand the meaning of variation and to extract the pertinent information from it. It is here that statistics comes into play. Engineers need to understand the basic principals of experimental design to fulfill their roles in points 3 and 5.

---

† "Out of the Crisis," W. Edwards Deming (1986) Massachusetts Institute of Technology, Cambridge, Mass. Reprinted by permission of the author.

The main thrust of the statistical approach is that, in order to improve quality, it is better to work upstream on the processes. That is, build quality into the product by concentrating on the equipment, components, and materials that go into making it.

The consumer also has a role in the new way of quality improvement. It has always been (1) design a product, (2) make it, and (3) market it. Now, there is a new fourth step, (4) find out the purchasers' reactions to the product. Also find out why others did not purchase. Statistical methods of sampling will provide a way of finding out what the consumer thinks. Changes can then be made in design and production to better match the product to the market. These four steps must be repeated over and over again in the search for continual product improvement.

14.2
STARTING A QUALITY-IMPROVEMENT PROGRAM

It is the prevailing wisdom that top management must be involved in any quality improvement. Once committed, it must take action and select initial processes to serve as flagship projects. It is good to start with processes that have a large potential for improvement and where the prospect for large financial gains is greatest. Even though this is management's decision, the most successful programs start with committees formed with employees from all levels. More enthusiasm can be generated when there is a consensus regarding the selection of the process. A modified Delphi technique can help groups reach unanimity. Each person writes down his or her top three choices. With 3 points for first place, 2 for second, and 1 for third, the totals for each candidate process are tabulated for all to see. Perhaps after some discussion, each person votes again and the process continues until a unanimous choice is reached.

Suppose the process selected concerns piston rings. The first step is to collect data. We will talk to those who run the process about causes and types of defects, but to start we want fresh data on all of the defectives that occur over a period of two weeks. This information is presented as a Pareto diagram in Figure 14.3.

Defect	*Number*
Height	30
Diameter	14
Cracks	4
Scratches	2
Other	5

We see that 30 out of 45 defective rings have incorrect heights.

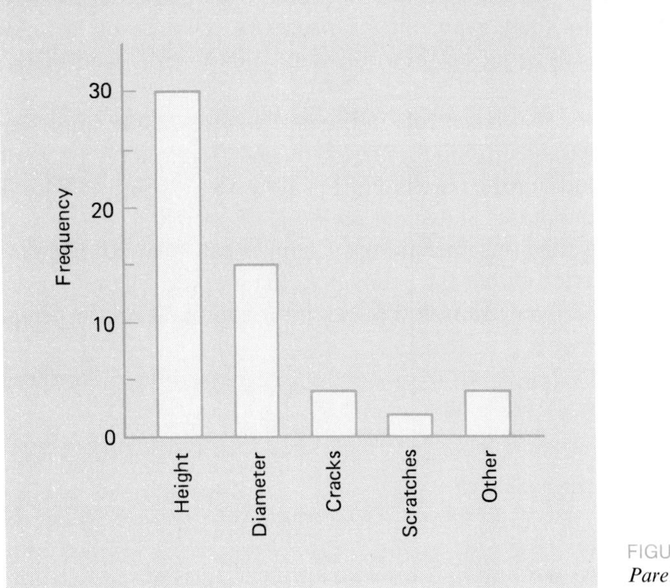

FIGURE 14.3
Pareto diagram for piston rings.

To proceed, we gather the engineers, supervisors, and operators who make the rings for a "brainstorming" session. They construct a list of the possible causes for the variation in height. These may be graphically displayed in a **cause-and-effect diagram**. It arranges causes and causes of causes, as shown in Figure 14.4 (see the book by K. Ishikawa, listed in the bibliography, for more examples). The cause-and-effect diagram, which resembles the skeleton of a fish, starts with a central horizontal line for a major problem such as incorrect height. Major factors that affect height are listed on diagonal lines attached to the central horizontal line. Factors that effect the major factors, as cooling time affects tempering, are labeled on horizontal lines connected to the diagonal lines. To proceed further, action must be taken on the system. A two-level factorial design was run with the suppliers as Factor A and two speeds for the grinder as Factor B. The response, the number of defectives out of 200 rings made at each condition, is recorded. The results confirm Deming's fourth point. Work with a single supplier.

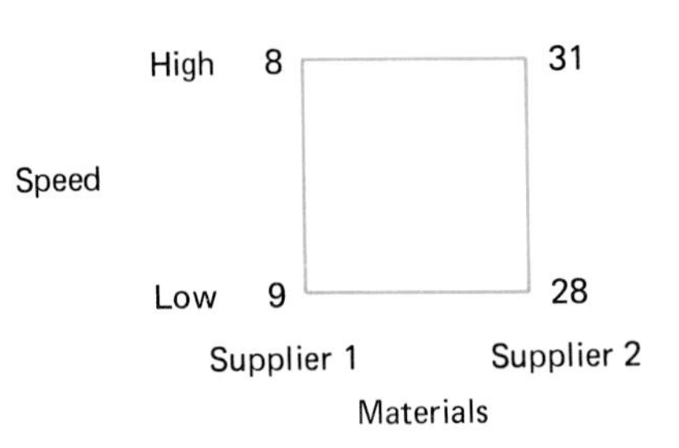

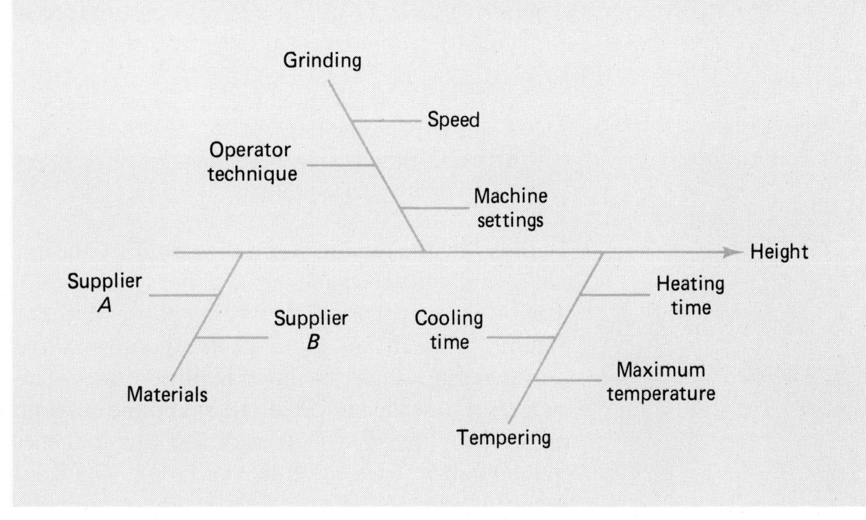

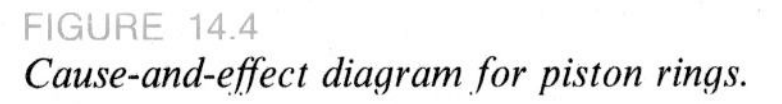

FIGURE 14.4
Cause-and-effect diagram for piston rings.

If the process is stable with the material from supplier 1, then it is time to make another Pareto diagram and continue the cycle of improvement.

One outgrowth of the Japanese way of working together has been the formation of small groups of employees called *quality circles.* These groups, consisting of employees at all levels, meet on a regular basis to discuss ways of continually improving their process.

With all the workers given some statistical training and engineers some training in experimental design, all the processes within the company can receive attention and be improved.

14.3 EXPERIMENTAL DESIGNS FOR QUALITY IMPROVEMENT

The modern emphasis, developed in Japan, has been to build quality into the product rather than waiting until the end of the line and trying to inspect bad quality out. The job of quality becomes a full-time job for everyone in the company, working as a team. They must learn about the process by observing and conducting statistically designed experiments.

In addition to the factorial designs discussed in the previous chapter, the Japanese, and Professor Genichi Taguchi in particular, have introduced good engineering ideas to produce new design procedures. Two of his major contributions involve using designed experiments to

(a) select one input variable to minimize variation while another input variable holds the response on target;
(b) create products that are not sensitive to variations in their components or environmental conditions.

To illustrate the minimization of variation concept, suppose a 2^3 factorial design is run to study the effects of initial concentration of Acid A, rag content, and digester time on the tear strength of writing paper. Rather than summarize the experiment in terms of the means $\bar{y}_{ijk.}$ at each experimental condition, the statistician G. E. P. Box suggests a chart of the individual values. This graph portrays both the level with respect to the indicated target and the amount of variation. From Figure 14.5, we see that the acid concentration can influence the mean level, whereas rag content can be used to reduce variation. The third factor, digester time, does not seem to have an effect. That is, we can manipulate rag content and initial concentration to both be on target and to reduce variation.

To illustrate the idea of making products that are insensitive to variation, suppose that the output voltage of a circuit is related to the value of a resistor, as in

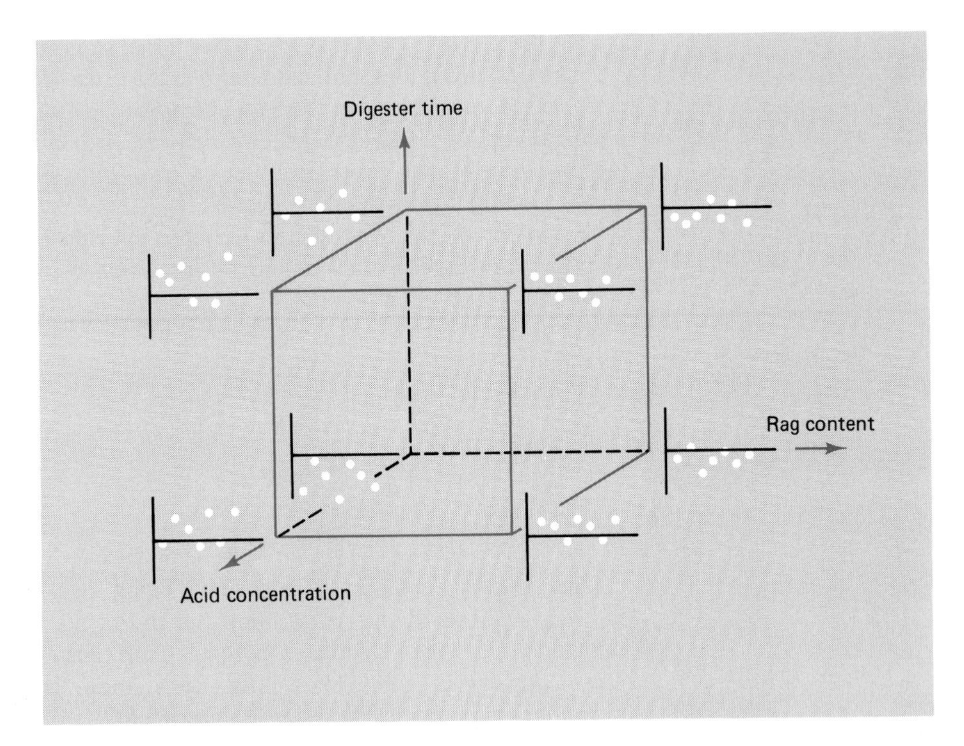

FIGURE 14.5

The summary of a design to study the effect on both mean response and variation.

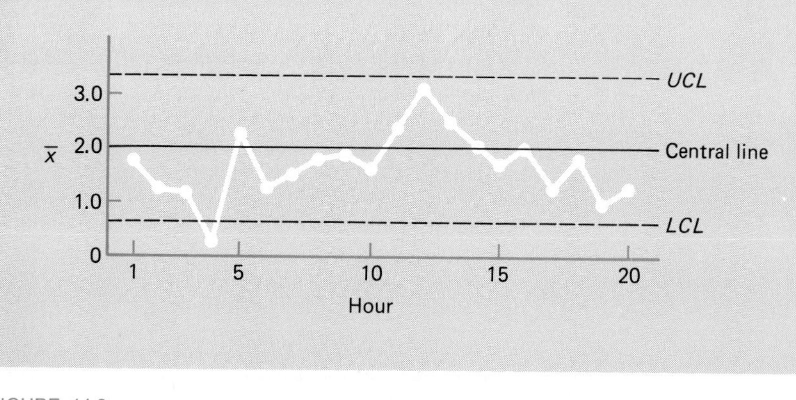

FIGURE 14.8
$\bar{x}$ chart.

In controlling a process, it may not be enough to monitor the population mean. Although an increase in process variability may become apparent from increased fluctuations of the $\bar{x}$'s, a more sensitive test of shifts in process variability is provided by a separate control chart, an R chart based on the sample ranges or a σ chart based on the sample standard deviations. An example of the latter may be found in Exercise 14.5 on page 513.

The central line and control limits of an R chart are based on the distribution of the range of samples of size n from a normal population. As we observed on page 258, the mean and the standard deviation of this sampling distribution are given by $d_2\sigma$ and $d_3\sigma$, respectively, when σ is known. Thus, three-sigma control limits for the range are given by $d_2\sigma \pm 3d_3\sigma$, and the complete set of control-chart values for an R chart (with σ known) is given by

Control-chart values for an R chart (σ known)

$$\textbf{central line} = d_2\sigma$$
$$UCL = D_2\sigma$$
$$LCL = D_1\sigma$$

Here $D_1 = d_2 - 3d_3$, and $D_2 = d_2 + 3d_3$, and values of these constants can be found in Table 13 for various values of n.

If σ is unknown, it is estimated from past data as previously described, and the control-chart values for an R chart (with σ unknown) are as follows:

Control-chart values for an R chart (σ unknown)

$$\textbf{central line} = \bar{R}$$
$$UCL = D_4\bar{R}$$
$$LCL = D_3\bar{R}$$

Here $D_3 = D_1/d_2$ and $D_4 = D_2/d_2$, and values of these constants can also be found in Table 13 for various values of n.

To illustrate the construction of an $\bar{x}$ chart and an R chart, suppose a manufacturer of a certain bearing knows from a preliminary record of 20 hourly samples of size 4 that for the diameters of these bearings $\bar{\bar{x}} = 0.9752$ and $\bar{R} = 0.0002$. Coding her data by means of the equation $\frac{x - 0.9750}{0.0001}$, that is, expressing each measurement as a deviation from 0.9750 in 0.0001 inch, she obtains

$\bar{x}$ *chart (coded)*		*R chart (coded)*	
central line	$\bar{\bar{x}} = 2.0$	central line	$\bar{R} = 2.0$
UCL	$\bar{\bar{x}} + A_2\bar{R} = 3.4$	*UCL*	$D_4\bar{R} = 4.6$
LCL	$\bar{\bar{x}} - A_2\bar{R} = 0.6$	*LCL*	$D_3\bar{R} = 0$

The values of $A_2 = 0.729$, $D_3 = 0$, and $D_4 = 2.282$ for samples of size 4 were obtained from Table 13. Graphically, these control charts are shown in Figures 14.8 and 14.9, where we have also indicated the results subsequently obtained in the following 20 samples:

Hour	*Coded sample values*				$\bar{x}$	*R*
1	1.7	2.2	1.9	1.2	1.75	1.0
2	0.8	1.5	2.1	0.9	1.32	1.3
3	1.0	1.4	1.0	1.3	1.18	0.4
4	0.4	−0.6	0.7	0.2	0.18	1.3
5	1.4	2.3	2.8	2.7	2.30	1.4
6	1.8	2.0	1.1	0.1	1.25	1.9
7	1.6	1.0	1.5	2.0	1.52	1.0
8	2.5	1.6	1.8	1.2	1.78	1.3
9	2.9	2.0	0.5	2.2	1.90	2.4
10	1.1	1.1	3.1	1.6	1.72	2.0
11	1.7	3.6	2.5	1.8	2.40	1.9
12	4.6	2.8	3.5	1.9	3.20	2.7
13	2.6	2.8	3.2	1.5	2.52	1.7
14	2.3	2.1	2.1	1.7	2.05	0.6
15	1.9	1.6	1.8	1.4	1.68	0.5
16	1.3	2.0	3.9	0.8	2.00	3.1
17	2.8	1.5	0.6	0.2	1.28	2.6
18	1.7	3.6	0.9	1.5	1.92	2.7
19	1.6	0.6	1.0	0.8	1.00	1.0
20	1.7	1.0	0.5	2.2	1.35	1.7

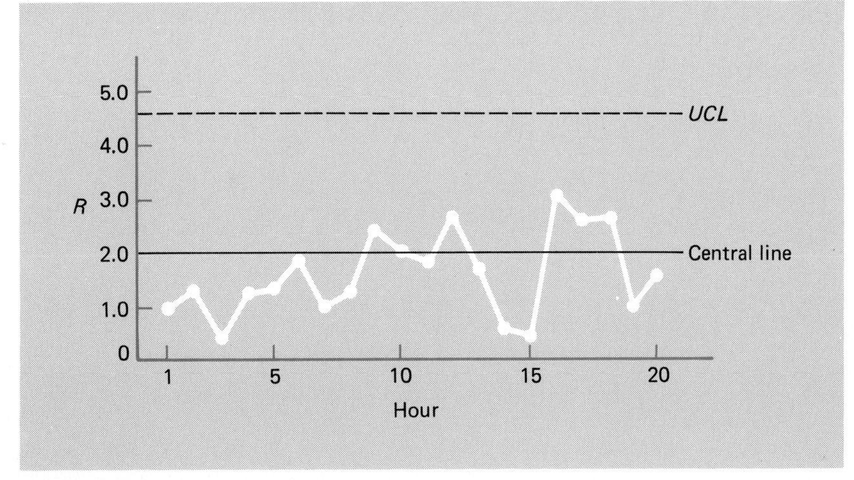

FIGURE 14.9
R chart.

Inspection of Figure 14.8 shows that only one of the points falls outside of the control limits, but it also shows that there may nevertheless have been a downward shift in the process average. Figure 14.9 shows a definite downward shift in the process variability; note especially that most of the sample ranges fall below the central line of the R chart.

The reader may have observed the close connection between the use of control charts and the testing of hypotheses. A point on an $\bar{x}$ chart that is out of control corresponds to a sample for which the null hypothesis that $\mu = \mu_0$ is rejected. To be more precise, we should say that control-chart techniques provide sequential, *temporally ordered* sets of tests. We are interested not only in the positions of individual points, but also in possible trends or other patterns exhibited by the points representing successive samples.

A different graph, called a **cumulative sum** (**cusum**) chart, is more effective for detecting small shifts in the mean. Consider the deviations *observation − target value*. In the context of our example, the target value could be 2.00. To construct the cusum chart, plot

$$S_1 = (1.75 - 2.00) = -0.25 \quad \text{versus} \quad 1$$
$$S_2 = (1.75 - 2.00) + (1.32 - 2.00) = -0.93 \quad \text{versus} \quad 2$$
$$S_3 = (-0.25) + (-0.68) + (1.18 - 2.00) = -1.75 \quad \text{versus} \quad 3$$
$$\vdots$$

In Figure 14.10(a) we see evidence of an initial constant downward trend masked somewhat by random variation. This behavior indicates that the level of the process is a fixed amount below the target value of 2.00. At observation 11 there

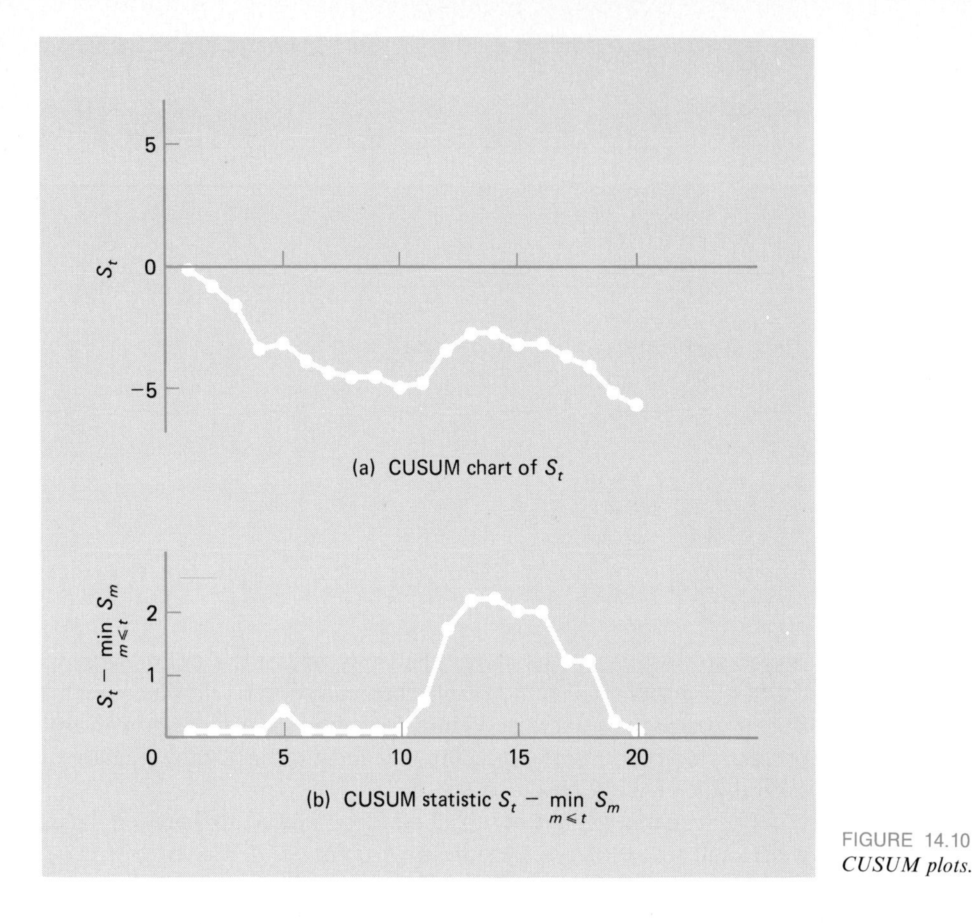

FIGURE 14.10
CUSUM plots.

is a distinct shift up in level, as if a change has been made, but after a few hours a shift to a level below 2.0 appears to occur.

The **cusum statistic** for detecting an increase is

$$\textit{current value of } S_t - \textit{minimum of } S_m \textit{ over all previous times}$$

This is plotted in Figure 14.10(b). It shows more vividly the temporary increase in level at observation 11.

There are theoretical reasons for creating separate charts to detect an increase and to detect a decrease. Separate centering constants may be used for each chart. If it is deemed important for engineering reasons to detect an increase to 2.50, we could then use the centering value $2.25 = 2.00 + (2.50 - 2.00)/2$ as recommended. (See the entry on cumulative sum control charts in the encyclopedia by S. Kotz and N. L. Johnson, listed in the bibliography for more details on critical values for cusum statistics.)

14.6
CONTROL CHARTS FOR ATTRIBUTES

Although more complete information can usually be gained from measurements made on a finished product, it is often quicker and cheaper to check the product against specifications on an "attribute" or "go, no-go" basis. For example, in checking the diameter and eccentricity of a ball bearing it is far simpler to determine whether it will pass through circular holes cut in a template than to make several measurements of the diameter with a micrometer. In this section we discuss two fundamental kinds of control charts used in connection with attribute sampling, the **fraction-defective chart**, also called a p chart, and the **number-of-defects chart**, also called a c chart. To clarify the distinction between "number defective" and "number of defects," note that a unit tested can have several **defects**, whereas, on the other hand, it is either defective or it is not. In many applications, a unit is referred to as **defective** if it has at least one defect.

Control limits for a fraction-defective chart are based on the sampling theory for proportions introduced in Section 9.1 and on the normal curve approximation to the binomial distribution. Thus, if a standard is given—that is, if the fraction defective should take on some preassigned value p—the central line is p and three-sigma control limits for the fraction defective in random samples of size n are given by

$$p \pm 3\sqrt{\frac{p(1-p)}{n}}$$

If no standard is given, which is more frequently the case in actual practice, p will have to be estimated from past data. If k samples are available, d_i is the number of defectives in the ith sample, and n_i is the number of observations in the ith sample, it is customary to estimate p as the proportion of defectives in the combined sample, namely, as

Proportion of defectives in combined sample

$$\bar{p} = \frac{d_1 + d_2 + \cdots + d_k}{n_1 + n_2 + \cdots + n_k}$$

Control-chart values for a fraction-defective chart

$$\textit{central line} = \bar{p}$$

$$UCL = \bar{p} + 3\sqrt{\frac{\bar{p}(1-\bar{p})}{n}}$$

$$LCL = \bar{p} - 3\sqrt{\frac{\bar{p}(1-\bar{p})}{n}}$$

Note that if p is small, as is often the case in practice, substitution in the formula for the lower control limit might yield a negative number. When this occurs, it is customary to regard the lower control limit as if it were zero and, in effect, to use only the upper control limit. Another complication that can arise if p is small is that the binomial distribution may not be adequately approximated by the normal distribution. Generally speaking, the use of the above control limits for p charts is unrealistic whenever n and p are such that the underlying binomial (or hypergeometric) distribution cannot be approximated by a normal curve (see page 148). In such cases it is best to use an upper control limit obtained directly from a table of binomial probabilities or, perhaps, use the Poisson approximation to the binomial distribution.

As an illustration of a p chart, suppose that it is desired to control the output of a certain integrated circuit production line to maintain a "yield" of 60 percent, that is, a proportion defective of 40 percent. To this end, daily samples of 100 units are checked to electrical specifications, with the following results:

Date	*Number of defectives*	*Date*	*Number of defectives*	*Date*	*Number of defectives*
3-12	24	3-26	44	4- 9	23
3-13	38	3-27	52	4-10	31
3-16	62	3-30	45	4-13	26
3-17	34	3-31	30	4-14	32
3-18	26	4- 1	34	4-15	35
3-19	36	4- 2	33	4-16	15
3-20	38	4- 3	22	4-17	24
3-23	52	4- 6	34	4-20	38
3-24	33	4- 7	43	4-21	21
3-25	44	4- 8	28	4-22	16

Since the standard is given as $p = 0.40$, the control-chart values are

$$\text{central line} = 0.40$$

$$UCL = 0.40 + 3\sqrt{\frac{(0.40)(0.60)}{100}} = 0.55$$

$$LCL = 0.40 - 3\sqrt{\frac{(0.40)(0.60)}{100}} = 0.25$$

The corresponding control chart with points for the 30 sample fractions defective is shown in Figure 14.11, and it exhibits some interesting characteristics. Note that there is only one point out of control on the high side, but there are seven points out of control on the low side. Most of these seven "low points" occurred after

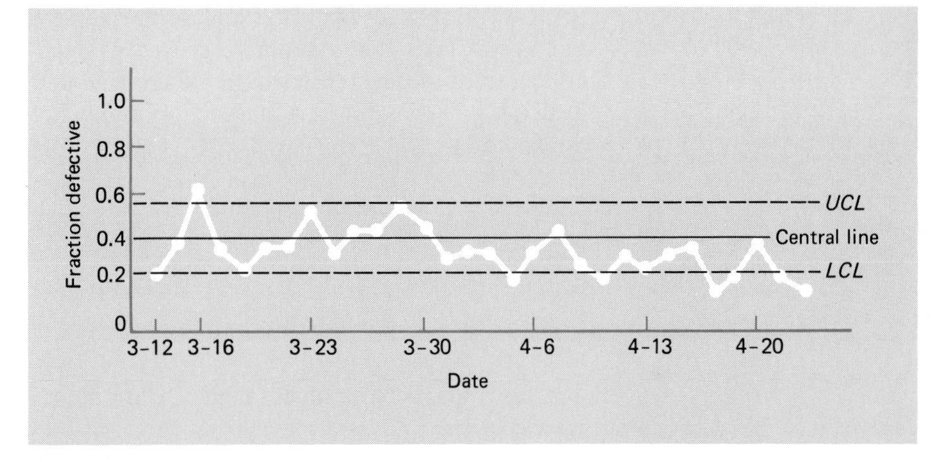

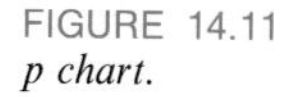

FIGURE 14.11
p chart.

April 1, and there appears to be a general downward trend. In fact, there is an unbroken run of eleven points below the central line after April 7. It would appear from this chart that the yield is not yet stabilized and that the process is potentially capable of maintaining a yield well above the nominal 60% value.

Equivalent to the p chart for the fraction defective is the control chart for the number of defectives. Instead of plotting the fraction defective in a sample of size n, one plots the number of defectives, and the control-chart values for this kind of chart are obtained by multiplying the above values for the central line and the control limits by n. Thus, if p is estimated by $\bar{p}$, the control-chart values for a **number-of-defectives** chart are as follows:

Control-chart values for a number-of-defectives chart

$$\text{central line} = n\bar{p}$$

$$UCL = n\bar{p} + 3\sqrt{n\bar{p}(1-\bar{p})}$$

$$LCL = n\bar{p} - 3\sqrt{n\bar{p}(1-\bar{p})}$$

There are situations where it is necessary to control the number of defects in a unit of product, rather than the fraction defective or the number of defectives. For example, in the production of carpeting it is important to control the number of defects per hundred yards; in the production of newsprint one may wish to control the number of defects per roll. These situations are similar to the one described in Section 4.7, which led to the Poisson distribution. Thus, if c is the number of defects per manufactured unit, c is taken to be a value of a random variable having the Poisson distribution.

It follows that the center line for a number-of-defects chart is the parameter λ of the corresponding Poisson distribution, and that three-sigma control limits can be based on the fact that the standard deviation of this distribution is $\sqrt{\lambda}$. If λ is unknown, that is, if no standard is given, its value is usually estimated from at least 20 values of c observed from past data. If k is the number of units of product available for estimating λ, and if c_i is the number of defects in the ith unit, then λ is estimated by

Mean number of defects

$$\bar{c} = \frac{1}{k}\sum_{i=1}^{k} c_i$$

and the control-chart values for the *c chart* are

Control-chart values for a number-of defects chart

$$\textit{central line} = \bar{c}$$

$$UCL = \bar{c} + 3\sqrt{\bar{c}}$$

$$LCL = \bar{c} - 3\sqrt{\bar{c}}$$

To illustrate this kind of control chart, suppose that it is known from past experience that on the average an aircraft assembly made by a certain company has $\bar{c} = 4$ missing rivets. The corresponding control chart for the number of missing rivets is shown in Figure 14.12, on which we have also plotted the results of inspections which revealed 4, 6, 5, 1, 2, 3, 5, 7, 1, 2, 2, 4, 6, 5, 3, 2, 4, 1, 8, 4, 5, 6, 3, 4, and 2 missing rivets in 25 assemblies.

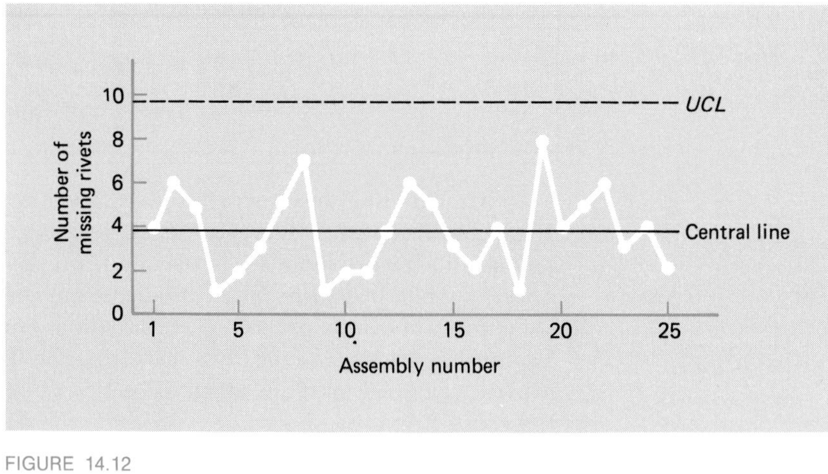

FIGURE 14.12
c chart.

EXERCISES

14.1 A plastics manufacturer extrudes blanks for use in the manufacture of eye-glass temples. Specifications require that the thickness of these blanks have $\mu = 0.150$ inch and $\sigma = 0.002$ inch.

(a) Use the specifications to calculate a central line and three-sigma control limits for an $\bar{x}$ chart with $n = 5$.

(b) Use the specifications to calculate a central line and three-sigma control limits for an R chart with $n = 5$.

(c) Plot the following means and ranges, obtained in 20 successive random samples of size 5, on charts based on the control-chart constants obtained in parts (a) and (b), and discuss the process.

Sample	$\bar{x}$	R	*Sample*	$\bar{x}$	R
1	0.152	0.004	11	0.149	0.003
2	0.147	0.006	12	0.153	0.004
3	0.153	0.004	13	0.150	0.005
4	0.153	0.002	14	0.152	0.001
5	0.151	0.003	15	0.149	0.003
6	0.148	0.002	16	0.146	0.002
7	0.149	0.006	17	0.154	0.004
8	0.144	0.001	18	0.152	0.005
9	0.149	0.003	19	0.151	0.002
10	0.152	0.005	20	0.149	0.004

14.2 Calculate $\bar{\bar{x}}$ and $\bar{R}$ for the data of part (c) of Exercise 14.1, and use these values to construct the central lines and three-sigma control limits for new $\bar{x}$ and R charts to be used in the control of the thickness of the extruded plastic blanks.

14.3 The following data give the means and ranges of 25 samples, each consisting of four compression test results on steel forgings, in thousands of pounds per square inch:

Sample	*1*	*2*	*3*	*4*	*5*	*6*	*7*	*8*
$\bar{x}$	45.4	48.1	46.2	45.7	41.9	49.4	52.6	54.5
R	2.7	3.1	5.0	1.6	2.2	5.7	6.5	3.6

Sample	*9*	*10*	*11*	*12*	*13*	*14*	*15*	*16*
$\bar{x}$	45.1	47.6	42.8	41.4	43.7	49.2	51.1	42.8
R	2.5	1.0	3.9	5.6	2.7	3.1	1.5	2.2

Sample	*17*	*18*	*19*	*20*	*21*	*22*	*23*	*24*	*25*
$\bar{x}$	51.1	52.4	47.9	48.6	53.3	49.7	48.2	51.6	52.3
R	1.4	4.3	2.2	2.7	3.0	1.1	2.1	1.6	2.4

(a) Use these data to find the central line and control limits for an $\bar{x}$ chart.

(b) Use these data to find the central line and control limits for an R chart.

(c) Plot the given data on $\bar{x}$ and R charts based on the control-chart constants computed in parts (a) and (b), and interpret the results.

(d) Using runs above and below the central line (similar to runs above and below the median discussed on page 315), test at a level of significance of 0.05 whether there is a trend in the $\bar{x}$ values.

(e) Would it be reasonable to use the control limits found in this exercise in connection with subsequent compression test measurements from the same process? Why?

14.4 Reverse-current readings (in nanoamperes) are made on a sample of 10 transistors every half hour. Since some of the units may prove to be "shorts" or "opens," it is not always possible to obtain 10 readings. The following table shows the number of readings made at the end of each half-hour interval during an 8-hour shift, and the mean reverse currents obtained.

Sample	*1*	*2*	*3*	*4*	*5*	*6*	*7*	*8*
n	10	6	9	8	8	10	7	9
$\bar{x}$	12.5	11.1	10.2	11.6	21.9	12.3	9.7	15.6

Sample	*9*	*10*	*11*	*12*	*13*	*14*	*15*	*16*
n	7	8	10	9	7	8	9	10
$\bar{x}$	16.7	9.8	11.6	17.2	10.1	9.5	13.1	14.2

(a) Find the central line for an $\bar{x}$ chart by taking the weighted mean of the 16 $\bar{x}$'s, weighting each value with the size of the corresponding sample.

(b) Construct a table showing the central line found in part (a) and three-sigma control limits corresponding to $n = 6, 7, 8, 9$, and 10. Use $R = 4.0$, a value based on prior data.

(c) Plot the data on a control chart like the one of Figure 14.13 and interpret the results.

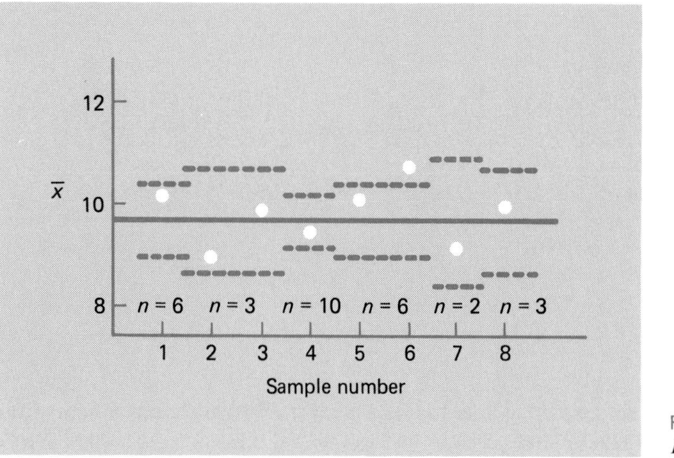

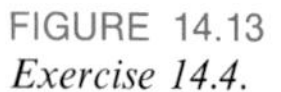
FIGURE 14.13
Exercise 14.4.

14.5 If the sample standard deviations instead of the sample ranges are used to estimate σ, the control limits for the resulting $\bar{x}$ chart are given by $\bar{\bar{x}} \pm A_1\bar{s}$, where $\bar{s}$ is the mean of the sample standard deviations obtained from given data, and A_1 can be found in Table 13. Note that in connection with problems of quality control the sample standard deviation is defined using the divisor n instead of $n - 1$. The corresponding R chart is replaced by a σ chart, having the central line $c_2\bar{s}$ and the upper and lower control limits $B_3\bar{s}$ and $B_4\bar{s}$, where B_3 and B_4 can be obtained from Table 13.

(a) Construct an $\bar{x}$ chart and a σ chart for 20 samples of size 3 which had $\bar{x}$ equal to 21.2, 19.4, 20.4, 20.4, 20.4, 19.0, 20.3, 21.1, 21.6, 22.1, 24.4, 23.9, 24.9, 24.1, 21.8, 19.5, 20.3, 22.5, 23.4, 23.3, and s equal to 2.0, 0.8, 1.1, 0.9, 1.0, 0.3, 1.3, 2.0, 0.8, 1.0, 1.5, 1.0, 1.5, 0.8, 1.3, 2.9, 4.3, 1.2, 0.3, 3.1.

(b) Would it be reasonable to use these control limits for subsequent data? Why?

14.6 In order to establish control charts for a boring process, 30 samples of five measurements of the inside diameters are taken, and the results are $\bar{\bar{x}} = 1.317$ inches and $\bar{s} = 0.002$ inch. Using the method of Exercise 14.5, construct an $\bar{x}$ chart for $n = 5$, and on it plot the following means obtained in 25 successive samples: 1.328, 1.330, 1.321, 1.325, 1.332, 1.340, 1.327, 1.321, 1.324, 1.325, 1.329, 1.326, 1.330, 1.324, 1.328, 1.322, 1.326, 1.327, 1.329, 1.325, 1.324, 1.329, 1.330, 1.321, and 1.329. Discuss the results.

14.7 Suppose that in the example of Exercise 14.6 it is desired to establish control also over the variability of the process. Using the method of Exercise 14.5 and the values of $\bar{\bar{x}}$ and $\bar{s}$ given in Exercise 14.6, calculate the central line and control limits for a σ chart with $n = 5$.

14.8 Thirty-five successive samples of 100 castings each, taken from a production line, contained, respectively, 3, 3, 5, 3, 5, 0, 3, 2, 3, 5, 6, 5, 9, 1, 2, 4, 5, 2, 0, 10, 3, 6, 3, 2, 5, 6, 3, 3, 2, 5, 1, 0, 7, 4, and 3 defectives. If the fraction defective is to be maintained at 0.02, construct a p chart for these data and state whether or not this standard is being met.

14.9 The data of Exercise 14.8 may be looked upon as evidence that the standard of 2% defectives is being exceeded.

(a) Use the data of Exercise 14.8 to construct new control limits for the fraction defective.

(b) Using the control limits found in part (a), continue the control of the process by plotting the following data on the number of defectives obtained in 20 subsequent samples of size $n = 100$: 2, 4, 2, 4, 7, 5, 3, 2, 2, 3, 5, 6, 4, 5, 8, 0, 5, 5, 4, and 2.

14.10 The specifications for a certain mass-produced valve prescribe a testing procedure according to which each valve can be classified as satisfactory or unsatisfactory (defective). Past experience has shown that the process can perform so that $\bar{p} = 0.03$. Construct a three-sigma control chart for the number of defectives obtained in samples of size 100, and on it plot the following numbers of defectives obtained in such samples randomly selected from 30 successive half-day's production: 3, 4, 2, 1, 5, 2, 1, 2, 3, 1, 3, 2, 2, 2, 1, 1, 2, 0, 4, 3, 1, 0, 2, 4, 0, 1, 5, 7, 3, and 2.

14.11 The standard for a process producing tin plate in a continuous strip is five defects in the form of pinholes or visual blemishes per hundred feet. Based on the following set of 25 observations, giving the number of defects per hundred feet, can it be concluded that the process is in control to this standard?

Inspection number	1	2	3	4	5	6	7	8	9	10	11	12
Number of defects	3	2	2	4	4	4	6	4	1	7	5	5

Inspection number	13	14	15	16	17	18	19	20	21	22	23	24	25
Number of defects	4	6	6	9	5	2	6	5	11	6	6	8	2

14.12 A process for the manufacture of 4-feet-by-8-feet woodgrained panels has performed in the past with an average of 2.7 imperfections per 100 panels. Construct a chart to be used in the inspection of the panels and discuss the control if 25 successive 100-panel lots contained, respectively, 4, 1, 0, 3, 5, 3, 5, 4, 1, 4, 0, 1, 4, 2, 3, 7, 4, 2, 1, 3, 0, 2, 6, 1, and 3 imperfections.

14.7 TOLERANCE LIMITS

Inherent in every phase of industrial quality control is the problem of comparing some quality characteristic or measurement of a finished product against given specifications. Sometimes the specifications, or **tolerance limits**, are so stated by the customer or by the design engineer that any appreciable departure will make the product unusable. There remains, however, the problem of producing the part so that an acceptably high proportion of units will fall within tolerance limits specified for the given quality characteristic. Also, if a product is made without prior specifications, or if modifications are made, it is desirable to know within what limits the process can hold a quality characteristic a reasonably high percentage of the time. We, thus, speak of "natural" tolerance limits; that is, we let the process establish its own limits which, according to experience, can be met in actual practice.

If reliable information is available about the distribution underlying the measurement in question, it is a relatively simple matter to find natural tolerance limits. For instance, if long experience with a product enables us to assume that a certain dimension is normally distributed with the mean μ and the standard deviation σ, it is easy to construct limits between which we can expect to find any given proportion P of the population. For $P = 0.90$ we have the tolerance limits $\mu \pm 1.645\sigma$, and for $P = 0.95$ we have $\mu \pm 1.96\sigma$, as can easily be verified from a table of normal-curve areas.

In most practical situations the true values of μ and σ are not known, and tolerance limits must be based on the mean $\bar{x}$ and the standard deviation s of a random sample. Whereas $\mu \pm 1.96\sigma$ are limits including 95% of a normal population, the same cannot be said for the limits $\bar{x} \pm 1.96s$. These limits are values of random variables and they may or may not include a given proportion of the population. Nevertheless, it is possible to determine a constant K so that *one can assert with* $(1 - \alpha)100\%$ *confidence that the proportion of the population contained between* $\bar{x} - Ks$ *and* $\bar{x} + Ks$ *is at least* P. Such values of K for random samples from normal populations are given in Table 14 for $P = 0.90$, 0.95, and 0.99, 95% and 99% levels of confidence, and selected values of n from 2 to 1,000.

To illustrate this technique, suppose that a manufacturer takes a sample of size $n = 100$ from a very large lot of mass-produced compression springs and that he obtains $\bar{x} = 1.507$ and $s = 0.004$ inch for the free lengths of the springs. Choosing the 99% level of confidence and a minimum proportion of $P = 0.95$, he obtains the tolerance limits $1.507 \pm (2.355)(0.004)$; in other words, the manufacturer can assert with 99% confidence that at least 95% of the springs in the entire lot have free lengths from 1.497 to 1.517 inches. Note that in problems like these the minimum proportion P as well as the degree of confidence $1 - \alpha$ must be specified; also note that the lower tolerance limit is rounded *down* and the upper tolerance limit is rounded *up*.

To avoid confusion, let us also point out that there is an essential difference between confidence limits and tolerance limits. Whereas confidence limits are used to estimate a parameter of a population, tolerance limits are used to indicate between what limits one can find a certain proportion of a population. This distinction is emphasized by the fact that when n becomes large the length of a confidence interval approaches zero, while the tolerance limits will approach the corresponding values for the population. Thus, for large n, K approaches 1.96 in the columns for $P = 0.95$ in Table 14.

The situation for one-sided tolerance bounds is different. In the context of strength of materials, it is the weaker specimens that break. Consequently, it is important for engineers to have an accurate estimate of the lower tail of the population of strengths. Recently engineers have realized that it is wiser to set specifications for strength in terms of a lower percentile η_β rather than the mean μ. It is the weaker specimens, not those of average strength that break. The lumber industry and many space-age materials groups specify that a 95% one-sided confidence bound be calculated for the 5*th* percentile $\eta_{0.05}$. That is, a lower bound $L(x_1, x_2, \ldots, x_n)$ is calculated from the observations and, prior to taking the observations,

$$P[L < \eta_{0.05}] = 0.95$$

But this one-sided confidence bound is just a **one-sided tolerance bound**, since the event *the bound* $L(x_1, x_2, \ldots, x_n)$ *is less than the population 0.05 point* $\eta_{0.05}$ is the same as the event as *at least 95% of the population is above* $L(x_1, x_2, \ldots, x_n)$.

For normal populations

$$L(x_1, x_2, \ldots, x_n) = \bar{x} - Ks$$

where K can be obtained from Table 14(b). (A more extensive table appears in the book by I. Guttman listed in the bibliography.)

EXAMPLE The cardboard industry is considering new standards for the cardboard used in boxes. One test involves placing weight on the box until it bursts. The burst strengths, in pounds per square inch, for 40 boxes are

```
210 234 216 232 262
183 227 197 248 218
256 218 244 259 263
185 218 196 235 223
212 237 275 240 217
263 240 247 253 269
231 254 248 261 268
262 247 292 238 215
```

Obtain a 95% tolerance bound that will be less than proportion 0.95 of the population of burst strengths.

Solution By computer, we determine that $\bar{x} = 237.32$ and $s = 25.10$. From Table 14.16(b), $K = 2.126$, so

$$L = \bar{x} - Ks = 237.32 - 2.126(25.10) = 183.0$$

We are 95% confident that a proportion 0.95 of the population of burst strengths, for cardboard boxes, is above 183.0 psi.

In Exercise 14.43 you are asked to verify that the strength measurements fail to exhibit departures from normality.

■

EXERCISES

14.13 To check the strength of carbon steel for use in chain links, the yield stress of a random sample of 25 pieces was measured, yielding a mean and a standard deviation of 52,800 psi and 4,600 psi, respectively. Establish tolerance limits with $\alpha = 0.05$ and $P = 0.99$, and express *in words* what these tolerance limits mean.

14.14 In a study designed to determine the number of turns required for an artillery-shell fuse to arm, 75 fuses, rotated on a turntable, averaged 38.7 turns with a standard deviation of 4.3 turns. Establish tolerance limits for which one can assert with 99% confidence that *at least 95%* of the fuses will arm within these limits.

14.15 In a random sample of 40 piston rings chosen from a production line, the mean edge width was 0.1063 inch, and the standard deviation was 0.0004 inch.

(a) Between what limits can it be said with 95% confidence that at least 90% of the edge widths of piston rings produced by this production line will lie?

(b) Find 95% confidence limits for the true mean edge width, and explain the difference between these limits and the tolerance limits found in part (a).

14.16 *Nonparametric tolerance limits* can be based on the extreme values in a random sample of size n taken from any continuous population. The following equation relates the quantities n, P, and α, where P is the minimum proportion of the population contained between the smallest and largest observations with $(1 - \alpha)100\%$ confidence:

$$nP^{n-1} - (n-1)P^n = \alpha$$

An approximate solution for n is given by

$$n \simeq \frac{1}{2} + \frac{1+P}{1-P} \cdot \frac{\chi_\alpha^2}{4}$$

where χ_α^2 is the value of chi square for 4 degrees of freedom that corresponds to a right-hand tail area α.

(a) How large a sample is required to be 95% certain that at least 90% of the population will be included between the extreme values of the sample?

(b) With 95% confidence, at least what proportion of the population can be expected to be included between the extreme values of a sample of size 100?

14.8 ACCEPTANCE SAMPLING

Manufactured goods are shipped to the purchaser in lots ranging in size from only a few to many thousands of individual items. Ideally, each lot should not contain any defectives, but practically speaking it is rarely possible to meet this goal. Recognizing the fact that some defective goods are bound to be delivered, even if each lot were to be inspected 100%, most consumers require that evidence, based on careful inspection, be given that the proportion of defectives in each lot is not excessive.

We must stress that this philosophy is strictly at odds with the new philosophy of quality improvement. There is no reason, in the long run, to settle for any fixed proportion of defectives. Instead, there must be continual effort to improve the process. However, because many contracts still specify the use of these plans, the engineer should be acquainted with the details of acceptance sampling.

A frequently used and highly effective method for providing such evidence is that of sampling inspection, where items are selected from each lot prior to shipment (or prior to acceptance by the consumer), and a decision is made on the basis of this sample whether to accept or reject the lot. Acceptance of a lot ordinarily implies that it can be shipped (or be accepted by the consumer), even though it may contain some defective items. Arrangements between the producer and the consumer may allow for some form of credit to be given for defectives subsequently discovered by the consumer. Rejection of a lot need not mean that it is to be scrapped; a rejected lot may be subjected to closer inspection with the aim of eliminating all defective items.

Since the cost of inspection is rarely negligible (sometimes it is nearly as high as or higher than the cost of production), it is seldom desirable to inspect each item in a lot. Thus, acceptance inspection usually involves **sampling**; more specifically, a random sample is selected from each lot and the lot is accepted if the number of defectives found in the sample does not exceed a given **acceptance number**. This procedure is equivalent to a test of the null hypothesis that the proportion defective p in the lot equals some specified value p_0 against the alternative that it equals p_1, where $p_1 > p_0$. In acceptance sampling the value p_0 is called the **acceptable quality level**, or **AQL**, and p_1 is called the **lot tolerance percent defective**, or **LTPD**. The probability of a Type I error, α, can be interpreted as an upper limit to the proportion of "good" lots (lots with $p \leq p_0$) that will be rejected, and in this context it is called the **producer's risk**. The probability of a Type II error, β, gives an upper bound to the proportion of "bad" lots (lots with $p \geq p_1$) that will be accepted, and it is called the **consumer's risk**.

A **single-sampling plan** is simply a specification of the sample size and the acceptance number to be used, and its choice is usually based on a specified AQL and (or) $LTPD$ in association with given producer's and (or) consumer's risks. A given sampling plan is best described by its operating characteristic or OC curve, which gives the probability of acceptance for each value that can be assumed by the lot proportion defective p. Thus, the OC curve describes the degree of protection offered by the sampling plan against incoming lots of various qualities. If a sample of size n is taken from a lot containing N units, and if the acceptance number is c, the probability of accepting a lot containing the proportion of defectives p (the lot contains Np defectives) can be calculated by using the hypergeometric distribution, as follows:

Probability of lot acceptance

$$L(p) = \sum_{x=0}^{c} h(x; n, Np, N)$$

To illustrate the calculation of points on an OC curve, suppose that the lot size is $N = 100$, the sample size is $n = 10$, and the acceptance number is $c = 1$. We have

$$L(p) = \frac{\binom{100p}{0}\binom{100(1-p)}{10} + \binom{100p}{1}\binom{100(1-p)}{9}}{\binom{100}{10}}$$

Since calculations involving the hypergeometric distribution are fairly tedious, especially when n and N are large, it is customary in acceptance sampling to approximate the hypergeometric distribution with the binomial distribution, as on page 98. For instance, for $p = 0.10$ the exact value is $L(0.10) = 0.739$, as the reader will be asked to verify in Exercise 14.21 on page 526, while Table 1 with $n = 10$ and $p = 0.10$ yields the binomial approximation

$$L(0.10) \simeq B(1; 10. 0.10) = 0.736$$

A sketch of the OC curve for the sampling plan $n = 10$, $c = 1$ can be made rapidly with the aid of Table 1, and the result is shown in Figure 14.14. From this curve it can be seen that the producer's risk is approximately 0.05 when the AQL is 0.04, and the consumer's risk is approximately 0.10 when the $LTPD$ is 0.34.

A sampling plan can also be described by means of its **average outgoing quality**, or ***AOQ*** curve. This curve describes the degree of protection offered by the sampling plan by showing the average quality of outgoing lots corresponding to each quality level of incoming lots (that is, lots prior to inspection). If incoming lots are of good quality, that is, if their proportion defective is smaller than the AQL, very few lots will be rejected and the average outgoing quality or AOQ will be good. If the incoming lots are of poor quality, that is, if their proportion defective is larger than the $LTPD$, most of them will be rejected. If all rejected lots are inspected 100% and all defective units are replaced by good units prior to acceptance of the lot, then the average outgoing quality will be good even though the average incoming

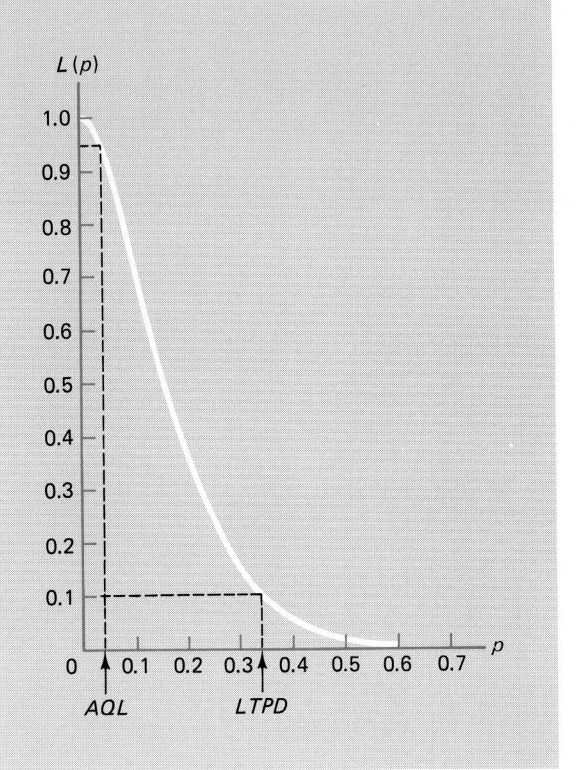

FIGURE 14.14
OC curve.

quality is poor. It is when the average incoming quality lies between the *AQL* and the *LTPD* that the poorest quality of lots will be shipped. In general, there will be a maximum *AOQ* over all values of incoming quality p, and this value is called the **average outgoing quality limit**, or ***AOQL***.

It is not difficult to derive a formula for finding the *AOQ* corresponding to a given incoming quality p under the assumption that all defectives in rejected lots are replaced by acceptable items prior to their final acceptance. If the incoming quality is p, the probability that a lot will be accepted is $L(p)$, and each such lot contains the proportion p of defectives. The proportion $1 - L(p)$ of lots that are eventually rejected contain no defectives, and it follows that the *AOQ* is given by $p \cdot L(p) + 0 \cdot [1 - L(p)]$, or

Average outgoing quality

$$AOQ = p \cdot L(p)$$

More common practice is to remove or replace defectives found in accepted as well as rejected lots, but the modification thus required in the *AOQ* is usually minor and it is customary to use the above formula regardless of the inspection procedure.

The *AOQ* curve for the sampling plan $n = 10$, $c = 1$ is shown in Figure 14.15, and it is apparent from this figure that the *AOQL* is approximately 0.081.

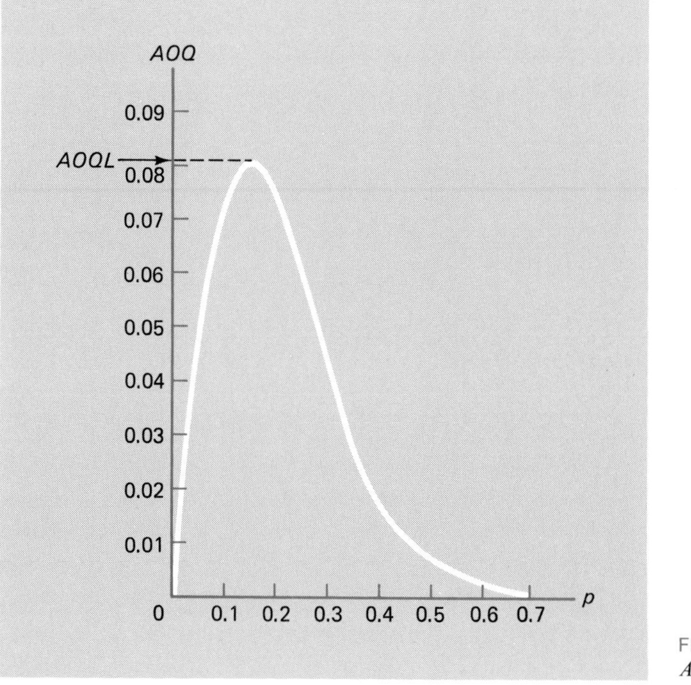

FIGURE 14.15
AOQ curve.

Sometimes, smaller samples (and hence, reductions in cost) can be achieved without sacrifice in the degree of protection by the use of **double** or **multiple sampling**. A double-sampling plan involves the selection of a random sample of size n_1 from a lot; if the sample contains c_1 or fewer defectives, the lot is accepted; if it contains c'_1 or more defectives ($c'_1 > c_1$) the lot is rejected; otherwise, a second sample of size n_2 is taken from the lot, and the lot is accepted unless the total number of defectives in the combined sample of size $n_1 + n_2$ exceeds c_2. A multiple-sampling plan is similar in nature to a double-sampling plan, but it involves more than two stages.

An illustration of a multiple-sampling plan is given in the following table:

		Combined samples		
Sample	*Sample size*	*Size*	*Acceptance number*	*Rejection number*
First	20	20		3
Second	20	40	1	4
Third	20	60	3	5
Fourth	20	80	3	6
Fifth	20	100	5	7
Sixth	20	120	6	8
Seventh	20	140	7	8

This table summarizes a procedure that can be described as follows. In the first step, the lot is rejected if there are 3 or more defectives; otherwise, sampling continues. In the second step, the lot is accepted if the combined sample contains at most 1 defective, it is rejected if there are 4 or more defectives; otherwise, sampling continues. This goes on, if necessary, until in the final step the lot is accepted if there are at most 7 defectives in the combined sample of size 140, and otherwise it is rejected.

By an appropriate choice of the sample sizes and the acceptance and rejection numbers it is possible to match the *OC* curve of a double- or multiple-sampling plan closely to that of an equivalent single-sampling plan. Thus, the degree of protection offered by a double- or multiple-sampling plan can be made essentially the same as that offered by an equivalent single-sampling plan.

The advantage of double or multiple sampling is that there is a high probability that a very good lot will be accepted or a very poor lot will be rejected on the basis of the first sample (or, at least, an early sample), thus reducing the required amount of inspection. On the other hand, if the lot quality is "intermediate," the total sample size required may actually be larger than that of the equivalent single-sampling plan.

To observe how a double- or multiple-sampling plan can conserve sampling effort, and to introduce the idea of the **average sample number (*ASN*) curve**, we consider the following double-sampling plan:

		Combined samples		
Sample	*Sample size*	*Size*	*Acceptance number*	*Rejection number*
First	15	15	1	5
Second	30	45	5	6

If the incoming lot quality is $p = 0.05$, the probability that a second sample will be required is the same as the probability that there will be 2, 3, or 4 defectives in a sample of size 15. According to Table 1, this probability is equal to

$$B(4;\, 15, 0.05) - B(1;\, 15, 0.05) = 0.170$$

Thus, on the average, it will require a sample of size $15 + (0.170)(30) = 20.1$ to decide whether to accept or reject an incoming lot of quality $p = 0.05$. Similar calculations enable us to find the average sample size required to inspect a lot having any given incoming quality p. A graph showing the relation between the average sample size (also called the **average sample number**) and the incoming lot quality is called an *ASN* curve; the *ASN* curve for the double-sampling plan described above is shown in Figure 14.16.

Several standard sampling plans have been published to facilitate the use of acceptance sampling (see the book by A. Duncan listed in the Bibliography). Among the most widely used standard plans are those contained in **Military Standard *105D*** (see the Bibliography). These plans stress the maintenance of a specified *AQL*, and they are designed to encourage the producer to offer only good products to the consumer. To accomplish this there are three general levels of inspection corresponding to different consumer's risks. (Inspection level II is normally chosen; level I uses smaller sample sizes and level III uses larger sample sizes than level II.) There are also three types of inspection: normal, tightened, and reduced. The type of inspection depends on whether the average proportion defectives for prior samples has been above or below the *AQL*, and it may be changed during the course of inspection. Under tightened inspection the producer's risk is increased and the consumer's risk is (slightly) decreased; under reduced inspection the consumer's risk is increased and the producer's risk is (slightly) decreased. Tables are available for single, double, and multiple sampling, and a brief portion of these tables is included in this book in Tables 15 and 16.

The procedure in using MIL-STD-105D for single sampling is first to find the sample size code letter corresponding to the lot size and the desired inspection level

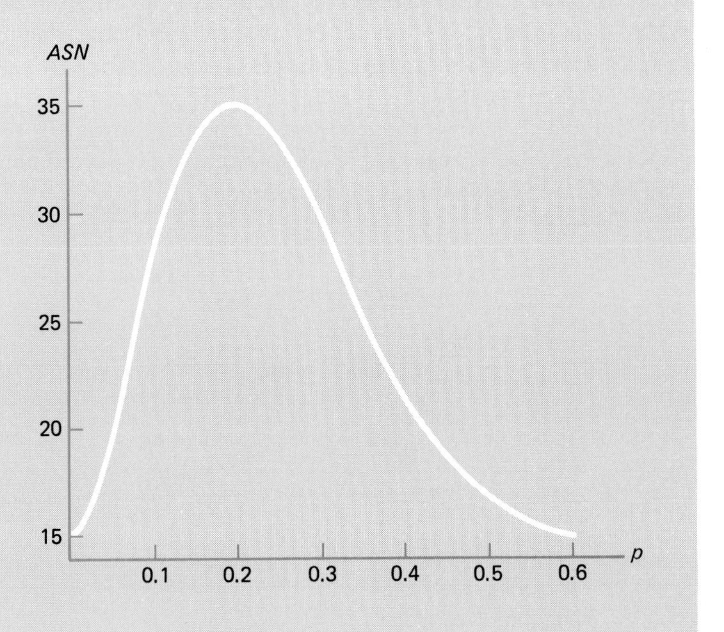

FIGURE 14.16
ASN curve.

and type. Then, using the sample size code letter thus obtained and the appropriate AQL, one finds the sample size and the acceptance number from the master table. A portion of the table for finding sample size code letters is given in Table 15, and a portion of the master table for normal inspection is included in Table 16.

To illustrate the use of MIL-STD-105D, suppose that incoming lots contain 2,000 items, and inspection level II is to be used in conjunction with normal inspection and an AQL of 0.025, or 2.5%. From Table 15 we find that the sample size code letter is K. Then, entering Table 16 in the row labeled K, we find that the sample size to be used is 125. Using the column labeled 2.5, we find that the acceptance number is 7 and the rejection number is 8. Thus, if a single sample of size 125, selected at random from a lot of 2,000 items, contains 7 or fewer defectives the lot is to be accepted; if the sample contains 8 or more defectives the lot is to be rejected.

The concept of multiple sampling is carried to its extreme in **sequential sampling**. A sampling procedure is said to be sequential if, after each observation, one of the following decisions is made: accept whatever hypothesis is being tested, reject the hypothesis, or take another observation. Although sequential procedures are used also in connection with other kinds of problems, we shall discuss this kind of sampling only in connection with acceptance sampling, where we shall thus decide after the inspection of each successive item whether to accept a lot, reject it, or continue sampling.

The construction of a sequential sampling plan consists of finding two sequences of numbers a_n and r_n, where n is the number of observations, so that the lot is accepted as soon as the number of defectives is less than or equal to a_n for some n, the lot is rejected as soon as the number of defectives is greater than or equal to r_n for some n, and sampling continues so long as the number of defectives in a sample of size n falls between a_n and r_n. If an acceptance plan is to have p_0 and p_1 as its AQL and $LTPD$, the producer's risk α, and the consumer's risk β, it can be shown (see the book by A. Wald in the bibliography) that the required values of a_n and r_n can be computed by means of the formulas

Acceptance and rejection numbers for sequential sampling

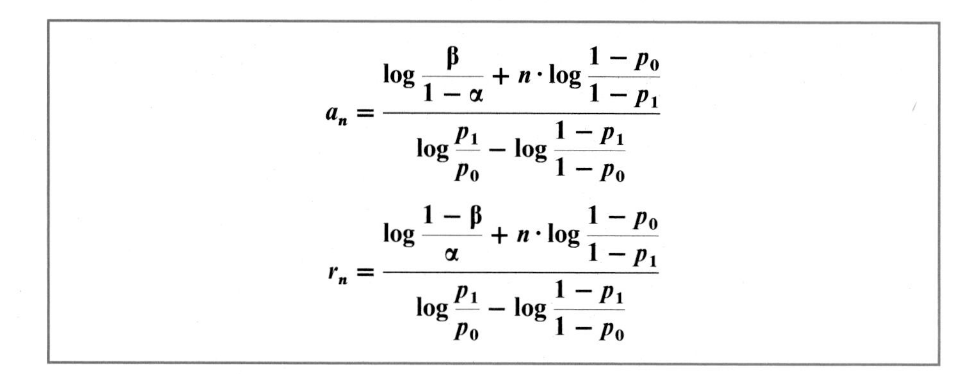

$$a_n = \frac{\log \dfrac{\beta}{1-\alpha} + n \cdot \log \dfrac{1-p_0}{1-p_1}}{\log \dfrac{p_1}{p_0} - \log \dfrac{1-p_1}{1-p_0}}$$

$$r_n = \frac{\log \dfrac{1-\beta}{\alpha} + n \cdot \log \dfrac{1-p_0}{1-p_1}}{\log \dfrac{p_1}{p_0} - \log \dfrac{1-p_1}{1-p_0}}$$

If a_n is not an integer, it is replaced by the largest integer less than a_n; if r_n is not an integer, it is replaced by the smallest integer greater than r_n.

To illustrate this procedure, let $p_0 = 0.05$, $p_1 = 0.20$, $\alpha = 0.05$, and $\beta = 0.10$. Substituting these values into the above formulas for a_n and r_n, we obtain

$$a_n = -1.45 + 0.11n$$

$$r_n = 1.86 + 0.11n$$

and, letting $n = 1, 2, 3, \ldots$, and 25, we get the acceptance and rejection numbers shown in the second and fourth columns of the following table:

Number of items inspected n	*Acceptance number* a_n	*Number of defectives* d_n	*Rejection number* r_n
1	—	0	—
2	—	0	—
3	—	0	3
4	—	0	3
5	—	0	3
6	—	0	3
7	—	0	3
8	—	1	3
9	—	1	3
10	—	1	3

(*continued*)

Number of items inspected n	*Acceptance number* a_n	*Number of defectives* d_n	*Rejection number* r_n
11	—	1	4
12	—	1	4
13	—	1	4
14	0	1	4
15	0	1	4
16	0	1	4
17	0	2	4
18	0	2	4
19	0	3	4
20	0	3	5
21	0	3	5
22	0	4	5
23	1	5	5
24	1		5
25	1		5

In the third column we have indicated the results obtained in an inspection where the 8th, 17th, 19th, 22nd, and 23rd items are defective, and where the inspection terminates with rejection of the lot after inspection of the 23rd item.

The tabular procedure illustrated can be replaced by an equivalent graphical procedure for carrying out sequential sampling inspection. Plotting the values of a_n and r_n obtained from the equations on page 524 (without rounding), we obtain two *straight lines* like those of Figure 14.17. Sampling terminates with rejection or

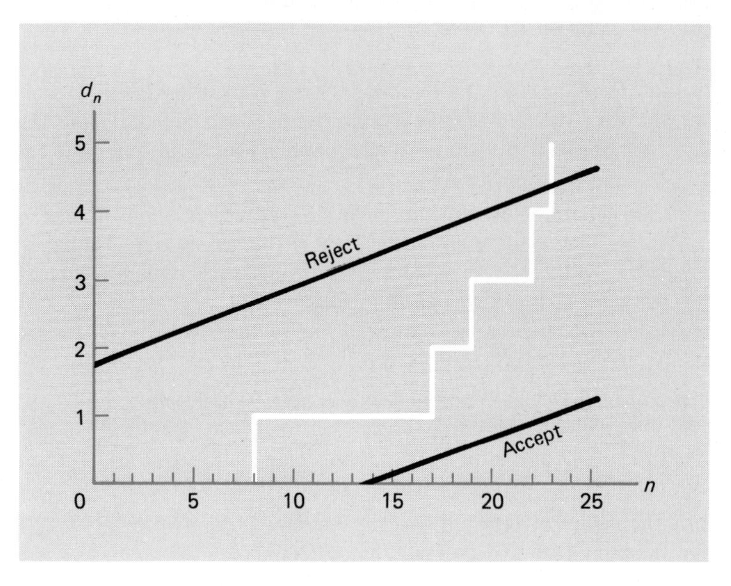

FIGURE 14.17
Graphical procedure for sequential sampling.

acceptance if the number of defectives observed falls above the line for r_n or below the line for a_n, respectively.

The main advantage of sequential sampling is that it can materially reduce the required amount of inspection. Studies have shown that the average decrease in sample size is often near 50% when compared with the sample size of equivalent single-sampling plans. The major disadvantage is that, in a sequential sampling plan, there is no upper limit to the number of items that might have to be inspected in order to reach a decision concerning the acceptance or rejection of a lot. In fact, the sample size is a random variable and its value will occasionally be very large. For this reason it is customary to *truncate* sequential sampling procedures by selecting a number N such that a decision must be made to accept or reject a lot prior to or at $n = N$.

EXERCISES

14.17 A sampling plan, calling for a sample of size $n = 50$, has the acceptance number $c = 3$. Assuming that the lot size is very large, calculate the probability of accepting a lot of incoming quality 15% defective and the probability of rejecting a lot of incoming quality 4% defective

(a) by calculating the corresponding binomial probabilities;

(b) by using the Poisson approximation to the binomial distribution.

14.18 A single-sampling plan calls for a sample of size 150.

(a) Use the normal approximation to find the acceptance number c if the *AQL* is to be 2.5% and the producer's risk is to be $\alpha = 0.05$.

(b) Using the acceptance number obtained in part (a), determine the consumer's risk if the *LTPD* is to be 6%.

14.19 Referring to the sampling plan of Exercise 14.17, use the Poisson approximation to calculate $L(p)$ for $p = 0.01, 0.02, \ldots, 0.19$, and 0.20. Sketch the *OC* curve for this sampling plan and read off the consumer's and producer's risks corresponding to an *AQL* of 4% and an *LTPD* of 14%.

14.20 A single-sampling plan has $n = 100$ and $c = 5$.

(a) Find the *AQL* if the producer's risk is to be 0.025.

(b) Find the *LTPD* if the consumer's risk is to be 0.05. [*Hint*: Use the normal approximation and set up equations leading to quadratic equations in p_0 and p_1, respectively.]

14.21 With reference to the example on page 518, verify that

(a) $L(0.10) = 0.739$, using the exact (hypergeometric) formula;

(b) $L(0.10) \simeq 0.736$, using the approximate (binomial) formula;

(c) the *OC* curve is as given in Figure 14.14;

(d) the *AOQ* curve is as given in Figure 14.15.

14.22 Calculate $L(p)$ for selected values of p and sketch the *OC* curve for the single-sampling plan $n = 120$, $c = 3$. [*Hint*: Assume a large lot size, and use the normal approximation to the binomial distribution.]

14.23 Sketch the *AOQ* curve for the sampling plan of Exercise 14.22, and estimate the *AOQL*.

14.24 Using the values of $L(p)$ obtained in Exercise 14.19, sketch the *AOQ* curve for the given sampling plan and estimate the *AOQL*.

14.25 Referring to the double-sampling plan on page 522, use Table 1 and the normal approximation (when necessary) to calculate the producer's risk α when the *AQL* is $p_0 = 0.10$.

14.26 Calculate the average sample size required for selected values of p and sketch the *ASN* curve for the following double-sampling plan:

		Combined samples		
Sample	*Sample size*	*Size*	*Acceptance number*	*Rejection number*
First	10	10	0	3
Second	35	45	3	4

14.27 A lot of 200 items is to be inspected at an *AQL* of 6.5%. If MIL-STD-105D is to be used, with normal inspection at general inspection level II, what single-sampling plan is required?

14.28 An incoming lot of 5,000 items is to be inspected using MIL-STD-105D, with normal inspection at general inspection level II and an *AQL* of 2.5%. What single-sampling plan should be used?

14.29 Find formulas for the acceptance and rejection numbers for the sequential sampling plan having an *AQL* of 0.10, an *LTPD* of 0.30, a producer's risk of 0.05, and a consumer's risk of 0.10. If a sample should contain defectives on the third, fifth, seventh, and eighth trials, would this plan accept or reject the corresponding lot prior to the tenth trial? If so, on which trial?

14.30 A sequential sampling plan is to have $p_0 = 0.01$, $p_1 = 0.10$, $\alpha = 0.05$, and $\beta = 0.20$.

(a) Determine the acceptance and rejection numbers for $n = 1, 2, \ldots,$ and 50.

(b) Use random numbers and the acceptance and rejection numbers obtained in part (a) to simulate the inspection of a very large lot containing 20% defectives.

14.9 REVIEW EXERCISES

14.31 The specifications require that the weight of castings have $\mu = 4.1$ ounces and $\sigma = 0.05$ ounces.

(a) Use the specifications to calculate a central line and three-sigma control limits for an $\bar{x}$ chart with $n = 5$.

(b) Use the specifications to calculate a central line and three-sigma control limits for an R chart with $n = 5$.

(c) Plot the following means and ranges, obtained in 20 successive random samples of size 5, on charts based on the control-chart constants obtained in part (a) and (b), and discuss the process.

Sample	$\bar{x}$	R
1	4.24	0.09
2	4.18	0.12
3	4.26	0.14
4	4.21	0.24
5	4.22	0.15
6	4.18	0.28
7	4.23	0.06
8	4.19	0.15
9	4.21	0.09
10	4.18	0.15
11	4.20	0.21
12	4.25	0.20
13	4.25	0.17
14	4.21	0.07
15	4.19	0.16
16	4.23	0.16
17	4.27	0.19
18	4.22	0.20
19	4.20	0.12
20	4.19	0.16

14.32 Calculate $\bar{\bar{x}}$ and $\bar{R}$ for the data of part (c) of Exercise 14.31, and use these values to construct the central lines and three-sigma control limits for new $\bar{x}$ and R charts to be used in the control of the weight of the castings.

14.33 Twenty-five successive samples of 200 switches, each taken from a production line, contained, respectively, 6, 7, 13, 7, 0, 9, 4, 6, 0, 4, 5, 11, 6, 18, 1, 4, 9, 8, 2, 17, 9, 12, 10, 5, and 4 defectives. If the fraction of defectives is to be maintained at 0.02, construct a p chart for these data and state whether or not this standard is being met.

14.34 The data of Exercise 14.33 may be looked upon as evidence that the standard of 2% defectives is being exceeded.

(a) Use the data of Exercise 14.33 to construct new control limits for the fraction defective.

(b) Using the limits found in part (a), continue the control of the process by plotting the following data on the next ten samples of size $n = 200$: 4, 7, 5, 3, 8, 3, 1, 4, 3, 9.

14.35 A process for the manufacture of film has performed in the past with an average of 0.8 imperfections per 10 linear feet.

(a) Construct a chart to be used in the inspection of 10 foot sections.

(b) Discuss the control if 20 successive 10-foot sections contained, respectively, 1, 0, 0, 1, 3, 1, 2, 1, 0, 2, 1, 3, 0, 0, 1, 1, 2, 0, 4, and 1 imperfections.

14.36 With reference to the aluminum alloy strength data on page 15, obtain a 95% tolerance limits on the proportion $P = 0.90$ of the population of strengths.

14.37 With reference to the interrequest time data on page 14, obtain a 95% tolerance limits on the proportion $P = 0.90$ of the population of inter-request times. Take logs, use the normal theory approach, and then transform back to the original scale.

14.38 A sampling plan, calling for a sample of size $n = 40$, has the acceptance number $c = 2$. Assuming the lot size is very large, calculate the probability of accepting a lot of incoming quality 15% defective and the probability of rejecting a lot of incoming quality 4% defective

(a) by calculating the corresponding binomial probabilities;

(b) by using the Poisson approximation to the binomial distribution.

14.39 Referring to the sampling plan of Exercise 14.38, use the Poisson approximation to calculate $L(p)$ for $p = 0.01, 0.02, \ldots, 0.19$, and 0.20. Sketch the *OC* curve for this sampling plan and read off the consumer's and producer's risks corresponding to an *AQL* of 4% and *LTPD* of 14%.

14.40 Using the values of $L(p)$ obtained in Exercise 14.39, sketch the *AOQ* curve for the given sampling plan and estimate the *AOQL*.

14.41 A lot of 150 items is to be inspected at an *AQL* of 4.0%. If MIL-STD-105*D* is to be used, with normal inspection at general inspection level II, what single-sampling plan is required?

14.42 With reference to the discussion on page 505, calculate the cusum using 2.25 in place of 2.00 as the centering value. Also make the cusum chart.

14.43 With reference to the example on page 516,

(a) verify the calculation of the tolerance bound L;

(b) if the confidence is decreased to 90%, calculate the new tolerance bound (use $K = 2.010$);

(c) check the cardboard strength data for departures from normality using a normal-scores plot.

14.44 Explain, from the perspective of quality improvement programs, why the $\bar{x}$, R, and fraction defective charts should be used to listen to the process and observe its natural variability, at any stage, rather than for the long-run control of the process.

14.10 CHECK LIST OF KEY TERMS (with page references)

Acceptable quality level (AQL) *518*
Acceptance number 518
Acceptance sampling 518
Assignable variation 500
Average outgoing quality (AOQ) *curve 519*
Average outgoing quality limit (AOQL) *520*
Average sample number (ASN) *curve 522*
Cause-and-effect diagram 496
Central line 500
Consumer's risk 518
Control chart 500
Control chart for attributes 500
Control chart for means 502
Control chart for measurements 500

Control limits 500
Cusum 505
Cusum statistic 506
Defect 507
Defective 507
Deming's 14 points 493
Double sampling 521
Fraction-defective chart 507
Lot tolerance percent defective (LTPD) 518
Military Standard 105D 522
Multiple sampling 521
Number-of-defects chart 507
Number-of-defectives chart 509
One-sided tolerance bound 515
Producer's risk 518
Quality assurance 491
Quality control 499
Quality improvement 491
R *chart* 503
Rejection number 524
Sequential sampling 523
σ *chart* 513
Single-sampling plan 518
Statistical control 500
Three-sigma limits 501
Tolerance limits 514
$\bar{x}$ *chart* 502

15 APPLICATIONS TO RELIABILITY AND LIFE TESTING

The task of designing and supervising the manufacture of a product has been made increasingly difficult by rapid strides in the sophistication of modern products and the severity of the environmental conditions under which they must perform. No longer can an engineer be satisfied if the operation of a product is technically feasible, or if it can be made to "work" under optimum conditions. In addition to such considerations as cost and ease of manufacture, increasing attention must now be paid to size and weight, ease of maintenance, and reliability. The magnitude of the problem of maintainability and reliability is illustrated by surveys which have uncovered the fact that frequently a high percentage of space-age electronic equipment has been in inoperative condition. Military surveys have further shown that maintenance and repair expenses for electronic equipment often exceed the original cost of procurement, even during the first year of operation.

In Section 15.1 we define the concept of reliability; we discuss and apply special probability distributions to the calculation of reliabilities in Sections 15.2 and 15.3. Some theory and applications relating to testing products for useful lifetime are introduced in Sections 15.4 and 15.5.

15.1 RELIABILITY

The problem of assuring and maintaining reliability has many facets, including original equipment design, control of quality during production, acceptance inspection, field trials, life testing, and design modifications. To complicate matters further, reliability competes directly or indirectly with a host of other engineering considerations, chiefly cost, complexity, size and weight, and maintainability. In spite of its complicated engineering aspects, it is possible to give a relatively simple mathematical definition of reliability. To motivate this definition, we call the reader's attention to the fact that a product may function satisfactorily under one set of conditions but not under other conditions, and that satisfactory performance for one purpose does not assure adequate performance for another purpose. For example, a microchip that is perfectly satisfactory for use in a home radio may be entirely unsatisfactory for use in the airborne guidance system of a missile. Accordingly, we shall define the **reliability** of a product as *the probability that it will function within specified limits for at least a specified period of time under specified environmental conditions.* Thus, the reliability of a "standard equipment" automobile tire is close to unity for 10,000 miles of normal operation on a passenger car, but it is virtually zero for use at the Indianapolis "500."

Since reliability has been defined as a probability, the theoretical treatment of this subject is based essentially on the material introduced in the early chapters of this book. Thus, the rules of probability introduced in Chapter 3 can be applied directly to the calculation of the reliability of a complex system, if the reliabilities of the individual components are known. (Estimates of the reliabilities of the individual components are usually obtained from statistical life tests, such as those discussed in Sections 15.4 and 15.5.)

Many systems can be considered to be series or parallel systems, or a combination of both. A **series system** is one in which all components are so interrelated that the entire system will fail if any one of its components fails; a **parallel system** is one that will fail only if all of its components fail.

Let us first discuss a system of n components connected in series, and let us suppose that the components are independent, namely, that the performance of any one part does not affect the reliability of the others. Under these conditions, the probability that the system will function is given by the special rule of multiplication for probabilities, and we have

Product law of reliabilities

$$R_s = \prod_{i=1}^{n} R_i$$

where R_i is the reliability of the ith component and R_s is the reliability of the series system. This simple **product law of reliabilities**, applicable to series systems of independent components, vividly demonstrates the effect of increased complexity on reliability.

EXAMPLE A system consists of five independent components in series, each having a reliability of 0.970. What is the reliability of the system? What happens to the system reliability if its complexity is increased so that it contains 10 similar components?

Solution The reliability of the five-component series system is

$$(0.970)^5 = 0.859$$

Increasing system complexity to 10 components will decrease the system reliability to

$$(0.970)^{10} = 0.738$$

Looking at the effect of increasing complexity in another way, we find that each of the components in the 10-component system would require a reliability of 0.985, instead of 0.970, for the 10-component system to have a reliability equal to that of the original 5-component system.

One way to increase the reliability of a system is to replace certain components by several similar components connected in parallel. If a system consists of n independent components connected in parallel, it will fail to function only if all n components fail. Thus, if $F_i = 1 - R_i$ is the "unreliability" of the ith component, we can again apply the special rule of multiplication for probabilities to obtain

$$F_p = \prod_{i=1}^{n} F_i$$

where F_p is the unreliability of the parallel system, and $R_p = 1 - F_p$ is the reliability of the parallel system. Thus, for parallel systems we have a **product law of unreliabilities** analogous to the product law of reliabilities for series systems. Writing this law in another way, we get

Product law of unreliabilities

$$\mathbf{R_p = 1 - \prod_{i=1}^{n} (1 - R_i)}$$

for the reliability of a parallel system.

EXAMPLE The two basic formulas for the reliability of series and parallel systems can be used in combination to calculate the reliability of a system having both series and parallel parts. To illustrate such a calculation, consider the system diagrammed in Figure 15.1, which consists of eight components having the reliabilities shown in that figure. Find the reliability of this system.

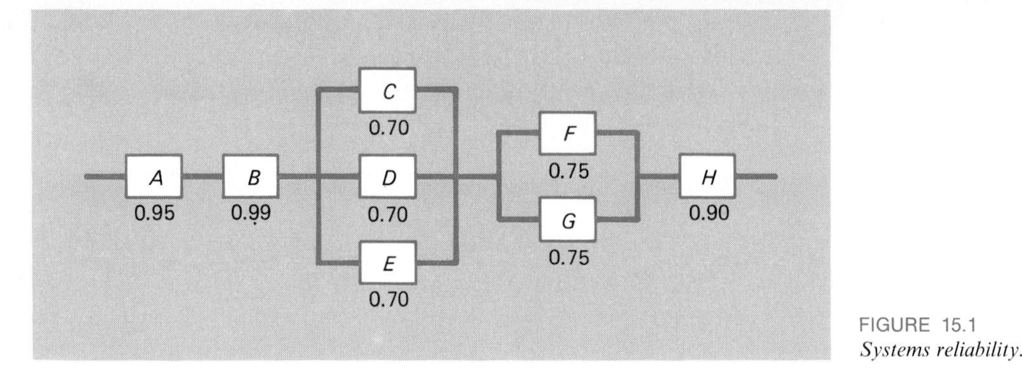

FIGURE 15.1
Systems reliability.

Solution The parallel assembly C, D, E can be replaced by an equivalent component C' having the reliability $1 - (1 - 0.70)^3 = 0.973$, without affecting the overall reliability of the system. Similarly, the parallel assembly F, G can be replaced by a single component F' having the reliability $1 - (1 - 0.75)^2 = 0.9375$. The resulting series system A, B, C', F', H, equivalent to the original system, has the reliability

$$(0.95)(0.99)(0.973)(0.9375)(0.90) = 0.772.$$

■

15.2 FAILURE-TIME DISTRIBUTIONS

According to the definition of reliability given in the preceding section, the reliability of a system or a component will often depend on the length of time it has been in service. Thus, of fundamental importance in reliability studies is the **failure-time distribution**, that is, the distribution of the time to failure of a component under given environmental conditions. A useful way to characterize this distribution is by means of its associated **instantaneous failure rate**. To develop this concept, first let $f(t)$ be the probability density of the time to failure of a given component, that is, the probability that the component will fail between times t and $t + \Delta t$ is given by $f(t) \cdot \Delta t$. Then, the probability that the component will fail on the interval from 0 to t is given by

$$F(t) = \int_0^t f(x)\, dx$$

and the **reliability function**, expressing the probability that it survives to time t, is given by

$$R(t) = 1 - F(t)$$

Thus, the probability that the component will fail in the interval from t to $t + \Delta t$ is $F(t + \Delta t) - F(t)$, and the conditional probability of failure in this interval, *given that the component survived to time t*, is expressed by

$$\frac{F(t + \Delta t) - F(t)}{R(t)}$$

Dividing by Δt, we find that the average rate of failure in the interval from t to $t + \Delta t$, given that the component survived to time t, is

$$\frac{F(t + \Delta t) - F(t)}{\Delta t} \cdot \frac{1}{R(t)}$$

Taking the limit as $\Delta t \to 0$, we then get the instantaneous failure rate, or simply the **failure rate**

$$Z(t) = \frac{F'(t)}{R(t)}$$

where $F'(t)$ is the derivative of $F(t)$ with respect to t. Finally, observing that $f(t) = F'(t)$ (see page 138) we get the relation

General equation for failure-rate function

$$\boldsymbol{Z(t) = \frac{f(t)}{R(t)} = \frac{f(t)}{1 - F(t)}}$$

which expresses the failure rate in terms of the failure-time distribution.

A failure-rate curve that is typical of many manufactured items is shown in Figure 15.2. The curve is conveniently divided into three parts. The first part is

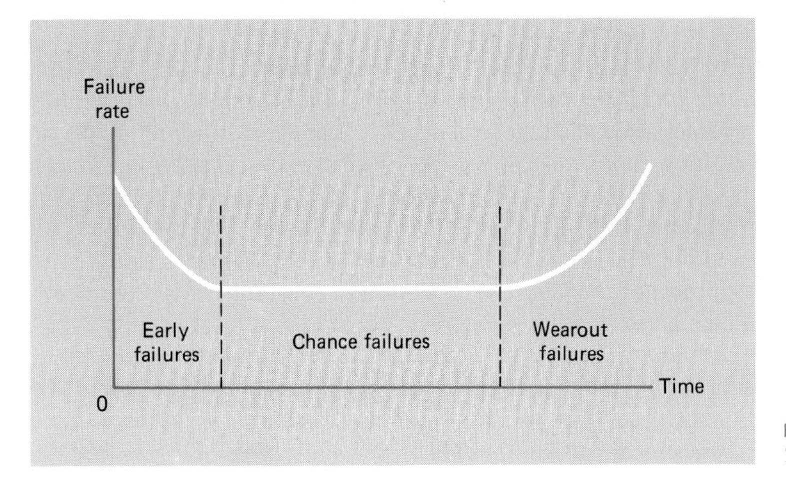

FIGURE 15.2
Typical failure-rate curve.

characterized by a decreasing failure rate and it represents the period during which poorly manufactured items are weeded out. (It is common in the electronics industry to "burn in" components prior to actual use in order to eliminate any early failures.) The second part, which is often characterized by a constant failure rate, is normally regarded as the period of useful life during which only chance failures occur. The third part is characterized by an increasing failure rate, and it is the period during which components fail primarily because they are worn out. Note that the same general failure-rate curve is typical of human mortality, where the first part represents infant mortality, and the third part corresponds to old-age mortality.

Let us now derive an important relationship expressing the failure-time density in terms of the failure-rate function. Making use of the fact that $R(t) = 1 - F(t)$ and, hence, that $F'(t) = -R'(t)$, we can write

$$Z(t) = -\frac{R'(t)}{R(t)} = -\frac{d[\ln R(t)]}{dt}$$

Solving this differential equation for $R(t)$, we obtain

$$R(t) = e^{-\int_0^t Z(x)\,dx}$$

and, making use of the relation $f(t) = Z(t) \cdot R(t)$, we finally get

General equation for failure-time distribution

$$\mathbf{f(t) = Z(t) \cdot e^{-\int_0^t Z(x)\,dx}}$$

As illustrated in Figure 15.2, it is often assumed that the failure rate is constant during the period of useful life of a component. Denoting this constant failure rate by α, where $\alpha > 0$, and substituting α for $Z(t)$ in the formula for $f(t)$, we obtain

$$f(t) = \alpha \cdot e^{-\alpha t} \qquad t > 0$$

Thus, we have an **exponential failure-time distribution** when it can be assumed that the failure rate is constant. For this reason, the assumption of constant failure rates is sometimes also called the *exponential assumption*. Interpreting the time to failure as a waiting time, we can use the results of Section 5.7 to conclude that the occurrence of failures is a Poisson process, if a component which fails is immediately replaced with a new one having the same constant failure rate α. As we observed on page 157, the mean waiting time between successive failures is $1/\alpha$, or the reciprocal of the failure rate. Thus, the constant $1/\alpha$ is often referred to as the **mean time between failures**, abbreviated $MTBF$.

There are situations in which the assumption of a constant failure rate is not realistic, and in many of these situations one assumes instead that the failure rate function increases or decreases "smoothly" with time. In other words, it is assumed that there are no discontinuities or turning points. This assumption would be consistent with the initial and last stages of the failure-rate curve shown in Figure 15.2.

A useful function that is often used to approximate such failure-rate curves is given by

$$Z(t) = \alpha\beta t^{\beta-1} \qquad t > 0$$

where α and β are positive constants. Note the generality of this function: If $\beta < 1$ the failure rate *decreases* with time, if $\beta > 1$ it *increases* with time, and if $\beta = 1$ the failure rate equals α. Note that the assumption of a constant failure rate, the exponential assumption, is thus included as a special case.

If we substitute the above expression for $Z(t)$ into the formula for $f(t)$ on page 536, we obtain

$$f(t) = \alpha\beta t^{\beta-1} e^{-\alpha t^{\beta}} \qquad t > 0$$

where α and β are positive constants. This density, or distribution, is the Weibull distribution, introduced in Section 5.9, and we discuss its application to problems of life testing in Section 15.5.

15.3
THE EXPONENTIAL MODEL IN RELIABILITY

If we make the exponential assumption about the distribution of failure times, some very useful results can be derived concerning the *MTBF*, the mean time between failure, of series and parallel systems. In order to use the product laws of Section 15.1, we shall first have to obtain a relation expressing the reliability of a component in terms of its service time t. Making use of the fact that

$$R(t) = 1 - F(t) = 1 - \int_0^t f(x)\, dx$$

we obtain

$$R(t) = 1 - \int_0^t \alpha e^{-\alpha x}\, dx = e^{-\alpha t}$$

for the reliability function of the exponential model. Thus, if a component has a failure rate of 0.05 per thousand hours, the probability that it will survive at least 10,000 hours of operation is $e^{-(0.05)10} = 0.607$.

Suppose now that a system consists of n components connected in *series*, and that these components have the respective failure rates $\alpha_1, \alpha_2, \ldots,$ and α_n. The product law of reliabilities can be written as

Product law of reliabilities—exponential case

$$R_s(t) = \prod_{i=1}^{n} e^{-\alpha_i t} = e^{-t\sum_{i=1}^{n}\alpha_i}$$

and it can be seen that the reliability function of the series system also satisfies the exponential assumption. The failure rate of the entire series system is readily identified as $\sum_{i=1}^{n} \alpha_i$, the *sum* of the failure rates of its components. Since the $MTBF$ is the reciprocal of the failure rate when each component which fails is replaced immediately with another having the identical failure rate, we obtain the formula

Mean time between failures—series system

$$\mu_s = \frac{1}{\dfrac{1}{\mu_1} + \dfrac{1}{\mu_2} + \cdots + \dfrac{1}{\mu_n}}$$

expressing the $MTBF$ μ_s of a series system in terms of the $MTBF$'s μ_i of its components. In the special case where all n components have the same failure rate α and hence the same $MTBF$ μ, the system failure rate is $n\alpha$, and the system $MTBF$ is $1/n\alpha = \mu/n$.

For parallel systems the results are not quite so simple. If a system consists of n components in parallel, having the respective failure rates $\alpha_1, \alpha_2, \ldots, \alpha_n$, the system "unreliability" to time t is given by

$$F_p(t) = \prod_{i=1}^{n} (1 - e^{-\alpha_i t})$$

Thus, the failure-time distribution of a parallel system is not exponential even when each of its components satisfies the exponential assumption. The system failure-rate function can be obtained by means of the formula $Z_p(t) = F'_p(t)/R_p(t)$, but the result is fairly complicated. Note, however, that the system failure rate is not constant, but depends on t, the "age" of the system.

The mean time to failure of a parallel system is also difficult to obtain in general, but in the special case where all components have the same failure rate α, an interesting and useful result can be obtained. In this special case the system reliability function becomes

$$\begin{aligned} R_p(t) &= 1 - (1 - e^{-\alpha t})^n \\ &= \binom{n}{1} e^{-\alpha t} - \binom{n}{2} e^{-2\alpha t} + \cdots + (-1)^{n-1} e^{-n\alpha t} \end{aligned}$$

after using the binomial theorem to expand $(1 - e^{-\alpha t})^n$. Then, making use of the fact that $f_p(t) = -R'_p(t)$, we obtain

$$f_p(t) = \alpha\binom{n}{1}e^{-\alpha t} - 2\alpha\binom{n}{2}e^{-2\alpha t} + \cdots + (-1)^{n-1}n\alpha e^{-n\alpha t}$$

and the mean of the failure-time distribution is given by

$$\begin{aligned}\mu_p &= \int_0^\infty t \cdot f_p(t)\,dt \\ &= \alpha\binom{n}{1}\int_0^\infty te^{-\alpha t}\,dt - 2\alpha\binom{n}{2}\int_0^\infty te^{-2\alpha t}\,dt + \cdots + (-1)^{n-1}n\alpha\int_0^\infty te^{-n\alpha t}\,dt \\ &= \frac{1}{\alpha}\binom{n}{1} - \frac{1}{2\alpha}\binom{n}{2} + \cdots + (-1)^{n-1}\frac{1}{n\alpha}\end{aligned}$$

It can be proved by induction that this expression is equivalent to

Mean time between failures—parallel system

$$\boldsymbol{\mu_p = \frac{1}{\alpha}\left(1 + \frac{1}{2} + \cdots + \frac{1}{n}\right)}$$

Thus, if a parallel system consists of n components having the identical failure rate α, the mean time between failures of the system equals $\left(1 + \frac{1}{2} + \cdots + \frac{1}{n}\right)$ times the common *MTBF* of its components, provided each defective component is replaced whenever the whole parallel system fails. Thus, if we use two parallel components rather than one, the mean time to failure of the pair exceeds that of the single component by 50 percent, rather than doubling it. In general, the above formula for μ_p expresses a rather severe law of diminishing returns for parallel redundancy.

The formulas derived in this section can be used in system design. To illustrate, we again consider the system diagrammed in Figure 15.1 on page 534. Assuming the exponential model and that the reliabilities are given for 10 hours of operation, we can calculate the failure rate of component A by solving the equation $0.95 = e^{-10\alpha}$ for α, and we obtain $\alpha = 5.1 \cdot 10^{-3}$ failures per hour, or 5.1 failures per thousand hours. The failure rates of all eight components (in failures per thousand hours) are as shown in the following table:

Component	*A*	*B*	*C*	*D*	*E*	*F*	*G*	*H*
Failure rate	5.1	1.0	35.7	35.7	35.7	28.8	28.8	10.5

To compute the mean time to failure for the entire system, we first obtain the mean failure times for the parallel assemblies C, D, E, and F, G, respectively. For C, D, E

we have $\mu_{CDE} = \frac{1}{35.7}\left(1 + \frac{1}{2} + \frac{1}{3}\right) = 0.051$ thousand hours, or 51 hours; for F, G we have $\mu_{FG} = \frac{1}{28.8}\left(1 + \frac{1}{2}\right) = 0.052$ thousand hours, or 52 hours. Although the two parallel assemblies do not have constant failure rates, we shall approximate their respective failure rates by $1/0.051 = 19.6$ and $1/0.052 = 19.2$ failures per thousand hours, and treat the entire system as a series system. Thus, the system failure rate is given approximately by $5.1 + 1.0 + 19.6 + 19.2 + 10.5 = 55.4$ failures per thousand hours, and the mean time to failure of the system is approximately $1/55.4 = 0.018$, or 18 hours.

EXERCISES

15.1 An old-fashioned string of holiday lights has eight bulbs connected in series. What would have to be the reliability of each bulb if there is to be a 95% chance of the string's lighting after a year's storage?

15.2 A system consists of five identical components connected in parallel. What must be the reliability of each component if the overall reliability of the system is to be 0.96?

15.3 A system consists of six components connected as in Figure 15.3. Find the overall reliability of the system, given that the reliabilities of A, B, C, D, E, and F are, respectively, 0.95, 0.80, 0.90, 0.99, 0.90, and 0.85.

15.4 Suppose that the flight of an aircraft is regarded as a system having the three main components A (aircraft), B (pilot), and C (airport). Suppose, furthermore, that component B can be regarded as a parallel subsystem consisting of B_1 (captain), B_2 (first officer), and B_3 (flight engineer); and C is a parallel subsystem consisting of C_1 (scheduled airport) and C_2 (alternate airport). Under given flight conditions, the reliabilities of components A, B_1, B_2, B_3, C_1, and C_2 (defined as the probabilities that they can contribute to the successful completion of the scheduled flight) are, respectively, 0.9999, 0.9995, 0.999, 0.20, 0.95, and 0.85.

(a) What is the reliability of the system?

(b) What is the effect on system reliability of having a flight engineer who is also a trained pilot, so that the reliability of B_3 is increased from 0.20 to 0.99?

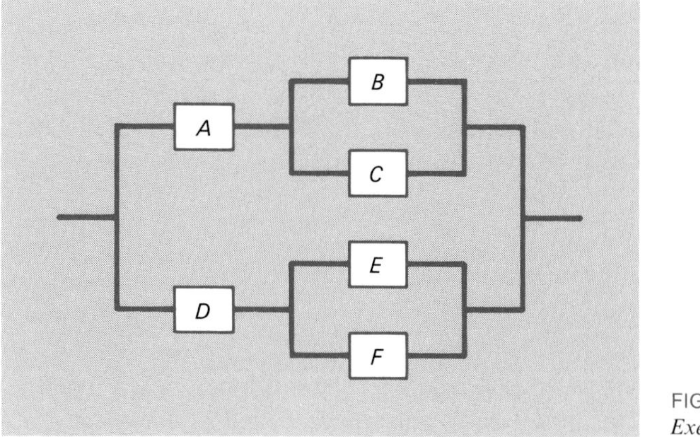

FIGURE
Exercise 15.3.

(c) If the flight crew did not have a first officer, what then would be the effect of increasing the reliability of B_3 from 0.20 to 0.99?

(d) What is the effect of adding a second alternate landing point, C_3, with reliability 0.80?

15.5 In some reliability problems we are concerned only with initial failures, treating a component as if (for all practical purposes) it never fails, once it has survived past a certain time $t = \alpha$. In a problem like this, it may be reasonable to use the failure rate

$$Z(t) = \begin{cases} \beta\left(1 - \dfrac{t}{\alpha}\right) & \text{for } 0 < t < \alpha \\ 0 & \text{elsewhere} \end{cases}$$

(a) Find expressions for $f(t)$ and $F(t)$.

(b) Show that the probability of an initial failure is given by

$$1 - e^{-\alpha\beta/2}$$

15.6 As has been indicated in the text, one often distinguishes between initial failures, random failures during the useful life of the product, and wear-out failures. Thus, suppose that for a given product the probability of an initial failure (a failure prior to time $t = \alpha$) is θ_1, the probability of a wear-out failure (a failure beyond time $t = \beta$) is θ_2, and that for the interval $\alpha \le t \le \beta$ the failure-time density is given by

$$f(t) = \frac{1 - \theta_1 - \theta_2}{\beta - \alpha}$$

(a) Find an expression for $F(t)$ for the interval $\alpha \le t \le \beta$.

(b) Show that for the interval $\alpha \le t \le \beta$ the failure rate is given by

$$Z(t) = \frac{1 - \theta_1 - \theta_2}{(\beta - \alpha)(1 - \theta_1) - (1 - \theta_1 - \theta_2)(t - \alpha)}$$

(c) Suppose that the failure of a color television set is considered to be an initial failure if it occurs during the first 100 hours of usage and a wear-out failure if it occurs after 15,000 hours. Assuming that the model given in this exercise holds and that θ_1 and θ_2 equal 0.05 and 0.75, respectively, sketch the graph of the failure-rate function from $t = 100$ to $t = 15{,}000$ hours.

15.7 An integrated-circuit chip has a constant failure rate of 0.02 per thousand hours.

(a) What is the probability that it will operate satisfactorily for at least 20,000 hours?

(b) What is the 5,000-hour reliability of a component consisting of four such chips connected in series?

15.8 After burn-in, the lifetime of a solar cell is modeled as an exponential distribution with failure rate $\alpha = 0.0005$ failures per day.

(a) What is the probability that the cell will fail within the first 365 days that it is in operation?

(b) What is the probability that two such cells, operating independently, will both survive the first 365 days they are in operation?

15.9 A system consisting of several identical components connected in parallel is to have a failure rate of at most 4×10^{-4} per hour. What is the least number of components that must be used if each has a constant failure rate of 9×10^{-4}.

15.10 A system consists of four different components connected in series. Find the *MTBF* of the system if the four components have exponential time-to-failure distributions with failure rates 1.9, 3.1, 1.4 and 2.2 per 10,000 hours, respectively.

15.11 A certain part has an exponential life distribution with a mean life, that is, an *MTBF*, of 1,000 hours.

(a) What is the probability that such a part will last at least 500 hours?

(b) What is the probability that among three such parts at least one will fail during the first 1,000 hours?

(c) What is the probability that among four such parts exactly two will fail during the first 600 hours?

15.12 If a component has the Weibull failure-time distribution with the parameters $\alpha = 0.005$ per hour and $\beta = 0.80$, find the probability that it will operate successfully for at least 5,000 hours.

15.13 Throughout Sections 15.2 and 15.3 we assumed that the products with which we were concerned were in continuous operation. Consequently, the models discussed in these sections do not apply if we want to investigate the ability of light bulbs to withstand successive voltage overloads, the performance of switches which are repeatedly turned on and off, etc. In each of these cases failure can occur on the xth trial ($x = 1, 2, 3, \ldots$), and it is often assumed that the probability of failure on the xth trial equals some constant p, provided the item has not failed prior to that trial.

(a) Show that the probability of failure on the xth trial is given by

$$f(x) = p(1 - p)^{x-1}$$

for $x = 1, 2, 3, \ldots$. This probability distribution is the *geometric distribution* introduced in Section 4.8.

(b) Find $F(x)$ for the probability distribution obtained in part (a).

(c) What is the probability that a switch will survive 2,000 cycles of operation if the preceding model holds and the constant probability of failure as the result of any one of the switching cycles is $p = 6 \times 10^{-4}$?

15.4
THE EXPONENTIAL MODEL IN LIFE TESTING

An effective and widely used method of handling problems of reliability is that of **life testing**. For the purpose of such tests, a random sample of n components is selected from a lot, put on test under specified environmental conditions, and the times to failure of the individual components are observed. If each component that fails is immediately replaced by a new one, the resulting life test is called a **replacement test**; otherwise, the life test is called a **nonreplacement test**. Whenever the mean lifetime of the components is so large that it is not practical, or economically feasible, to test each component to failure, the life test may be **truncated**, that is, it may be terminated after the first r failures have occurred ($r \leq n$), or after a fixed period of time has elapsed.

A special method that is often used when early results are required in connection with very-high-reliability components is that of **accelerated life testing**. In an accelerated life test the components are put on test under environmental conditions that are far more severe than those normally encountered in practice. This causes the components to fail more quickly, and it can drastically reduce both the time required for the test and the number of components that must be placed on test. Accelerated life testing can be used to compare two or more types of components for the purpose of obtaining a rapid assessment of which is the most reliable. Sometimes, preliminary experimentation is carried out to determine the relationship between the proportion of failures that can be expected under nominal conditions and under various levels of accelerated environmental conditions. The methods of Sections 11.4 and 13.2 can be applied in this connection to determine "derating curves," relating the reliability of the component to the severity of the environmental conditions under which it is to operate.

In the remainder of this section we shall assume that the exponential model holds, namely, that the failure-time distribution of each component is given by

$$f(t) = \alpha \cdot e^{-\alpha t} \qquad t > 0, \alpha > 0$$

In what follows, we shall assume that n components are put on test, life testing is discontinued after a fixed number, $r(r \leq n)$, of components have failed, and that the observed failure times are $t_1 \leq t_2 \leq \cdots \leq t_r$. We shall be concerned with estimating and testing hypotheses about the mean life of the component, namely, $\mu = 1/\alpha$.

Using theory developed in the article by B. Epstein mentioned in the bibliography, it can be shown that unbiased estimates of the mean life of the component are given by

Estimate of mean life

$$\hat{\mu} = \frac{T_r}{r}$$

where T_r is the accumulated life on test until the rth failure occurs, and hence

Accumulated life to r failures—nonreplacement test

$$T_r = \sum_{i=1}^{r} t_i + (n - r)t_r$$

for nonreplacement tests and

Accumulated life to r failures—replacement test

$$T_r = nt_r$$

if the test is with replacement. Note that if the test is without replacement and $r = n$, $\hat{\mu}$ is simply the mean of the observed times to failure.

To make inferences concerning the mean life μ of the component, we use the fact that $2T_r/\mu$ is a value of a random variable having the chi-square distribution with $2r$ degrees of freedom (see reference to B. Epstein in the Bibliography). With the appropriate expression substituted for T_r, this is true regardless of whether the test is conducted with or without replacement. Thus, in either case a two-sided $(1 - \alpha)100\%$ confidence interval for μ is given by

Confidence interval for mean life

$$\frac{2T_r}{\chi_2^2} < \mu < \frac{2T_r}{\chi_1^2}$$

where χ_1^2 and χ_2^2 cut off left- and right-hand tails of area $\alpha/2$ under the chi-square distribution with $2r$ degrees of freedom. (See Exercise 15.19 on page 552.)

Tests of the null hypothesis that $\mu = \mu_0$ can also be based on the sampling distribution of $2T_r/\mu$, using the appropriate expression for T_r depending on whether the test is with or without replacement. Thus, if the alternative hypothesis is $\mu > \mu_0$, we reject the null hypothesis at the level of significance α when $2T_r/\mu_0$ exceeds χ_α^2, or

Critical region for testing H_0: $\mu = \mu_0$ against H_1: $\mu > \mu_0$

$$T_r > \tfrac{1}{2}\mu_0\chi_\alpha^2$$

where χ_α^2, to be determined for $2r$ degrees of freedom, is as defined on page 200. In Exercises 15.14 and 15.17 on page 551 the reader is asked to construct and perform similar tests corresponding to the alternative hypotheses $\mu < \mu_0$ and $\mu \neq \mu_0$.

An alternate life-testing procedure consists of discontinuing the test after a fixed accumulated amount of lifetime T has elapsed and treating the observed number of failures k as the value of a random variable. In the important special case where n items are tested with replacement for a length of time t^*, we have $T = nt^*$. Regardless of whether the test is with or without replacement, an *approximate* $1 - \alpha$ confidence interval for the mean life of the component is given by

$$\frac{2T}{\chi_4^2} < \mu < \frac{2T}{\chi_3^2}$$

Here χ_4^2 cuts off a right-hand tail of area $\alpha/2$ under the chi-square distribution with $2k + 2$ degrees of freedom, while χ_3^2 cuts off a left-hand tail of area $\alpha/2$ under the chi-square distribution with $2k$ degrees of freedom.

EXAMPLE

Suppose that 50 units are placed on life test (without replacement) and that the test is to be truncated after $r = 10$ of them have failed. We shall suppose, furthermore, that the first 10 failure times are 65, 110, 380, 420, 505, 580, 650, 840, 910, and 950 hours. Estimate the mean life of the component, its failure rate, and calculate a 0.90 confidence interval for μ.

Solution Since $n = 50$, $r = 10$,

$$T_{10} = (65 + 110 + \cdots + 950) + (50 - 10)950$$
$$= 43{,}410 \text{ hours}$$

and we estimate the mean life of the component as $\hat{\mu} = \frac{43{,}410}{10} = 4{,}341$ hours. The failure rate α is estimated by $1/\hat{\mu} = 0.00023$ failure per hour, or 0.23 failure per thousand hours. Also, a 0.90 confidence interval for μ is given by

$$\frac{2(43{,}410)}{31.410} < \mu < \frac{2(43{,}410)}{10.851}$$

or

$$2{,}764 < \mu < 8{,}001$$

EXAMPLE Using the data of the preceding example, test whether the failure rate is 0.40 failure per thousand hours against the alternative that the failure rate is less. Use the 0.05 level of significance.

Solution

1. *Null hypothesis*: $\mu = \frac{1{,}000}{0.40} = 2{,}500$ hours

   Alternative hypothesis: $\mu > 2{,}500$ hours
2. *Level of significance*: $\alpha = 0.05$
3. *Criterion*: Reject the null hypothesis if $T_r > \frac{1}{2}\mu_0\chi^2_{0.05}$ where $\chi^2_{0.05} = 31.410$ is the value of $\chi^2_{0.05}$ having 20 degrees of freedom.
4. *Calculations*: Substituting $r = 10$ and $\mu_0 = 2{,}500$, we find the critical value for this test to be

$$\tfrac{1}{2}\mu_0\chi^2_{0.05} = \tfrac{1}{2}(2{,}500)(31.410) = 39{,}263$$

5. *Decision*: Since $T_{10} = 43{,}410$ exceeds the critical value, we must reject the null hypothesis, concluding that the mean life-time exceeds 2,500 hours, or, equivalently, that the failure rate is less than 0.40 failure per thousand hours.

Because of the simplicity of the statistical procedures, the exponential model is frequently considered. Before making inferences, it is imperative that this model be checked for adequacy. We recommend making a **total time on test plot**. Plot the total time on test until the ith failure, T_i, divided by the total time on test through the last (rth) observed failure, against i/r. If the population is exponential, we would expect to see a straight line along the 45-degree line. When this straight-line pattern occurs, we conclude that no violations of the exponential model are evident over the range of failure times. If the plot is a curve above the 45-degree line, the evidence favors an increasing hazard rate model.

We illustrate the total time on test plot using the data of the example on page 544. For $t_1 = 65$, we calculate the total time on test

$$T_1 = 65 + (50 - 1)65 = 3{,}250$$

Next, for $t_2 = 110$,

$$T_2 = 65 + 110 + (50 - 2)110 = 5{,}455$$

Continuing, we obtain all the values

3,250	5,455	18,415	20,295	24,205
27,580	30,660	38,830	41,770	43,410

so the total time on test until the last, $r = 10$, failure is $T_r = 43{,}410$. The first ratio $T_1/T_{10} = 3{,}250/43{,}410 = 0.0749$ is plotted against $1/10 = 0.10$. The ratios for all ten failures are plotted in Figure 15.4. Over the range of failure times observed, the plot does not exhibit any marked departures from the assumed exponential model.

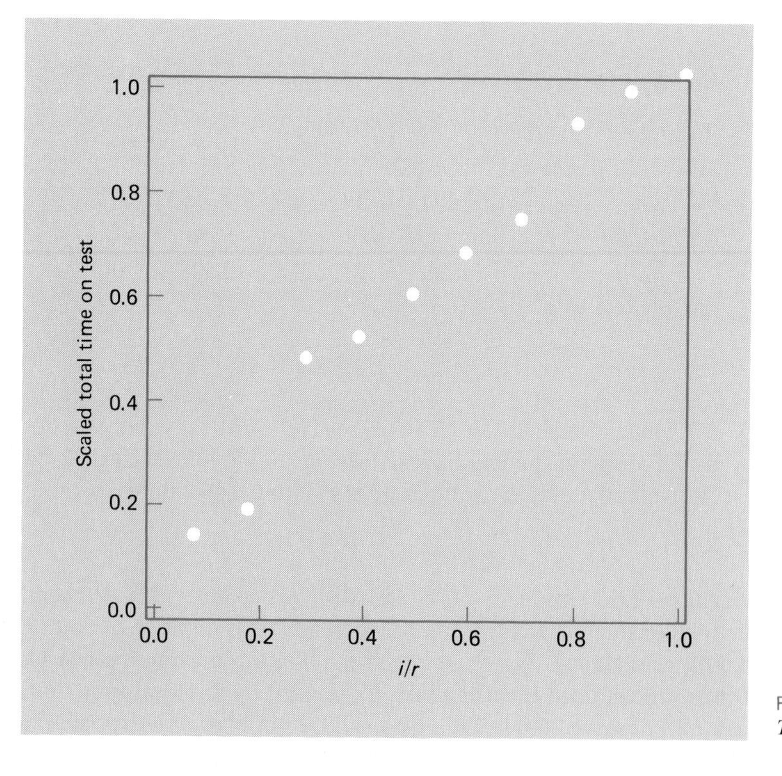

FIGURE 15.4
The total time on test plot.

15.5
THE WEIBULL MODEL IN LIFE TESTING

Although life testing of components during the period of useful life is generally based on the exponential model, we have already pointed out that the failure rate of a component may not be constant throughout a period under investigation. In some instances the period of initial failure may be so long that the component's main use is during this period, and in other instances the main purpose of life testing may be that of determining the time to wear-out failure rather than chance failure. In such cases the exponential model generally does not apply, and it is necessary to substitute a more general assumption for that of a constant failure rate.

As we observed on page 537, the Weibull distribution adequately describes the failure times of components when their failure rate either increases or decreases with time. It has the parameters α and β, its formula is given by

Weibull distribution

$$f(t) = \alpha\beta t^{\beta-1}e^{-\alpha t^{\beta}} \qquad t > 0, \alpha > 0, \beta > 0$$

and it follows (see Exercise 15.24 on page 552) that the reliability function associated with the **Weibull failure-time distribution** is given by

Weibull reliability function

$$R(t) = e^{-\alpha t^{\beta}}$$

We already showed on page 537 that the failure rate leading to the Weibull distribution is given by

Weibull failure-rate function

$$Z(t) = \alpha\beta t^{\beta-1}$$

The range of shapes a graph of the Weibull density can take on is very broad, depending primarily on the value of the parameter β. As illustrated in Figure 15.5, the Weibull curve is asymptotic to both axes and highly skewed to the right for values of β less than 1; it is identical to that of the exponential density for $\beta = 1$, and it is "bell-shaped" but skewed for values of β greater than 1.

The mean of the Weibull distribution having the parameters α and β may be obtained by evaluating the integral

$$\mu = \int_0^{\infty} t \cdot \alpha\beta t^{\beta-1}e^{-\alpha t^{\beta}}\,dt$$

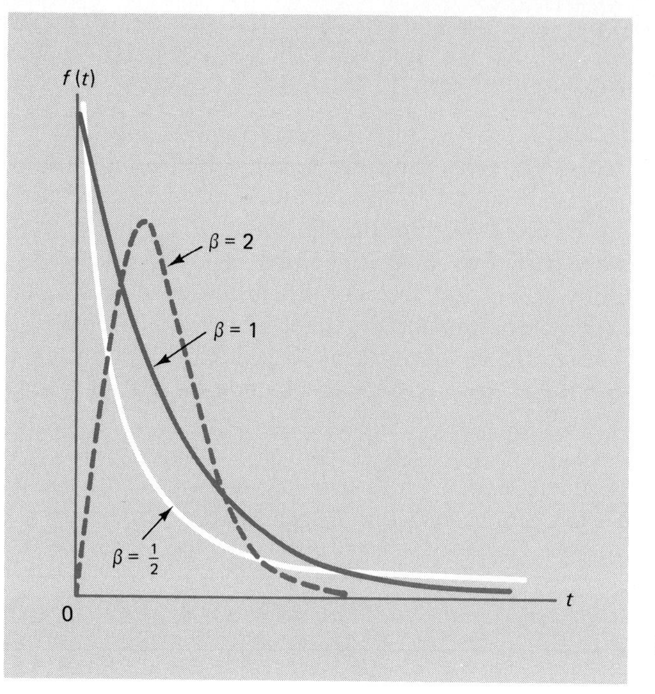

FIGURE 15.5
Weibull density functions ($\alpha = 1$).

Making the change of variable $u = \alpha t^{\beta}$, we get

$$\mu = \alpha^{-1/\beta} \int_0^{\infty} u^{1/\beta} e^{-u} \, du$$

Recognizing the integral as $\Gamma\left(1 + \frac{1}{\beta}\right)$, we find that the mean time to failure for the Weibull model is

Mean time to failure—Weibull model

$$\boldsymbol{\mu = \alpha^{-1/\beta}\Gamma\left(1 + \frac{1}{\beta}\right)}$$

The reader will be asked to show in Exercise 15.25 on page 552 that the variance of this distribution is given by

Variance of Weibull model

$$\boldsymbol{\sigma^2 = \alpha^{-2/\beta}\left\{\Gamma\left(1 + \frac{2}{\beta}\right) - \left[\Gamma\left(1 + \frac{1}{\beta}\right)\right]^2\right\}}$$

Estimates of the parameters α and β of the Weibull distribution are somewhat difficult to obtain.

The most widely accepted approach, the *maximum likelihood* method (see Exercise 7.23), maximizes the likelihood. Because the partial derivatives with respect to α and β must vanish at the maximum, the method selects the solution to these two equations as the estimates of α and β. If the lifetimes are censored at the rth failure, or uncensored so $r = n$, the equations are

$$\frac{\sum_{i=1}^{r} t_i^{\beta} \ln t_i + (n-r) t_r^{\beta} \ln t_r}{\sum_{i=1}^{r} t_i^{\beta} + (n-r) t_r^{\beta}} - \frac{1}{\beta} - \frac{1}{r} \sum_{i=1}^{r} \ln t_i = 0$$

$$\alpha = \frac{1}{\frac{1}{r}\left[\sum_{i=1}^{r} t_i^{\beta} + (n-r) t_r^{\beta}\right]}$$

The first equation is solved for $\hat{\beta}$ by numerical techniques. Then the second yields the estimate of $\hat{\alpha}$. These are easy computer calculations.

If the lifetimes are time censored at time T_0, the terms with a factor $n - r$ are modified by replacing each t_r by T_0.

A graphical method provides a check on the adequacy of the Weibull model. This method is based on the fact that the reliability function of the Weibull distribution can be transformed into a linear function of $\ln t$ by means of a double-logarithmic transformation. Taking the natural logarithm of $R(t)$, we obtain

$$\ln R(t) = -\alpha t^{\beta} \quad \text{or} \quad \ln \frac{1}{R(t)} = \alpha t^{\beta}$$

Again taking logarithms, we have

$$\ln \ln \frac{1}{R(t)} = \ln \alpha + \beta \cdot \ln t$$

and it can be seen that the right-hand side is linear in $\ln t$.

The usual experimental procedure is to place n units on life test and observe their failure times. If the ith unit fails at time t_i, we estimate $F(t_i) = 1 - R(t_i)$ by the same method used for the normal-scores plot on page 171, namely,

$$\widehat{F(t_i)} = \frac{i}{n+1}$$

To construct a **Weibull plot**, we plot $\ln \ln \dfrac{1}{1 - \widehat{F(t_i)}}$ versus $\ln t_i$. If the points fall reasonably close to a straight line, it can be assumed that the underlying failure-time distribution is of the Weibull type.

As an illustration of this procedure, suppose that a sample of 100 components is put on life test for 500 hours and that the times to failure of the 12 components that failed during the test are as follows: 6, 21, 50, 84, 95, 130, 205, 260, 270, 370, 440, and 480 hours. Setting

$$x_i = \ln t_i$$

$$y_i = \ln \ln \frac{1}{1 - \widehat{F(t_i)}}$$

we obtain

$\widehat{F(t_i)}$	t_i	x_i	y_i
0.010	6	1.79	−4.61
0.020	21	3.04	−3.91
0.030	50	3.91	−3.50
0.040	84	4.43	−3.21
0.050	95	4.55	−2.98
0.059	130	4.87	−2.79
0.069	205	5.32	−2.63
0.079	260	5.56	−2.49
0.089	270	5.60	−2.37
0.099	370	5.91	−2.26
0.109	440	6.09	−2.16
0.119	480	6.17	−2.07

The points (x_i, y_i) are plotted in Figure 15.6, and it can be seen that they fall fairly close to a straight line. After checking the adequacy of the Weibull distribution, we obtain the maximum-likelihood estimators defined on page 549. Our computer calculations yield $\hat{\alpha} = 0.001505$ and $\hat{\beta} = 0.7148$. It follows that the mean time to failure is estimated as

$$\hat{\mu} = (0.001505)^{-1/0.7148}\Gamma\left(1 + \frac{1}{0.7148}\right)$$

which equals approximately 11,000 hours. Also, values of the failure-rate function may be obtained by substituting for t into

$$\widehat{Z(t)} = (0.001505)(0.7148)t^{0.7148-1} = 0.00108t^{-0.2852}$$

Since $\hat{\beta} < 1$, the failure rate is decreasing with time. After 1 hour ($t = 1$), units are failing at the rate of 0.00108 unit per hour, and after 1,000 hours the failure rate has decreased to $0.00108(1000)^{-0.2852} = 0.00015$ unit per hour.

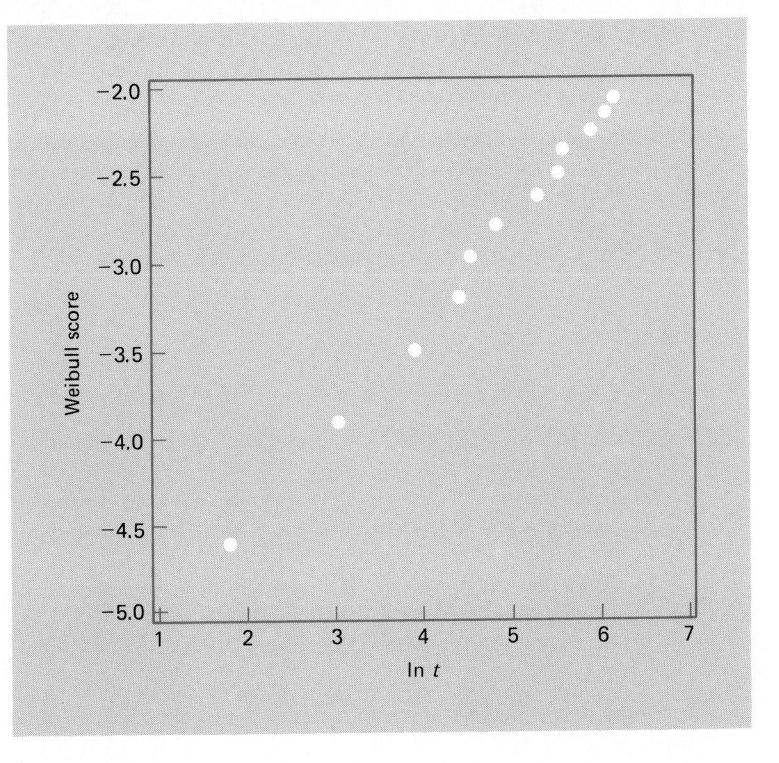

FIGURE 15.6
A Weibull plot of failure times.

EXERCISES

15.14 Suppose that 50 units are put on life test, each unit that fails is immediately replaced, and the test is discontinued after 8 units have failed. If the eighth failure occurred at 760 hours, assuming an exponential model,

(a) construct a 95% confidence interval for the mean life of such units;

(b) test at the 0.05 level of significance whether or not the mean life is less than 10,000 hours.

15.15 In a life test with replacement, 35 space heaters were put into continuous operation, and the first five failures occurred after 250, 380, 610, 980, and 1,250 hours.

(a) Assuming the exponential model, construct a 99% confidence interval for the mean life of this kind of space heater.

(b) To check the manufacturer's claim that the mean life of these heaters is at least 5,000 hours, test the null hypothesis $\mu = 5{,}000$ against an appropriate alternative, so that the burden of proof is put on the manufacturer. Use $\alpha = 0.05$.

15.16 With reference to the data in Exercise 15.15, make a total time on test plot.

15.17 To investigate the average time to failure of a certain weld subjected to continuous vibration, seven welded pieces were subjected to specified frequencies and amplitudes of vibration and their times to failure were 211, 350, 384, 510, 539, 620, and 715 thousand cycles.

(a) Assuming the exponential model, construct a 95% confidence interval for the mean life (in thousands of cycles) of such a weld under the given vibration conditions.

(b) Assuming the exponential model, test the null hypothesis that the mean life of the weld under the given vibration conditions is 500,000 cycles against the two-sided alternative $\mu \neq 500{,}000$. Use the level of significance 0.10.

15.18 In life testing we are sometimes interested in establishing tolerance limits for the life of a component (see Section 14.7); in particular, we may be interested in a one-sided tolerance limit t^*, for which we can assert with a $(1 - \alpha)100\%$ confidence that at least $100 \cdot P$ percent of the components have a life exceeding t^*. Using the exponential model, it can be shown that a good approximation is given by

$$t^* = \frac{-2T_r(\ln P)}{\chi_\alpha^2}$$

where T_r is as defined on page 543 and the value of χ_α^2 is to be obtained from Table 5 with $2r$ degrees of freedom.

(a) Using the data of Exercise 15.15, establish a lower tolerance limit for which one can assert with 95% confidence that it is exceeded by at least 80% of the lifetimes of the heaters.

(b) Using the data of Exercise 15.17, establish a lower tolerance limit for which one can assert with 99% confidence that it is exceeded by at least 90% of the lifetimes of the given welds.

15.19 Using the fact that $2T_r/\mu$ is a value of a random variable having the chi-square distribution with $2r$ degrees of freedom, derive the confidence interval for μ given on page 544.

15.20 One hundred devices are put on life test and the times to failure (in hours) of the first 10 that fail are as follows: 7.0, 14.1, 18.9, 31.6, 52.8, 80.0, 164.5, 355.4, 451.0, and 795.1. Assuming a Weibull failure-time distribution, estimate the parameters α and β as well as the failure rate at 1,000 hours. How does this value of the failure rate compare with the value we would obtain if we assumed the exponential model?

15.21 A sample of 200 switches was placed on life test for 3000 on-off cycles and there were no failures; the test was then terminated. Find a 95% *lower* confidence limit for the mean life, in number of cycles, of the switches.

15.22 A sample of 60 diaphragm valves, used in the control system of a chemical process, are placed on life test (without replacement). The first nine failures are observed after 3.6, 6.9, 9.5, 15.7, 27.3, 41.2, 81.7, 178.3, 227.1 hours. Using the Weibull model, estimate the mean life of this valve. How does this value compare with the mean life that would have been obtained under the exponential assumption?

15.23 Using the estimates of the parameter of the Weibull model obtained in Exercise 15.22, estimate the probability that this kind of diaphragm valve will perform satisfactorily for at least 150 hours.

15.24 Show that the reliability function associated with the Weibull failure-time distribution is given by

$$R(t) = e^{-\alpha t^\beta}$$

15.25 Derive the formula for the variance of the Weibull distribution given on page 548.

15.6

REVIEW EXERCISES

15.26 A system consists of seven identical components connected in parallel. What must be the reliability of each component if the overall reliability of the system is to be 0.90?

15.27 A certain component has an exponential life distribution with a failure rate of $\alpha = 0.0045$ failures per hour.

(a) What is the probability that the component will fail during the first 250 hours it is in operation?

(b) What is the probability that two such components will both survive the first 100 hours of operation?

15.28 A system containing several identical components in parallel is to have a failure rate of at most $5 \cdot 10^{-5}$ per hour. What is the least number of components that must be used if each has a constant failure rate of $2.0 \cdot 10^{-4}$ per hour?

15.29 A system consists of six different components connected in series. Find the *MTBF* of the system if the six components have exponential time-to-failure distributions with failure rates of 1.8, 2.4, 2.0, 1.3, 3.0, and 1.5 per 1,000 hours, respectively.

15.30 Fifteen assemblies are put on accelerated life test without replacement, and the test is truncated after four failures. If the first four failures occurred at 16.5, 19.2, 20.8, and 37.3 hours, assuming an exponential model,

(a) find a 90% confidence interval for the failure rate of such assemblies under these accelerated conditions;

(b) test the null hypothesis that the failure rate is 0.004 failure per hour against the alternative that it is less than 0.004, using the 0.01 level of significance.

15.31 A sample of 300 high-reliability capacitors was placed on life test for 2,000 hours and there were no failures; the test was then terminated. Find a 0.95 *lower* confidence limit for the mean life of the capacitors assuming an exponential model.

15.32 To investigate the performance of a logic circuit for a small electronic calculator, a laboratory puts 75 of the circuits on life test (without replacement) under specified environmental conditions, and the first 10 failures are observed after 28, 46, 50, 63, 81, 101, 116, 137, 159, and 175 hours. Using the Weibull model, estimate the mean life of such a circuit. How does this value compare with the mean life that would have been obtained under the exponential assumption?

15.33 Using the estimates of the parameters of the Weibull model obtained in Exercise 15.32, estimate the probability that this kind of circuit will perform satisfactorily for at least 100 hours.

15.34 With reference to Exercise 15.32, make

(a) a total time on test plot; (b) a Weibull plot.

15.35 (*Stress-strength models* for *reliability*) An alternative model used in reliability treats the environmental stress as a random variable X and the strength on the component to withstand this stress as an independent random variable Y. Then, the reliability is defined as

$$R = P[Y > X] = \int_{-\infty}^{\infty} \int_{-\infty}^{y} f(x, y)\, dx\, dy = \int_{-\infty}^{\infty} F(y) g(y)\, dy$$

where $F(x)$ is the distribution function of X and $g(y)$ is the probability density of Y. Evaluate this reliability when

(a) X has an exponential distribution with $\alpha = 0.01$ and Y has an exponential distribution with failure rate 0.005;

(b) X has an exponential distribution with $\alpha = 0.005$ and Y has an exponential distribution with failure rate 0.005;

(c) $\ln X$ has a normal distribution with $\mu = 60$ and $\sigma = 5$ and $\ln Y$ has a normal distribution with $\mu = 80$ and $\sigma = 5$.

15.7 CHECK LIST OF KEY TERMS (with page references)

BIBLIOGRAPHY

1. ENGINEERING STATISTICS

BROWNLEE, K. A., *Statistical Theory and Methodology in Science and Engineering*, 2d ed. New York: John Wiley & Sons, Inc., 1965.

GUTTMAN, I., WILKS, S. S., and HUNTER, J. S., *Introductory Engineering Statistics*, 3d ed. New York: John Wiley & Sons, Inc., 1982.

HALD, A., *Statistical Theory with Engineering Applications.* New York: John Wiley & Sons, Inc., 1952.

WALPOLE, R. E., and MYERS, R. H., *Probability and Statistics for Engineers and Scientists*, 3d ed. New York: Macmillan Publishing Company, 1985.

2. THEORETICAL STATISTICS

FREUND, J. E., and WALPOLE, R., *Mathematical Statistics*, 4th ed. Englewood Cliffs, N.J.: Prentice-Hall, Inc., 1987.

HOEL, P., *Introduction to Mathematical Statistics*, 5th ed. New York: John Wiley & Sons, Inc., 1984.

HOGG, R. V., and CRAIG, A. T., *Introduction to Mathematical Statistics*, 4th ed. New York: Macmillan Publishing Company, 1978.

KENDALL, M. G., and STUART, A., *The Advanced Theory of Statistics*, Vol. 1, 4th ed., Vol. 2, 4th ed., Vol. 3, 3d ed. New York: Hafner Press, 1977, 1979, and 1983.

MOOD, A. M., GRAYBILL, F. A., and BOES, D. C., *Introduction to the Theory of Statistics*, 3d ed. New York: McGraw-Hill Book Company, 1973.

3. EXPERIMENTAL DESIGN AND ANALYSIS OF VARIANCE

BOX, G. E. P., HUNTER, W. G., and HUNTER, J. S., *Statistics for Experimenters*, New York: John Wiley and Sons, Inc., 1978.

COCHRAN, W. G., and COX, G. M., *Experimental Designs*, 2d ed. New York: John Wiley & Sons, Inc., 1957.

DAVIES, O. L., ED., *The Design and Analysis of Industrial Experiments*, 2d rev. ed. New York: Hafner Press, 1967.

HICKS, C. R., *Fundamental Concepts in the Design of Experiments*, 3d ed. New York: Holt Rinehart and Winston, 1982.

LIPSON, C., and SHETH, N. J., *Statistical Design and Analysis of Engineering Experiments*. New York: McGraw-Hill Book Company, 1973.

SCHEFFE, H., *The Analysis of Variance*. New York: John Wiley & Sons, Inc., 1959.

SNEDECOR, G. W., and COCHRAN, W. G., *Statistical Methods*, 7th ed. Ames, Iowa: Iowa University Press, 1980.

4. QUALITY IMPROVEMENT AND ASSURANCE

DEMING, W. E., *Out of the Crisis*, Cambridge, Mass.: MIT., 1986.

DUNCAN, A. J., *Quality Control and Industrial Statistics*, 5th ed. Homewood, Ill.: Richard D. Irwin, Inc., 1986.

GRANT, E. L., and LEAVENWORTH, R., *Statistical Quality Control*, 5th ed. New York: McGraw-Hill Book Company, 1979.

ISHIKAWA, K., *Guide to Quality Control*, Tokyo: Asian Productivity Organization, 1976.

TAGUCHI, G., and WU, Y., *Introduction to Off-Line Quality Control*, Nagoya, Japan: Central Japan Quality Control Association.

5. SPECIAL TOPICS

BARLOW, R. E., and PROSCHAN, F., *Statistical Theory of Reliability and Life Testing: Probability Models*. New York: Holt, Rinehart and Winston, 1975.

BOX, G. E. P., and DRAPER, N. R., *Empirical Model-Building and Response Surfaces*, New York: John Wiley and Sons, Inc., 1987.

EPSTEIN, B., "Statistical Life Test Acceptance Procedures," *Technometrics*, Vol. 2, November, 1960.

GUTTMAN, I., *Statistical Tolerance Regions: Classical and Bayesian*. Darien, Conn.: Hafner, 1970.

KALBFLEISCH, J. D., and PRENTICE, R. L., *The Statistical Analysis of Failure Time Data*. New York: John Wiley & Sons, Inc., 1980.

LAWLESS, J. F., *Statistical Models and Methods for Lifetime Data*. New York: John Wiley & Sons, Inc., 1981.

MILLER, R. G., Jr., *Survival Analysis.* New York: John Wiley & Sons, Inc., 1981.

NELSON, W., *Applied Life Data Analysis.* New York: John Wiley & Sons, Inc., 1981.

WALD, A., *Sequential Analysis.* New York: John Wiley & Sons, Inc., 1947.

6. GENERAL REFERENCE WORKS AND TABLES

JOHNSON, R. A., and BHATTACHARYYA, G. K., *Statistics: Principles and Methods.* New York: John Wiley & Sons, Inc., 1986.

KENDALL, M. G., and BUCKLAND, W. R., *A Dictionary of Statistical Terms,* 3d ed. New York: John Wiley and Sons, Inc., 1983.

KOTZ, S., and JOHNSON, N. L., *Encyclopedia of Statistical Sciences,* Vol. 2., New York: John Wiley & Sons, 1982.

Military Standard 105D. Washington, D.C.: U.S. Government Printing Office, 1963.

National Bureau of Standards Handbook 91: Experimental Statistics. Washington, D.C.: U.S. Government Printing Office, 1963.

National Bureau of Standards: Tables of the Binomial Distribution. Washington, D.C.: U.S. Government Printing Office, 1950.

PEARSON, E. S., and HARTLEY, H. O., *Biometrika Tables for Statisticians,* 3d ed. Cambridge: Cambridge University Press, 1966.

RAND Corporation, *A Million Random Digits with 100,000 Normal Deviates.* New York: Macmillan Publishing Company, third printing 1966.

ROMIG, H. G., *50–100 Binomial Tables.* New York: John Wiley & Sons, Inc., 1953.

STATISTICAL TABLES

Table 1 Binomial Distribution Function

$$B(x; n, p) = \sum_{k=0}^{x} \binom{n}{k} p^k (1-p)^{n-k}$$

n	x	p = 0.05	0.10	0.15	0.20	0.25	0.30	0.35	0.40	0.45	0.50	0.55	0.60	0.65	0.70	0.75	0.80	0.85	0.90	0.95
2	0	0.9025	0.8100	0.7225	0.6400	0.5625	0.4900	0.4225	0.3600	0.3025	0.2500	0.2025	0.1600	0.1225	0.0900	0.0625	0.0400	0.0225	0.0100	0.0025
	1	0.9975	0.9900	0.9775	0.9600	0.9375	0.9100	0.8775	0.8400	0.7975	0.7500	0.6975	0.6400	0.5775	0.5100	0.4375	0.3600	0.2775	0.1900	0.0975
3	0	0.8574	0.7290	0.6141	0.5120	0.4219	0.3430	0.2746	0.2160	0.1664	0.1250	0.0911	0.0640	0.0429	0.0270	0.0156	0.0080	0.0034	0.0010	0.0001
	1	0.9927	0.9720	0.9393	0.8960	0.8438	0.7840	0.7183	0.6480	0.5748	0.5000	0.4252	0.3520	0.2818	0.2160	0.1563	0.1040	0.0607	0.0280	0.0073
	2	0.9999	0.9990	0.9966	0.9920	0.9844	0.9730	0.9571	0.9360	0.9089	0.8750	0.8336	0.7840	0.7254	0.6570	0.5781	0.4880	0.3859	0.2710	0.1426
4	0	0.8145	0.6561	0.5220	0.4096	0.3164	0.2401	0.1785	0.1296	0.0915	0.0625	0.0410	0.0256	0.0150	0.0081	0.0039	0.0016	0.0005	0.0001	0.0000
	1	0.9860	0.9477	0.8905	0.8192	0.7383	0.6517	0.5630	0.4752	0.3910	0.3125	0.2415	0.1792	0.1265	0.0837	0.0508	0.0272	0.0120	0.0037	0.0005
	2	0.9995	0.9963	0.9880	0.9728	0.9492	0.9163	0.8735	0.8208	0.7585	0.6875	0.6090	0.5248	0.4370	0.3483	0.2617	0.1808	0.1095	0.0523	0.0140
	3	1.0000	0.9999	0.9995	0.9984	0.9961	0.9919	0.9850	0.9744	0.9590	0.9375	0.9085	0.8704	0.8215	0.7599	0.6836	0.5904	0.4780	0.3439	0.1855
5	0	0.7738	0.5905	0.4437	0.3277	0.2373	0.1681	0.1160	0.0778	0.0503	0.0313	0.0185	0.0102	0.0053	0.0024	0.0010	0.0003	0.0001	0.0000	0.0000
	1	0.9774	0.9185	0.8352	0.7373	0.6328	0.5282	0.4284	0.3370	0.2562	0.1875	0.1312	0.0870	0.0540	0.0308	0.0156	0.0067	0.0022	0.0005	0.0000
	2	0.9988	0.9914	0.9734	0.9421	0.8965	0.8369	0.7648	0.6826	0.5931	0.5000	0.4069	0.3174	0.2352	0.1631	0.1035	0.0579	0.0266	0.0086	0.0012
	3	1.0000	0.9995	0.9978	0.9933	0.9844	0.9692	0.9460	0.9130	0.8688	0.8125	0.7438	0.6630	0.5716	0.4718	0.3672	0.2627	0.1648	0.0815	0.0226
	4	1.0000	1.0000	0.9999	0.9997	0.9990	0.9976	0.9947	0.9898	0.9815	0.9688	0.9497	0.9222	0.8840	0.8319	0.7627	0.6723	0.5563	0.4095	0.2262
6	0	0.7351	0.5314	0.3771	0.2621	0.1780	0.1176	0.0754	0.0467	0.0277	0.0156	0.0083	0.0041	0.0018	0.0007	0.0002	0.0001	0.0000	0.0000	0.0000
	1	0.9672	0.8857	0.7765	0.6554	0.5339	0.4202	0.3191	0.2333	0.1636	0.1094	0.0692	0.0410	0.0223	0.0109	0.0046	0.0016	0.0004	0.0001	0.0000
	2	0.9978	0.9841	0.9527	0.9011	0.8306	0.7443	0.6471	0.5443	0.4415	0.3438	0.2553	0.1792	0.1174	0.0705	0.0376	0.0170	0.0059	0.0013	0.0001
	3	0.9999	0.9987	0.9941	0.9830	0.9624	0.9295	0.8826	0.8208	0.7447	0.6563	0.5585	0.4557	0.3529	0.2557	0.1694	0.0989	0.0473	0.0158	0.0022
	4	1.0000	0.9999	0.9996	0.9984	0.9954	0.9891	0.9777	0.9590	0.9308	0.8906	0.8364	0.7667	0.6809	0.5798	0.4661	0.3446	0.2235	0.1143	0.0328
	5	1.0000	1.0000	1.0000	0.9999	0.9998	0.9993	0.9982	0.9959	0.9917	0.9844	0.9723	0.9533	0.9246	0.8824	0.8220	0.7379	0.6229	0.4686	0.2649
7	0	0.6983	0.4783	0.3206	0.2097	0.1335	0.0824	0.0490	0.0280	0.0152	0.0078	0.0037	0.0016	0.0006	0.0002	0.0001	0.0000	0.0000	0.0000	0.0000
	1	0.9556	0.8503	0.7166	0.5767	0.4449	0.3294	0.2338	0.1586	0.1024	0.0625	0.0357	0.0188	0.0090	0.0038	0.0013	0.0004	0.0001	0.0000	0.0000
	2	0.9962	0.9743	0.9262	0.8520	0.7564	0.6471	0.5323	0.4199	0.3164	0.2266	0.1529	0.0963	0.0556	0.0288	0.0129	0.0047	0.0012	0.0002	0.0000
	3	0.9998	0.9973	0.9879	0.9667	0.9294	0.8740	0.8002	0.7102	0.6083	0.5000	0.3917	0.2898	0.1998	0.1260	0.0706	0.0333	0.0121	0.0027	0.0002
	4	1.0000	0.9998	0.9988	0.9953	0.9871	0.9712	0.9444	0.9037	0.8471	0.7734	0.6836	0.5801	0.4677	0.3529	0.2436	0.1480	0.0738	0.0257	0.0038
	5	1.0000	1.0000	0.9999	0.9996	0.9987	0.9962	0.9910	0.9812	0.9643	0.9375	0.8976	0.8414	0.7662	0.6706	0.5551	0.4233	0.2834	0.1497	0.0444
	6	1.0000	1.0000	1.0000	1.0000	0.9999	0.9998	0.9994	0.9984	0.9963	0.9922	0.9848	0.9720	0.9510	0.9176	0.8665	0.7903	0.6794	0.5217	0.3017
8	0	0.6634	0.4305	0.2725	0.1678	0.1001	0.0576	0.0319	0.0168	0.0084	0.0039	0.0017	0.0007	0.0002	0.0001	0.0000	0.0000	0.0000	0.0000	0.0000
	1	0.9428	0.8131	0.6572	0.5033	0.3671	0.2553	0.1691	0.1064	0.0632	0.0352	0.0181	0.0085	0.0036	0.0013	0.0004	0.0001	0.0000	0.0000	0.0000
	2	0.9942	0.9619	0.8948	0.7969	0.6785	0.5518	0.4278	0.3154	0.2201	0.1445	0.0885	0.0498	0.0253	0.0113	0.0042	0.0012	0.0002	0.0000	0.0000
	3	0.9996	0.9950	0.9786	0.9437	0.8862	0.8059	0.7064	0.5941	0.4770	0.3633	0.2604	0.1737	0.1061	0.0580	0.0273	0.0104	0.0029	0.0004	0.0000
	4	1.0000	0.9996	0.9971	0.9896	0.9727	0.9420	0.8939	0.8263	0.7396	0.6367	0.5230	0.4059	0.2936	0.1941	0.1138	0.0563	0.0214	0.0050	0.0004
	5	1.0000	1.0000	0.9998	0.9988	0.9958	0.9887	0.9747	0.9502	0.9115	0.8555	0.7799	0.6846	0.5722	0.4482	0.3215	0.2031	0.1052	0.0381	0.0058
	6	1.0000	1.0000	1.0000	0.9999	0.9996	0.9987	0.9964	0.9915	0.9819	0.9648	0.9368	0.8936	0.8309	0.7447	0.6329	0.4967	0.3428	0.1869	0.0572
	7	1.0000	1.0000	1.0000	1.0000	1.0000	0.9999	0.9998	0.9993	0.9983	0.9961	0.9916	0.9832	0.9681	0.9424	0.8999	0.8322	0.7275	0.5695	0.3366

n	x	0.05	0.10	0.15	0.20	0.25	0.30	0.35	0.40	0.45	0.50	0.55	0.60	0.65	0.70	0.75	0.80	0.85	0.90	0.95
		p																		
9	0	0.6302	0.3874	0.2316	0.1342	0.0751	0.0404	0.0207	0.0101	0.0046	0.0020	0.0008	0.0003	0.0001	0.0000	0.0000	0.0000	0.0000	0.0000	0.0000
	1	0.9288	0.7748	0.5995	0.4362	0.3003	0.1960	0.1211	0.0705	0.0385	0.0195	0.0091	0.0038	0.0014	0.0004	0.0001	0.0000	0.0000	0.0000	0.0000
	2	0.9916	0.9470	0.8591	0.7382	0.6007	0.4628	0.3373	0.2318	0.1495	0.0898	0.0498	0.0250	0.0112	0.0043	0.0013	0.0003	0.0000	0.0000	0.0000
	3	0.9994	0.9917	0.9661	0.9144	0.8343	0.7297	0.6089	0.4826	0.3614	0.2539	0.1658	0.0994	0.0536	0.0253	0.0100	0.0031	0.0006	0.0001	0.0000
	4	1.0000	0.9991	0.9944	0.9804	0.9511	0.9012	0.8283	0.7334	0.6214	0.5000	0.3786	0.2666	0.1717	0.0988	0.0489	0.0196	0.0056	0.0009	0.0000
	5	1.0000	0.9999	0.9994	0.9969	0.9900	0.9747	0.9464	0.9006	0.8342	0.7461	0.6386	0.5174	0.3911	0.2703	0.1657	0.0856	0.0339	0.0083	0.0006
	6	1.0000	1.0000	1.0000	0.9997	0.9987	0.9957	0.9888	0.9750	0.9502	0.9102	0.8505	0.7682	0.6627	0.5372	0.3993	0.2618	0.1409	0.0530	0.0084
	7	1.0000	1.0000	1.0000	1.0000	0.9999	0.9996	0.9986	0.9962	0.9909	0.9805	0.9615	0.9295	0.8789	0.8040	0.6997	0.5638	0.4005	0.2252	0.0712
	8	1.0000	1.0000	1.0000	1.0000	1.0000	1.0000	0.9999	0.9997	0.9992	0.9980	0.9954	0.9899	0.9793	0.9596	0.9249	0.8658	0.7684	0.6126	0.3698
10	0	0.5987	0.3487	0.1969	0.1074	0.0563	0.0282	0.0135	0.0060	0.0025	0.0010	0.0003	0.0001	0.0000	0.0000	0.0000	0.0000	0.0000	0.0000	0.0000
	1	0.9139	0.7361	0.5443	0.3758	0.2440	0.1493	0.0860	0.0464	0.0233	0.0107	0.0045	0.0017	0.0005	0.0001	0.0000	0.0000	0.0000	0.0000	0.0000
	2	0.9885	0.9298	0.8202	0.6778	0.5256	0.3828	0.2616	0.1673	0.0996	0.0547	0.0274	0.0123	0.0048	0.0016	0.0004	0.0001	0.0000	0.0000	0.0000
	3	0.9990	0.9872	0.9500	0.8791	0.7759	0.6496	0.5138	0.3823	0.2660	0.1719	0.1020	0.0548	0.0260	0.0106	0.0035	0.0009	0.0001	0.0000	0.0000
	4	0.9999	0.9984	0.9901	0.9672	0.9219	0.8497	0.7515	0.6331	0.5044	0.3770	0.2616	0.1662	0.0949	0.0473	0.0197	0.0064	0.0014	0.0001	0.0000
	5	1.0000	0.9999	0.9986	0.9936	0.9803	0.9527	0.9051	0.8338	0.7384	0.6230	0.4956	0.3669	0.2485	0.1503	0.0781	0.0328	0.0099	0.0016	0.0001
	6	1.0000	1.0000	0.9999	0.9991	0.9965	0.9894	0.9740	0.9452	0.8980	0.8281	0.7340	0.6177	0.4862	0.3504	0.2241	0.1209	0.0500	0.0128	0.0010
	7	1.0000	1.0000	1.0000	0.9999	0.9996	0.9984	0.9952	0.9877	0.9726	0.9453	0.9004	0.8327	0.7384	0.6172	0.4744	0.3222	0.1798	0.0702	0.0115
	8	1.0000	1.0000	1.0000	1.0000	1.0000	0.9999	0.9995	0.9983	0.9955	0.9893	0.9767	0.9536	0.9140	0.8507	0.7560	0.6242	0.4557	0.2639	0.0861
	9	1.0000	1.0000	1.0000	1.0000	1.0000	1.0000	1.0000	0.9999	0.9997	0.9990	0.9975	0.9940	0.9865	0.9718	0.9437	0.8926	0.8031	0.6513	0.4013
11	0	0.5688	0.3138	0.1673	0.0859	0.0422	0.0198	0.0088	0.0036	0.0014	0.0005	0.0002	0.0000	0.0000	0.0000	0.0000	0.0000	0.0000	0.0000	0.0000
	1	0.8981	0.6974	0.4922	0.3221	0.1971	0.1130	0.0606	0.0302	0.0139	0.0059	0.0022	0.0007	0.0002	0.0000	0.0000	0.0000	0.0000	0.0000	0.0000
	2	0.9848	0.9104	0.7788	0.6174	0.4552	0.3127	0.2001	0.1189	0.0652	0.0327	0.0148	0.0059	0.0020	0.0006	0.0001	0.0000	0.0000	0.0000	0.0000
	3	0.9984	0.9815	0.9306	0.8389	0.7133	0.5696	0.4256	0.2963	0.1911	0.1133	0.0610	0.0293	0.0122	0.0043	0.0012	0.0002	0.0000	0.0000	0.0000
	4	0.9999	0.9972	0.9841	0.9496	0.8854	0.7897	0.6683	0.5328	0.3971	0.2744	0.1738	0.0994	0.0501	0.0216	0.0076	0.0020	0.0003	0.0000	0.0000
	5	1.0000	0.9997	0.9973	0.9883	0.9657	0.9218	0.8513	0.7535	0.6331	0.5000	0.3669	0.2465	0.1487	0.0782	0.0343	0.0117	0.0027	0.0003	0.0000
	6	1.0000	1.0000	0.9997	0.9980	0.9924	0.9784	0.9499	0.9006	0.8262	0.7256	0.6029	0.4672	0.3317	0.2103	0.1146	0.0504	0.0159	0.0028	0.0001
	7	1.0000	1.0000	1.0000	0.9998	0.9988	0.9957	0.9878	0.9707	0.9390	0.8867	0.8089	0.7037	0.5744	0.4304	0.2867	0.1611	0.0694	0.0185	0.0016
	8	1.0000	1.0000	1.0000	1.0000	0.9999	0.9994	0.9980	0.9941	0.9852	0.9673	0.9348	0.8811	0.7999	0.6873	0.5448	0.3826	0.2212	0.0896	0.0152
	9	1.0000	1.0000	1.0000	1.0000	1.0000	1.0000	0.9998	0.9993	0.9978	0.9941	0.9861	0.9698	0.9394	0.8870	0.8029	0.6779	0.5078	0.3026	0.1019
	10	1.0000	1.0000	1.0000	1.0000	1.0000	1.0000	1.0000	1.0000	0.9998	0.9995	0.9986	0.9964	0.9912	0.9802	0.9578	0.9141	0.8327	0.6862	0.4312
12	0	0.5404	0.2824	0.1422	0.0687	0.0317	0.0138	0.0057	0.0022	0.0008	0.0002	0.0001	0.0000	0.0000	0.0000	0.0000	0.0000	0.0000	0.0000	0.0000
	1	0.8816	0.6590	0.4435	0.2749	0.1584	0.0850	0.0424	0.0196	0.0083	0.0032	0.0011	0.0003	0.0001	0.0000	0.0000	0.0000	0.0000	0.0000	0.0000
	2	0.9804	0.8891	0.7358	0.5583	0.3907	0.2528	0.1513	0.0834	0.0421	0.0193	0.0079	0.0028	0.0008	0.0002	0.0000	0.0000	0.0000	0.0000	0.0000
	3	0.9978	0.9744	0.9078	0.7946	0.6488	0.4925	0.3467	0.2253	0.1345	0.0730	0.0356	0.0153	0.0056	0.0017	0.0004	0.0001	0.0000	0.0000	0.0000
	4	0.9998	0.9957	0.9761	0.9274	0.8424	0.7237	0.5833	0.4382	0.3044	0.1938	0.1117	0.0573	0.0255	0.0095	0.0028	0.0006	0.0001	0.0000	0.0000
	5	1.0000	0.9995	0.9954	0.9806	0.9456	0.8822	0.7873	0.6652	0.5269	0.3872	0.2607	0.1582	0.0846	0.0386	0.0143	0.0039	0.0007	0.0001	0.0000
	6	1.0000	0.9999	0.9993	0.9961	0.9857	0.9614	0.9154	0.8418	0.7393	0.6128	0.4731	0.3348	0.2127	0.1178	0.0544	0.0194	0.0046	0.0005	0.0000
	7	1.0000	1.0000	0.9999	0.9994	0.9972	0.9905	0.9745	0.9427	0.8883	0.8062	0.6956	0.5618	0.4167	0.2763	0.1576	0.0726	0.0239	0.0043	0.0002
	8	1.0000	1.0000	1.0000	0.9999	0.9996	0.9983	0.9944	0.9847	0.9644	0.9270	0.8655	0.7747	0.6533	0.5075	0.3512	0.2054	0.0922	0.0256	0.0022
	9	1.0000	1.0000	1.0000	1.0000	1.0000	0.9998	0.9992	0.9972	0.9921	0.9807	0.9579	0.9166	0.8487	0.7472	0.6093	0.4417	0.2642	0.1109	0.0196
	10	1.0000	1.0000	1.0000	1.0000	1.0000	1.0000	0.9999	0.9997	0.9989	0.9968	0.9917	0.9804	0.9576	0.9150	0.8416	0.7251	0.5565	0.3410	0.1184
	11	1.0000	1.0000	1.0000	1.0000	1.0000	1.0000	1.0000	1.0000	0.9999	0.9998	0.9992	0.9978	0.9943	0.9862	0.9683	0.9313	0.8578	0.7176	0.4596

n	x	0.05	0.10	0.15	0.20	0.25	0.30	0.35	0.40	0.45	0.50	0.55	0.60	0.65	0.70	0.75	0.80	0.85	0.90	0.95
		p																		
13	0	0.5133	0.2542	0.1209	0.0550	0.0238	0.0097	0.0037	0.0013	0.0004	0.0001	0.0000	0.0000	0.0000	0.0000	0.0000	0.0000	0.0000	0.0000	0.0000
	1	0.8646	0.6213	0.3983	0.2336	0.1267	0.0637	0.0296	0.0126	0.0049	0.0017	0.0005	0.0001	0.0000	0.0000	0.0000	0.0000	0.0000	0.0000	0.0000
	2	0.9755	0.8661	0.6920	0.5017	0.3326	0.2025	0.1132	0.0579	0.0269	0.0112	0.0041	0.0013	0.0003	0.0001	0.0000	0.0000	0.0000	0.0000	0.0000
	3	0.9969	0.9658	0.8820	0.7473	0.5843	0.4206	0.2783	0.1686	0.0929	0.0461	0.0203	0.0078	0.0025	0.0007	0.0001	0.0000	0.0000	0.0000	0.0000
	4	0.9997	0.9935	0.9658	0.9009	0.7940	0.6543	0.5005	0.3530	0.2279	0.1334	0.0698	0.0321	0.0126	0.0040	0.0010	0.0002	0.0000	0.0000	0.0000
	5	1.0000	0.9991	0.9925	0.9700	0.9198	0.8346	0.7159	0.5744	0.4268	0.2905	0.1788	0.0977	0.0462	0.0182	0.0056	0.0012	0.0002	0.0000	0.0000
	6	1.0000	0.9999	0.9987	0.9930	0.9757	0.9376	0.8705	0.7712	0.6437	0.5000	0.3563	0.2288	0.1295	0.0624	0.0243	0.0070	0.0013	0.0001	0.0000
	7	1.0000	1.0000	0.9998	0.9988	0.9944	0.9818	0.9538	0.9023	0.8212	0.7095	0.5732	0.4256	0.2841	0.1654	0.0802	0.0300	0.0075	0.0009	0.0000
	8	1.0000	1.0000	1.0000	0.9998	0.9990	0.9960	0.9874	0.9679	0.9302	0.8666	0.7721	0.6470	0.4995	0.3457	0.2060	0.0991	0.0342	0.0065	0.0003
	9	1.0000	1.0000	1.0000	1.0000	0.9999	0.9993	0.9975	0.9922	0.9797	0.9539	0.9071	0.8314	0.7217	0.5794	0.4157	0.2527	0.1180	0.0342	0.0031
	10	1.0000	1.0000	1.0000	1.0000	1.0000	0.9999	0.9997	0.9987	0.9959	0.9888	0.9731	0.9421	0.8868	0.7975	0.6674	0.4983	0.3080	0.1339	0.0245
	11	1.0000	1.0000	1.0000	1.0000	1.0000	1.0000	1.0000	0.9999	0.9995	0.9983	0.9951	0.9874	0.9704	0.9363	0.8733	0.7664	0.6017	0.3787	0.1354
	12	1.0000	1.0000	1.0000	1.0000	1.0000	1.0000	1.0000	1.0000	1.0000	0.9999	0.9996	0.9987	0.9963	0.9903	0.9762	0.9450	0.8791	0.7458	0.4867
14	0	0.4877	0.2288	0.1028	0.0440	0.0178	0.0068	0.0024	0.0008	0.0002	0.0001	0.0000	0.0000	0.0000	0.0000	0.0000	0.0000	0.0000	0.0000	0.0000
	1	0.8470	0.5846	0.3567	0.1979	0.1010	0.0475	0.0205	0.0081	0.0029	0.0009	0.0003	0.0001	0.0000	0.0000	0.0000	0.0000	0.0000	0.0000	0.0000
	2	0.9699	0.8416	0.6479	0.4481	0.2811	0.1608	0.0839	0.0398	0.0170	0.0065	0.0022	0.0006	0.0001	0.0000	0.0000	0.0000	0.0000	0.0000	0.0000
	3	0.9958	0.9559	0.8535	0.6982	0.5213	0.3552	0.2205	0.1243	0.0632	0.0287	0.0114	0.0039	0.0011	0.0002	0.0000	0.0000	0.0000	0.0000	0.0000
	4	0.9996	0.9908	0.9533	0.8702	0.7415	0.5842	0.4227	0.2793	0.1672	0.0898	0.0426	0.0175	0.0060	0.0017	0.0003	0.0000	0.0000	0.0000	0.0000
	5	1.0000	0.9985	0.9885	0.9561	0.8883	0.7805	0.6405	0.4859	0.3373	0.2120	0.1189	0.0583	0.0243	0.0083	0.0022	0.0004	0.0000	0.0000	0.0000
	6	1.0000	0.9998	0.9978	0.9884	0.9617	0.9067	0.8164	0.6925	0.5461	0.3953	0.2586	0.1501	0.0753	0.0315	0.0103	0.0024	0.0003	0.0000	0.0000
	7	1.0000	1.0000	0.9997	0.9976	0.9897	0.9685	0.9247	0.8499	0.7414	0.6047	0.4539	0.3075	0.1836	0.0933	0.0383	0.0116	0.0022	0.0002	0.0000
	8	1.0000	1.0000	1.0000	0.9996	0.9978	0.9917	0.9757	0.9417	0.8811	0.7880	0.6627	0.5141	0.3595	0.2195	0.1117	0.0439	0.0115	0.0015	0.0000
	9	1.0000	1.0000	1.0000	1.0000	0.9997	0.9983	0.9940	0.9825	0.9574	0.9102	0.8328	0.7207	0.5773	0.4158	0.2585	0.1298	0.0467	0.0092	0.0004
	10	1.0000	1.0000	1.0000	1.0000	1.0000	0.9998	0.9989	0.9961	0.9886	0.9713	0.9368	0.8757	0.7795	0.6448	0.4787	0.3018	0.1465	0.0441	0.0042
	11	1.0000	1.0000	1.0000	1.0000	1.0000	1.0000	0.9999	0.9994	0.9978	0.9935	0.9830	0.9602	0.9161	0.8392	0.7189	0.5519	0.3521	0.1584	0.0301
	12	1.0000	1.0000	1.0000	1.0000	1.0000	1.0000	1.0000	0.9999	0.9997	0.9991	0.9971	0.9919	0.9795	0.9525	0.8990	0.8021	0.6433	0.4154	0.1530
	13	1.0000	1.0000	1.0000	1.0000	1.0000	1.0000	1.0000	1.0000	1.0000	0.9999	0.9998	0.9992	0.9976	0.9932	0.9822	0.9560	0.8972	0.7712	0.5123
15	0	0.4633	0.2059	0.0874	0.0352	0.0134	0.0047	0.0016	0.0005	0.0001	0.0000	0.0000	0.0000	0.0000	0.0000	0.0000	0.0000	0.0000	0.0000	0.0000
	1	0.8290	0.5490	0.3186	0.1671	0.0802	0.0353	0.0142	0.0052	0.0017	0.0005	0.0001	0.0000	0.0000	0.0000	0.0000	0.0000	0.0000	0.0000	0.0000
	2	0.9638	0.8159	0.6042	0.3980	0.2361	0.1268	0.0617	0.0271	0.0107	0.0037	0.0011	0.0003	0.0001	0.0000	0.0000	0.0000	0.0000	0.0000	0.0000
	3	0.9945	0.9444	0.8227	0.6482	0.4613	0.2969	0.1727	0.0905	0.0424	0.0176	0.0063	0.0019	0.0005	0.0001	0.0000	0.0000	0.0000	0.0000	0.0000
	4	0.9994	0.9873	0.9383	0.8358	0.6865	0.5155	0.3519	0.2173	0.1204	0.0592	0.0255	0.0093	0.0028	0.0007	0.0001	0.0000	0.0000	0.0000	0.0000
	5	0.9999	0.9977	0.9832	0.9389	0.8516	0.7216	0.5643	0.4032	0.2608	0.1509	0.0769	0.0338	0.0124	0.0037	0.0008	0.0001	0.0000	0.0000	0.0000
	6	1.0000	0.9997	0.9964	0.9819	0.9434	0.8689	0.7548	0.6098	0.4522	0.3036	0.1818	0.0950	0.0422	0.0152	0.0042	0.0008	0.0001	0.0000	0.0000
	7	1.0000	1.0000	0.9994	0.9958	0.9827	0.9500	0.8868	0.7869	0.6535	0.5000	0.3465	0.2131	0.1132	0.0500	0.0173	0.0042	0.0006	0.0000	0.0000
	8	1.0000	1.0000	0.9999	0.9992	0.9958	0.9848	0.9578	0.9050	0.8182	0.6964	0.5478	0.3902	0.2452	0.1311	0.0566	0.0181	0.0036	0.0003	0.0000
	9	1.0000	1.0000	1.0000	0.9999	0.9992	0.9963	0.9876	0.9662	0.9231	0.8491	0.7392	0.5968	0.4357	0.2784	0.1484	0.0611	0.0168	0.0022	0.0001
	10	1.0000	1.0000	1.0000	1.0000	0.9999	0.9993	0.9972	0.9907	0.9745	0.9408	0.8796	0.7827	0.6481	0.4845	0.3135	0.1642	0.0617	0.0127	0.0006
	11	1.0000	1.0000	1.0000	1.0000	1.0000	0.9999	0.9995	0.9981	0.9937	0.9824	0.9576	0.9095	0.8273	0.7031	0.5387	0.3518	0.1773	0.0556	0.0055
	12	1.0000	1.0000	1.0000	1.0000	1.0000	1.0000	0.9999	0.9997	0.9989	0.9963	0.9893	0.9729	0.9383	0.8732	0.7639	0.6020	0.3958	0.1841	0.0362
	13	1.0000	1.0000	1.0000	1.0000	1.0000	1.0000	1.0000	1.0000	0.9999	0.9995	0.9983	0.9948	0.9858	0.9647	0.9198	0.8329	0.6814	0.4510	0.1710
	14	1.0000	1.0000	1.0000	1.0000	1.0000	1.0000	1.0000	1.0000	1.0000	1.0000	0.9999	0.9995	0.9984	0.9953	0.9866	0.9648	0.9126	0.7941	0.5367

n	x	p 0.05	0.10	0.15	0.20	0.25	0.30	0.35	0.40	0.45	0.50	0.55	0.60	0.65	0.70	0.75	0.80	0.85	0.90	0.95
16	0	0.4401	0.1853	0.0743	0.0281	0.0100	0.0033	0.0010	0.0003	0.0001	0.0000	0.0000	0.0000	0.0000	0.0000	0.0000	0.0000	0.0000	0.0000	0.0000
	1	0.8108	0.5147	0.2839	0.1407	0.0635	0.0261	0.0098	0.0033	0.0010	0.0003	0.0001	0.0000	0.0000	0.0000	0.0000	0.0000	0.0000	0.0000	0.0000
	2	0.9571	0.7892	0.5614	0.3518	0.1971	0.0994	0.0451	0.0183	0.0066	0.0021	0.0006	0.0001	0.0000	0.0000	0.0000	0.0000	0.0000	0.0000	0.0000
	3	0.9930	0.9316	0.7899	0.5981	0.4050	0.2459	0.1339	0.0651	0.0281	0.0106	0.0035	0.0009	0.0002	0.0000	0.0000	0.0000	0.0000	0.0000	0.0000
	4	0.9991	0.9830	0.9209	0.7982	0.6302	0.4499	0.2892	0.1666	0.0853	0.0384	0.0149	0.0049	0.0013	0.0003	0.0000	0.0000	0.0000	0.0000	0.0000
	5	0.9999	0.9967	0.9765	0.9183	0.8103	0.6598	0.4900	0.3288	0.1976	0.1051	0.0486	0.0191	0.0062	0.0016	0.0003	0.0000	0.0000	0.0000	0.0000
	6	1.0000	0.9995	0.9944	0.9733	0.9204	0.8247	0.6881	0.5272	0.3660	0.2272	0.1241	0.0583	0.0229	0.0071	0.0016	0.0002	0.0000	0.0000	0.0000
	7	1.0000	0.9999	0.9989	0.9930	0.9729	0.9256	0.8406	0.7161	0.5629	0.4018	0.2559	0.1423	0.0671	0.0257	0.0075	0.0015	0.0002	0.0000	0.0000
	8	1.0000	1.0000	0.9998	0.9985	0.9925	0.9743	0.9329	0.8577	0.7441	0.5982	0.4371	0.2839	0.1594	0.0744	0.0271	0.0070	0.0011	0.0001	0.0000
	9	1.0000	1.0000	1.0000	0.9998	0.9984	0.9929	0.9771	0.9417	0.8759	0.7728	0.6340	0.4728	0.3119	0.1753	0.0796	0.0267	0.0056	0.0005	0.0000
	10	1.0000	1.0000	1.0000	1.0000	0.9997	0.9984	0.9938	0.9809	0.9514	0.8949	0.8024	0.6712	0.5100	0.3402	0.1897	0.0817	0.0235	0.0033	0.0001
	11	1.0000	1.0000	1.0000	1.0000	1.0000	0.9997	0.9987	0.9951	0.9851	0.9616	0.9147	0.8334	0.7108	0.5501	0.3698	0.2018	0.0791	0.0170	0.0009
	12	1.0000	1.0000	1.0000	1.0000	1.0000	1.0000	0.9998	0.9991	0.9965	0.9894	0.9719	0.9349	0.8661	0.7541	0.5950	0.4019	0.2101	0.0684	0.0070
	13	1.0000	1.0000	1.0000	1.0000	1.0000	1.0000	1.0000	0.9999	0.9994	0.9979	0.9934	0.9817	0.9549	0.9006	0.8029	0.6482	0.4386	0.2108	0.0429
	14	1.0000	1.0000	1.0000	1.0000	1.0000	1.0000	1.0000	1.0000	0.9999	0.9997	0.9990	0.9967	0.9902	0.9739	0.9365	0.8593	0.7161	0.4853	0.1892
	15	1.0000	1.0000	1.0000	1.0000	1.0000	1.0000	1.0000	1.0000	1.0000	1.0000	0.9999	0.9997	0.9990	0.9967	0.9900	0.9719	0.9257	0.8147	0.5599
17	0	0.4181	0.1668	0.0631	0.0225	0.0075	0.0023	0.0007	0.0002	0.0000	0.0000	0.0000	0.0000	0.0000	0.0000	0.0000	0.0000	0.0000	0.0000	0.0000
	1	0.7922	0.4818	0.2525	0.1182	0.0501	0.0193	0.0067	0.0021	0.0006	0.0001	0.0000	0.0000	0.0000	0.0000	0.0000	0.0000	0.0000	0.0000	0.0000
	2	0.9497	0.7618	0.5198	0.3096	0.1637	0.0774	0.0327	0.0123	0.0041	0.0012	0.0003	0.0001	0.0000	0.0000	0.0000	0.0000	0.0000	0.0000	0.0000
	3	0.9912	0.9174	0.7556	0.5489	0.3530	0.2019	0.1028	0.0464	0.0184	0.0064	0.0019	0.0005	0.0001	0.0000	0.0000	0.0000	0.0000	0.0000	0.0000
	4	0.9988	0.9779	0.9013	0.7582	0.5739	0.3887	0.2348	0.1260	0.0596	0.0245	0.0086	0.0025	0.0006	0.0001	0.0000	0.0000	0.0000	0.0000	0.0000
	5	0.9999	0.9953	0.9681	0.8943	0.7653	0.5968	0.4197	0.2639	0.1471	0.0717	0.0301	0.0106	0.0030	0.0007	0.0001	0.0000	0.0000	0.0000	0.0000
	6	1.0000	0.9992	0.9917	0.9623	0.8929	0.7752	0.6188	0.4478	0.2902	0.1662	0.0826	0.0348	0.0120	0.0032	0.0006	0.0001	0.0000	0.0000	0.0000
	7	1.0000	0.9999	0.9983	0.9891	0.9598	0.8954	0.7872	0.6405	0.4743	0.3145	0.1834	0.0919	0.0383	0.0127	0.0031	0.0005	0.0000	0.0000	0.0000
	8	1.0000	1.0000	0.9997	0.9974	0.9876	0.9597	0.9006	0.8011	0.6626	0.5000	0.3374	0.1989	0.0994	0.0403	0.0124	0.0026	0.0003	0.0000	0.0000
	9	1.0000	1.0000	1.0000	0.9995	0.9969	0.9873	0.9617	0.9081	0.8166	0.6855	0.5257	0.3595	0.2128	0.1046	0.0402	0.0109	0.0017	0.0001	0.0000
	10	1.0000	1.0000	1.0000	0.9999	0.9994	0.9968	0.9880	0.9652	0.9174	0.8338	0.7098	0.5522	0.3812	0.2248	0.1071	0.0377	0.0083	0.0008	0.0000
	11	1.0000	1.0000	1.0000	1.0000	0.9999	0.9993	0.9970	0.9894	0.9699	0.9283	0.8529	0.7361	0.5803	0.4032	0.2347	0.1057	0.0319	0.0047	0.0001
	12	1.0000	1.0000	1.0000	1.0000	1.0000	0.9999	0.9994	0.9975	0.9914	0.9755	0.9404	0.8740	0.7652	0.6113	0.4261	0.2418	0.0987	0.0221	0.0012
	13	1.0000	1.0000	1.0000	1.0000	1.0000	1.0000	0.9999	0.9995	0.9981	0.9936	0.9816	0.9536	0.8972	0.7981	0.6470	0.4511	0.2444	0.0826	0.0088
	14	1.0000	1.0000	1.0000	1.0000	1.0000	1.0000	1.0000	0.9999	0.9997	0.9988	0.9959	0.9877	0.9673	0.9226	0.8363	0.6904	0.4802	0.2382	0.0503
	15	1.0000	1.0000	1.0000	1.0000	1.0000	1.0000	1.0000	1.0000	1.0000	0.9999	0.9994	0.9979	0.9933	0.9807	0.9499	0.8818	0.7475	0.5182	0.2078
	16	1.0000	1.0000	1.0000	1.0000	1.0000	1.0000	1.0000	1.0000	1.0000	1.0000	1.0000	0.9998	0.9993	0.9977	0.9925	0.9775	0.9369	0.8332	0.5819
18	0	0.3972	0.1501	0.0536	0.0180	0.0056	0.0016	0.0004	0.0001	0.0000	0.0000	0.0000	0.0000	0.0000	0.0000	0.0000	0.0000	0.0000	0.0000	0.0000
	1	0.7735	0.4503	0.2241	0.0991	0.0395	0.0142	0.0046	0.0013	0.0003	0.0001	0.0000	0.0000	0.0000	0.0000	0.0000	0.0000	0.0000	0.0000	0.0000
	2	0.9419	0.7338	0.4797	0.2713	0.1353	0.0600	0.0236	0.0082	0.0025	0.0007	0.0001	0.0000	0.0000	0.0000	0.0000	0.0000	0.0000	0.0000	0.0000
	3	0.9891	0.9018	0.7202	0.5010	0.3057	0.1646	0.0783	0.0328	0.0120	0.0038	0.0010	0.0002	0.0000	0.0000	0.0000	0.0000	0.0000	0.0000	0.0000
	4	0.9985	0.9718	0.8794	0.7164	0.5187	0.3327	0.1886	0.0942	0.0411	0.0154	0.0049	0.0013	0.0003	0.0000	0.0000	0.0000	0.0000	0.0000	0.0000
	5	0.9998	0.9936	0.9581	0.8671	0.7175	0.5344	0.3550	0.2088	0.1077	0.0481	0.0183	0.0058	0.0014	0.0003	0.0000	0.0000	0.0000	0.0000	0.0000
	6	1.0000	0.9988	0.9882	0.9487	0.8610	0.7217	0.5491	0.3743	0.2258	0.1189	0.0537	0.0203	0.0062	0.0014	0.0002	0.0000	0.0000	0.0000	0.0000
	7	1.0000	0.9998	0.9973	0.9837	0.9431	0.8593	0.7283	0.5634	0.3915	0.2403	0.1280	0.0576	0.0212	0.0061	0.0012	0.0002	0.0000	0.0000	0.0000
	8	1.0000	1.0000	0.9995	0.9957	0.9807	0.9404	0.8609	0.7368	0.5778	0.4073	0.2527	0.1347	0.0597	0.0210	0.0054	0.0009	0.0001	0.0000	0.0000
	9	1.0000	1.0000	0.9999	0.9991	0.9946	0.9790	0.9403	0.8653	0.7473	0.5927	0.4222	0.2632	0.1391	0.0596	0.0193	0.0043	0.0005	0.0000	0.0000
	10	1.0000	1.0000	1.0000	0.9998	0.9988	0.9939	0.9788	0.9424	0.8720	0.7597	0.6085	0.4366	0.2717	0.1407	0.0569	0.0163	0.0027	0.0002	0.0000

n	x	0.05	0.10	0.15	0.20	0.25	0.30	0.35	0.40	0.45	0.50	0.55	0.60	0.65	0.70	0.75	0.80	0.85	0.90	0.95
	11	1.0000	1.0000	1.0000	1.0000	0.9998	0.9986	0.9938	0.9797	0.9463	0.8811	0.7742	0.6257	0.4509	0.2783	0.1390	0.0513	0.0118	0.0012	0.0000
	12	1.0000	1.0000	1.0000	1.0000	1.0000	0.9997	0.9986	0.9942	0.9817	0.9519	0.8923	0.7912	0.6450	0.4656	0.2825	0.1329	0.0419	0.0064	0.0002
	13	1.0000	1.0000	1.0000	1.0000	1.0000	1.0000	0.9997	0.9987	0.9951	0.9846	0.9589	0.9058	0.8114	0.6673	0.4813	0.2836	0.1206	0.0282	0.0015
	14	1.0000	1.0000	1.0000	1.0000	1.0000	1.0000	1.0000	0.9998	0.9990	0.9962	0.9880	0.9672	0.9217	0.8354	0.6943	0.4990	0.2798	0.0982	0.0109
	15	1.0000	1.0000	1.0000	1.0000	1.0000	1.0000	1.0000	1.0000	0.9999	0.9993	0.9975	0.9918	0.9764	0.9400	0.8647	0.7287	0.5203	0.2662	0.0581
	16	1.0000	1.0000	1.0000	1.0000	1.0000	1.0000	1.0000	1.0000	1.0000	0.9999	0.9997	0.9987	0.9954	0.9858	0.9605	0.9009	0.7759	0.5497	0.2265
	17	1.0000	1.0000	1.0000	1.0000	1.0000	1.0000	1.0000	1.0000	1.0000	1.0000	1.0000	0.9999	0.9996	0.9984	0.9944	0.9820	0.9464	0.8499	0.6028
19	0	0.3774	0.1351	0.0456	0.0144	0.0042	0.0011	0.0003	0.0001	0.0000	0.0000	0.0000	0.0000	0.0000	0.0000	0.0000	0.0000	0.0000	0.0000	0.0000
	1	0.7547	0.4203	0.1985	0.0829	0.0310	0.0104	0.0031	0.0008	0.0002	0.0000	0.0000	0.0000	0.0000	0.0000	0.0000	0.0000	0.0000	0.0000	0.0000
	2	0.9335	0.7054	0.4413	0.2369	0.1113	0.0462	0.0170	0.0055	0.0015	0.0004	0.0001	0.0000	0.0000	0.0000	0.0000	0.0000	0.0000	0.0000	0.0000
	3	0.9868	0.8850	0.6841	0.4551	0.2631	0.1332	0.0591	0.0230	0.0077	0.0022	0.0005	0.0001	0.0000	0.0000	0.0000	0.0000	0.0000	0.0000	0.0000
	4	0.9980	0.9648	0.8556	0.6733	0.4654	0.2822	0.1500	0.0696	0.0280	0.0096	0.0028	0.0006	0.0001	0.0000	0.0000	0.0000	0.0000	0.0000	0.0000
	5	0.9998	0.9914	0.9463	0.8369	0.6678	0.4739	0.2968	0.1629	0.0777	0.0318	0.0109	0.0031	0.0007	0.0001	0.0000	0.0000	0.0000	0.0000	0.0000
	6	1.0000	0.9983	0.9837	0.9324	0.8251	0.6655	0.4812	0.3081	0.1727	0.0835	0.0342	0.0116	0.0031	0.0006	0.0001	0.0000	0.0000	0.0000	0.0000
	7	1.0000	0.9997	0.9959	0.9767	0.9225	0.8180	0.6656	0.4878	0.3169	0.1796	0.0871	0.0352	0.0114	0.0028	0.0005	0.0000	0.0000	0.0000	0.0000
	8	1.0000	1.0000	0.9992	0.9933	0.9713	0.9161	0.8145	0.6675	0.4940	0.3238	0.1841	0.0885	0.0347	0.0105	0.0023	0.0003	0.0000	0.0000	0.0000
	9	1.0000	1.0000	0.9999	0.9984	0.9911	0.9674	0.9125	0.8139	0.6710	0.5000	0.3290	0.1861	0.0875	0.0326	0.0089	0.0016	0.0001	0.0000	0.0000
	10	1.0000	1.0000	1.0000	0.9997	0.9977	0.9895	0.9653	0.9115	0.8159	0.6762	0.5060	0.3325	0.1855	0.0839	0.0287	0.0067	0.0008	0.0000	0.0000
	11	1.0000	1.0000	1.0000	1.0000	0.9995	0.9972	0.9886	0.9648	0.9129	0.8204	0.6831	0.5122	0.3344	0.1820	0.0775	0.0233	0.0041	0.0003	0.0000
	12	1.0000	1.0000	1.0000	1.0000	0.9999	0.9994	0.9969	0.9884	0.9658	0.9165	0.8273	0.6919	0.5188	0.3345	0.1749	0.0676	0.0163	0.0017	0.0000
	13	1.0000	1.0000	1.0000	1.0000	1.0000	0.9999	0.9993	0.9969	0.9891	0.9682	0.9223	0.8371	0.7032	0.5261	0.3322	0.1631	0.0537	0.0086	0.0002
	14	1.0000	1.0000	1.0000	1.0000	1.0000	1.0000	0.9999	0.9994	0.9972	0.9904	0.9720	0.9304	0.8500	0.7178	0.5346	0.3267	0.1444	0.0352	0.0020
	15	1.0000	1.0000	1.0000	1.0000	1.0000	1.0000	1.0000	0.9999	0.9995	0.9978	0.9923	0.9770	0.9409	0.8668	0.7369	0.5449	0.3159	0.1150	0.0132
	16	1.0000	1.0000	1.0000	1.0000	1.0000	1.0000	1.0000	1.0000	0.9999	0.9996	0.9985	0.9945	0.9830	0.9538	0.8887	0.7631	0.5587	0.2946	0.0665
	17	1.0000	1.0000	1.0000	1.0000	1.0000	1.0000	1.0000	1.0000	1.0000	1.0000	0.9998	0.9992	0.9969	0.9896	0.9690	0.9171	0.8015	0.5797	0.2453
	18	1.0000	1.0000	1.0000	1.0000	1.0000	1.0000	1.0000	1.0000	1.0000	1.0000	1.0000	0.9999	0.9997	0.9989	0.9958	0.9856	0.9544	0.8649	0.6226
20	0	0.3585	0.1216	0.0388	0.0115	0.0032	0.0008	0.0002	0.0000	0.0000	0.0000	0.0000	0.0000	0.0000	0.0000	0.0000	0.0000	0.0000	0.0000	0.0000
	1	0.7358	0.3917	0.1756	0.0692	0.0243	0.0076	0.0021	0.0005	0.0001	0.0000	0.0000	0.0000	0.0000	0.0000	0.0000	0.0000	0.0000	0.0000	0.0000
	2	0.9245	0.6769	0.4049	0.2061	0.0913	0.0355	0.0121	0.0036	0.0009	0.0002	0.0000	0.0000	0.0000	0.0000	0.0000	0.0000	0.0000	0.0000	0.0000
	3	0.9841	0.8670	0.6477	0.4114	0.2252	0.1071	0.0444	0.0160	0.0049	0.0013	0.0003	0.0000	0.0000	0.0000	0.0000	0.0000	0.0000	0.0000	0.0000
	4	0.9974	0.9568	0.8298	0.6296	0.4148	0.2375	0.1182	0.0510	0.0189	0.0059	0.0015	0.0003	0.0000	0.0000	0.0000	0.0000	0.0000	0.0000	0.0000
	5	0.9997	0.9887	0.9327	0.8042	0.6172	0.4164	0.2454	0.1256	0.0553	0.0207	0.0064	0.0016	0.0003	0.0000	0.0000	0.0000	0.0000	0.0000	0.0000
	6	1.0000	0.9976	0.9781	0.9133	0.7858	0.6080	0.4166	0.2500	0.1299	0.0577	0.0214	0.0065	0.0015	0.0003	0.0000	0.0000	0.0000	0.0000	0.0000
	7	1.0000	0.9996	0.9941	0.9679	0.8982	0.7723	0.6010	0.4159	0.2520	0.1316	0.0580	0.0210	0.0060	0.0013	0.0002	0.0000	0.0000	0.0000	0.0000
	8	1.0000	0.9999	0.9987	0.9900	0.9591	0.8867	0.7624	0.5956	0.4143	0.2517	0.1308	0.0565	0.0196	0.0051	0.0009	0.0001	0.0000	0.0000	0.0000
	9	1.0000	1.0000	0.9998	0.9974	0.9861	0.9520	0.8782	0.7553	0.5914	0.4119	0.2493	0.1275	0.0532	0.0171	0.0039	0.0006	0.0000	0.0000	0.0000
	10	1.0000	1.0000	1.0000	0.9994	0.9961	0.9829	0.9468	0.8725	0.7507	0.5881	0.4086	0.2447	0.1218	0.0480	0.0139	0.0026	0.0002	0.0000	0.0000
	11	1.0000	1.0000	1.0000	0.9999	0.9991	0.9949	0.9804	0.9435	0.8692	0.7483	0.5857	0.4044	0.2376	0.1133	0.0409	0.0100	0.0013	0.0001	0.0000
	12	1.0000	1.0000	1.0000	1.0000	0.9998	0.9987	0.9940	0.9790	0.9420	0.8684	0.7480	0.5841	0.3990	0.2277	0.1018	0.0321	0.0059	0.0004	0.0000
	13	1.0000	1.0000	1.0000	1.0000	1.0000	0.9997	0.9985	0.9935	0.9786	0.9423	0.8701	0.7500	0.5834	0.3920	0.2142	0.0867	0.0219	0.0024	0.0000
	14	1.0000	1.0000	1.0000	1.0000	1.0000	1.0000	0.9997	0.9984	0.9936	0.9793	0.9447	0.8744	0.7546	0.5836	0.3828	0.1958	0.0673	0.0113	0.0003
	15	1.0000	1.0000	1.0000	1.0000	1.0000	1.0000	1.0000	0.9997	0.9985	0.9941	0.9811	0.9490	0.8818	0.7625	0.5852	0.3704	0.1702	0.0432	0.0026
	16	1.0000	1.0000	1.0000	1.0000	1.0000	1.0000	1.0000	1.0000	0.9997	0.9987	0.9951	0.9840	0.9556	0.8929	0.7748	0.5886	0.3523	0.1330	0.0159
	17	1.0000	1.0000	1.0000	1.0000	1.0000	1.0000	1.0000	1.0000	1.0000	0.9998	0.9991	0.9964	0.9879	0.9645	0.9087	0.7939	0.5951	0.3231	0.0755
	18	1.0000	1.0000	1.0000	1.0000	1.0000	1.0000	1.0000	1.0000	1.0000	1.0000	0.9999	0.9995	0.9979	0.9924	0.9757	0.9308	0.8244	0.6083	0.2642
	19	1.0000	1.0000	1.0000	1.0000	1.0000	1.0000	1.0000	1.0000	1.0000	1.0000	1.0000	1.0000	0.9998	0.9992	0.9968	0.9885	0.9612	0.8784	0.6415

Table 2 *Poisson Distribution Function**

$$F(x; \lambda) = \sum_{k=0}^{x} e^{-\lambda} \frac{\lambda^k}{k!}$$

λ \ x	0	1	2	3	4	5	6	7	8	9
0.02	0.980	1.000								
0.04	0.961	0.999	1.000							
0.06	0.942	0.998	1.000							
0.08	0.923	0.997	1.000							
0.10	0.905	0.995	1.000							
0.15	0.861	0.990	0.999	1.000						
0.20	0.819	0.982	0.999	1.000						
0.25	0.779	0.974	0.998	1.000						
0.30	0.741	0.963	0.996	1.000						
0.35	0.705	0.951	0.994	1.000						
0.40	0.670	0.938	0.992	0.999	1.000					
0.45	0.638	0.925	0.989	0.999	1.000					
0.50	0.607	0.910	0.986	0.998	1.000					
0.55	0.577	0.894	0.982	0.998	1.000					
0.60	0.549	0.878	0.977	0.997	1.000					
0.65	0.522	0.861	0.972	0.996	0.999	1.000				
0.70	0.497	0.844	0.966	0.994	0.999	1.000				
0.75	0.472	0.827	0.959	0.993	0.999	1.000				
0.80	0.449	0.809	0.953	0.991	0.999	1.000				
0.85	0.427	0.791	0.945	0.989	0.998	1.000				
0.90	0.407	0.772	0.937	0.987	0.998	1.000				
0.95	0.387	0.754	0.929	0.984	0.997	1.000				
1.00	0.368	0.736	0.920	0.981	0.996	0.999	1.000			
1.1	0.333	0.699	0.900	0.974	0.995	0.999	1.000			
1.2	0.301	0.663	0.879	0.966	0.992	0.998	1.000			
1.3	0.273	0.627	0.857	0.957	0.989	0.998	1.000			
1.4	0.247	0.592	0.833	0.946	0.986	0.997	0.999	1.000		
1.5	0.223	0.558	0.809	0.934	0.981	0.996	0.999	1.000		
1.6	0.202	0.525	0.783	0.921	0.976	0.994	0.999	1.000		
1.7	0.183	0.493	0.757	0.907	0.970	0.992	0.998	1.000		
1.8	0.165	0.463	0.731	0.891	0.964	0.990	0.997	0.999	1.000	
1.9	0.150	0.434	0.704	0.875	0.956	0.987	0.997	0.999	1.000	
2.0	0.135	0.406	0.677	0.857	0.947	0.983	0.995	0.999	1.000	

* *Reprinted by kind permission from E. C. Molina, Poisson's Exponential Binomial Limit,* D. Van Nostrand Company, Inc., Princeton, N.J., 1947.

Table 2 *Poisson Distribution Function (Continued)*

λ \ x	0	1	2	3	4	5	6	7
2.2	0.111	0.355	0.623	0.819	0.928	0.975	0.993	0.998
2.4	0.091	0.308	0.570	0.779	0.904	0.964	0.988	0.997
2.6	0.074	0.267	0.518	0.736	0.877	0.951	0.983	0.995
2.8	0.061	0.231	0.469	0.692	0.848	0.935	0.976	0.992
3.0	0.050	0.199	0.423	0.647	0.815	0.916	0.966	0.988
3.2	0.041	0.171	0.380	0.603	0.781	0.895	0.955	0.983
3.4	0.033	0.147	0.340	0.558	0.744	0.871	0.942	0.977
3.6	0.027	0.126	0.303	0.515	0.706	0.844	0.927	0.969
3.8	0.022	0.107	0.269	0.473	0.668	0.816	0.909	0.960
4.0	0.018	0.092	0.238	0.433	0.629	0.785	0.889	0.949
4.2	0.015	0.078	0.210	0.395	0.590	0.753	0.867	0.936
4.4	0.012	0.066	0.185	0.359	0.551	0.720	0.844	0.921
4.6	0.010	0.056	0.163	0.326	0.513	0.686	0.818	0.905
4.8	0.008	0.048	0.143	0.294	0.476	0.651	0.791	0.887
5.0	0.007	0.040	0.125	0.265	0.440	0.616	0.762	0.867
5.2	0.006	0.034	0.109	0.238	0.406	0.581	0.732	0.845
5.4	0.005	0.029	0.095	0.213	0.373	0.546	0.702	0.822
5.6	0.004	0.024	0.082	0.191	0.342	0.512	0.670	0.797
5.8	0.003	0.021	0.072	0.170	0.313	0.478	0.638	0.771
6.0	0.002	0.017	0.062	0.151	0.285	0.446	0.606	0.744

λ \ x	10	11	12	13	14	15	16
2.8	1.000						
3.0	1.000						
3.2	1.000						
3.4	0.999	1.000					
3.6	0.999	1.000					
3.8	0.998	0.999	1.000				
4.0	0.997	0.999	1.000				
4.2	0.996	0.999	1.000				
4.4	0.994	0.998	0.999	1.000			
4.6	0.992	0.997	0.999	1.000			
4.8	0.990	0.996	0.999	1.000			
5.0	0.986	0.995	0.998	0.999	1.000		
5.2	0.982	0.993	0.997	0.999	1.000		
5.4	0.977	0.990	0.996	0.999	1.000		
5.6	0.972	0.988	0.995	0.998	0.999	1.000	
5.8	0.965	0.984	0.993	0.997	0.999	1.000	
6.0	0.957	0.980	0.991	0.996	0.999	0.999	1.000

Table 2 Poisson Distribution Function (Continued)

λ \ x	0	1	2	3	4	5	6	7	8	9
6.2	0.002	0.015	0.054	0.134	0.259	0.414	0.574	0.716	0.826	0.902
6.4	0.002	0.012	0.046	0.119	0.235	0.384	0.542	0.687	0.803	0.886
6.6	0.001	0.010	0.040	0.105	0.213	0.355	0.511	0.658	0.780	0.869
6.8	0.001	0.009	0.034	0.093	0.192	0.327	0.480	0.628	0.755	0.850
7.0	0.001	0.007	0.030	0.082	0.173	0.301	0.450	0.599	0.729	0.830
7.2	0.001	0.006	0.025	0.072	0.156	0.276	0.420	0.569	0.703	0.810
7.4	0.001	0.005	0.022	0.063	0.140	0.253	0.392	0.539	0.676	0.788
7.6	0.001	0.004	0.019	0.055	0.125	0.231	0.365	0.510	0.648	0.765
7.8	0.000	0.004	0.016	0.048	0.112	0.210	0.338	0.481	0.620	0.741
8.0	0.000	0.003	0.014	0.042	0.100	0.191	0.313	0.453	0.593	0.717
8.5	0.000	0.002	0.009	0.030	0.074	0.150	0.256	0.386	0.523	0.653
9.0	0.000	0.001	0.006	0.021	0.055	0.116	0.207	0.324	0.456	0.587
9.5	0.000	0.001	0.004	0.015	0.040	0.089	0.165	0.269	0.392	0.522
10.0	0.000	0.000	0.003	0.010	0.029	0.067	0.130	0.220	0.333	0.458

λ \ x	10	11	12	13	14	15	16	17	18	19
6.2	0.949	0.975	0.989	0.995	0.998	0.999	1.000			
6.4	0.939	0.969	0.986	0.994	0.997	0.999	1.000			
6.6	0.927	0.963	0.982	0.992	0.997	0.999	0.999	1.000		
6.8	0.915	0.955	0.978	0.990	0.996	0.998	0.999	1.000		
7.0	0.901	0.947	0.973	0.987	0.994	0.998	0.999	1.000		
7.2	0.887	0.937	0.967	0.984	0.993	0.997	0.999	0.999	1.000	
7.4	0.871	0.926	0.961	0.980	0.991	0.996	0.998	0.999	1.000	
7.6	0.854	0.915	0.954	0.976	0.989	0.995	0.998	0.999	1.000	
7.8	0.835	0.902	0.945	0.971	0.986	0.993	0.997	0.999	1.000	
8.0	0.816	0.888	0.936	0.966	0.983	0.992	0.996	0.998	0.999	1.000
8.5	0.763	0.849	0.909	0.949	0.973	0.986	0.993	0.997	0.999	0.999
9.0	0.706	0.803	0.876	0.926	0.959	0.978	0.989	0.995	0.998	0.999
9.5	0.645	0.752	0.836	0.898	0.940	0.967	0.982	0.991	0.996	0.998
10.0	0.583	0.697	0.792	0.864	0.917	0.951	0.973	0.986	0.993	0.997

λ \ x	20	21	22
8.5	1.000		
9.0	1.000		
9.5	0.999	1.000	
10.0	0.998	0.999	1.000

Table 2 *Poisson Distribution Function (Continued)*

λ \ x	0	1	2	3	4	5	6	7	8	9
10.5	0.000	0.000	0.002	0.007	0.021	0.050	0.102	0.179	0.279	0.397
11.0	0.000	0.000	0.001	0.005	0.015	0.038	0.079	0.143	0.232	0.341
11.5	0.000	0.000	0.001	0.003	0.011	0.028	0.060	0.114	0.191	0.289
12.0	0.000	0.000	0.001	0.002	0.008	0.020	0.046	0.090	0.155	0.242
12.5	0.000	0.000	0.000	0.002	0.005	0.015	0.035	0.070	0.125	0.201
13.0	0.000	0.000	0.000	0.001	0.004	0.011	0.026	0.054	0.100	0.166
13.5	0.000	0.000	0.000	0.001	0.003	0.008	0.019	0.041	0.079	0.135
14.0	0.000	0.000	0.000	0.000	0.002	0.006	0.014	0.032	0.062	0.109
14.5	0.000	0.000	0.000	0.000	0.001	0.004	0.010	0.024	0.048	0.088
15.0	0.000	0.000	0.000	0.000	0.001	0.003	0.008	0.018	0.037	0.070
	10	11	12	13	14	15	16	17	18	19
10.5	0.521	0.639	0.742	0.825	0.888	0.932	0.960	0.978	0.988	0.994
11.0	0.460	0.579	0.689	0.781	0.854	0.907	0.944	0.968	0.982	0.991
11.5	0.402	0.520	0.633	0.733	0.815	0.878	0.924	0.954	0.974	0.986
12.0	0.347	0.462	0.576	0.682	0.772	0.844	0.899	0.937	0.963	0.979
12.5	0.297	0.406	0.519	0.628	0.725	0.806	0.869	0.916	0.948	0.969
13.0	0.252	0.353	0.463	0.573	0.675	0.764	0.835	0.890	0.930	0.957
13.5	0.211	0.304	0.409	0.518	0.623	0.718	0.798	0.861	0.908	0.942
14.0	0.176	0.260	0.358	0.464	0.570	0.669	0.756	0.827	0.883	0.923
14.5	0.145	0.220	0.311	0.413	0.518	0.619	0.711	0.790	0.853	0.901
15.0	0.118	0.185	0.268	0.363	0.466	0.568	0.664	0.749	0.819	0.875
	20	21	22	23	24	25	26	27	28	29
10.5	0.997	0.999	0.999	1.000						
11.0	0.995	0.998	0.999	1.000						
11.5	0.992	0.996	0.998	0.999	1.000					
12.0	0.988	0.994	0.997	0.999	0.999	1.000				
12.5	0.983	0.991	0.995	0.998	0.999	0.999	1.000			
13.0	0.975	0.986	0.992	0.996	0.998	0.999	1.000			
13.5	0.965	0.980	0.989	0.994	0.997	0.998	0.999	1.000		
14.0	0.952	0.971	0.983	0.991	0.995	0.997	0.999	0.999	1.000	
14.5	0.936	0.960	0.976	0.986	0.992	0.996	0.998	0.999	0.999	1.000
15.0	0.917	0.947	0.967	0.981	0.989	0.994	0.997	0.998	0.999	1.000

Table 2 Poisson Distribution Function (Continued)

λ \ x	4	5	6	7	8	9	10	11	12	13
16	0.000	0.001	0.004	0.010	0.022	0.043	0.077	0.127	0.193	0.275
17	0.000	0.001	0.002	0.005	0.013	0.026	0.049	0.085	0.135	0.201
18	0.000	0.000	0.001	0.003	0.007	0.015	0.030	0.055	0.092	0.143
19	0.000	0.000	0.001	0.002	0.004	0.009	0.018	0.035	0.061	0.098
20	0.000	0.000	0.000	0.001	0.002	0.005	0.011	0.021	0.039	0.066
21	0.000	0.000	0.000	0.000	0.001	0.003	0.006	0.013	0.025	0.043
22	0.000	0.000	0.000	0.000	0.001	0.002	0.004	0.008	0.015	0.028
23	0.000	0.000	0.000	0.000	0.000	0.001	0.002	0.004	0.009	0.017
24	0.000	0.000	0.000	0.000	0.000	0.000	0.001	0.003	0.005	0.011
25	0.000	0.000	0.000	0.000	0.000	0.000	0.001	0.001	0.003	0.006

λ \ x	14	15	16	17	18	19	20	21	22	23
16	0.368	0.467	0.566	0.659	0.742	0.812	0.868	0.911	0.942	0.963
17	0.281	0.371	0.468	0.564	0.655	0.736	0.805	0.861	0.905	0.937
18	0.208	0.287	0.375	0.469	0.562	0.651	0.731	0.799	0.855	0.899
19	0.150	0.215	0.292	0.378	0.469	0.561	0.647	0.725	0.793	0.849
20	0.105	0.157	0.221	0.297	0.381	0.470	0.559	0.644	0.721	0.787
21	0.072	0.111	0.163	0.227	0.302	0.384	0.471	0.558	0.640	0.716
22	0.048	0.077	0.117	0.169	0.232	0.306	0.387	0.472	0.556	0.637
23	0.031	0.052	0.082	0.123	0.175	0.238	0.310	0.389	0.472	0.555
24	0.020	0.034	0.056	0.087	0.128	0.180	0.243	0.314	0.392	0.473
25	0.012	0.022	0.038	0.060	0.092	0.134	0.185	0.247	0.318	0.394

λ \ x	24	25	26	27	28	29	30	31	32	33
16	0.978	0.987	0.993	0.996	0.998	0.999	0.999	1.000		
17	0.959	0.975	0.985	0.991	0.995	0.997	0.999	0.999	1.000	
18	0.932	0.955	0.972	0.983	0.990	0.994	0.997	0.998	0.999	1.000
19	0.893	0.927	0.951	0.969	0.980	0.988	0.993	0.996	0.998	0.999
20	0.843	0.888	0.922	0.948	0.966	0.978	0.987	0.992	0.995	0.997
21	0.782	0.838	0.883	0.917	0.944	0.963	0.976	0.985	0.991	0.994
22	0.712	0.777	0.832	0.877	0.913	0.940	0.959	0.973	0.983	0.989
23	0.635	0.708	0.772	0.827	0.873	0.908	0.936	0.956	0.971	0.981
24	0.554	0.632	0.704	0.768	0.823	0.868	0.904	0.932	0.953	0.969
25	0.473	0.553	0.629	0.700	0.763	0.818	0.863	0.900	0.929	0.950

λ \ x	34	35	36	37	38	39	40	41	42	43
19	0.999	1.000								
20	0.999	0.999	1.000							
21	0.997	0.998	0.999	0.999	1.000					
22	0.994	0.996	0.998	0.999	0.999	1.000				
23	0.998	0.993	0.996	0.997	0.999	0.999	1.000			
24	0.979	0.987	0.992	0.995	0.997	0.998	0.999	0.999		
25	0.966	0.978	0.985	0.991	0.994	0.997	0.998	0.999	1.000	

Table 3 *Normal Distribution Function*

$$F(z) = \frac{1}{\sqrt{2\pi}} \int_{-\infty}^{z} e^{-t^2/2}\, dt$$

z	0.00	0.01	0.02	0.03	0.04	0.05	0.06	0.07	0.08	0.09
0.0	0.5000	0.5040	0.5080	0.5120	0.5160	0.5199	0.5239	0.5279	0.5319	0.5359
0.1	0.5398	0.5438	0.5478	0.5517	0.5557	0.5596	0.5636	0.5675	0.5714	0.5753
0.2	0.5793	0.5832	0.5871	0.5910	0.5948	0.5987	0.6026	0.6064	0.6103	0.6141
0.3	0.6179	0.6217	0.6255	0.6293	0.6331	0.6368	0.6406	0.6443	0.6480	0.6517
0.4	0.6554	0.6591	0.6628	0.6664	0.6700	0.6736	0.6772	0.6808	0.6844	0.6879
0.5	0.6915	0.6950	0.6985	0.7019	0.7054	0.7088	0.7123	0.7157	0.7190	0.7224
0.6	0.7257	0.7291	0.7324	0.7357	0.7389	0.7422	0.7454	0.7486	0.7517	0.7549
0.7	0.7580	0.7611	0.7642	0.7673	0.7704	0.7734	0.7764	0.7794	0.7823	0.7852
0.8	0.7881	0.7910	0.7939	0.7967	0.7995	0.8023	0.8051	0.8078	0.8106	0.8133
0.9	0.8159	0.8186	0.8212	0.8238	0.8264	0.8289	0.8315	0.8340	0.8365	0.8389
1.0	0.8413	0.8438	0.8461	0.8485	0.8508	0.8531	0.8554	0.8577	0.8599	0.8621
1.1	0.8643	0.8665	0.8686	0.8708	0.8729	0.8749	0.8770	0.8790	0.8810	0.8830
1.2	0.8849	0.8869	0.8888	0.8907	0.8925	0.8944	0.8962	0.8980	0.8997	0.9015
1.3	0.9032	0.9049	0.9066	0.9082	0.9099	0.9115	0.9131	0.9147	0.9162	0.9177
1.4	0.9192	0.9207	0.9222	0.9236	0.9251	0.9265	0.9279	0.9292	0.9306	0.9319
1.5	0.9332	0.9345	0.9357	0.9370	0.9382	0.9394	0.9406	0.9418	0.9429	0.9441
1.6	0.9452	0.9463	0.9474	0.9484	0.9495	0.9505	0.9515	0.9525	0.9535	0.9545
1.7	0.9554	0.9564	0.9573	0.9582	0.9591	0.9599	0.9608	0.9616	0.9625	0.9633
1.8	0.9641	0.9649	0.9656	0.9664	0.9671	0.9678	0.9686	0.9693	0.9699	0.9706
1.9	0.9713	0.9719	0.9726	0.9732	0.9738	0.9744	0.9750	0.9756	0.9761	0.9767
2.0	0.9772	0.9778	0.9783	0.9788	0.9793	0.9798	0.9803	0.9808	0.9812	0.9817
2.1	0.9821	0.9826	0.9830	0.9834	0.9838	0.9842	0.9846	0.9850	0.9854	0.9857
2.2	0.9861	0.9864	0.9868	0.9871	0.9875	0.9878	0.9881	0.9884	0.9887	0.9890
2.3	0.9893	0.9896	0.9898	0.9901	0.9904	0.9906	0.9909	0.9911	0.9913	0.9916
2.4	0.9918	0.9920	0.9922	0.9925	0.9927	0.9929	0.9931	0.9932	0.9934	0.9936
2.5	0.9938	0.9940	0.9941	0.9943	0.9945	0.9946	0.9948	0.9949	0.9951	0.9952
2.6	0.9953	0.9955	0.9956	0.9957	0.9959	0.9960	0.9961	0.9962	0.9963	0.9964
2.7	0.9965	0.9966	0.9967	0.9968	0.9969	0.9970	0.9971	0.9972	0.9973	0.9974
2.8	0.9974	0.9975	0.9976	0.9977	0.9977	0.9978	0.9979	0.9979	0.9980	0.9981
2.9	0.9981	0.9982	0.9982	0.9983	0.9984	0.9984	0.9985	0.9985	0.9986	0.9986
3.0	0.9987	0.9987	0.9987	0.9988	0.9988	0.9989	0.9989	0.9989	0.9990	0.9990
3.1	0.9990	0.9991	0.9991	0.9991	0.9992	0.9992	0.9992	0.9992	0.9993	0.9993
3.2	0.9993	0.9993	0.9994	0.9994	0.9994	0.9994	0.9994	0.9995	0.9995	0.9995
3.3	0.9995	0.9995	0.9995	0.9996	0.9996	0.9996	0.9996	0.9996	0.9996	0.9997
3.4	0.9997	0.9997	0.9997	0.9997	0.9997	0.9997	0.9997	0.9997	0.9997	0.9998
3.5	0.9998									
4.0	0.99997									
5.0	0.9999997									
6.0	0.999999999									

Table 4 Values of t_α*

ν	$\alpha = 0.10$	$\alpha = 0.05$	$\alpha = 0.025$	$\alpha = 0.01$	$\alpha = 0.005$	ν
1	3.078	6.314	12.706	31.821	63.657	1
2	1.886	2.920	4.303	6.965	9.925	2
3	1.638	2.353	3.182	4.541	5.841	3
4	1.533	2.132	2.776	3.747	4.604	4
5	1.476	2.015	2.571	3.365	4.032	5
6	1.440	1.943	2.447	3.143	3.707	6
7	1.415	1.895	2.365	2.998	3.499	7
8	1.397	1.860	2.306	2.896	3.355	8
9	1.383	1.833	2.262	2.821	3.250	9
10	1.372	1.812	2.228	2.764	3.169	10
11	1.363	1.796	2.201	2.718	3.106	11
12	1.356	1.782	2.179	2.681	3.055	12
13	1.350	1.771	2.160	2.650	3.012	13
14	1.345	1.761	2.145	2.624	2.977	14
15	1.341	1.753	2.131	2.602	2.947	15
16	1.337	1.746	2.120	2.583	2.921	16
17	1.333	1.740	2.110	2.567	2.898	17
18	1.330	1.734	2.101	2.552	2.878	18
19	1.328	1.729	2.093	2.539	2.861	19
20	1.325	1.725	2.086	2.528	2.845	20
21	1.323	1.721	2.080	2.518	2.831	21
22	1.321	1.717	2.074	2.508	2.819	22
23	1.319	1.714	2.069	2.500	2.807	23
24	1.318	1.711	2.064	2.492	2.797	24
25	1.316	1.708	2.060	2.485	2.787	25
26	1.315	1.706	2.056	2.479	2.779	26
27	1.314	1.703	2.052	2.473	2.771	27
28	1.313	1.701	2.048	2.467	2.763	28
29	1.311	1.699	2.045	2.462	2.756	29
inf.	1.282	1.645	1.960	2.326	2.576	inf.

Table 5 Values of χ^2_α *

ν	$\alpha = 0.995$	$\alpha = 0.99$	$\alpha = 0.975$	$\alpha = 0.95$	$\alpha = 0.05$	$\alpha = 0.025$	$\alpha = 0.01$	$\alpha = 0.005$	ν
1	0.0000393	0.000157	0.000982	0.00393	3.841	5.024	6.635	7.879	1
2	0.0100	0.0201	0.0506	0.103	5.991	7.378	9.210	10.597	2
3	0.0717	0.115	0.216	0.352	7.815	9.348	11.345	12.838	3
4	0.207	0.297	0.484	0.711	9.488	11.143	13.277	14.860	4
5	0.412	0.554	0.831	1.145	11.070	12.832	15.056	16.750	5
6	0.676	0.872	1.237	1.635	12.592	14.449	16.812	18.548	6
7	0.989	1.239	1.690	2.167	14.067	16.013	18.475	20.278	7
8	1.344	1.646	2.180	2.733	15.507	17.535	20.090	21.955	8
9	1.735	2.088	2.700	3.325	16.919	19.023	21.666	23.589	9
10	2.156	2.558	3.247	3.940	18.307	20.483	23.209	25.188	10
11	2.603	3.053	3.816	4.575	19.675	21.920	24.725	26.757	11
12	3.074	3.571	4.404	5.226	21.026	23.337	26.217	28.300	12
13	3.565	4.107	5.009	5.892	22.362	24.736	27.688	29.819	13
14	4.075	4.660	5.629	6.571	23.685	26.119	29.141	31.319	14
15	4.601	5.229	6.262	7.261	24.996	27.488	30.578	32.801	15
16	5.142	5.812	6.908	7.962	26.296	28.845	32.000	34.267	16
17	5.697	6.408	7.564	8.672	27.587	30.191	33.409	35.718	17
18	6.265	7.015	8.231	9.390	28.869	31.526	34.805	37.156	18
19	6.844	7.633	8.907	10.117	30.144	32.852	36.191	38.582	19
20	7.434	8.260	9.591	10.851	31.410	34.170	37.566	39.997	20
21	8.034	8.897	10.283	11.591	32.671	35.479	38.932	41.401	21
22	8.643	9.542	10.982	12.338	33.924	36.781	40.289	42.796	22
23	9.260	10.196	11.689	13.091	35.172	38.076	41.638	44.181	23
24	9.886	10.856	12.401	13.484	36.415	39.364	42.980	45.558	24
25	10.520	11.524	13.120	14.611	37.652	40.646	44.314	46.928	25
26	11.160	12.198	13.844	15.379	38.885	41.923	45.642	48.290	26
27	11.808	12.879	14.573	16.151	40.113	43.194	46.963	49.645	27
28	12.461	13.565	15.308	16.928	41.337	44.461	48.278	50.993	28
29	13.121	14.256	16.047	17.708	42.557	45.772	49.588	52.336	29
30	13.787	14.953	16.791	18.493	43.773	46.979	50.892	53.672	30
40	20.706	22.164	24.433	26.509	55.758	59.342	63.691	66.766	40
50	27.991	29.707	32.357	34.764	67.505	71.420	76.154	79.490	50
60	35.535	37.485	40.482	43.118	79.082	83.298	88.379	91.952	60
70	43.275	45.442	48.758	51.739	90.531	95.023	100.425	104.215	70
80	51.172	53.540	57.153	60.391	101.879	106.629	112.329	116.321	80
90	59.196	61.754	65.646	69.126	113.145	118.136	124.116	128.299	90
100	67.328	70.065	74.222	77.929	124.342	129.561	135.807	140.169	100

* This table is based on Table 8 of *Biometrika Tables for Statisticians*, Vol. 1, by permission of the *Biometrika* trustees.

Table 6(*a*) Values of $F_{0.05}$*

ν_2 = Degrees of freedom for denominator	ν_1 = Degrees of freedom for numerator																		
	1	2	3	4	5	6	7	8	9	10	12	15	20	24	30	40	60	120	∞
1	161	200	216	225	230	234	237	239	241	242	244	246	248	249	250	251	252	253	254
2	18.50	19.00	19.20	19.20	19.30	19.30	19.40	19.40	19.40	19.40	19.40	19.40	19.40	19.50	19.50	19.50	19.50	19.50	19.50
3	10.10	9.55	9.28	9.12	9.01	8.94	8.89	8.85	8.81	8.79	8.74	8.70	8.66	8.64	8.62	8.59	8.57	8.55	8.53
4	7.71	6.94	6.59	6.39	6.26	6.16	6.09	6.04	6.00	5.96	5.91	5.86	5.80	5.77	5.75	5.72	5.69	5.66	5.63
5	6.61	5.79	5.41	5.19	5.05	4.95	4.88	4.82	4.77	4.74	4.68	4.62	4.56	4.53	4.50	4.46	4.43	4.40	4.37
6	5.99	5.14	4.76	4.53	4.39	4.28	4.21	4.15	4.10	4.06	4.00	3.94	3.87	3.84	3.81	3.77	3.74	3.70	3.67
7	5.59	4.74	4.35	4.12	3.97	3.87	3.79	3.73	3.68	3.64	3.57	3.51	3.44	3.41	3.38	3.34	3.30	3.27	3.23
8	5.32	4.46	4.07	3.84	3.69	3.58	3.50	3.44	3.39	3.35	3.28	3.22	3.15	3.12	3.08	3.04	3.01	2.97	2.93
9	5.12	4.26	3.86	3.63	3.48	3.37	3.29	3.23	3.18	3.14	3.07	3.01	2.94	2.90	2.86	2.83	2.79	2.75	2.71
10	4.96	4.10	3.71	3.48	3.33	3.22	3.14	3.07	3.02	2.98	2.91	2.85	2.77	2.74	2.70	2.66	2.62	2.58	2.54
11	4.84	3.98	3.59	3.36	3.20	3.09	3.01	2.95	2.90	2.85	2.79	2.72	2.65	2.61	2.57	2.53	2.49	2.45	2.40
12	4.75	3.89	3.49	3.26	3.11	3.00	2.91	2.85	2.80	2.75	2.69	2.62	2.54	2.51	2.47	2.38	2.38	2.30	2.30
13	4.67	3.81	3.41	3.18	3.03	2.92	2.83	2.77	2.71	2.67	2.60	2.53	2.46	2.42	2.38	2.34	2.30	2.25	2.21
14	4.60	3.74	3.34	3.11	2.96	2.85	2.76	2.70	2.65	2.60	2.53	2.46	2.39	2.35	2.31	2.27	2.22	2.18	2.13
15	4.54	3.68	3.29	3.06	2.90	2.79	2.71	2.64	2.59	2.54	2.48	2.40	2.33	2.29	2.25	2.20	2.16	2.11	2.07
16	4.49	3.63	3.24	3.01	2.85	2.74	2.66	2.59	2.54	2.49	2.42	2.35	2.28	2.24	2.19	2.15	2.11	2.06	2.01
17	3.45	3.59	3.20	2.96	2.81	2.70	2.61	2.55	2.49	2.45	2.38	2.31	2.23	2.19	2.15	2.10	2.06	2.01	1.96
18	4.41	3.55	3.16	2.93	2.77	2.66	2.58	2.51	2.46	2.41	2.34	2.27	2.19	2.15	2.11	2.06	2.02	1.97	1.93
19	4.38	3.52	3.13	2.90	2.74	2.63	2.54	2.48	2.42	2.38	2.31	2.23	2.16	2.11	2.07	2.03	1.98	1.93	1.88
20	4.35	3.49	3.10	2.87	2.71	2.60	2.51	2.45	2.39	2.35	2.28	2.20	2.12	2.08	2.04	1.99	1.95	1.90	1.84
21	4.32	3.47	3.07	2.84	2.68	2.57	2.49	2.42	2.37	2.32	2.25	2.18	2.10	2.05	2.01	1.96	1.92	1.87	1.81
22	4.30	3.44	3.05	2.82	2.66	2.55	2.46	2.40	2.34	2.30	2.23	2.15	2.07	2.03	1.98	1.94	1.89	1.84	1.78
23	4.28	3.42	3.03	2.80	2.64	2.53	2.44	2.37	2.32	2.27	2.20	2.13	2.05	2.01	1.96	1.91	1.86	1.81	1.76
24	4.26	3.40	3.01	2.78	2.62	2.51	2.42	2.36	2.30	2.25	2.18	2.11	2.03	1.98	1.94	1.89	1.84	1.79	1.73
25	4.24	3.39	2.99	2.76	2.60	2.49	2.40	2.34	2.28	2.24	2.16	2.09	2.01	1.96	1.92	1.87	1.82	1.77	1.71
30	4.17	3.32	2.92	2.69	2.53	2.42	2.33	2.27	2.21	2.16	2.09	2.01	1.93	1.89	1.84	1.79	1.74	1.68	1.62
40	4.08	3.23	2.84	2.61	2.45	2.34	2.25	2.18	2.12	2.08	2.00	1.92	1.84	1.79	1.74	1.69	1.64	1.58	1.51
60	4.00	3.15	2.76	2.53	2.37	2.25	2.17	2.10	2.04	1.99	1.92	1.84	1.75	1.70	1.65	1.59	1.53	1.47	1.39
120	3.92	3.07	2.68	2.45	2.29	2.18	2.09	2.02	1.96	1.91	1.83	1.75	1.66	1.61	1.55	1.50	1.43	1.35	1.25
∞	3.84	3.00	2.60	2.37	2.21	2.10	2.01	1.94	1.88	1.83	1.75	1.67	1.57	1.52	1.46	1.39	1.32	1.22	1.00

* This table is reproduced from M. Merrington and C. M. Thompson, "Tables of percentage points of the inverted beta (F) distribution", *Biometrika*, Vol. 33 (1943), by permission of the *Biometrika* trustees.

Table 6(*b*) Values of $F_{0.01}$*

ν_2 = Degrees of freedom for denominator	ν_1 = Degrees of freedom for numerator																		
	1	2	3	4	5	6	7	8	9	10	12	15	20	24	30	40	60	120	∞
1	4,052	5,000	5,403	5,625	5,764	5,859	5,928	5,982	6,023	6,056	6,106	6,157	6,209	6,235	6,261	6,287	6,313	6,339	6,366
2	98.50	99.00	99.20	99.20	99.30	99.30	99.40	99.40	99.40	99.40	99.40	99.40	99.40	99.50	99.50	99.50	99.50	99.50	99.50
3	34.10	30.80	29.50	28.70	28.20	27.90	27.70	27.50	27.30	27.20	27.10	26.90	26.70	26.60	26.50	26.40	26.30	26.20	26.10
4	21.20	18.00	16.70	16.00	15.50	15.20	15.00	14.80	14.70	14.50	14.40	14.20	14.00	13.90	13.80	13.70	13.70	13.60	13.50
5	16.30	13.30	12.10	11.40	11.00	10.70	10.50	10.30	10.20	10.10	9.89	9.72	9.55	9.47	9.38	9.29	9.20	9.11	9.02
6	13.70	10.90	9.78	9.15	8.75	8.47	8.26	8.10	7.98	7.87	7.72	7.56	7.40	7.31	7.23	7.14	7.06	6.97	6.88
7	12.20	9.55	8.45	7.85	7.46	7.19	6.99	6.84	6.72	6.62	6.47	6.31	6.16	6.07	5.99	5.91	5.82	5.74	5.65
8	11.30	8.65	7.59	7.01	6.63	6.37	6.18	6.03	5.91	5.81	5.67	5.52	5.36	5.28	5.20	5.12	5.03	4.95	4.83
9	10.60	8.02	6.99	6.42	6.06	5.80	5.61	5.47	5.35	5.26	5.11	4.96	4.81	4.73	4.65	4.57	4.48	4.40	4.31
10	10.00	7.56	6.55	5.99	5.64	5.39	5.20	5.06	4.94	4.85	4.71	4.56	4.41	4.33	4.25	4.17	4.08	4.00	3.91
11	9.65	7.21	6.22	5.67	5.32	5.07	4.89	4.74	4.63	4.54	4.40	4.25	4.10	4.02	3.94	3.86	3.78	3.69	3.60
12	9.33	6.93	5.95	5.41	5.06	4.82	4.64	4.50	4.39	4.30	4.16	4.01	3.86	3.78	3.70	3.62	3.54	3.45	3.36
13	9.07	6.70	5.74	5.21	4.86	4.62	4.44	4.30	4.19	4.10	3.96	3.82	3.66	3.59	3.51	3.43	3.34	3.25	3.17
14	8.86	6.51	5.56	5.04	4.70	4.46	4.28	4.14	4.03	3.94	3.80	3.66	3.51	3.43	3.35	3.27	3.18	3.09	3.00
15	8.68	6.36	5.42	4.89	4.56	4.32	4.14	4.00	3.89	3.80	3.67	3.52	3.37	3.29	3.21	3.13	3.05	2.96	2.87
16	8.53	6.23	5.29	4.77	4.44	4.20	4.03	3.89	3.78	3.69	3.55	3.41	3.26	3.18	3.10	3.02	2.93	2.84	2.75
17	8.40	6.11	5.19	4.67	4.34	4.10	3.93	3.79	3.68	3.59	3.46	3.31	3.16	3.08	3.00	2.92	2.83	2.75	2.65
18	8.29	6.01	5.09	4.58	4.25	4.01	3.84	3.71	3.60	3.51	3.37	3.23	3.08	3.00	2.92	2.84	2.75	2.66	2.57
19	8.19	5.93	5.01	4.50	4.17	3.94	3.77	3.63	3.52	3.43	3.30	3.15	3.00	2.92	2.84	2.76	2.67	2.58	2.49
20	8.10	5.85	4.94	4.43	4.10	3.87	3.70	3.56	3.46	3.37	3.23	3.09	2.94	2.86	2.78	2.69	2.61	2.52	2.42
21	8.02	5.78	4.87	4.37	4.04	3.81	3.64	3.51	3.40	3.31	3.17	3.03	2.88	2.80	2.72	2.64	2.55	2.46	2.36
22	7.95	5.72	4.82	4.31	3.99	3.76	3.59	3.45	3.35	3.26	3.12	2.98	2.83	2.75	2.67	2.58	2.50	2.40	2.31
23	7.88	5.66	4.76	4.26	3.94	3.71	3.54	3.41	3.30	3.21	3.07	2.93	2.78	2.70	2.62	2.54	2.45	2.35	2.26
24	7.82	5.61	4.72	4.22	3.90	3.67	3.50	3.36	3.26	3.17	3.03	2.89	2.74	2.66	2.58	2.49	2.40	2.31	2.21
25	7.77	5.57	4.68	4.18	3.86	3.63	3.46	3.32	3.22	3.13	2.99	2.85	2.70	2.62	2.53	2.45	2.36	2.27	2.17
30	7.56	5.39	4.51	4.02	3.70	3.47	3.30	3.17	3.07	2.98	2.84	2.70	2.55	2.47	2.39	2.30	2.21	2.11	2.01
40	7.31	5.18	4.31	3.83	3.51	3.29	3.12	2.99	2.89	2.80	2.66	2.52	2.37	2.29	2.20	2.11	2.02	1.92	1.80
60	7.08	4.98	4.13	3.65	3.34	3.12	2.95	2.82	2.72	2.63	2.50	2.35	2.20	2.12	2.03	1.94	1.84	1.73	1.60
120	6.85	4.79	3.95	3.48	3.17	2.96	2.79	2.66	2.56	2.47	2.34	2.19	2.03	1.95	1.86	1.76	1.66	1.53	1.38
∞	6.63	4.61	3.78	3.32	3.02	2.80	2.64	2.51	2.41	2.32	2.18	2.04	1.88	1.79	1.70	1.59	1.47	1.32	1.00

* This table is reproduced from M. Merrington and C. M. Thompson, "Tables of percentage points of the inverted beta (*F*) distribution," *Biometrika*. Vol. 33 (1943), by permission of the *Biometrika* trustees.

Table 7 Random Digits*

1306	1189	5731	3968	5606	5084	8947	3897	1636	7810
0422	2431	0649	8085	5053	4722	6598	5044	9040	5121
6597	2022	6168	5060	8656	6733	6364	7649	1871	4328
7965	6541	5645	6243	7658	6903	9911	5740	7824	8520
7695	6937	0406	8894	0441	8135	9797	7285	5905	9539
5160	7851	8464	6789	3938	4197	6511	0407	9239	2232
2961	0551	0539	8288	7478	7565	5581	5771	5442	8761
1428	4183	4312	5445	4854	9157	9158	5218	1464	3634
3666	5642	4539	1561	7849	7520	2547	0756	1206	2033
6543	6799	7454	9052	6689	1946	2574	9386	0304	7945
9975	6080	7423	3175	9377	6951	6519	8287	8994	5532
4866	0956	7545	7723	8085	4948	2228	9583	4415	7065
8239	7068	6694	5168	3117	1568	0237	6160	9585	1133
8722	9191	3386	3443	0434	4586	4150	1224	6204	0937
1330	9120	8785	8382	2929	7089	3109	6742	2468	7025
2296	2952	4764	9070	6356	9192	4012	0618	2219	1109
3582	7052	3132	4519	9250	2486	0830	8472	2160	7046
5872	9207	7222	6494	8973	3545	6967	8490	5264	9821
1134	6324	6201	3792	5651	0538	4676	2064	0584	7996
1403	4497	7390	8503	8239	4236	8022	2914	4368	4529
3393	7025	3381	3553	2128	1021	8353	6413	5161	8583
1137	7896	3602	0060	7850	7626	0854	6565	4260	6220
7437	5198	8772	6927	8527	6851	2709	5992	7383	1071
8414	8820	3917	7238	9821	6073	6658	1280	9643	7761
8398	5224	2749	7311	5740	9771	7826	9533	3800	4553
0995	8935	2939	3092	2496	0359	0318	4697	7181	4035
6657	0755	9685	4017	6581	7292	5643	5064	1142	1297
8875	8369	7868	0190	9278	1709	4253	9346	4335	3769
8399	6702	0586	6428	7985	2979	4513	1970	1989	3105
6703	1024	2064	0393	6815	8502	1375	4171	6970	1201
4730	1653	9032	9855	0957	7366	0325	5178	7959	5371
8400	6834	3187	8688	1079	1480	6776	9888	7585	9998
3647	8002	6726	0877	4552	3238	7542	7804	3933	9475
6789	5197	8037	2354	9262	5497	0005	3986	1767	7981
2630	2721	2810	2185	6323	5679	4931	8336	6662	3566
1374	8625	1644	3342	1587	0762	6057	8011	2666	3759
1519	7625	9110	4409	0239	7059	3415	5537	2250	7292
9678	2877	7579	4935	0449	8119	6969	5383	1717	6719
0882	6781	3538	4090	3092	2365	6001	3446	9985	6007
0006	4205	2389	4365	1981	8158	7784	6256	3842	5603

4611	9861	7916	9305	2074	9462	0254	4827	9198	3974
1093	3784	4190	6332	1175	8599	9735	8584	6581	7194
3374	3545	6865	8819	3342	1676	2264	6014	5012	2458
3650	9676	1436	4374	4716	5548	8276	6235	6742	2154
7292	5749	7977	7602	9205	3599	3880	9537	4423	2330
2353	8319	2850	4026	3027	1708	3518	7034	7132	6903
1094	2009	8919	5676	7283	4982	9642	9235	8167	3366
0568	4002	0587	7165	1094	2006	7471	0940	4366	9554
5606	4070	5233	4339	6543	6695	5799	5821	3953	9458
8285	7537	1181	2300	5294	6892	1627	3372	1952	3028
2444	9039	4803	8568	1590	2420	2547	2470	8179	4617
5748	7767	2800	6289	2814	8281	1549	9519	3341	1192
7761	8583	0852	5619	6864	8506	9643	7763	9611	1289
6838	9280	2654	0812	3988	2146	5095	0150	8043	9079
6440	2631	3033	9167	4998	7036	0133	7428	9702	1376
8829	0094	2887	3802	5497	0318	5168	6377	9216	2802
9845	4796	2951	4449	1999	2691	5328	7674	7004	6212
5072	9000	3887	5739	7920	6074	4715	3681	2721	2701
9035	0553	1272	2600	3828	8197	8852	9092	8027	6144
5562	1080	2222	0336	1411	0303	7424	3713	9278	1918
2757	2650	8727	3953	9579	2442	8041	9869	2887	3933
6397	1848	1476	0787	4990	4666	1208	2769	3922	1158
9208	7641	3575	4279	1282	1840	5999	1806	7809	5885
2418	9289	6120	8141	3908	5577	3590	2317	8975	4593
7300	9006	5659	8258	3662	0332	5369	3640	0563	7939
6870	2535	8916	3245	2256	4350	6064	2438	2002	1272
2914	7309	4045	7513	3195	4166	0878	5184	6680	2655
0868	8657	8118	6340	9452	7460	3291	5778	1167	0312
7994	6579	6461	2292	9554	8309	5036	0974	9517	8293
8587	0764	6687	9150	1642	2050	4934	0027	1376	5040
8016	8345	2257	5084	8004	7949	3205	3972	7640	3478
5581	5775	7517	9076	4699	8313	8401	7147	9416	7184
2015	3364	6688	2631	2152	2220	1637	8333	4838	5699
7327	8987	5741	0102	1173	7350	7080	7420	1847	0741
3589	1991	1764	8355	9684	9423	7101	1063	4151	4875
2188	6454	7319	1215	0473	6589	2355	9579	7004	6209
2924	0472	9878	7966	2491	5662	5635	2789	2564	1249
1961	1669	2219	1113	9175	0260	4046	8142	4432	2664
2393	9637	0410	7536	0972	5153	0708	1935	1143	1704
7585	4424	2648	6728	2233	3518	7267	1732	1926	3833

Table 7 Random Digits* (continued)

0197	4021	9207	7327	9212	7017	8060	6216	1942	6817
9719	5336	5532	8537	2980	8252	4971	0110	6209	1556
8866	4785	6007	8006	9043	4109	5570	9249	9905	2152
5744	3957	8786	9023	1472	7275	1014	1104	0832	7680
7149	5721	1389	6581	7196	7072	6360	3084	7009	0239
7710	8479	9345	7773	9086	1202	8845	3163	7937	6163
5246	5651	0432	8644	6341	9661	2361	8377	8673	6098
3576	0013	7381	0124	8559	9813	9080	6984	0926	2169
3026	1464	2671	4691	0353	5289	8754	2442	7799	8983
6591	4365	8717	2365	5686	8377	8675	9798	7745	6360
0402	3257	0480	5038	1998	2935	1306	1190	2406	2596
7105	7654	4745	4482	8471	1424	2031	7803	4367	6816
7181	4140	1046	0885	1264	7755	1653	8924	5822	4401
3655	3282	2178	8134	3291	7262	8229	2866	7065	4806
5121	6717	3117	1901	5184	6467	8954	3884	0279	8635
3618	3098	9208	7429	1578	1917	7927	2696	3704	0833
0166	3638	4947	1414	4799	9189	2459	5056	5982	6154
6187	9653	3658	4730	1652	8096	8288	9368	5531	7788
1234	1448	0276	7290	1667	2823	3755	5642	4854	8844
8949	8731	4875	5724	2962	1182	2930	7539	4526	7252
4357	4146	8353	9952	8004	7945	1530	5207	4730	1967
5339	7325	6862	7584	8634	3485	2278	5832	0612	8118
6583	8433	0717	0606	9284	2719	1888	2889	0285	2765
6564	3526	2171	3809	3428	5523	9078	0648	7768	3326
4811	1933	3763	6265	8931	0649	8085	6177	4450	2139
6931	7236	1230	0441	4013	1352	6563	1499	7332	3068
8755	3390	6120	7825	9005	7012	1643	9934	4044	7022
6742	2260	3443	0190	9278	1816	7697	7933	0067	2906
6655	3930	9014	6032	7574	1685	5258	3100	5358	1929
8514	4806	4124	9286	0449	5051	4772	4651	0038	1580
8135	5004	7299	8981	4689	1950	2271	2201	8344	3852
4414	6855	0127	5489	5157	6386	7492	3736	7164	0498
3727	7959	5056	5983	8021	0204	7616	4325	7454	5039
5434	7342	0314	7525	0067	2800	6292	4706	3454	6881
7195	8828	9869	2785	3186	8375	7414	7232	0401	2483
2705	8245	6251	9611	1077	0641	0195	7024	6202	3899
1547	8981	4972	1280	4286	5678	0338	8096	8284	7010
3424	1435	1354	7631	7260	7361	0151	8903	9056	8684
8969	7551	3695	4915	7921	2913	3840	9031	9747	9735
5225	8720	8898	2478	3342	9200	8836	7269	2992	6284

Table 7 Random Digits* (continued)

6432	9861	1516	2849	2539	2208	4595	8616	6170	5865
3085	5903	8319	2744	0814	7318	8619	7614	3265	5999
0264	1246	3687	9759	6995	6565	3949	1012	0179	0059
8710	2419	6065	0036	9650	2027	6042	5467	1839	5577
5736	9001	3132	4521	9973	5070	8078	4150	2276	5059
7529	1339	4802	5751	3785	7125	4922	8877	9530	6499
5133	7995	8030	7408	2186	0725	5554	5664	6791	9677
3170	9915	6960	2621	6718	4059	9919	1007	6469	5410
3024	0680	1127	8088	0200	5868	0084	6362	6808	3727
4398	3121	7749	8191	2087	8270	5233	3980	6774	8522
0082	5419	7659	2061	2506	7573	1157	3979	2309	0811
4351	6516	6814	5898	3973	8103	3616	2049	7843	0568
3268	0086	7580	1337	3884	5679	4830	4509	9587	2184
4391	8487	4884	1488	2249	6661	5774	7205	2717	7030
7328	0705	0652	9424	7082	8579	5647	5571	9667	8555
3835	2938	2671	4691	0559	8382	2825	4928	5379	8635
8731	4980	8674	4506	7262	8127	2022	2178	7463	4842
2995	7868	0683	3768	0625	9887	7060	0514	0034	8600
5597	9028	5660	5006	8325	9677	2169	3196	0357	7811
3081	5876	8150	1360	1868	9265	3277	8465	7502	6458
7406	4439	5683	6877	2920	9588	3002	2869	3746	3690
5969	9442	7696	7510	1620	4973	1911	1288	6160	9797
4765	9647	4364	1037	4975	1998	1359	1346	6125	5078
3219	2532	7577	2815	8696	9248	9410	9282	6572	3940
6906	8859	5044	8826	6218	3206	9034	0843	9832	2703
7993	3141	0103	4528	7988	4635	8478	9094	9077	5306
2549	3737	7686	0723	4505	6841	1379	6460	1869	5700
3672	7033	4844	0149	7412	6370	1884	0717	5740	8477
2217	0293	3978	5933	1032	5192	1732	2137	9357	5941
3162	9968	6369	1258	0416	4326	7840	6525	2608	5255
1758	1489	2774	6033	9813	1052	1816	7484	1699	7350
6430	8803	0478	4157	5626	1603	1339	4666	1207	2135
4893	8857	1717	1533	6572	8408	2173	4754	0272	1305
1516	2733	7326	8674	9233	1799	5281	0797	0885	0947
4950	3171	5756	3036	9047	8719	8498	1312	7124	4787
0549	6775	9360	6639	0990	0037	7309	4702	0812	4195
1018	7027	7569	7549	2539	2315	8030	7663	3881	8264
2241	9965	9729	7092	4891	9239	0738	1804	3025	1030
1602	0708	2201	9848	6241	1084	8142	8555	7291	5016
5840	8381	1549	9902	6935	3681	6420	0214	8489	5911

Table 7 *Random Digits* (continued)*

1676	0367	7484	1595	5693	3008	9816	7311	6162	1024
6048	4175	8940	9029	8306	8892	4127	1709	4043	6591
5549	9621	2563	0515	0560	9021	0632	4309	4044	7010
5317	4584	9418	4600	0640	9668	6379	6515	6310	7916
2532	7784	6469	4793	5957	4123	6555	3237	6915	6960
2300	5412	3106	4877	6936	4109	8060	1896	6881	7028
1499	8699	4534	5367	7557	2701	2587	2521	2159	6991
6201	3791	2946	2863	5684	5517	7448	2227	8991	7505
6839	9736	8312	8068	7339	5395	9559	3416	6169	5484
0092	5537	1933	3186	8482	6680	2656	1864	4535	2193
1862	3253	6515	6299	2929	2219	9145	7511	2146	4962
9886	6744	3097	8894	0446	3494	8211	1723	6138	3181
5289	9071	1231	0651	9109	7448	2228	9700	0224	4595
2685	7104	7193	5506	2993	7028	4830	3866	8698	0277
6055	7092	4786	6847	4543	7448	2017	4114	8385	3625
4092	4995	0280	9371	3375	3503	4496	8642	5388	1831
5951	4937	3670	5797	5030	6524	2265	7748	7875	6976
7687	3849	4821	2373	1157	4208	3623	9399	7349	6663
9886	3463	2055	4872	2702	1807	9056	8576	4237	8757
3193	3011	8899	6721	0086	2623	7977	7578	4024	1997
9181	7365	9135	1669	2007	7784	6363	6913	6017	6588
9459	2175	5728	4933	0111	6703	1234	2410	1620	4859
9874	5278	2849	3163	6372	2600	9887	7060	3919	1111
7729	2099	7513	2774	6030	8260	9023	1368	9513	6122
4699	8102	3001	7947	1659	6571	1969	7152	7356	1062
1872	7244	3954	7422	2688	8649	0156	1965	5012	2461
9636	0123	2438	1757	4204	1650	2486	0002	4724	7412
6403	9054	7632	7469	8973	3332	0294	5062	0303	7315
4433	3293	2314	7431	2389	4094	5062	0118	0046	6070
2361	3933	8026	0431	8012	3214	8927	7355	0585	7638
4077	8463	6580	6983	0181	3327	6812	1755	3387	4569
6678	0006	3686	8478	9187	2291	9032	9852	1450	7940
6499	2582	5207	4627	0456	4245	3583	8996	1006	5839
6663	9021	0319	7908	1241	9977	7042	6923	2539	2103
9999	4503	6105	5525	1068	6272	7036	0200	6291	2841
9048	6982	3845	6865	9029	8700	0349	3416	8236	1129
5136	9653	3654	2863	5565	8923	5596	8389	9927	9092
9906	1070	5693	3012	1218	7309	4361	3041	4327	8423
4198	7035	8182	6270	6461	2079	2998	5507	9605	6734
2030	5878	8989	6789	4359	1820	5063	9199	7751	6337

* From Donald B. Owen, *Handbook of Statistical Tables*. Reading, Mass.: Addison Wesley, 1962.

Table 8 Operating Characteristic Curves for Tests About Means*

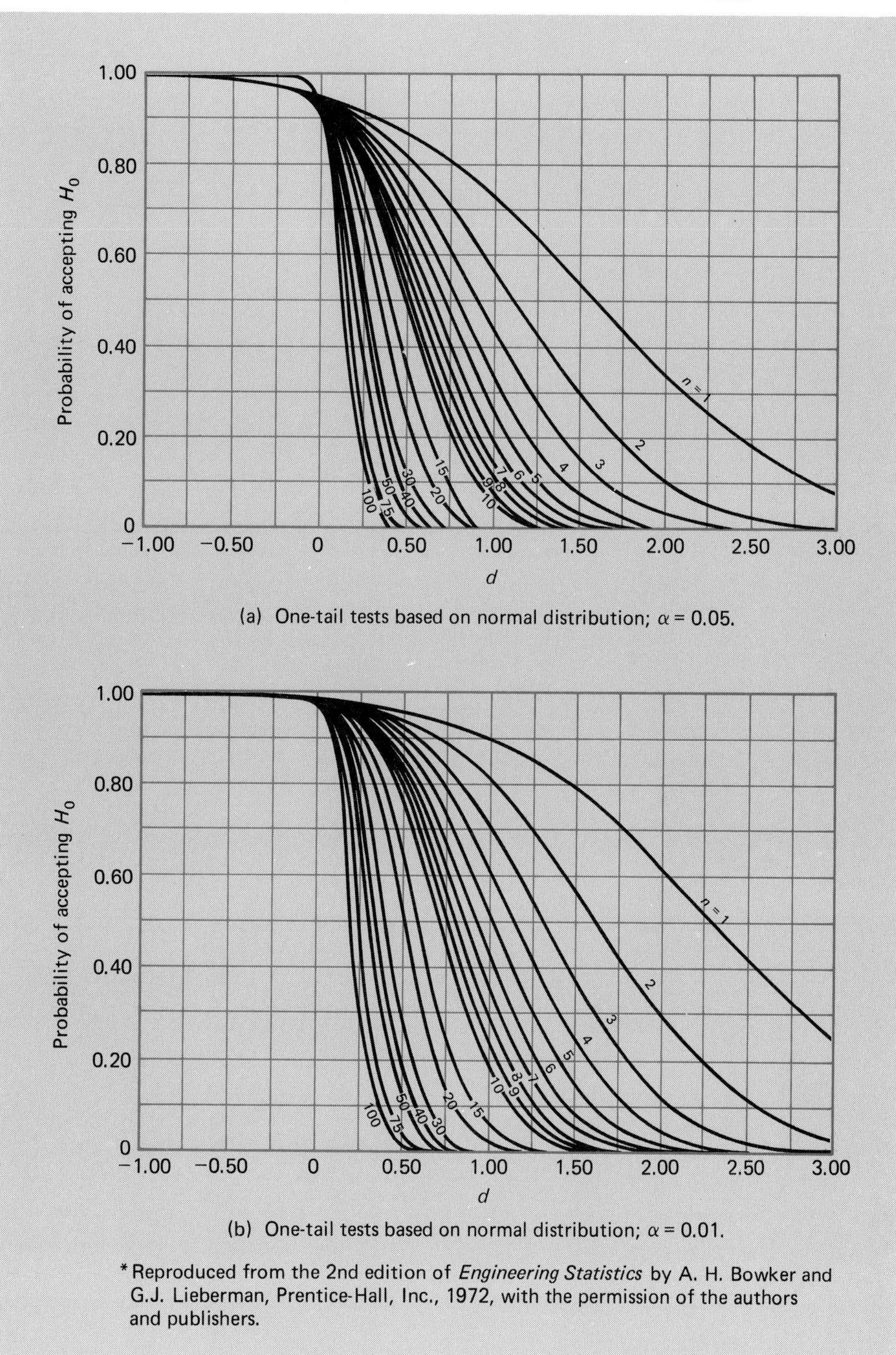

(a) One-tail tests based on normal distribution; $\alpha = 0.05$.

(b) One-tail tests based on normal distribution; $\alpha = 0.01$.

*Reproduced from the 2nd edition of *Engineering Statistics* by A. H. Bowker and G.J. Lieberman, Prentice-Hall, Inc., 1972, with the permission of the authors and publishers.

Table 8 Operating Characteristic Curves for Tests About Means (Continued)

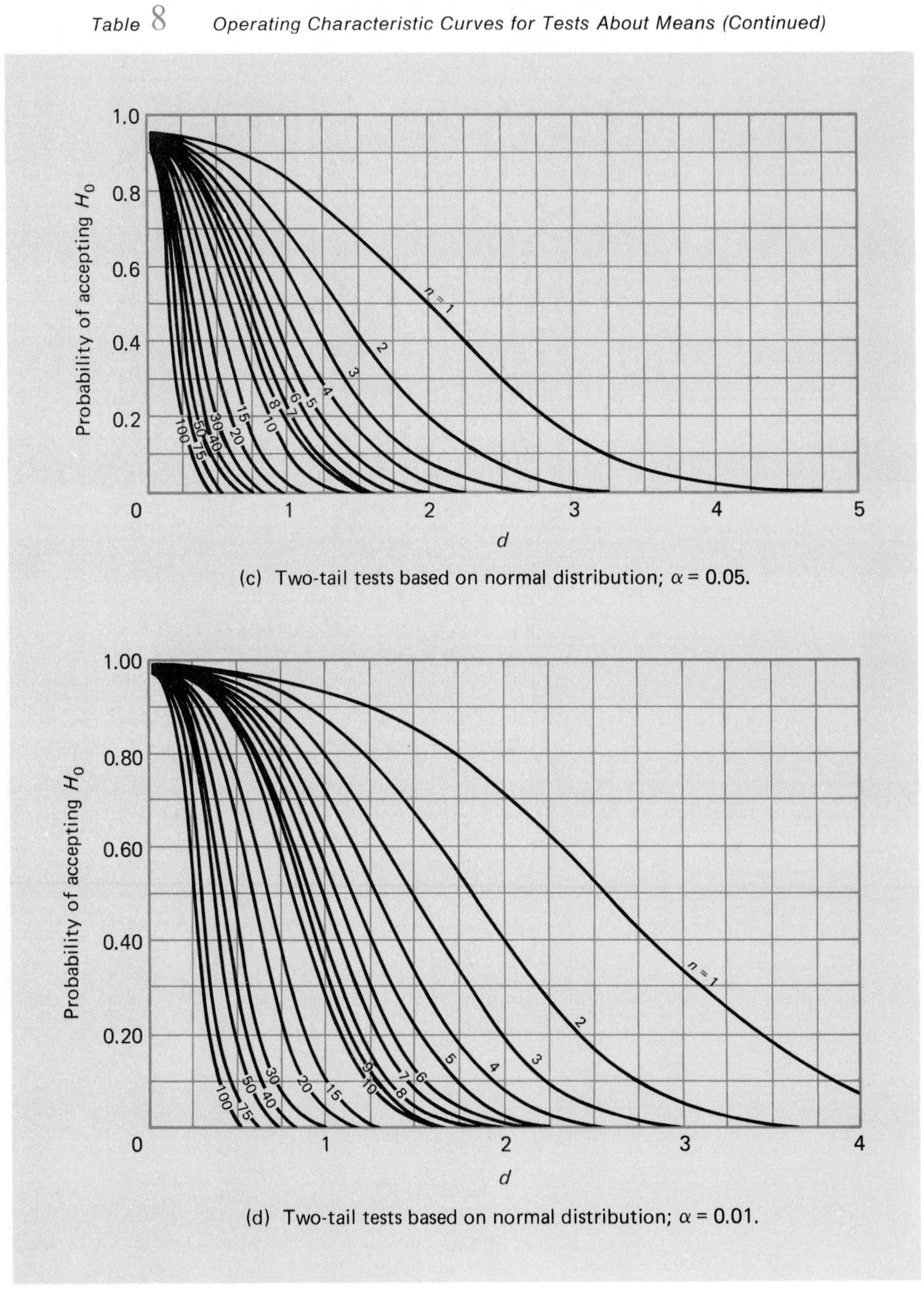

(c) Two-tail tests based on normal distribution; $\alpha = 0.05$.

(d) Two-tail tests based on normal distribution; $\alpha = 0.01$.

Table 9(*a*) *95% Confidence Intervals for Proportions**

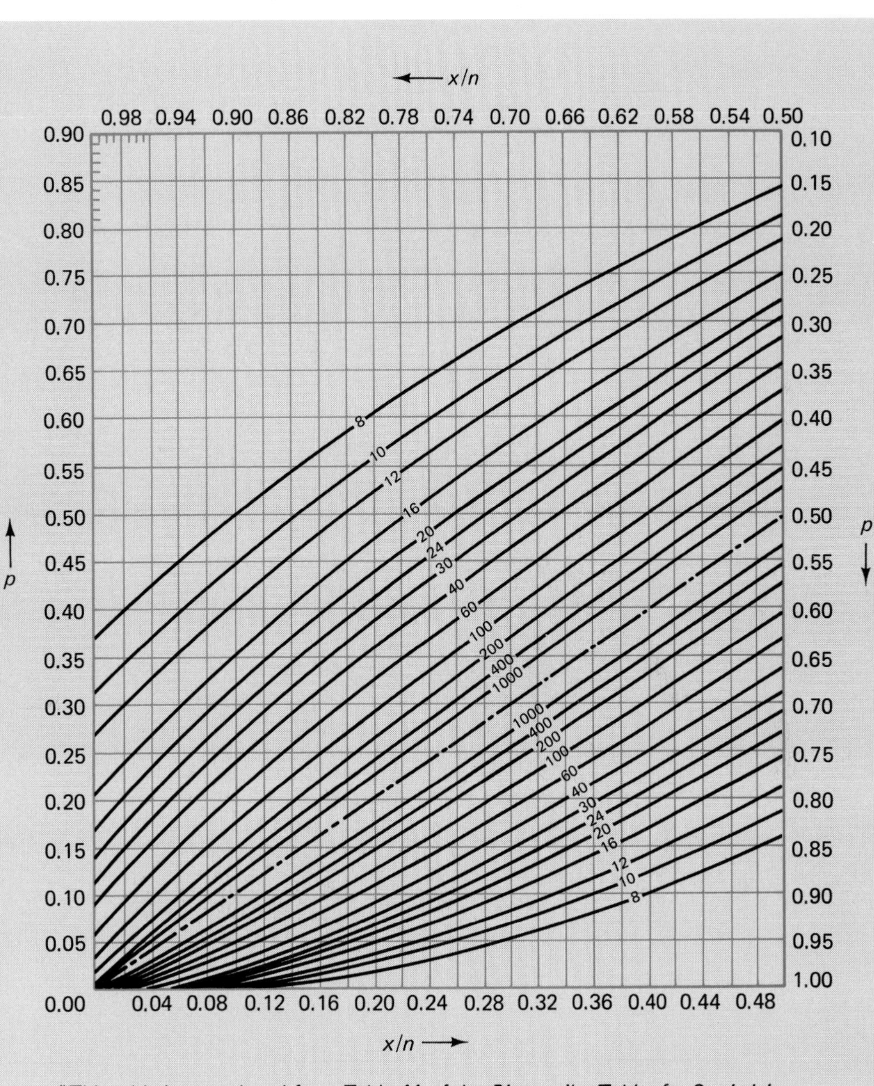

*This table is reproduced from Table 41 of the *Biometrika Tables for Statisticians*, Vol. I (New York: Cambridge University Press, 1954) by permission of the *Biometrika* trustees.

Table 9(b) 99% Confidence Intervals for Proportions*

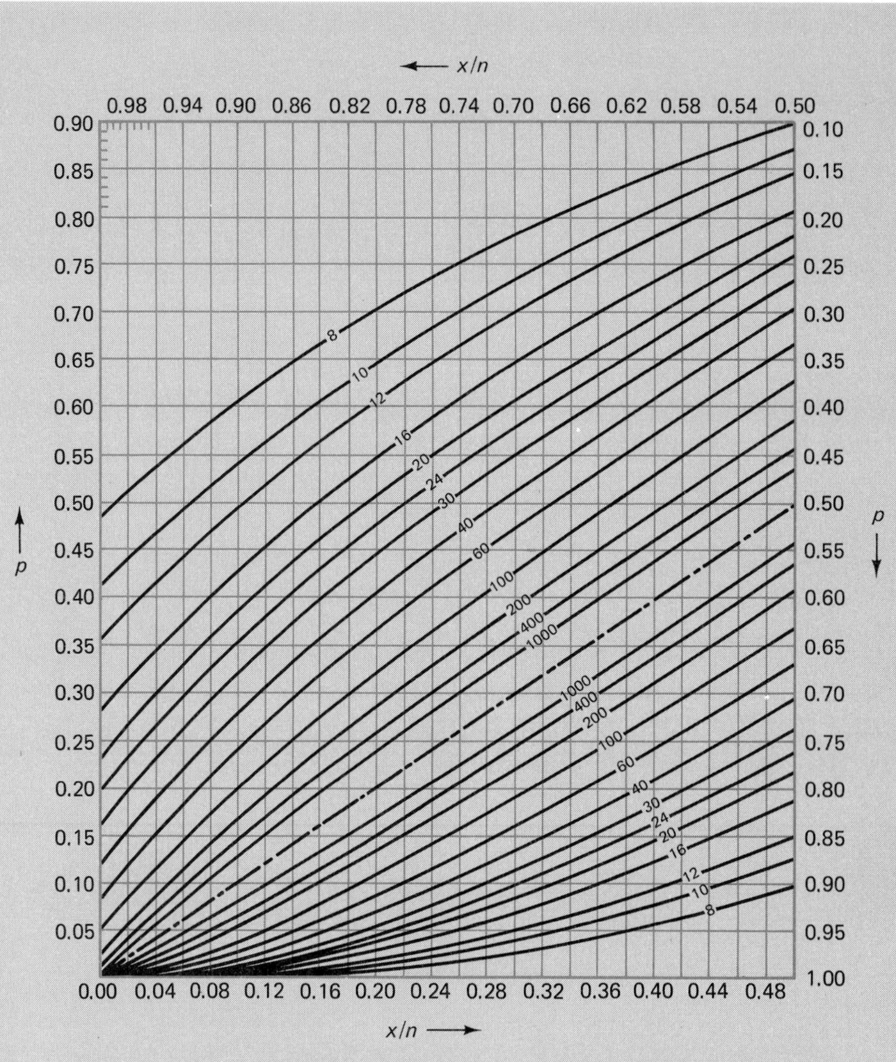

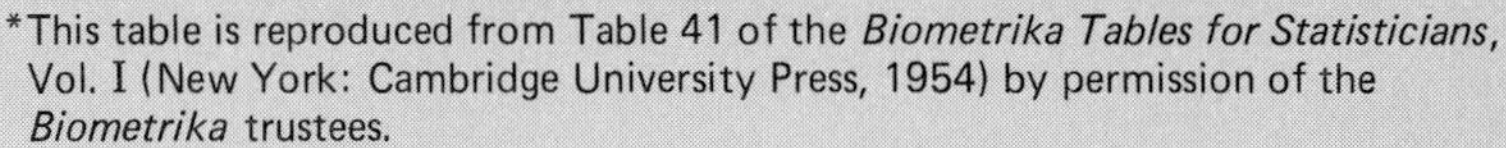
*This table is reproduced from Table 41 of the *Biometrika Tables for Statisticians*, Vol. I (New York: Cambridge University Press, 1954) by permission of the *Biometrika* trustees.

Table 10 Critical Values of D*

Sample size n	$D_{.10}$	$D_{.05}$	$D_{.01}$
1	0.950	0.975	0.995
2	0.776	0.842	0.929
3	0.642	0.708	0.828
4	0.564	0.624	0.733
5	0.510	0.565	0.669
6	0.470	0.521	0.618
7	0.438	0.486	0.577
8	0.411	0.457	0.543
9	0.388	0.432	0.514
10	0.368	0.410	0.490
11	0.352	0.391	0.468
12	0.338	0.375	0.450
13	0.325	0.361	0.433
14	0.314	0.349	0.418
15	0.304	0.338	0.404
16	0.295	0.328	0.392
17	0.286	0.318	0.381
18	0.278	0.309	0.371
19	0.272	0.301	0.363
20	0.264	0.294	0.356
25	0.24	0.27	0.32
30	0.22	0.24	0.29

* Adapted from F. J. Massey, Jr., "The Kolgomorov-Smirnov test for goodness of fit," *J. Amer. Statist. Ass.*, Vol. 46 (1951), p. 70, with the kind permission of the author and publisher.

Table 11* Values of $Z = \frac{1}{2} \ln \frac{1+r}{1-r}$

r	0.00	0.01	0.02	0.03	0.04	0.05	0.06	0.07	0.08	0.09
0.0	0.000	0.010	0.020	0.030	0.040	0.050	0.060	0.070	0.080	0.090
0.1	0.100	0.110	0.121	0.131	0.141	0.151	0.161	0.172	0.182	0.192
0.2	0.203	0.213	0.224	0.234	0.245	0.255	0.266	0.277	0.288	0.299
0.3	0.310	0.321	0.332	0.343	0.354	0.365	0.377	0.388	0.400	0.412
0.4	0.424	0.436	0.448	0.460	0.472	0.485	0.497	0.510	0.523	0.536
0.5	0.549	0.563	0.576	0.590	0.604	0.618	0.633	0.648	0.662	0.678
0.6	0.693	0.709	0.725	0.741	0.758	0.775	0.793	0.811	0.829	0.848
0.7	0.867	0.887	0.908	0.929	0.950	0.973	0.996	1.020	1.045	1.071
0.8	1.099	1.127	1.157	1.188	1.221	1.256	1.293	1.333	1.376	1.422
0.9	1.472	1.528	1.589	1.658	1.738	1.832	1.946	2.092	2.298	2.647

* For negative values of r put a minus sign in front of the corresponding Z's, and vice versa.

Table 12(a) Values of r_p for $\alpha = 0.05$*

d.f. \ p	2	3	4	5	6	7	8	9	10
1	17.97								
2	6.09	6.09							
3	4.50	4.52	4.52						
4	3.93	4.01	4.03	4.03					
5	3.64	3.75	3.80	3.81	3.81				
6	3.46	3.59	3.65	3.68	3.69	3.70			
7	3.34	3.48	3.55	3.59	3.61	3.62	3.63		
8	3.26	3.40	3.48	3.52	3.55	3.57	3.57	3.58	
9	3.20	3.34	3.42	3.47	3.50	3.52	3.54	3.54	3.55
10	3.15	3.29	3.38	3.43	3.47	3.49	3.51	3.52	3.52
11	3.11	3.26	3.34	3.40	3.44	3.46	3.48	3.49	3.50
12	3.08	3.23	3.31	3.37	3.41	3.44	3.46	3.47	3.48
13	3.06	3.20	3.29	3.35	3.39	3.42	3.46	3.46	3.47
14	3.03	3.18	3.27	3.33	3.37	3.40	3.43	3.44	3.46
15	3.01	3.16	3.25	3.31	3.36	3.39	3.41	3.43	3.45
16	3.00	3.14	3.23	3.30	3.34	3.38	3.40	3.42	3.44
17	2.98	3.13	3.22	3.28	3.33	3.37	3.39	3.41	3.43
18	2.97	3.12	3.21	3.27	3.32	3.36	3.38	3.40	3.42
19	2.96	3.11	3.20	3.26	3.31	3.35	3.38	3.40	3.41
20	2.95	3.10	3.19	3.25	3.30	3.34	3.37	3.39	3.41
24	2.92	3.07	3.16	3.23	3.28	3.31	3.35	3.37	3.39
30	2.89	3.03	3.13	3.20	3.25	3.29	3.32	3.35	3.37
40	2.86	3.01	3.10	3.17	3.22	3.27	3.30	3.33	3.35
60	2.83	2.98	3.07	3.14	3.20	3.24	3.28	3.31	3.33
120	2.80	2.95	3.04	3.12	3.17	3.22	3.25	3.29	3.31
∞	2.77	2.92	3.02	3.09	3.15	3.19	3.23	3.27	3.29

* This table is reproduced from H. L. Harter, "Critical values for Duncan's new multiple range test." It contains some corrected values to replace those given by D. B. Duncan in "Multiple Range and Multiple *F* Tests," *Biometrics*, Vol. 11 (1955). The above table is reproduced with the permission of the author and the editor of *Biometrics*.

Table 12(b) Values of r_p for $\alpha = 0.01$*

d.f. \ p	2	3	4	5	6	7	8	9	10
1	90.02								
2	14.04	14.04							
3	8.26	8.32	8.32						
4	6.51	6.68	6.74	6.76					
5	5.70	5.90	5.99	6.04	6.07				
6	5.24	5.44	5.55	5.62	5.66	5.68			
7	4.95	5.15	5.26	5.33	5.38	5.42	5.44		
8	4.74	4.94	5.06	5.13	5.19	5.23	5.26	5.28	
9	4.60	4.79	4.91	4.99	5.04	5.09	5.12	5.14	5.16
10	4.48	4.67	4.79	4.88	4.93	4.98	5.01	5.04	5.06
11	4.39	4.58	4.70	4.78	4.84	4.89	4.92	4.95	4.97
12	4.32	4.50	4.62	4.71	4.77	4.81	4.85	4.88	4.91
13	4.26	4.44	4.56	4.64	4.71	4.75	4.79	4.82	4.85
14	4.21	4.39	4.51	4.59	4.66	4.70	4.74	4.77	4.80
15	4.17	4.34	4.46	4.55	4.61	4.66	4.70	4.73	4.76
16	4.13	4.31	4.43	4.51	4.57	4.62	4.66	4.70	4.72
17	4.10	4.27	4.39	4.47	4.54	4.59	4.63	4.66	4.69
18	4.07	4.25	4.36	4.45	4.51	4.56	4.60	4.64	4.66
19	4.05	4.22	4.33	4.42	4.48	4.53	4.57	4.61	4.64
20	4.02	4.20	4.31	4.40	4.46	4.51	4.55	4.59	4.62
24	3.96	4.13	4.24	4.32	4.39	4.44	4.48	4.52	4.55
30	3.89	4.06	4.17	4.25	4.31	4.36	4.41	4.45	4.48
40	3.82	3.99	4.10	4.18	4.24	4.29	4.33	4.38	4.41
60	3.76	3.92	4.03	4.11	4.18	4.23	4.37	4.31	4.34
120	3.70	3.86	3.97	4.04	4.11	4.16	4.20	4.24	4.27
∞	3.64	3.80	3.90	3.98	4.04	4.09	4.13	4.17	4.21

* This table is reproduced from H. L. Harter, "Critical values for Duncan's new multiple range test." It contains some corrected values to replace those given by D. B. Duncan in "Multiple Range and Multiple *F* Tests," *Biometrics*, Vol. 11 (1955). The above table is reproduced with the permission of the author and the editor of *Biometrics*.

Table 13 Control Chart Constants*

Number of observations in sample, n	Chart for averages			Chart for standard deviations					Chart for ranges				
	Factors for control limits			Factor for central line	Factors for control limits				Factor for central line	Factors for control limits			
	A	A_1	A_2	c_2	B_1	B_2	B_3	B_4	d_2	D_1	D_2	D_3	D_4
2	2.121	3.760	1.880	0.5642	0	1.843	0	3.267	1.128	0	3.686	0	3.267
3	1.732	2.394	1.023	0.7236	0	1.858	0	2.568	1.693	0	4.358	0	2.575
4	1.500	1.880	0.729	0.7979	0	1.808	0	2.266	2.059	0	4.698	0	2.282
5	1.342	1.596	0,577	0.8407	0	1.756	0	2.089	2.326	0	4.918	0	2.115
6	1.225	1.410	0.483	0.8686	0.026	1.711	0.030	1.970	2.534	0	5.078	0	2.004
7	1.134	1.277	0.419	0.8882	0.105	1.672	0.118	1.882	2.704	0.205	5.203	0.076	1.924
8	1.061	1.175	0.373	0.9027	0.167	1.638	0.185	1.815	2.847	0.387	5.307	0.136	1.864
9	1.000	1.094	0.337	0.9139	0.219	1.609	0.239	1.761	2.970	0.546	5.394	0.184	1.816
10	0.949	1.028	0.308	0.9227	0.262	1.584	0.284	1.716	3.078	0.687	5.469	0.223	1.777
11	0.905	0.973	0.285	0.9300	0.299	1.561	0.321	1.679	3.173	0.812	5.534	0.256	1.744
12	0.866	0.925	0.266	0.9359	0.331	1.541	0.354	1.646	3.258	0.924	5.592	0.284	1.716
13	0.832	0.884	0.249	0.9410	0.359	1.523	0.382	1.618	3.336	1.026	5.646	0.308	1.692
14	0.802	0.848	0.235	0.9453	0.384	1.507	0.406	1.594	3.407	1.121	5.693	0.329	1.671
15	0.775	0.816	0.223	0.9490	0.406	1.492	0.428	1.572	3.472	1.207	5.737	0.348	1.652

* Reprinted with permission from the American Society for Testing and Materials, 1916 Race Street, Philadelphia, Pa. 19103.

Table 14(a) *Factors for Two-Sided Tolerance Limits**

	$1-\alpha = 0.95$			$1-\alpha = 0.99$		
n \ P	0.90	0.95	0.99	0.90	0.95	0.99
2	32.019	37.674	48.430	160.193	188.491	242.300
3	8.380	9.916	12.861	18.930	22.401	29.055
4	5.369	6.370	8.299	9.398	11.150	14.527
5	4.275	5.079	6.634	6.612	7.855	10.260
6	3.712	4.414	5.775	5.337	6.345	8.301
7	3.369	4.007	5.248	4.613	5.488	7.187
8	3.136	3.732	4.891	4.147	4.936	6.468
9	2.967	3.532	4.631	3.822	4.550	5.966
10	2.839	3.379	4.433	3.582	4.265	5.594
11	2.737	3.259	4.277	3.397	4.045	5.308
12	2.655	3.162	4.150	3.250	3.870	5.079
13	2.587	3.081	4.044	3.130	3.727	4.893
14	2.529	3.012	3.955	3.029	3.608	4.737
15	2.480	2.954	3.878	2.945	3.507	4.605
16	2.437	2.903	3.812	2.872	3.421	4.492
17	2.400	2.858	3.754	2.808	3.345	4.393
18	2.366	2.819	3.702	2.753	3.279	4.307
19	2.337	2.784	3.656	2.703	3.221	4.230
20	2.310	2.752	3.615	2.659	3.168	4.161
25	2.208	2.631	3.457	2.494	2.972	3.904
30	2.140	2.549	3.350	2.385	2.841	3.733
35	2.090	2.490	3.272	2.306	2.748	3.611
40	2.052	2.445	3.213	2.247	2.677	3.518
45	2.021	2.408	3.165	2.200	2.621	3.444
50	1.996	2.379	3.126	2.162	2.576	3.385
55	1.976	2.354	3.094	2.130	2.538	3.335

(continued)

Table 14(a) *Factors for Two-Sided Tolerance Limits** (*Continued*)

	$1-\alpha = 0.95$			$1-\alpha = 0.99$		
n \ P	0.90	0.95	0.99	0.90	0.95	0.99
60	1.958	2.333	3.066	2.103	2.506	3.293
65	1.943	2.315	3.042	2.080	2.478	3.257
70	1.929	2.299	3.021	2.060	2.454	3.225
75	1.917	2.285	3.002	2.042	2.433	3.197
80	1.907	2.272	2.986	2.026	2.414	3.173
85	1.897	2.261	2.971	2.012	2.397	3.150
90	1.889	2.251	2.958	1.999	2.382	3.130
95	1.881	2.241	2.945	1.987	2.368	3.112
100	1.874	2.233	2.934	1.977	2.355	3.096
150	1.825	2.175	2.859	1.905	2.270	2.983
200	1.798	2.143	2.816	1.865	2.222	2.921
250	1.780	2.121	2.788	1.839	2.191	2.880
300	1.767	2.106	2.767	1.820	2.169	2.850
400	1.749	2.084	2.739	1.794	2.138	2.809
500	1.737	2.070	2.721	1.777	2.117	2.783
600	1.729	2.060	2.707	1.764	2.102	2.763
700	1.722	2.052	2.697	1.755	2.091	2.748
800	1.717	2.046	2.688	1.747	2.082	2.736
900	1.712	2.040	2.682	1.741	2.075	2.726
1000	1.709	2.036	2.676	1.736	2.068	2.718
∞	1.645	1.960	2.576	1.645	1.960	2.576

* Adapted by permission from *Techniques of Statistical Analysis* by Columbia University Statistical Research Group, Copyright 1947, McGraw-Hill Book Company, Inc., New York.

Table 14(b) Factors for One-Sided Tolerance Limits

	$1 - \alpha = 0.95$			$1 - \alpha = 0.99$		
n \ P	0.90	0.95	0.99	0.90	0.95	0.99
2	20.581	26.260	37.094	103.029	131.426	185.617
3	6.156	7.656	10.553	13.995	17.370	23.896
4	4.162	5.144	7.042	7.380	9.083	12.387
5	3.407	4.203	5.741	5.362	6.578	8.939
6	3.006	3.708	5.062	4.411	5.406	7.335
7	2.756	3.400	4.642	3.859	4.728	6.412
8	2.582	3.187	4.354	3.497	4.285	5.812
9	2.454	3.031	4.143	3.241	3.972	5.389
10	2.355	2.911	3.981	3.048	3.738	5.074
11	2.275	2.815	3.852	2.898	3.556	4.829
12	2.210	2.736	3.747	2.777	3.410	4.633
13	2.155	2.671	3.659	2.677	3.290	4.472
14	2.109	2.615	3.585	2.593	3.189	4.337
15	2.068	2.566	3.520	2.522	3.102	4.222
16	2.033	2.524	3.464	2.460	3.028	4.123
17	2.002	2.486	3.414	2.405	2.963	4.037
18	1.974	2.453	3.370	2.357	2.905	3.960
19	1.949	2.423	3.331	2.314	2.854	3.892
20	1.926	2.396	3.295	2.276	2.808	3.832
25	1.838	2.292	3.158	2.129	2.633	3.601
30	1.777	2.220	3.064	2.030	2.516	3.447
35	1.732	2.167	2.995	1.957	2.430	3.334
40	1.697	2.126	2.941	1.902	2.364	3.249
45	1.669	2.092	2.898	1.857	2.312	3.180
50	1.646	2.065	2.863	1.821	2.269	3.125
55	1.626	2.042	2.833	1.790	2.233	3.078
60	1.609	2.022	2.807	1.764	2.202	3.038
65	1.594	2.005	2.785	1.741	2.176	3.004
70	1.581	1.990	2.765	1.722	2.153	2.974
75	1.570	1.976	2.748	1.704	2.132	2.947
80	1.559	1.965	2.733	1.688	2.114	2.924
85	1.550	1.954	2.719	1.674	2.097	2.902
90	1.542	1.944	2.706	1.661	2.082	2.883
95	1.534	1.935	2.695	1.650	2.069	2.866
100	1.527	1.927	2.684	1.639	2.056	2.850
150	1.478	1.870	2.611	1.566	1.971	2.741
200	1.450	1.837	2.570	1.524	1.923	2.679
250	1.431	1.815	2.542	1.496	1.891	2.638
300	1.417	1.800	2.522	1.476	1.868	2.608
∞	1.282	1.645	2.326	1.282	1.645	2.326

Table 15 *Sample Size Code Letters from MIL-STD-105D*

Lot or batch size			*General inspection levels*		
			I	II	III
2	to	8	A	A	B
9	to	15	A	B	C
16	to	25	B	C	D
26	to	50	C	D	E
51	to	90	C	E	F
91	to	150	D	F	G
151	to	280	E	G	H
281	to	500	F	H	J
501	to	1,200	G	J	K
1,201	to	3,200	H	K	L
3,201	to	10,000	J	L	M
10,001	to	35,000	K	M	N
35,001	to	150,000	L	N	P
150,001	to	500,000	M	P	Q
500,001	and	over	N	Q	R

Table 16 *Master Table for Single Sampling (Normal Inspection) from MIL-STD-105D*

Sample size code letter	Sample size	Acceptable quality levels (normal inspection) 0.010	0.015	0.025	0.040	0.065	0.10	0.15	0.25	0.40	0.65	1.0	1.5	2.5	4.0	6.5	10	15	25	40	65	100	150	250	400	650	1000
		Ac Re	Ac Re	Ac Re	Ac Re	Ac Re	Ac Re	Ac Re	Ac Re	Ac Re	Ac Re	Ac Re	Ac Re	Ac Re	Ac Re	Ac Re	Ac Re	Ac Re	Ac Re	Ac Re	Ac Re	Ac Re	Ac Re	Ac Re	Ac Re	Ac Re	Ac Re
A	2	↓	↓	↓	↓	↓	↓	↓	↓	↓	↓	↓	↓	↓	↓	0 1	↓	↓	1 2	2 3	3 4	5 6	7 8	10 11	14 15	21 22	30 31
B	2	↓	↓	↓	↓	↓	↓	↓	↓	↓	↓	↓	↓	↓	0 1	↑	↓	1 2	2 3	3 4	5 6	7 8	10 11	14 15	21 22	30 31	44 45
C	5	↓	↓	↓	↓	↓	↓	↓	↓	↓	↓	↓	↓	0 1	↑	↓	1 2	2 3	3 4	5 6	7 8	10 11	14 15	21 22	30 31	44 45	↑
D	8	↓	↓	↓	↓	↓	↓	↓	↓	↓	↓	↓	0 1	↑	↓	1 2	2 3	3 4	5 6	7 8	10 11	14 15	21 22	30 31	44 45	↑	↑
E	13	↓	↓	↓	↓	↓	↓	↓	↓	↓	↓	0 1	↑	↓	1 2	2 3	3 4	5 6	7 8	10 11	14 15	21 22	30 31	44 45	↑	↑	↑
F	20	↓	↓	↓	↓	↓	↓	↓	↓	↓	0 1	↑	↓	1 2	2 3	3 4	5 6	7 8	10 11	14 15	21 22	↑	↑	↑	↑	↑	↑
G	32	↓	↓	↓	↓	↓	↓	↓	↓	0 1	↑	↓	1 2	2 3	3 4	5 6	7 8	10 11	14 15	21 22	↑	↑	↑	↑	↑	↑	↑
H	50	↓	↓	↓	↓	↓	↓	↓	0 1	↑	↓	1 2	2 3	3 4	5 6	7 8	10 11	14 15	21 22	↑	↑	↑	↑	↑	↑	↑	↑
J	80	↓	↓	↓	↓	↓	↓	0 1	↑	↓	1 2	2 3	3 4	5 6	7 8	10 11	14 15	21 22	↑	↑	↑	↑	↑	↑	↑	↑	↑
K	125	↓	↓	↓	↓	↓	0 1	↑	↓	1 2	2 3	3 4	5 6	7 8	10 11	14 15	21 22	↑	↑	↑	↑	↑	↑	↑	↑	↑	↑
L	200	↓	↓	↓	↓	0 1	↑	↓	1 2	2 3	3 4	5 6	7 8	10 11	14 15	21 22	↑	↑	↑	↑	↑	↑	↑	↑	↑	↑	↑
M	315	↓	↓	↓	0 1	↑	↓	1 2	2 3	3 4	5 6	7 8	10 11	14 15	21 22	↑	↑	↑	↑	↑	↑	↑	↑	↑	↑	↑	↑
N	500	↓	↓	0 1	↑	↓	1 2	2 3	3 4	5 6	7 8	10 11	14 15	21 22	↑	↑	↑	↑	↑	↑	↑	↑	↑	↑	↑	↑	↑
P	800	↓	0 1	↑	↓	1 2	2 3	3 4	5 6	7 8	10 11	14 15	21 22	↑	↑	↑	↑	↑	↑	↑	↑	↑	↑	↑	↑	↑	↑
Q	1250	0 1	↑	↓	1 2	2 3	3 4	5 6	7 8	10 11	14 15	21 22	↑	↑	↑	↑	↑	↑	↑	↑	↑	↑	↑	↑	↑	↑	↑
R	2000	↑	↑	1 2	2 3	3 4	5 6	7 8	10 11	14 15	21 22	↑	↑	↑	↑	↑	↑	↑	↑	↑	↑	↑	↑	↑	↑	↑	↑

↓ = Use first sampling plan below arrow. If sample size equals, or exceeds, lot or batch size, do 100 percent inspection.

↑ = Use first sampling plan above arrow.

Ac = Acceptance number.
Re = Rejection number.

ANSWERS TO ODD-NUMBERED EXERCISES

CHAPTER 2

2.5 (a) 10.95, 11.95, 12.95, 13.95; (b) 10.45, 11.45, 12.45, 13.45, and 14.45; (c) 1.

2.7 The class frequencies are 3, 15, 24, 12, and 6.

2.9 (a) 11.5, 20.5, 29.5, 38.5, 47.5, 56.5, 65.5; (b) 12–20, 21–29, 30–38, 39–47, 48–56, 57–65.

2.11 The cumulative "less than" percentages are 0.00, 12.50, 25.00, 36.25, 50.00, 65.00, 77.50, 85.00, 90.00, 95.00, 96.25, 98.75, and 100%.

2.13 The cumulative "less than" percentages are 0, 4, 17, 35, 60, 80, 94, and 100%.

2.15 The cumulative "or more" frequencies are 60, 45, 33, 22, 15, 7, 2, and 0.

2.17 No, because we tend to compare areas; the large sack should be modified so that its area is double that of the small sack.

2.21

2**	67, 88, 95
3**	55, 70, 91, 83, 17
4**	05, 19, 34, 62
5**	40, 08
6**	12

2.23

2*	2 1
2.	8 6
3*	2 4 3 4
3.	6 6 7 9 5 5 5 5 8
4*	3 3 0 0 4 2 1 1
4.	9 6 8 5 8 5 5 5 7 7
5*	0 1 2 3 1 4 1 3 2 2 0 3 4 0
5.	6 5 6 7 7 9 6 5 5 9 8
6*	2 0 2 3 4 0 1 1 0 2 1
6.	5 7 8 5 8 7 9 8 5
7*	2 3 4 0 4 3 4 0
7.	9 8 6 5 6 7
8*	4 2 0 2
8.	8 5

2.25 (a) 35; (b) 34.5.

2.27 No, the total earnings are only $375,000.

2.29 (a) 8; (b) 9.
2.33 97.43.
2.37 (a) $\bar{x} = 30.25$ and $s = 11.15$; (b) $\bar{x} = 30.0$ and $s = 10.99$.
2.39 $\bar{x} = 5.18$, $s = 2.684$ and $v = 51.8\%$.
2.41 $v = 2.63\%$ for the price of copper and $v = 7.85\%$ for the price of coal; the price of coal is relatively more variable.
2.45 (a) 11.375; (b) 27.45; (c) 4.995.
2.47 $Q_1 = 14.95$, $Q_3 = 22.83$, and the interquartile range is 7.88; (b) $Q_1 = 7.33$, $Q_3 = 16.06$.
2.49 (a) 73.0; (b) 43.64%.
2.51 (a) $Q_1 = 1712$, $Q_2 = 1863$, $Q_3 = 2061$; (c) $Q_1 = 69.5$, $Q_2 = 70.55$, $Q_3 = 71.80$.
2.53 (a) The class frequencies are 1, 8, 19, 17, 9, 3, and 1.
2.55 (a) $\bar{x} = 5.4835$ and $s = 0.19042$; (b) median is 5.46 $Q_1 = 5.34$ and $Q_3 = 5.63$; (c) there is no apparent trend.
2.57 (a) median = 0.40, maximum = 0.57, minimum = 0.32 and the range = 0.25; (b) median = 0.51, maximum = 0.63, minimum = 0.47 and the range = 0.16.
2.59 (a) $Q_1 = 18.0$, $Q_2 = 27.0$, $Q_3 = 30.0$; (b) minimum = 12, maximum = 48, range = 36 and the interquartile range = 12.0.
2.61 (a) $Q_1 = 19$, $Q_2 = 28$, $Q_3 = 55$; (b) minimum = 12, maximum = 63, range = 51 and the interquartile range = 36.
2.65 (a) 1160–1179, 1180–1199, 1200–1219, 1220–1239, 1240–1259, 1260–1279, 1280–1299, 1300–1319; (b) 1159.5, 1179.5, 1199.5, 1219.5, 1239.5, 1259.5, 1279.5, 1299.5, 1319.5; (c) 1169.5, 1189.5, 1209.5, 1229.5, 1249.5, 1269.5, 1289.5, 1309.5; (d) 20.
2.67 The coefficient of variation is 5.22% for the first student and 6.73% for the second student, so the first student is more consistent.

CHAPTER 3

3.1 (b) $R = \{(0, 0), (1, 1), (2, 2)\}$. $T = \{(0, 0), (1, 0), (2, 0), (3, 0)\}$. $U = \{(0, 1), (0, 2), (1, 2)\}$.
3.3 (a) $R \cup U = \{(0, 0), (1, 1), (2, 2), (0, 1), (0, 2), (1, 2)\}$ is the event that at least as many carbon dioxide lasers are suitable as solid crystal lasers; (b) $R \cap T = \{(0, 0)\}$ is the event that none of the lasers is suitable; (c) $T' = \{(0, 1), (1, 1), (2, 1), (3, 1), (0, 2), (1, 2), (2, 2), (3, 2)\}$ is the event that at least one of the carbon dioxide lasers is suitable.
3.5 (a) $A \cup B = \{2, 3, 4\}$ is the event that work on the car is easy, average or difficult; (b) $A \cap B = \{3\}$ is the event that work on the car is average; (c) $A \cup B' = \{1, 3, 4, 5\}$ is the event that work on the car is not easy; (d) $C' = \{1, 2, 3\}$ is the event that work on the car is very easy, easy or average.
3.7 (b) B is the event that three graduate assistants are present, C is the event that as many professors as graduate assistants are present, D is the event that altogether three professors or graduate assistants are present; (c) $C \cup D = \{(1, 1), (1, 2), (2, 1), (2, 2)\}$, is the event that at most two graduate assistants are present; (d) B and D are mutually exclusive.
3.9 Region 1 represents the event that the ore contains copper and uranium; region 2 represents the event that the ore contains copper but not unranium; region 3 represents the event that the ore contains uranium but not copper; region 4 represents the event that the ore contains neither uranium nor copper.
3.11 (a) Region 5 represents the event that the windings are improper, but the shaft size is not too large and the electrical connections are satisfactory; (b) regions 4 and 6 together represent the event that the electrical connections are unsatisfactory, but the

windings are proper; (c) regions 7 and 8 together represent the event that the windings are proper and the electrical connections are satisfactory; (d) regions 1, 2, 3, and 5 together represent the event that the windings are improper.

3.17 72.

3.19 (a) 1,680; (b) 4,096.

3.21 720.

3.23 105.

3.25 (a) 55; (b) 165.

3.27 4,200.

3.29 (a) 1/6; (b) 1/18; (c) 2/9; (d) 1/18; (e) 1/18; (f) 1/9.

3.31 0.368.

3.33 45.

3.35 (a) Yes; (b) no, sum exceeds 1; (c) no, $P(C)$ is negative; (d) no, sum is less than 1; (e) yes.

3.37 (b) 27/112, 45/112 and 5/14; (c) 25/56, 5/16 and 27/112.

3.41 (a) 0.71; (b) 0.72; (c) 0.29; (d) 0.28.

3.43 (a) 0.43; (b) 0.67; (c) 0.11; (d) 0.59.

3.45 (a) 15/32; (b) 13/32; (c) 5/32; (d) 23/32; (e) 8/32; (f) 9/32.

3.47 (a) 0.29; (b) 0.18.

3.51 (a) 4 to 3; (b) 19 to 1 against it; (c) 4 to 1.

3.53 (a) 0.60; (b) $0.75 \leq p < 0.80$.

3.55 $P(I|D) = 2/3$; $P(I|D') = 4/97$.

3.57 (a) 62/85; (b) 74/84; (c) 29/51.

3.59 (a) 0.4333; (b) 0.5000; (c) 0.4375.

3.65 (a) 33/59; (b) 7/118; (c) 45/118.

3.67 Yes.

3.69 (a) 1/256; (b) 1/648; (c) 1/243.

3.71 0.758.

3.73 0.068.

3.75 (a) 0.60; (b) 0.20; (c) 0.70.

3.77 (a) 0.686; (b) 0.171; (c) 0.0286.

3.79 $0.20.

3.81 0.

3.83 93/16.

3.87 $1.

3.89 (a) $p > 1/6$; (b) $p < 1/6$; (c) $p = 1/6$.

3.91 3/4.

3.93 Discontinue the operation.

3.95 (a) FF costs $6.25, LWR costs $10.53 per kilowatt per year of useful life; (b) if FF cost exceeds the LWR cost by $4.28, the LWR plant is best; (c) LWR now costs $11.40 per kilowatt per year. The FF fuel cost must now exceed 1.82 times LWR fuel cost.

3.97 (a) $X' = \{(0, 0), (0, 1), (0, 2), (0, 3), (1, 0), (1, 1), (1, 2), (2, 0), (2, 1), (3, 0)\}$ is the event that the salesman will not visit all four customers; (b) $X \cup Y = \{(4, 0), (3, 1), (2, 2), (1, 3), (0, 4), (1, 0), (2, 0), (2, 1), (3, 0)\}$ is the event that the salesman will visit all four customers or more on the first day than on the second day; (c) $X \cap Z = \{(1, 3), (0, 4)\}$ is the event that he will visit all four customers but at most one on the first day; (d) $X' \cap Y = \{(1, 0), (2, 0), (2, 1), (3, 0)\}$ is the event that he will visit at most three of the customers and more on the first day than on the second day.

3.101 21.

3.103 (a) 0.56; (b) 0.26; (c) 0.06; (d) 0.76; (e) no.

3.105 There is a contradiction in his claim.
3.109 Purposeful action is most likely.
3.111 (a) First job; (b) second job.

CHAPTER 4

4.1 1/12, 2/12, 3/12, 3/12, 2/12, and 1/12.
4.3 (a) No, sum exceeds 1; (b) yes; (c) no, $f(3)$ is negative.
4.5 $k = 16/31$.
4.9 .1176, .3025, .3241, .1852, .0595, .0102, .0007.
4.11 (a) 0.3169; (b) 0.1442; (c) 0.0861; (d) .0746; (e) .4862; (f) .7052.
4.13 27/128.
4.15 (a) 0.2969; (b) 0.0152; (c) 0.2061.
4.17 (a) 0.1501; (b) 0.7338; (c) 0.0982.
4.19 (a) 0.9571; (b) 0.7892; (c) 0.5614; (d) 0.3518.
4.21 0.3167.
4.23 (a) 0.4032; (b) 0.4536; (c) 0.1432.
4.25 (a) 0.15; (b) 0.0625.
4.27 (a) 0.1470; (b) 0.1468.
4.31 $\sigma^2 = 1.0$.
4.33 $\sigma^2 = 1.8$.
4.35 (a) $\mu = 2.8$ and $\sigma^2 = 0.84$.
4.37 (a) $\sigma^2 = 1.25$.
4.39 (a) $\mu = 338$ and $\sigma = 13$; (b) $\mu = 120$ and $\sigma = 10$; (c) $\mu = 24$ and $\sigma = 4.8$; (d) $\mu = 520$ and $\sigma = 13.49$.
4.43 The probability of getting more than 95 or less than 49 correct is less than or equal to 1/16.
4.49 0.0498, 0.1494, 0.2240, 0.2240, 0.1680, 0.1008, 0.0504, 0.0216, 0.0081, 0.0027.
4.51 (a) 0.242; (b) 0.087; (c) 0.937.
4.53 (a) 0.182; (b) 0.857; (c) 0.648.
4.55 0.007.
4.57 (a) 0.981; (b) 0.577; (c) 0.978.
4.59 (a) 0.0314; (b) 0.0121.
4.61 (a) 0.3010.
4.63 (a) 1.5; (b) 0.9; (c) 3.0.
4.65 80 dollars for 8 hour day. Not worthwhile.
4.73 0.117.
4.75 (b) 0.195.
4.79 000–040, 041–170, 171–379, 380–602, 603–780, 781–894, 895–954, 955–982, 983–993, 994–997, 998–999.
4.83 (a) Yes; (b) Yes; (c) No, sum exceeds 1.
4.85 (a) 0.1468; (b) 0.1468.
4.87 (a) 0.5238; (b) 0.4190; (c) 0.0571.
4.89 (a) 0.72; (b) 0.72.
4.91 0.2707.
4.93 Probability at least 0.96.
4.95 0.5488.
4.97 (a) 0000–2465, 2466–5917, 5918–8334, 8335–9462, 9463–9857, 9858–9968, 9969–9994, 9995–9999.

CHAPTER 5

5.3 (a) 0.5904; (b) 0.024.
5.5 (a) 0.02; (b) 0.84.
5.7 (a) 0.556; (b) 0.09.
5.9 (a) 0.707; (b) 0.1339.
5.11 0.0916.
5.13 $\mu = 0.8$ and $\sigma^2 = 0.0267$.
5.15 $\mu = 4$ and σ^2 does not exist.
5.17 20,000 miles.
5.19 (a) 0.9332; (b) 0.1151; (c) 0.0154; (d) 0.9599.
5.21 (a) 2.37; (b) 1.23; (c) -0.37; (d) -2.02; (e) 1.81.
5.25 $\sigma = 19.88$.
5.27 (a) 0.7580; (b) 0.6578.
5.29 (a) 0.9568; (b) 0.7475; (c) 0.7201.
5.31 83.15%.
5.33 $\mu = 2.984$.
5.35 (a) 0.1212; (b) 0.4097.
5.37 0.1841.
5.39 0.0808.
5.45 $F(x) = 0$ for $x \le \alpha$, $F(x) = \dfrac{x-\alpha}{\beta-\alpha}$ for $\alpha < x < \beta$, and $F(x) = 1$ for $x \ge \beta$.
5.47 50%.
5.49 0.2646.
5.51 (a) 0.0807; (b) 0.0960.
5.53 0.5940.
5.55 (a) 0.049; (b) 0.799.
5.57 No relative maximum when $0 < \alpha < 1$; relative maximum at $x = 0$ when $\alpha = 1$.
5.59 (a) 25.9%; (b) 22.3%.
5.61 $e^{-\alpha t}$.
5.65 (a) $\mu = 0.2$; (b) 0.3164.
5.67 0.6321.
5.69 0.2057.
5.71 (a) 1/4; (b) 1/24.
5.73 $F(x_1, x_2) = 0$ for $x_1 \le 0$ or $x_2 \le 0$, $F(x_1, x_2) = \frac{1}{4}x_1^2x_2^2$ for $0 < x_1 < 1$ and $0 < x_2 < 2$, $F(x_1, x_2) = x_1^2$ for $0 < x_1 < 1$ and $x_2 \ge 2$, $F(x_1, x_2) = \frac{1}{4}x_2^2$ for $x_1 \ge 1$ and $0 < x_2 < 2$, and $F(x_1, x_2) = 1$ for $x_1 \ge 1$ and $x_2 \ge 2$; $F_1(x_1) = 0$ for $x_1 \le 0$, $F_1(x_1) = x_1^2$ for $0 < x_1 < 1$, and $F_1(x_1) = 1$ for $x_1 \ge 1$; $F_2(x_2) = 0$ for $x_2 \le 0$, $F_2(x_2) = \frac{1}{4}x_2^2$ for $0 < x_2 < 1$, and $F_2(x_2) = 1$ for $x_2 \ge 2$; they are independent.
5.75 $F(x, y) = 0$ for $x \le 0$ or $y \le 0$, $F(x, y) = \frac{3}{5}x^2y + \frac{2}{5}xy^3$ for $0 < x < 1$ and $0 < y < 1$, $F(x, y) = \frac{3}{5}x^2 + \frac{2}{5}x$ for $0 < x < 1$ and $y \ge 1$, $F(x, y) = \frac{3}{5}y + \frac{2}{5}y^3$ for $x \ge 1$ and $0 < y < 1$, and $F(x, y) = 1$ for $x \ge 1$ and $y \ge 1$.
5.77 (a) $f_1(x|y) = (x + y^2)/(5 + y^2)$ for $0 < x < 1$ and $f_1(x|y) = 0$ elsewhere; (b) $f_1(x|\frac{1}{2}) = \frac{1}{3}(4x + 1)$ for $0 < x < 1$ and $f_1(x|y) = 0$ elsewhere; (c) 11/18.
5.79 (a) 1/3; (b) $5/(6e) = 0.3066$.
5.81 (a) 0.3264; (b) 0.4712.
5.83 2.
5.85 $\mu = LW$ and $\sigma^2 = \frac{1}{12}(a^2W^2 + b^2L^2 + \frac{1}{12}a^2b^2)$.
5.91 2.45400, 5.42158, 1.59896.

5.93 (a) From the simulated data, time of the first failure is 1.74915 and the time of the fifth failure 5.13457.

5.97 (a) 0.104043, 0.978997, 0.059544, 0.889298, 0.367487, 0.870607, 0.660003, 0.135508; (b) 2.1973, 77.2619, 1.2278, 44.0183, 9.1611, 40.8981, 21.5764, 2.9123.

5.101 (a) 0.1465; (b) 0.3125.

5.103 (a) 1; (b) 0.25.

5.105 (a) 0.4938; (b) 0.1018; (c) 0.2789; (d) 0.9093.

5.109 (a) 0.997; (b) 0.0012.

5.113 $n = 25$; maximum profit is 125 dollars.

5.115 (a) $f_1(x) = 0.2e^{-.2x}$ for $x > 0$, $f_1(x) = 0$ elsewhere; $f_2(y) = 0.2e^{-.2y}$ for $y > 0$, $f_2(y) = 0$ elsewhere; $E(X) = E(Y) = 5$; (b) $E(X + Y) = 10$.

CHAPTER 6

6.5 (a) 15; (b) 300.

6.7 (a) $\mu = 0$ and $\sigma^2 = 26/3$.

6.9 The samples are 1 and 1, 1 and 2, 1 and 3, 1 and 4, 2 and 1, 2 and 2, 2 and 3, 2 and 4, 3 and 1, 3 and 2, 3 and 3, 3 and 4, 4 and 1, 4 and 2, 4 and 3, 4 and 4; the probabilities that $\bar{x}$ equals 1, 1.5, 2, 2.5, 3, 3.5, or 4 are 1/16, 1/8, 3/16, 1/4, 3/16, 1/8, and 1/16.

6.11 (a) It is divided by 2; (b) it is divided by 3/2; (c) it is multiplied by 3; (d) it is multiplied by 4.

6.15 approximately 0.628.

6.17 approximately 0.111.

6.21 $t = 1.15$; since $t_{0.10} = 1.476$ for 5 degrees of freedom, the data fail to reject the claim.

6.23 0.025.

6.25 0.02.

6.27 0.2873.

6.29 0.3125.

6.31 (a) 28 (b) 190.

6.33 (a) 6/7 (b) 18/19.

6.35 0.9876.

6.37 (a) It is divided by $\sqrt{2}$; (b) it is multiplied by $\sqrt{2/3}$; (c) it is multiplied by 2.

CHAPTER 7

7.3 (a) The medians are 4, 5, 4, 3, 5, 2, 3, 5, 3, 2, 3, and 4; the means are 4, 4.3, 4, 3.3, 4, 2, 2.7, 4, 3.3, 3, 3, and 4; (b) the frequencies are 2, 4, 3, and 3 for the medians, and 1, 5, 6, and 0 for the means.

7.5 $7899.4 < \mu < 15690.6$.

7.7 87.6%.

7.9 84.7%.

7.11 $n = 208$.
7.13 (a) $E = 22.14$; (b) $E = 4.46$.
7.15 99.82%.
7.17 $E = 1.16$ minutes.
7.19 $.5031 < \mu < .5089$.
7.21 $-0.34 < \mu < 8.34$, that is $\mu < 8.34$.
7.23 (a) $\frac{1}{n}\sum_{i=1}^{n} x_i$; (b) $\frac{1}{n}\sum_{i=1}^{n} x_i$.
7.25 (a) .6548; (b) .9632.
7.27 Type I; Type II.
7.29 (a) .1056; (b) .1056.
7.31 (a) .014; (b) .2327.
7.33 (a) 0.0087; (b) $\alpha \leq 0.0087$.
7.35 $n = 47$.
7.37 (a) $\mu \neq 1250$; (b) $\mu < 1250$; (c) $\mu > 1250$.
7.39 $z = 2.73$; reject H_0.
7.41 $z = 2.08$; reject H_0.
7.43 $z = 1.52$; cannot reject H_0.
7.45 $t = 4.79$; reject H_0.
7.47 $t = 3.087$; reject H_0.
7.49 $t = 2.11$; cannot reject H_0. The paradox is explained by the standard deviation, which has greatly increased.
7.51 (a) .80; (b) .09.
7.53 (a) .26; (b) .15; (c) .05.
7.55 (a) .92; (b) .79; (c) .56; (d) .30; (e) .12; (f) .05.
7.57 (a) .59; (b) .32; (c) .12; (d) .03; (e) .00.
7.59 n should be 40.
7.61 (a) yes; (b) $17.843 < \mu < 19.949$; (c) $70.22 < \mu < 71.17$.
7.63 (a) $\mu = .006$ and $\sigma = .0036$; (b) .0475.
7.65 $z = 2.07$; the data is not in favor of that the men earn $20/week more than women.
7.67 $t = .96$; cannot reject H_0.
7.69 $t = 1.03$; cannot reject H_0.
7.71 $t = 2.205$; cannot reject H_0.
7.73 $0.0508 < \delta < 0.0552$.
7.75 (a) select three amplifiers by random drawing; (b) flip coin for each of the six amplifiers. If heads, channel 1 gets the modified power source.
7.77 randomly select 25 cars, and install the modified air pollution device. The other 25 cars use the current device.
7.79 $70.224 < \mu < 71.169$.
7.81 $24.92 < \mu < 27.88$.
7.83 $0.992 < \mu < 1.048$.
7.85 $n = 11$.
7.87 $-0.181 < \delta < -0.036$.
7.89 $0.007 < \delta < 0.150$.
7.91 (a) new dividing line $z = 2.210$; (b) .9999, .9984, .9864, .9296, .7689, .50, .2311, .0704, .0136, .0016, .0001.
7.95 n should be 25.
7.97 (a) Randomly select 10 cars to use the modified spark plugs. The other 10 cars use the regular spark plugs; (b) select 7 specimens, by random drawing, to try in the old oven.
7.99 $t = -1.99$ we would not reject $\mu_A = \mu_B$ at the level of significance $\alpha = .05$.

CHAPTER 8

8.1 (a) $s = 4.195$; (b) 4.341.
8.3 (a) 1.787; (b) 2.144.
8.5 $.082 < \sigma < .695$
8.7 $\chi^2 = 5.832$; cannot reject H_0.
8.9 $\chi^2 = 115.886$; reject H_0.
8.11 $\chi^2 = 10.89$; cannot reject H_0.
8.13 $F = 1.496$; we cannot reject H_0.
8.15 $F = 2.42$ we cannot reject H_0.
8.17 (a) No, samples are not normal and variances are unequal; (b) base test on the logarithms of the observations.
8.19 $1.52 < \sigma_I < 2.20$ and $1.91 < \sigma_{II} < 3.32$.
8.21 $\chi^2 = 19.21$; cannot reject H_0.
8.23 $\chi^2 = 72.22$; reject H_0.
8.25 $F = 2.797$ we cannot reject H_0.

CHAPTER 9

9.1 (a) $0.35 < p < 0.49$; (b) $0.352 < p < 0.488$.
9.3 (a) $0.52 < p < 0.64$; (b) $0.514 < p < 0.642$.
9.5 (a) $0.319 < p < 0.445$; (b) $E = 0.06$.
9.7 $E = 0.0885$.
9.9 $0.617 < p < 0.743$.
9.11 $n = 201$.
9.13 $n = 1{,}300$.
9.15 $0.513 < p < 0.639$.
9.17 $p < 0.026$.
9.19 (a) 0.75 and 0.25; (b) 0.0052 and 0.9948.
9.21 0.238.
9.23 $z = 2.19$; reject H_0.
9.25 $z = -1.83$; cannot reject H_0.
9.27 $z = 1.489$; cannot reject H_0.
9.29 At most three or at least twelve heads; 0.0352.
9.31 $\chi^2 = 2.37$; cannot reject H_0.
9.33 $\chi^2 = 9.39$; reject H_0.
9.37 $-0.419 < p_1 - p_2 < -0.031$.
9.39 $0.170 < p_1 - p_2 < 0.374$.
9.43 $\chi^2 = 3.457$; cannot reject H_0.
9.45 $\chi^2 = 54.328$; reject H_0.
9.47 $\chi^2 = 0.657$; cannot reject H_0.
9.49 $\chi^2 = 9.185$; cannot reject H_0.
9.51 (a) 0.0375, 0.1071, 0.2223, 0.2811, 0.2163, 0.1013 and 0.0344; (b) 3, 8.6, 17.8, 22.5, 17.3, 8.1 and 2.8; (c) $\chi^2 = 1.264$; good fit.
9.55 (a) $0.115 < p < 0.275$; (b) $0.105 < p < 0.255$.
9.57 (a) $0.084 < p < 0.243$; (b) $0.077 < p < 0.223$.
9.59 (a) $0.075 < p < 0.217$; (b) $0.064 < p < 0.196$.
9.61 $p < 0.00424$.

9.63 $z = 0.807$; cannot reject H_0.
9.65 $z = -1.746$; reject H_0.
9.67 (a) $\chi^2 = 8.190$; reject H_0; (b) $0.256 < p_1 < 0.611$; $0.108 < p_2 < 0.425$; $0.461 < p_3 < 0.806$.
9.69 $\chi^2 = 10.481$; cannot reject H_0.
9.71 $\chi^2 = 47.862$; reject H_0.

CHAPTER 10

10.1 $P(10\ or\ more) = 0.2272$; cannote reject H_0.
10.3 $P(6\ or\ fewer) = 0.3036$; cannot reject H_0.
10.5 $z = -3.91$; reject H_0.
10.7 $P(3\ or\ fewer) = 0.1719$; cannot reject H_0.
10.9 $z = -0.10$; cannot reject H_0.
10.11 $z = 2.92$; difference is significant.
10.13 $z = 1.62$; cannot reject H_0.
10.15 $H = 26.0$; the populations are not identical.
10.17 $z = -0.244$; cannot reject H_0.
10.21 $z = -4.00$; reject H_0.
10.23 Maximum difference is about 0.22; cannot reject H_0.
10.25 $P(6\ or\ more) = 0.0625$; reject H_0.
10.27 $z = -1.814$; reject H_0.
10.29 $H = 0.904$; cannot reject H_0.
10.31 $z = -1.797$; reject H_0.
10.33 $W_1 = 30$ so $U_1 = 2$; reject H_0.
10.35 $z = -2.248$; reject H_0.

CHAPTER 11

11.1 (b) $\hat{y} = 39.05 + 0.764x$; $\hat{y} = 65.8$.
11.3 (b) $\hat{y} = 1.13 + 14.49x$; $\hat{y} = 51.8$.
11.5 39.97 – 63.71.
11.7 $t = -1.212$; cannot reject H_0.
11.9 (a) $\hat{y} = 0.0 + 2.0x$; (b) $\hat{y} = 7.0$.
11.11 $t = 0.0$; cannot reject H_0.
11.13 7.23 – 12.39.
11.15 $22.37 < \beta < 27.49$.
11.17 (a) $\sum xy / \sum x^2$; (b) 14.75.
11.19 $12.72 < \alpha < 33.08$.
11.21 (a) $\hat{y} = 3.39 + 0.679x$; $\hat{y} = 6.11$; (b) $\hat{y} = 2.95 + 0.242x$; $\hat{y} = 5.61$.
11.27 (b) $\ln \hat{y} = 11.150 + 1.3899x$ or $\hat{y} = 69559.2(1.149)^x$; (c) 1,121,073.
11.29 44.86
11.31 $\hat{y} = exp[exp(0.00041x + 1.21)]$
11.33 $\hat{\alpha} = 0.240$.
11.35 (a) $t = -2.28$; cannot reject $\beta_1 = 0$; (b) $F = 38.59$; reject $\beta_2 = 0$.
11.39 $\hat{y} = 64.1$.

11.41 $\hat{y} = 2.266 + 0.225x_1 + 0.0623x_2$; $\hat{y} = 8.37$.
11.45 (a) slight increase in variance; (b) U-shaped pattern.
11.47 A serious violation, time trend.
11.49 $z = 4.89$; reject H_0: $\rho = 0$.
11.51 $z = 5.15$; reject H_0: $\rho = 0$.
11.53 $.717 < \rho < .976$.
11.55 The first linear relationship is 1.9 times stronger.
11.57 (a) $0.395 < \rho < 0.885$; (b) $-0.052 < \rho < 0.655$; (c) $0.314 < \rho < .749$.
11.63 (a) 2812.4; (b) 233.9; (c) 0.958.
11.65 (a) $r_s = 0.697$; (b) $r_s = 0.811$; (c) $r_s = 0.915$.
11.67 $r_s = 0.39$; not significant.
11.71 $-57.2 < \alpha < 37.2$.
11.73 (a) 19.9 to 128.1 hours; (b) 0.0 to 173.1 hours.
11.75 $1.69 < \alpha < 2.11$.
11.77 $\hat{\gamma} = 1.499$.
11.79 $\hat{\tau} = 0.9284$.
11.81 (a) 3.99 to 8.01 hours; (b) 1.07 to 10.93 hours.
11.83 The first linear relationship is twice as strong.
11.85 (a) $0.446 < \rho < 0.923$; (b) $-0.797 < \rho < -0.346$; (c) $-0.173 < \rho < 0.476$.
11.87 (a) $\hat{y} = -0.0206 + 0.462x$; (b) $0.43 < \beta < 0.50$; (c) $t = -0.32$, cannot reject H_0. (d) The variance appears to increase somewhat with x.

CHAPTER 12

12.5 $F = 0.68$, not significant at the 0.01 level.
12.7 (a) $SS(Tr) = 204$, with 3 degrees of freedom; $SSE = 34$, with 11 degrees of freedom; $SST = 238$, with 14 degrees of freedom. (b) $F = 22.0$, significant at the 0.05 level.
12.9 $F = 11.3$, significant at the 0.01 level.
12.11 $F = 15.7$, significant at the 0.05 level.
12.19 For technicians, $F = 0.64$, not significant at the 0.01 level; for days, $F = 0.77$, not significant at the 0.01 level.
12.21 (b) $SS(Tr) = 70$ with 3 degrees of freedom; $SS(Bl) = 136$, with 4 degrees of freedom; $SSE = 66$, with 12 degrees of freedom; $SST = 272$, with 19 degrees of freedom. (c) For treatments, $F = 4.24$, significant at the 0.05 level.
12.23 $F = 1.92$, not significant at the 0.05 level.
12.25 For machines, $F = 0.064$, not significant at the 0.05 level; for workmen, $F = 1.346$, not significant at the 0.05 level; for replicates, $F = 6.217$, significant at the 0.05 level.

12.29

Tr 1	Tr 4	Tr 3	Tr 2
8	9	10	13

12.31

Method A	Method C	Method B
72	76	86

12.33

Thread 1	Thread 5	Thread 3	Thread 4	Thread 2
20.675	20.9	23.525	23.7	25.65

12.35 For Laboratories, $F = 6.55$, significant at the 0.05 level; for "along the rolling direction," $F = 1.29$, not significant at the 0.05 level; for "across the rolling direction," $F = 26.83$, significant at the 0.05 level.

12.37 For automobile size classes, $F = 414$, significant at the 0.01 level; for barrier impact speeds, $F = 99$, significant at the 0.01 level; for impact angles, $F = 17.5$, significant at the 0.01 level; for restraint systems, $F = 0.72$, not signficant at the 0.05 level.

12.41 For track designs, $F = 6.44$, significant at the 0.01 level. The estimated effect of usage on breakage resistance is 0.39.

12.43 $F = 2.80$, not significant at the 0.05 level.

12.45 (b) $SS(Tr) = 56$, with 2 degrees of freedom; $SS(Bl) = 138$, with 3 degress of freedom; $SSE = 32$, with 6 degrees of freedom; $SST = 226$, with 11 degrees of freedom. (c) For treatments, $F = 5.25$, significant at the 0.05 level; for blocks, $F = 8.63$, significant at the 0.05 level.

12.47 (a) For fuels, $F = 0.35$, not significant at the 0.05 level. (b) For launchers, $F = 0.003$, not significant at the 0.05 level;

12.49 For flow, $F = 7.8$, significant at the 0.01 level; for precipitators, $F = 8.1$, significant at the 0.01 level.

12.51

Site E	Site B	Site C	Site A	Site D
3.70	6.77	14.20	21.30	26.73

12.53 For golf-ball designs, $F = 66.6$, significant at the 0.01 level; for golf pros, $F = 24.4$, significant at the 0.01 level; for drivers, $F = 2.7$, not significant at the 0.05 level; for fairways, $F = 73.3$, significant at the 0.01 level.

12.55 $F = 56.4$, significant at the 0.05 level.

CHAPTER 13

13.1 Significant effects are temperature ($F = 11.5$, $\alpha = 0.01$), concentration ($F = 48.0$, $\alpha = 0.01$), and their interaction ($F = 3.8$, $\alpha = 0.05$). Optimal conditions for reflectivity are sulfone concentration at 10 grams per liter and bath temperature at 75°F. The 95% confidence interval for reflectivity is 37.32 to 46.02.

13.3 For detergents, $F = 0.05$, not significant at the 0.05 level; for machines, $F = 6.89$, significant at the 0.05 level; for interaction, $F = 0.81$, not significant at the 0.05 level.

13.5 For racket face, $F = 0.2$, not significant at the 0.05 level; for tension, $F = 14.0$, significant at the 0.01 level; for interaction, $F = 4.5$, significant at the 0.05 level; for replicates, $F = 1.2$, not significant at the 0.05 level. Medium string tension is preferred with large racket face.

13.7 Significant effects at the 0.01 level are specimen preparation (factor C, $F = 12.67$), and twisting temperature (factor D, $F = 86.51$). No other effects are significant at the 0.05 level. The F-values for ingot locations (factor A) and slab positions (factor B) are 0.56 and 0.0008 respectively. For two-factor interactions AB, AC, AD, BC, BD, and CD, the corresponding F-values are 0.09, 1.12, 1.20, 1.04, 0.09 and 0.98 respectively. The F-values for three-factor interactions ABC, ABD, ACD and BCD are 0.28, 0.53, 1.76, and 0.27 respectively. The F-value for interaction $ABCD$ is 0.68.

13.11 (a) $SST = 110.4375$, $SS(Tr) = 103.9375$, $SS(Rep) = 1.5625$, $SSE = 4.9375$. (b) $[A] = -1$, $[B] = -13$, $[C] = -25$, $[AB] = 3$, $[AC] = 3$, $[BC] = -29$, $[ABC] = 3$. (c) $SSA = 0.0625$, $SSB = 10.5625$, $SSC = 39.0625$, $SS(AB) = .5625$, $SS(AC) = .5625$, $SS(BC) = 52.5625$, $SS(ABC) = .5625$. (e) B, C and BC are significant at the 0.01 level with corresponding F-values of 14.9, 55.0 and 74.0.

13.13 The replication effect is significant at the 0.01 level with an F-value of 9.1. The main effect B and the three-factor interaction ABC are significant at the 0.05 level with corresponding F-values of 8.1 and 5.6 respectively. No other effects are significant at the 0.05 level.

13.21 (a) $SSE = SST - SS(Tr) - SSR = 138 - 118 - 2 = 18$, with 3 degrees of freedom. $MSE = 6$.

13.23 The 95% confidence intervals for effects A, B and interaction AB are $7 \pm 12{,}86$, 3 ± 12.86 and -1 ± 12.86 respectively. All the confidence intervals cover zero.

13.25 The 95% confidence intervals for oil viscosity and temperature are 0.69 to 1.71 and -1.41 to -0.39 respectively. The 95% confidence intervals of all other effects cover zero. Conclusion: (1) changing oil viscosity level from low to high increases the engine wear; (2) changing temperature level from low to high decreases the engine wear; (3) the special additive has no effect on engine wear; neither do the two-factor or three-factor interactions.

13.27 (a) Block 1: *a, b, c, abc, d, abd, acd, bcd*;
Block 2: 1, *ab, ac, bc, ad, bd, cd, abdc*;
(b) Block 1: 1, *bc, abd, acd*;
Block 2: *a, abc, bd, cd*;
Block 3: *b, c, ad, abcd*;
Block 4: *ab, ac, d, bcd*.

13.29 The B and C main effects are significant at the 0.01 level with F-values of 11.9 and 27.8; the AD interaction is significant at the 0.05 level with an F-value of 8.1; use combination of drugs A, B and C.

13.33 1, *ab, ac, ad, ae, af, bc, bd, be, bf, cd, ce, cf, de, df, ef, cdef, bdef, bcef, bcdf, bcde, adef, acef, acdf, acde, abef, abdf, abde, abcf, abce, abcd* and *abcdef*. Each effect is aliased with its generalized interaction with $ABCDF$.

13.35 (a) 1, *af, bf, ab, cf, ac, bc, abcf, df, ad, bd, abdf, cd, acdf, bcdf, abcd, ef, ae, be, abef, ce, acef, bcef, abce, de, adef, bdef, abde, cdef, acde, bcde, bcde, abcdef*.

13.39 The required half replicate includes treatment combinations 1, *ab, ac, ad, bc, bd, cd*, and *abcd*. The MSE is 32.896 with 4 degrees of freedom. No effect is significant at the 0.05 level. The F-values are 1.26 for $A = BCD$, 1.32 for $B = ACD$, 3.02 for $C = ABD$, 5.32 for $D = ABC$, 0.04 for $AB = CD$, 0.06 for $AC = BD$ and 0.46 for $BC = AD$.

13.41 (a) For treatments, $F = 97.3$, significant at the 0.01 level; (c) significant effects at the 0.01 level are tenderizers ($F = 486.4$), cooking times ($F = 51.4$), cooking temperature ($F = 28.5$), and tenderizer-time interaction ($F = 6.0$).

13.43 The AB interaction is significant at the 0.01 level with an F-value of 19.0. No other effects are significant at the 0.05 level.

13.45 Changing temperature or ph or both from low to high increase the yield substantially. The 95% confidence intervals for factors A, B and interaction AB effects are 1.28 to 11.06, 0.94 to 10.72 and -6.39 to 3.39 respectively.

13.47 Significant effects at the 0.05 level are A and B with F-values 9.5 and 8.5 respectively.

13.49 The largest number of blocks in which one can perform a 2^6 factorial experiment without confounding any main effect is 32. For example, we could confound on AB, BC,CD, DE and EF.

13.51 (a) Block 1: 1, *bc, abd, acd, abe, ace, de, bcde*;
Block 2: *ab, ac, d, bcd, e, bce, abde, acde*;
Block 3: *b, c, ad, abcd, ae, abce, bde, cde*;
Block 4: *a, abc, bd, cd, be, ce, ade, abcde*.
(b) The intrablock SSE equals 401.875; A and E are significant at the 0.01 level with F-values 12.7 and 17.3; B, C, AD and ACD are significant at the .05 level with F-values 5.6, 6.0, 6.3 and 4.7.

13.53 We could use half the replicate that includes experimental condition 1. A and B are significant at the 0.01 level with F-values of 72.8 and 23.3. C is significant at the 0.05 level with an F-value of 10.0.

CHAPTER 14

14.1 (a) Central line = 0.150, $UCL = 0.153$, $LCL = 0.147$;
(b) central line = 0.005, $UCL = 0.010$, $LCL = 0$;
(c) $\bar{x}$: eighth, sixteenth, and seventeenth sample values outside limits; R: all sample values within limits.

14.3 (a) Central line = 48.1, $UCL = 50.3$, $LCL = 46.0$;
(b) central line = 2.9, $UCL = 6.7$, $LCL = 0$;
(c) process mean out of control, process variability in control;
(d) $z = -2.12$, there is a trend;
(e) no, process is not in control.

14.5 (a) $\bar{x}$: central line = 21.7, $UCL = 25.2$, $LCL = 18.2$; σ: central line = 1.05, $UCL =$ 3.74, $LCL = 0$;
(b) yes, process is in control.

14.7 Central line = 0.0017, $UCL = 0.0042$, $LCL = 0$;

14.9 (a) Central line = 0.037, $UCL = 0.093$, $LCL = 0$;

14.11 Yes, central line for c chart is 4.9, $UCL = 11.6$, and $LCL = 0$.

14.13 We can assert with 95% confidence that 99% of the pieces will have yield strength between 36,897 and 68,703 *psi*.

14.15 (a) 0.1063 ± 0.0008; (b) 0.1063 ± 0.0001.

14.17 (a) 0.046, 0.139; (b) 0.059, 0.143.

14.19 Producer's risk = 0.143, consumer's risk = 0.082.

14.23 $AOQL \approx 0.016$.

14.25 $\alpha = 0.15$.

14.27 Sample size is 32, acceptance number is 5, rejection number is 6.

14.29 $a_n = -1.67 + 0.19n$, $r_n = 2.14 + 0.19n$; reject on eighth trial.

14.31 (a) Central line = 4.1, $UCL = 4.17$, $LCL = 4.03$;
(b) Central line = 0.12, $UCL = 0.25$, $LCL = 0$;
(c) $\bar{x}$: All sample values are outside limits; R: sixth sample value outside limits.

14.33 The standard is being met.

14.35 (a) $UCL = 3.48$, $LCL = 0$; (b) all the 10 foot sections are under control except the 19-th section, which is out of the limits.

14.37 We can assert with 95% confidence that 90% of the inter-request times will be between 889 and 54,176 microseconds.

14.39 .992, .953, .879, .783, .677, .570, .469, .380, .303, .238, .185, .143, .109, .082, .062, .046, .034, .025, .019, .014; producer's risk = .217, consumer's risk = .082.

14.41 Sample size 20, acceptance number is 2, rejection number is 3.

14.43 (b) $L = 186.87$; (c) the cardboard strength data seems to be sampled from a normal distribution.

CHAPTER 15

15.1 $R = .9936$.

15.3 $R = .9983$.

15.5 (a)

$$f(t) = \begin{cases} \beta(1 - t/\alpha)\, exp[-\beta(t - t^2/(2\alpha))] & \text{for } 0 < t < \alpha. \\ 0 & \text{elsewhere.} \end{cases}$$

$$F(t) = \begin{cases} 1 - exp[-\beta(t - t^2/(2\alpha))] & \text{for } 0 < t < \alpha. \\ 1 - exp[-\alpha\beta/2] & \text{for } t > \alpha. \end{cases}$$

15.7 (a) 0.6703; (b) 0.6703

15.9 $n = 5$ and the failure rate is 3.94×10^{-4}.

15.11 (a) 0.6065; (b) 0.9502; (c) 0.3679.

15.13 (b) $F(x) = p\dfrac{1 - (1 - p)^x}{p} = 1 - (1 - p)^x$ for $x = 1, 2, 3, \ldots$; (c) .301

15.15 (a) $3{,}253.1 < \mu < 38{,}005.6$ (b) $T_r = 40{,}970 < 45{,}767.5$; cannot reject H_0.

15.17 (a) $254.9 < \mu < 1{,}182.8$ (b) $T_r = 3{,}329$ and $1{,}642.7 < T_r < 5{,}921.3$, we cannot reject H_0.

15.21 2.003×10^5.

15.23 0.8758.

15.27 (a) 0.6753; (b) 0.4066

15.29 83.3 hours.

15.31 2.003×10^5.

15.33 0.9520.

15.35 (a) 0.6667; (b) 0.50; (c) 0.9977.

INDEX

Difference between proportions (large samples)

$$z = \frac{\dfrac{x_1}{n_1} - \dfrac{x_2}{n_2}}{\sqrt{\hat{p}(1 - \hat{p})\left(\dfrac{1}{n_1} + \dfrac{1}{n_2}\right)}} \quad \text{with} \quad \hat{p} = \frac{x_1 + x_2}{n_1 + n_2}$$

Differences among proportions, contingency tables, or goodness of fit

$$\chi^2 = \sum \frac{(o - e)^2}{e}$$

Mean (large sample—known or estimated by s)

$$z = \frac{\bar{x} - \mu_o}{\sigma/\sqrt{n}}$$

Mean (small sample)

$$t = \frac{\bar{x} - \mu_o}{s/\sqrt{n}}$$

Proportion (large sample)

$$z = \frac{x - np_o}{\sqrt{np_o(1 - p_o)}}$$

REGRESSION

Coefficient of correlation

$$r = \frac{S_{xy}}{\sqrt{S_{xx} \cdot S_{yy}}}$$

where $S_{xx} = \sum x^2 - \frac{1}{n}(\sum x)^2$, $S_{yy} = \sum y^2 - \frac{1}{n}(\sum y)^2$,

and $S_{xy} = \sum xy - \frac{1}{n}(\sum x)(\sum y)$

Least squares line, $\hat{y} = a + bx$

$$b = \frac{S_{xy}}{S_{xx}} \quad \text{and} \quad a = \frac{\sum y - b(\sum x)}{n}$$